Springer Series on Atomic, Optical, and Plasma Physics

The Springer Series on Atomic, Optical, and Plasma Physics covers in a comprehensive manner theory and experiment in the entire field of atoms and molecules and their interaction with electromagnetic radiation. Books in the series provide a rich source of new ideas and techniques with wide applications in fields such as chemistry, materials science, astrophysics, surface science, plasma technology, advanced optics, aeronomy, and engineering. Laser physics is a particular connecting theme that has provided much of the continuing impetus for new developments in the field, such as quantum computation and Bose-Einstein condensation. The purpose of the series is to cover the gap between standard undergraduate textbooks and the research literature with emphasis on the fundamental ideas, methods, techniques, and results in the field.

More information about this series at http://www.springer.com/series/411

Babak Shokri • Anri A. Rukhadze

Electrodynamics of Conducting Dispersive Media

Springer

Babak Shokri
Physics Department
Shahid Beheshti University
Tehran, Iran

Anri A. Rukhadze
Fizika plazmy
Prokhorov General Physics Institute
Moscow, Russia

ISSN 1615-5653 ISSN 2197-6791 (electronic)
Springer Series on Atomic, Optical, and Plasma Physics
ISBN 978-3-030-28970-6 ISBN 978-3-030-28968-3 (eBook)
https://doi.org/10.1007/978-3-030-28968-3

This Springer imprint is published by the registered company Springer Nature Switzerland AG
The registered company address is: Gewerbestrasse 11, 6330 Cham, Switzerland

Preface

It has been more than 55 years since the Russian edition of the book *Electromagnetic Properties of Plasma and Plasma-like Media* by V. Silin and A. Rukhadze appeared. At that time, this book was one of the world's first book publications on the electrodynamics of plasma and was the first in which both the electrodynamics of plasma and plasma-like media such as gaseous plasma, metals, semiconductors, and molecular crystals were expounded in a common approach. It was impossible to be confined only to the classical approach and so the authors presented the quantum description for the first time in monograph form.

Despite the great popularity of the book in many countries, the book has not yet been translated to any other language and so only the Russian edition is used. There appears a new opportunity to gather this book with a significant expansion to make it more popular for students and researchers.

Over the years, electrodynamics of plasma-like media achieved significant developments, particularly in respect of the applications. The methods and applications outlined in the present monograph of electrodynamic properties have generally found their relevance and can rightfully become the table book for theorists working in this field of science, especially for young physicists.

The authors have written this book using the books *Electrodynamics of Plasma and Plasma-like Media* (by Silin and Rukhadze in the Russian language) and *Principles of Plasma Electrodynamics* (by Alexandrov, Bogdankevich, and Rukhadze in the English language) supplemented by new subjects. Furthermore, each chapter has its own problems introducing some concepts and models. Some of these problems have been selected from the books *Principles of Plasma Electrodynamics* (by Alexandrov, Bogdankevich, and Rukhadze) and *Electromagnetic Phenomena in Matter: Statistical and Quantum Approaches* (by I.N. Toptygin).

In this book, a sequential representation of the electrodynamics of conducting media with dispersion in three vector representations $(\vec{E}, \vec{B}, \vec{D})$ is given. In addition to the general formalism of such a representation, specific media such as classical nondegenerate plasma, degenerate metal plasma, magnetoactive anisotropic plasma, atomic hydrogen gas, semiconductors, and molecular crystals are considered. Each chapter is equipped with a sufficient number of problems with solutions that have academic and applied importance.

This book is designed for post-graduate students, as well as scientific professionals involved in relevant fields of science.

We hope that the content of this book significantly expands the circle of readers of the book, which is designed not only for plasma physicist but also for the professionals in material media with temporal and spatial dispersions, senior and graduate students, and postdoc researchers.

Here it is worthy to respect the memory of Professor Anri A. Rukhadze, who passed away in March 2018 and left me alone in the final preparation of this book. He was an outstanding scientist and teacher. May his memory be eternal.

Tehran, Iran Babak Shokri
Moscow, Russia Anri A. Rukhadze

Contents

List of Symbols

ρ	Charge density
$\vec{j}$	Current density
$\vec{E}$	Electric field
$\vec{D}$	Electric induction
$\vec{B}$	Magnetic induction
$\vec{M}$	Magnetization vector
$\vec{H}$	Magnetic field strength
$\vec{P}$	Polarization vector
$\vec{j}_{\text{cond}}$	Conduction current
σ_{ij}	Conductivity tensor
ε_{ij}	Dielectric permittivity tensor
μ_{ij}	Magnetic permittivity tensor
$\vec{i}$	Surface current density
$\vec{n}$	Surface normal vector and unit vector
ε_0	Static dielectric constant
σ_0	Electrostatic conductivity
σ	Surface charge density
ω	Frequency
$\vec{k}$	Wave vector
$\imath$	Imaginary unit
$\vec{r}$	Position vector
r_{scr}	Radius of screening
$*$	Complex conjugate
$\varepsilon' \equiv \text{Re}\,\varepsilon$	Real part of ε
$\varepsilon'' = \text{Im}\,\varepsilon$	Imaginary part of ε
ε^{tr}	Transverse dielectric permittivity

ε^l	Longitudinal dielectric permittivity
ε	Dielectric permittivity
μ	Magnetic permittivity
δ_{ij}	Kronecker delta
e_{ijl}	Completely anti-symmetric unit tensor of rank 3
χ	Magnetic susceptibility
$\mathcal{P}$	Principle part
$\delta(x)$	Delta function
ϕ	Electric potential
$n(\omega)$	Refractive index
ϵ	Energy
$\vec{p}$	Momentum
κ	Boltzmann's constant
μ_α	Chemical potential
F	Free energy
N_α	Number of α particles
$K_n(x)$	MacDonald function of order n
M	Mass of ion
m	Mass of electron
$\prod_{ij}^{(\alpha)}\left(\vec{r},t\right)$	Momentum flux density tensor of α particles, pressure tensor
$W^{(\alpha)}\left(\vec{r},t\right)$	Mass density of α particles
$\vec{V}^{(\alpha)}$	Hydrodynamic velocity of α particles
ν_{eff}	Effective collision frequency
$\nu_{\alpha n}$	Collision frequency of α particles with neutrals
v_s	Acoustic velocity
v_A	Alfven velocity
Ω_α	Larmor frequency of α particles
$J_s(z)$	Bessel function of order s
$I_s(z)$	Modified Bessel function of order s
$I_+(x)$	Tabulated function
P	Pressure
D_α	Diffusion coefficient of α particles
$\hat{H}$	Hamiltonian
$\vec{M}\left(\vec{r},t\right)$	Magnetization density
$\langle 1 \rangle$	Normalization of the derivative of the distribution function
λ_{ijkl}	Elasticity tensor
Λ_{ij}	Tensor characterizing the energy change due to the lattice deformation
β_{ijk}	Piezoelectric tensor
ρ_m	Lattice density
e_α	Charge of α particles

Chapter 1
Principles of Electrodynamics of Media with Spatial and Temporal Dispersion

1.1 Equations of Electromagnetic Fields

The electrodynamics of material media differs from the electrodynamics of vacuum due to the presence of charge and current densities ρ and $\vec{j}$ induced in the medium under the action of external field sources $\vec{j}_0$ and ρ_0 in the field equations

$$\begin{cases} \nabla \cdot \vec{E} = 4\pi(\rho + \rho_0), & \nabla \times \vec{E} = -\dfrac{1}{c}\dfrac{\partial \vec{B}}{\partial t}, \\ \nabla \times \vec{B} = \dfrac{1}{c}\dfrac{\partial \vec{E}}{\partial t} + \dfrac{4\pi}{c}\left(\vec{j} + \vec{j}_0\right), & \nabla \cdot \vec{B} = 0. \end{cases} \tag{1.1}$$

The induced charge and current densities satisfy the continuity equation, which expresses the conservation law of the electric charge:

$$\frac{\partial \rho}{\partial t} + \nabla \cdot \vec{j} = 0. \tag{1.2}$$

The physical meanings of electric field strength $\vec{E}$ and magnetic induction $\vec{B}$ of Eq. (1.1) are defined in vacuum as well as within the media by the force exerting on a test point charge e moving with the velocity $\vec{v}$, i.e., the Lorentz force:

$$\vec{F} = e\left\{\vec{E} + \frac{1}{c}\left(\vec{v} \times \vec{B}\right)\right\}. \tag{1.3}$$

The system of Eqs. (1.1) is not closed, unless the relations of the induced current densities with the electric field $\vec{E}$ and magnetic induction $\vec{B}$ are given. Such relations, which determine the electromagnetic properties of material media, are

B. Shokri, A. A. Rukhadze, *Electrodynamics of Conducting Dispersive Media*, Springer Series on Atomic, Optical, and Plasma Physics 111,
https://doi.org/10.1007/978-3-030-28968-3_1

called material equations. However, material equations and, consequently, the corresponding system of the field equations are not unique. The electromagnetic field equations in a medium are usually written as [1, 2]

$$\begin{cases} \nabla \cdot \vec{D} = 4\pi\rho_0, & \nabla \times \vec{E} = -\dfrac{1}{c}\dfrac{\partial \vec{B}}{\partial t}, \\ \nabla \times \vec{H} = \dfrac{1}{c}\dfrac{\partial \vec{D}}{\partial t} + \dfrac{4\pi}{c}\vec{j}_0, & \nabla \cdot \vec{B} = 0, \end{cases} \tag{1.4}$$

where $\vec{H}$, the strength of the magnetic field, and $\vec{D}$, the electric induction, are related to the induced current densities:

$$\vec{j} = \frac{\partial \vec{P}}{\partial t} + c\nabla \times \vec{M}, \tag{1.5}$$

$$\vec{H} = \vec{B} - 4\pi\vec{M}, \tag{1.6}$$

$$\vec{D} = \vec{E} + 4\pi\vec{P}. \tag{1.7}$$

Here, $\vec{M}$ is the magnetization vector and $\vec{P}$ is the polarization vector of the medium. By substituting Eq. (1.5) into the system of Eqs. (1.1) and making use of relations (1.6) and (1.7) we obtain Eq. (1.4).

It must be noticed that the system of Eqs. (1.4) includes two additional vector quantities $\vec{D}$ and $\vec{H}$ instead of one vector quantity $\vec{j}$ characterizing the medium. This indicates that there exist some arbitrary forms of field equations depending on convenience. For example, sometimes, one can single out the conduction current from the polarization current as

$$\frac{\partial \vec{P}}{\partial t} \to \frac{\partial \vec{P}'}{\partial t} + \vec{j}_{\text{cond}}.$$

Such decomposition makes sense only for slowly varying fields, because for static fields we have

$$\frac{\partial \vec{P}'}{\partial t} = 0,$$

and

$$\vec{j} = \vec{j}_{\text{cond}} + c\nabla \times \vec{M}.$$

In this case, the conduction current density is determined in such a way that its integral over a certain surface crossing the medium is equal to the total current. As a result, the quantity $\vec{M}$ (magnetization) gives a non-zero current density even if the total current flowing through each surface is zero.[1]

In general, for the varying fields it is quite difficult to justify the division of the induced current in some parts and also the setting off the displacement current $(1/4\pi)\partial\vec{E}/\partial t$ from it. A more appropriate way is the introduction of only a single vector quantity, called the vector of electric induction,[2] through

$$\vec{D}'\left(\vec{r},t\right) = \vec{E}\left(\vec{r},t\right) + 4\pi\int_{-\infty}^{t} dt'\,\vec{j}\left(\vec{r},t'\right). \tag{1.8}$$

This vector quantity combines the densities of induced charges and currents with the displacement current. Then, the field equations (1.1) can be written as [2]:

$$\begin{cases} \nabla\cdot\vec{D}' = 4\pi\rho_0, & \nabla\times\vec{E} = -\dfrac{1}{c}\dfrac{\partial\vec{B}}{\partial t}, \\ \nabla\times\vec{B} = \dfrac{1}{c}\dfrac{\partial\vec{D}'}{\partial t} + \dfrac{4\pi}{c}\vec{j}_0, & \nabla\cdot\vec{B} = 0. \end{cases} \tag{1.9}$$

This system of equations should be completed by the material equation determined by the electric induction $\vec{D'}$. In contrast to the system (1.4), this system of field equations does not include the magnetic field $\vec{H}$ at all. As it was mentioned above, such a form of field equations is preferable for description of fast varying processes. However, the form of Eqs. (1.4) is more common and, therefore, in the discussion of material equations we begin with this form of field equations.

In linear electrodynamics, material equations are linear relations. In this case, for the constant electric and magnetic fields the material equations corresponding to Eqs. (1.4) can be written as:

$$D_i = \varepsilon_{ij}E_j, \qquad B_i = \mu_{ij}H_j.$$

Here, ε_{ij} and μ_{ij} are the tensors of electric permittivity and magnetic permeability, respectively. Determined by the concrete properties of the medium, they characterize its electromagnetic properties. These material equations are valid only for the sufficiently slowly varying fields. Another situation takes place for the fast varying fields, when the field variation time is much less than the characteristic relaxation time in the medium or the period of characteristic oscillations of the medium. Then,

---

[1]Then, instead of definition (1.7), one must introduce the electric induction as $\vec{E} + 4\pi\vec{p'}$.

[2]The quantity D' is called as the electric induction vector as the quantity $\vec{D}(r,t)$ in system (1.4). We hope that this does not become a reason for confusion.

the induced current $\vec{j}(t)$ should depend on the field not only at moment t, but also at all the previous moments before t. For example, the relaxation process arising in a medium under the action of a field originated at moment t and terminated for a while much less than the relaxation time will continue to be after the field termination. Therefore, for high-frequency fields we should use the following material equations:

$$\begin{cases} D_i(t) = \int\limits_{-\infty}^{t} dt' \hat{\varepsilon}_{ij}(t-t') E_j(t'), \\ B_i(t) = \int\limits_{-\infty}^{t} dt' \hat{\mu}_{ij}(t-t') H_j(t'). \end{cases} \tag{1.10}$$

This relation takes into account the memory effect on the electromagnetic properties of the medium. In this case, there is the frequency or temporal dispersion of the medium.

Quite natural is the following question: why the electric induction in Eq. (1.10), for example, depends only on the electric field and is not associated with the magnetic induction. One can answer this question by using the Faraday's law

$$\nabla \times \vec{E} = -\frac{1}{c}\frac{\partial \vec{B}}{\partial t},$$

which eliminates the magnetic induction $\vec{B}$ and expresses it in terms of $\vec{E}$. In this case, in the material equation, the spatial derivatives of $\vec{E}$ should arise from $\vec{D}$. Hence, we can conclude that material equations (1.10) are only valid for those fields that vary slowly in space or, in other words, when the terms with spatial derivatives can be neglected.

It is obvious that for the fields varying sharply in space, it is necessary to consider the field impacts of the far points on the electromagnetic properties of the medium at the given space point. In fact, because of the transport processes, the state of a certain point of the medium is determined not only by the field value at this point, but also by the fields of all regions of medium from which the influence of fields is transported as a result of matter transport. In other words, owing to the transport processes, for example, the action of the field can be transported from one space point to the other one. Therefore, instead of material equations (1.10), one must use the spatially non-local relations, which account for not only temporal (frequency) but also spatial dispersions. For the homogeneous isotropic and non-gyrotropic media, such relations can be written as:[3]

[3]About material equations for isotropic and gyrotropic media see Sect. 1.7.

$$\begin{cases} \vec{D}\left(\vec{r},t\right) = \int\limits_{-\infty}^{t} dt' \int d\vec{r'} \hat{\varepsilon}\left(t-t', |\vec{r}-\vec{r'}|\right) \vec{E}\left(\vec{r'},t'\right), \\ \vec{B}\left(\vec{r},t\right) = \int\limits_{-\infty}^{t} dt' \int d\vec{r'} \hat{\mu}\left(t-t', |\vec{r}-\vec{r'}|\right) \vec{H}\left(\vec{r'},t'\right). \end{cases} \tag{1.11}$$

From these relations, it follows that the electromagnetic properties of such media are determined by two functions of $\vec{r}$ and t.

It must be noted that for Eqs. (1.11) we restricted ourselves only to the isotropic media. The direct generalization of Eqs. (1.10) leads to a large number of functions (two tensors of $\vec{r}$, $\vec{r'}$, and $t - t'$) for description of the electromagnetic properties of the anisotropic media. At the same time, the material equation for $\vec{D}'\left(\vec{r},t\right)$ consists of one tensor function only. One can be easily convinced by writing the material equation for $\vec{D}'\left(\vec{r},t\right)$ in the linear electrodynamics along with taking into account temporal and spatial dispersions in the form of

$$D_i'\left(\vec{r},t\right) = \int\limits_{-\infty}^{t} dt' \int d\vec{r'} \hat{\varepsilon}_{ij}\left(t-t', \vec{r}, \vec{r'}\right) E_j\left(\vec{r'},t'\right). \tag{1.12}$$

The dependence of the kernel of this integral relation on $t - t'$ is the result of time homogeneity.[4] Also for the spatially homogeneous media, the kernel of Eq. (1.12) depends only on $\vec{r} - \vec{r'}$.

From Eq. (1.12) it follows that the linear electromagnetic properties of anisotropic media can be described by one tensor function $\hat{\varepsilon}_{ij}$ of $\vec{r}$, $\vec{r'}$, $t - t'$. Therefore, two tensors arising from the direct generalization of Eqs. (1.10) cannot be independent in the case of spatial dispersion. This indicates that the introduction of the magnetic field, apart from the electric induction, for the anisotropic media with spatial dispersion is not appropriate.

The field equations must be supplemented by boundary conditions. They follow from the field equations by integrating them over an infinitely thin layer enclosing the boundary. Let us consider a homogeneous surface between media 1 and 2 and assume that the surface normal $\vec{n}$ is directed from 1 to 2. Then, the continuity of the normal components of the magnetic induction vector, $\vec{B}_{1n} = \vec{B}_{2n}$, follows from $\nabla \cdot \vec{B} = 0$. Integration of $\nabla \times \vec{E} = -(1/c)\partial \vec{B}/\partial t$ leads to the continuity condition of the tangential components of the electric field on the interface, $\vec{E}_{1t} = \vec{E}_{2t}$. These two boundary conditions as a result of the field equations do not depend on the

[4]Of course, in the absence of time homogeneity, when the properties of the medium vary with time, the kernel of the integral relation (1.12) becomes a more complicated function, $\hat{\varepsilon}_{ij}\left(t, t'; \vec{r}, \vec{r'}\right)$.

specific properties of the media and consequently they are valid for both forms of field equations (1.4) and (1.9).

The complication of boundary conditions arises when the field equations contain the induced charges and currents. The integration of

$$\vec{\nabla} \times \vec{B} = \frac{1}{c}\frac{\partial \vec{E}}{\partial t} + \frac{4\pi}{c}\left(\vec{j} + \vec{j}_0\right),$$

results in the boundary condition $\left[\vec{n} \times \left(\vec{B}_2 - \vec{B}_1\right)\right] = 4\pi/c\left(\vec{i}' + \vec{i}_0\right)$ where $\vec{i}_0$ is the surface current density of external sources, and

$$\vec{i}' = -\frac{1}{4\pi}\int_1^2 dl \frac{\partial \vec{D}'}{\partial t} = -\int_1^2 dl\left(\frac{1}{4\pi}\frac{\partial \vec{E}}{\partial t} + \vec{j}\right),$$

describes the induced surface current density. Integration is performed over a pill box of infinitesimal thickness. For the form of field equations (1.4), when one uses the induced current expression (1.5), we find

$$\vec{i}' = -\frac{1}{4\pi}\int_1^2 dl \frac{\partial \vec{D}}{\partial t} + c\left[\vec{n} \times \left(\vec{M}_2 - \vec{M}_1\right)\right].$$

Then, for the tangential components of the magnetic field, we obtain the following boundary condition:

$$\left[\vec{n} \times \left(\vec{H}_2 - \vec{H}_1\right)\right] = \frac{4\pi}{c}\left(\vec{i} + \vec{i}_0\right),$$

where

$$\vec{i} = -\frac{1}{4\pi}\int_1^2 dl \frac{\partial \vec{D}}{\partial t}.$$

Thus, the tangential components of both the magnetic induction and the magnetic field strength undergo discontinuity on the interface of two media even in the absence of external field surface sources. The last boundary condition, which is a consequence of $\nabla \cdot \vec{E} = 4\pi(\rho + \rho_0)$ for the equation system (1.9), looks as

$$D'_{2n} - D'_{1n} = 4\pi(\sigma' + \sigma_0),$$

where σ_0 is the surface charge density of the external sources and

$$\sigma' = \frac{1}{4\pi}\int_1^2 dl\nabla \cdot \left[\vec{n} \times \left(\vec{D}' \times n\right)\right].$$

Another quite similar boundary condition can be written for the system (1.4).

It should be noted that to solve the system of field equations, it is necessary to have surface current (and charges) density. In other words, to solve the problems of electrodynamics for bounded systems, the knowledge of surface material equations is also necessary (see Sect. 2.9). Thus, field equations (1.9) supplemented by the material equation (1.12) and the surface material equation by taking into account the boundary conditions on the interface

$$\begin{cases} B_{1n} = B_{2n} & D'_{2n} - D'_{1n} = 4\pi(\sigma_0 + \sigma'), \\ \vec{E}_{1t} = \vec{E}_{2t}, & \left[\vec{n} \times \left(\vec{B}_2 - \vec{B}_1\right)\right] = \frac{4\pi}{c}\left(\vec{i}_0 + \vec{i}'\right), \end{cases} \tag{1.13}$$

alongside with the conditions at infinity allow us to uniquely determine the electromagnetic fields in space.[5]

1.2 Tensor of Complex Dielectric Permittivity

By applying the Fourier-series expansion, the electromagnetic field in a medium can be expressed as a sum of monochromatic components of the type exp($-\imath\omega t$). Such expansion is known as spectral resolution as well. For the monochromatic electromagnetic field with frequency ω the material equation (1.12) becomes

$$D'_i\left(\vec{r}\right) = \int_{-\infty}^{\infty} d\omega \int dr' \varepsilon_{ij}\left(\omega, \vec{r}, \vec{r}'\right)\hat{E}_j\left(\vec{r}'\right), \tag{1.14}$$

where

[5]When the material equation, which determines $\vec{D}'$, could be written in terms of high derivatives of the electric field, it is necessary to use the supplementary boundary conditions, which are related to the order increase of the differential field equations. However, for non-local integral relation, such supplementary boundary conditions are not necessary.

$$\varepsilon_{ij}\left(\omega, \vec{r}, \vec{r'}\right) = \int_{0}^{\infty} dt \exp\left(\imath\omega t\right)\widehat{\varepsilon}_{ij}\left(t, \vec{r}, \vec{r'}\right). \tag{1.15}$$

The tensor $\widehat{\varepsilon}_{ij}\left(t, \vec{r}, \vec{r'}\right)$ is a real function of its variables, since it connects the real quantities $\vec{D'}\left(\vec{r}, t\right)$ and $\vec{E}\left(\vec{r}, t\right)$. Tensor $\varepsilon_{ij}\left(\omega, \vec{r}, \vec{r'}\right)$, however, appears as a complex function even when the variable ω is real. If one represents

$$\varepsilon_{ij}\left(\omega, \vec{r}, \vec{r'}\right) = \varepsilon'_{ij}\left(\omega, \vec{r}, \vec{r'}\right) + \imath\varepsilon''_{ij}\left(\omega, \vec{r}, \vec{r'}\right),$$

then from Eq. (1.15) and the reality of $\widehat{\varepsilon}_{ij}\left(t, \vec{r}, \vec{r'}\right)$ it follows that

$$\begin{cases} \varepsilon_{ij}\left(\omega, \vec{r}, \vec{r'}\right) = \varepsilon^{*}_{ij}\left(-\omega, \vec{r}, \vec{r'}\right), \\ \varepsilon'_{ij}\left(\omega, \vec{r}, \vec{r'}\right) = \varepsilon'_{ij}\left(-\omega, \vec{r}, \vec{r'}\right), \\ \varepsilon''_{ij}\left(\omega, \vec{r}, \vec{r'}\right) = -\varepsilon''_{ij}\left(-\omega, \vec{r}, \vec{r'}\right). \end{cases} \tag{1.16}$$

If the medium is homogeneous and unbounded in space, then the kernel of the integral relation (1.12) depends on $\vec{r} - \vec{r'}$:

$$D'_{i}\left(\vec{r}, t\right) = \int_{-\infty}^{t} dt' \int d\vec{r'}\,\widehat{\varepsilon}_{ij}\left(t - t', \vec{r} - \vec{r'}\right) E_{j}\left(\vec{r'}, t\right). \tag{1.17}$$

In this case, by applying the Fourier integral expansion, the electromagnetic field can be represented as the sum of the plane monochromatic waves of the type $\sim \exp\left(-\imath\omega t + \imath\vec{k}\cdot\vec{r}\right)$. For such waveforms, Eq. (1.17) reduces to

$$D'_{i} = \varepsilon_{ij}\left(\omega, \vec{k}\right) E_{j}, \tag{1.18}$$

where

$$\begin{aligned} \varepsilon_{ij}\left(\omega, \vec{k}\right) &= \int d\vec{r} \exp\left(-\imath\vec{k}\cdot\vec{r}\right)\varepsilon_{ij}\left(\omega, \vec{r}\right) \\ &= \int_{0}^{\infty} dt \int d\vec{r}\,\widehat{\varepsilon}_{ij}\left(t, \vec{r}\right) \exp\left(\imath\omega t - \imath\vec{k}\cdot\vec{r}\right). \end{aligned} \tag{1.19}$$

In the following, the quantity $\varepsilon_{ij}\left(\omega, \vec{k}\right)$ is called the tensor of dielectric permittivity of the medium. The dependence of this tensor on ω determines frequency dispersion, whereas the dependence of $\varepsilon_{ij}\left(\omega, \vec{k}\right)$ on the wave vector $\vec{k}$, stipulated by non-locality of the material equation (1.17), characterizes spatial dispersion.[6]

In general, the quantity $\varepsilon_{ij}\left(\omega, \vec{k}\right)$ is a complex function of real variables ω and $\vec{k}$. Based on Eq. (1.19), by considering $\widehat{\varepsilon}_{ij}\left(t, \vec{r}\right)$ as a real function, we obtain the following relations for the real and imaginary parts of the complex tensor $\varepsilon_{ij}\left(\omega, \vec{k}\right)$:

$$\begin{cases} \varepsilon_{ij}\left(\omega, \vec{k}\right) = \varepsilon_{ij}^{*}\left(-\omega, -\vec{k}\right), \\ \varepsilon_{ij}^{\prime}\left(\omega, \vec{k}\right) = \varepsilon_{ij}^{\prime}\left(-\omega, -\vec{k}\right), \\ \varepsilon_{ij}^{\prime\prime}\left(\omega, \vec{k}\right) = -\varepsilon_{ij}^{\prime\prime}\left(-\omega, -\vec{k}\right). \end{cases} \tag{1.20}$$

For spatially homogeneous and unbounded media, it is convenient to introduce one more additional quantity, which characterizes their electromagnetic properties. Let us represent the electromagnetic field, in this case, by means of Fourier integral expansion in terms of $\exp\left(\imath\vec{k}\cdot\vec{r}\right)$. For such fields, material equation (1.17) takes the form of

$$D_i^{\prime}(t) = \int_{-\infty}^{t} dt^{\prime}\, \varepsilon_{ij}\left(t - t^{\prime}, \vec{k}\right) E_j(t^{\prime}), \tag{1.21}$$

where

$$\varepsilon_{ij}\left(t, \vec{k}\right) = \int d\vec{r}\, \exp\left(-\imath\vec{k}\cdot\vec{r}\right) \widehat{\varepsilon}_{ij}\left(t, \vec{r}\right). \tag{1.22}$$

Tensor $\varepsilon_{ij}\left(t, \vec{k}\right)$, similar to $\varepsilon_{ij}\left(\omega, \vec{k}\right)$, as a complex function of the real $\vec{k}$ has the following properties:

$$\begin{cases} \varepsilon_{ij}^{\prime}\left(t, \vec{k}\right) = \varepsilon_{ij}^{\prime}\left(t, -\vec{k}\right), \\ \varepsilon_{ij}^{\prime\prime}\left(t, \vec{k}\right) = -\varepsilon_{ij}^{\prime\prime}\left(t, -\vec{k}\right). \end{cases} \tag{1.23}$$

[6]Term "spatial dispersion" was introduced first by Gertsentein [3].

The dependence of $\varepsilon_{ij}\left(\omega, \vec{k}\right)$ on the wave vector $\vec{k}$ results in this fact that even in an isotropic and non-gyrotropic medium, the tensor form of dielectric permittivity is conserved. In fact, due to the rotational invariance, in the case of the isotropic and non-gyrotropic media, the tensor $\varepsilon_{ij}\left(\omega, \vec{k}\right)$ can be composed of δ_{ij} and $k_i k_j$ [3, 4]:

$$\varepsilon_{ij}\left(\omega, \vec{k}\right) = \left(\delta_{ij} - \frac{k_i k_j}{k^2}\right)\varepsilon^{\mathrm{tr}}(\omega, k) + \frac{k_i k_j}{k^2}\varepsilon^{l}(\omega, k). \tag{1.24}$$

Coefficients $\varepsilon^{\mathrm{tr}}(\omega, k)$ and $\varepsilon^{l}(\omega, k)$ are called transverse and longitudinal dielectric permittivities of the isotropic media, respectively. They are complex functions of real frequency and wave vector. Furthermore, based on Eq. (1.20), they satisfy the following relations

$$\begin{cases} \varepsilon^{\mathrm{tr}'}(\omega, k) = \varepsilon^{\mathrm{tr}'}(-\omega, k), \\ \varepsilon^{\mathrm{tr}''}(\omega, k) = -\varepsilon^{\mathrm{tr}''}(-\omega, k), \\ \varepsilon^{l'}(\omega, k) = \varepsilon^{l'}(-\omega, k), \\ \varepsilon^{l''}(\omega, k) = -\varepsilon^{l''}(-\omega, k). \end{cases} \tag{1.25}$$

Earlier, we discussed two different forms of field equations in media. Now, it is appropriate to discuss the relation between material equations (1.11) and (1.12) for isotropic and non-gyrotropic homogeneous media as well as the relation between Maxwell's equations (1.4) and (1.9). However, beforehand let us consider some consequences arising from material equations (1.11).

For the monochromatic electromagnetic field with frequency ω, from Eqs. (1.11) it follows that

$$\begin{cases} \vec{D}\left(\vec{r}\right) = \int d\vec{r'}\,\varepsilon\left(\omega, \vec{r}, \vec{r'}\right)\vec{E}\left(\vec{r'}\right), \\ \vec{B}\left(\vec{r}\right) = \int d\vec{r'}\,\mu\left(\omega, \vec{r}, \vec{r'}\right)\vec{H}\left(\vec{r'}\right), \end{cases} \tag{1.26}$$

where

$$\begin{cases} \varepsilon\left(\omega, \vec{r}, \vec{r'}\right) = \int\limits_0^\infty dt \exp(\imath\omega t)\hat{\varepsilon}\left(t, \vec{r}, \vec{r'}\right), \\ \mu\left(\omega, \vec{r}, \vec{r'}\right) = \int\limits_0^\infty dt \exp(\imath\omega t)\hat{\mu}\left(t, \vec{r}, \vec{r'}\right), \end{cases} \tag{1.27}$$

The reality of fields $\vec{D}\left(\vec{r}, t\right)$, $\vec{E}\left(\vec{r}, t\right)$, $\vec{B}\left(\vec{r}, t\right)$, and $\vec{H}\left(\vec{r}, t\right)$ clearly results in the reality of functions $\hat{\varepsilon}\left(t, \vec{r}, \vec{r'}\right)$ and $\hat{\mu}\left(t, \vec{r}, \vec{r'}\right)$. Then, quite analogous to relations (1.15), from (1.27) it follows that the real parts of $\varepsilon\left(\omega, \vec{r}, \vec{r'}\right)$ and

$\mu\left(\omega, \vec{r}, \vec{r'}\right)$ are even functions of ω, whereas their imaginary parts are odd functions of ω.

In the case of homogeneous unbounded media, the functions of $\varepsilon\left(\omega, \vec{r}, \vec{r'}\right)$ and $\mu\left(\omega, \vec{r}, \vec{r'}\right)$ depend only on $|\vec{r} - \vec{r'}|$. It allows us to represent the electromagnetic fields as a sum of plane waves of the type $\exp\left(-\imath\omega t + \imath\vec{k}\cdot\vec{r}\right)$ and obtain from the material Eqs. (1.11) the following relations between the Fourier components:

$$\begin{cases} \vec{D} = \varepsilon(\omega, k)\vec{E}, \\ \vec{B} = \mu(\omega, k)\vec{H}, \end{cases} \tag{1.28}$$

where

$$\begin{cases} \varepsilon(\omega, k) = \int\limits_0^\infty dt \int d\vec{r} \exp\left(-\imath\vec{k}\cdot\vec{r} + \imath\omega t\right)\hat{\varepsilon}\left(t, \vec{r}\right) = \int d\vec{r} \exp\left(-\imath\vec{k}\cdot\vec{r}\right)\varepsilon\left(\omega, \vec{r}\right), \\ \mu(\omega, k) = \int\limits_0^\infty dt \int d\vec{r} \exp\left(-\imath\vec{k}\cdot\vec{r} + \imath\omega t\right)\hat{\mu}\left(t, \vec{r}\right) = \int d\vec{r} \exp\left(-\imath\vec{k}\cdot\vec{r}\right)\mu\left(\omega, \vec{r}\right). \end{cases} \tag{1.29}$$

The latter quantities are called dielectric and magnetic permittivities, correspondingly. Quite analogous to Eqs. (1.25), from Eqs. (1.29) follows the relations between real and imaginary parts of these quantities

$$\begin{cases} \varepsilon'\left(\omega, \vec{k}\right) = \varepsilon'\left(-\omega, \vec{k}\right), & \varepsilon''(\omega, k) = -\varepsilon''\left(-\omega, \vec{k}\right), \\ \mu'\left(\omega, \vec{k}\right) = \mu'\left(-\omega, \vec{k}\right), & \mu''(\omega, k) = -\mu''\left(-\omega, \vec{k}\right). \end{cases} \tag{1.30}$$

Now, let us establish a connection between the field equations (1.4) and (1.9). For the plane waves of the type $\exp\left(-\imath\omega t + \imath\vec{k}\cdot\vec{r}\right)$, Eqs. (1.4) can be rewritten as

$$\begin{cases} \imath\vec{k}\cdot\vec{E}\varepsilon\left(\omega, \vec{k}\right) = 4\pi\rho_0\left(\omega, \vec{k}\right), & \vec{k}\times\vec{E} = \dfrac{\omega}{c}\vec{B}, \\ \dfrac{\imath}{\mu(\omega, k)}\vec{k}\times\vec{B} = -\dfrac{\imath\omega}{c}\varepsilon(\omega, k)\vec{E} + \dfrac{4\pi}{c}\vec{j}_0\left(\omega, \vec{k}\right), & \vec{k}\cdot\vec{B} = 0. \end{cases} \tag{1.31}$$

Here, $\rho_0\left(\omega, \vec{k}\right)$ and $\vec{j}_0\left(\omega, \vec{k}\right)$ are the Fourier components of the charge and current densities of the external field sources.

In the same case, from the field Eqs. (1.9) in view of Eq. (1.24), we obtain

$$\begin{cases} \imath \vec{k}\cdot\vec{E}\varepsilon^{l}(\omega,k)=4\pi\rho_0\left(\omega,\vec{k}\right), & \vec{k}\times\vec{E}=\frac{\omega}{c}\vec{B} \\ \imath\left(\vec{k}\times\vec{B}\right)_i=-\frac{\imath\omega}{c}\left\{\left(\delta_{ij}-\frac{k_ik_j}{k^2}\right)\varepsilon^{\mathrm{tr}}(\omega,k)+\frac{k_ik_j}{k^2}\varepsilon^{l}(\omega,k)\right\}E_j+\frac{4\pi}{c}j_{0_i}\left(\omega,\vec{k}\right), & \vec{k}\cdot\vec{B}=0. \end{cases} \tag{1.32}$$

Comparison of the first equations of relations (1.31) and (1.32) yields

$$\varepsilon^{l}(\omega,k)=\varepsilon(\omega,k). \tag{1.33}$$

Now, let us establish the relation between $\mu(\omega,k)$, $\varepsilon^{\mathrm{tr}}(\omega,k)$, and $\varepsilon^{l}(\omega,k)$. For this aim, we must eliminate the magnetic induction $\vec{B}$ from the third equations of relations (1.31) and (1.32) by making use of the field equation

$$\vec{B}=\frac{c}{\omega}\vec{k}\times\vec{E}, \tag{1.34}$$

and the external current density $\vec{j}_0\left(\omega,\vec{k}\right)$. After this operation, taking into account the Eq. (1.33), we obtain

$$\frac{\omega}{c}\left(\delta_{ij}-\frac{k_ik_j}{k^2}\right)\left(\varepsilon^{l}-\varepsilon^{\mathrm{tr}}\right)E_j=\left(1-\frac{1}{\mu}\right)\frac{c}{\omega}\left[\vec{k}\times\left(\vec{k}\times\vec{E}\right)\right]_i.$$

From this relation, we find

$$1-\frac{1}{\mu(\omega,k)}=\frac{\omega^2}{c^2k^2}\left[\varepsilon^{\mathrm{tr}}(\omega,k)-\varepsilon^{l}(\omega,k)\right], \tag{1.35}$$

which expresses the magnetic permeability of an isotropic medium $\mu(\omega,k)$ in terms of longitudinal and transverse dielectric permittivities. For isotropic and non-gyrotropic media, using Eq. (1.24), one can rewrite material equation (1.18) as

$$\vec{D}'=\varepsilon^{\mathrm{tr}}(\omega,k)\vec{E}-\left[\varepsilon^{\mathrm{tr}}(\omega,k)-\varepsilon^{l}(\omega,k)\right]\frac{\vec{k}\times\left(\vec{k}\times\vec{E}\right)}{k^2}. \tag{1.36}$$

Now, one can use relations (1.33), (1.35) and field equation (1.34) to rewrite this equation in the form

$$\vec{D}'=\varepsilon(\omega,k)\vec{E}-4\pi\frac{c}{\omega}\frac{\chi(\omega,k)}{\mu(\omega,k)}\left(\vec{k}\times\vec{B}\right), \tag{1.37}$$

where

$$\chi(\omega, k) = \frac{1}{4\pi}[\mu(\omega, k) - 1]. \tag{1.38}$$

This quantity is known as magnetic susceptibility of the isotropic medium. Of course, material equations (1.36) and (1.37) are completely equivalent, since each of them consists of two functions of frequency and wave vector, which describe the electromagnetic properties of an isotropic and non-gyrotropic medium.

In conclusion, let us introduce the concept of the complex tensor of conductivity. It must be noticed that it is more convenient to write the material equation as a relation between $\vec{j}\left(\vec{r}, t\right)$ and $\vec{E}\left(\vec{r}, t\right)$ instead of material equation (1.12):

$$j_i\left(\vec{r}, t\right) = \int\limits_{-\infty}^{t} dt' \int d\vec{r'} \widehat{\sigma}_{ij}\left(t - t', \vec{r}, \vec{r'}\right) E_j\left(\vec{r'}, t'\right). \tag{1.39}$$

Then, from Eqs. (1.8) and (1.12), one can easily obtain

$$\widehat{\varepsilon}_{ij}\left(t, \vec{r}, \vec{r'}\right) = \delta(t)\delta\left(\vec{r} - \vec{r'}\right)\delta_{ij} + 4\pi \int_{-\infty}^{t} dt' \widehat{\sigma}_{ij}\left(t', \vec{r}, \vec{r'}\right). \tag{1.40}$$

For a monochromatic electromagnetic field of the type $\exp(-\imath\omega t)$, formula (1.39) takes the form of

$$j_i\left(\vec{r}\right) = \int d\vec{r'} \sigma_{ij}\left(\omega, \vec{r}, \vec{r'}\right) E_j\left(\vec{r'}\right), \tag{1.41}$$

where

$$\sigma_{ij}\left(\omega, \vec{r}, \vec{r'}\right) = \int\limits_{0}^{\infty} dt \exp\left(\imath\omega t\right) \widehat{\sigma}_{ij}\left(t, \vec{r}, \vec{r'}\right). \tag{1.42}$$

From relations (1.14) and (1.41), in view of Eq. (1.8) or Eq. (1.40), it follows that[7]

---

[7]To drive Eq. (1.43) we have used the relation

$$\delta_+(\omega) = \frac{1}{2\pi} \int\limits_{0}^{\infty} dt \exp\left(\imath\omega t\right) = \frac{1}{2}\delta(\omega) + \mathcal{P}\frac{\imath}{2\pi\omega}$$

where $\mathcal{P}$ denotes the principal value for singularity $\omega = 0$.

$$\varepsilon_{ij}\left(\omega, \vec{r}, \vec{r'}\right) = \delta\left(\vec{r} - \vec{r'}\right)\delta_{ij} + 8\pi^2\sigma_{ij}\left(\omega, \vec{r}, \vec{r'}\right)\delta_+(\omega). \tag{1.43}$$

For a homogeneous and unbounded medium, from Eq. (1.39) one can obtain the following relation between Fourier components of induced current density $\vec{j}$ and electric field $\vec{E}$:

$$j_i = \sigma_{ij}\left(\omega, \vec{k}\right)E_j, \tag{1.44}$$

where

$$\sigma_{ij}\left(\omega, \vec{k}\right) = \int_0^\infty dt \int d\vec{r}\, \exp\left(\imath\omega t - \imath\vec{k}.\vec{r}\right)\widehat{\sigma}_{ij}\left(t, \vec{r}\right)$$
$$= \int d\vec{r}\, \exp\left(-\imath\vec{k}.\vec{r}\right)\sigma_{ij}\left(\omega, \vec{r}\right). \tag{1.45}$$

The quantity $\sigma_{ij}\left(\omega, \vec{k}\right)$ is called the complex conductivity tensor of the medium. Formula (1.43) leads to a relation between the tensors of dielectric permittivity and conductivity:

$$\varepsilon_{ij}\left(\omega, \vec{k}\right) = \delta_{ij} + 8\pi^2\sigma_{ij}\left(\omega, \vec{k}\right)\delta_+(\omega). \tag{1.46}$$

If one denotes the real and imaginary parts of $\sigma_{ij}\left(\omega, \vec{k}\right)$ as $\sigma'_{ij}\left(\omega, \vec{k}\right)$ and $\sigma''_{ij}\left(\omega, \vec{k}\right)$, respectively, then from Eq. (1.46) we obtain

$$\begin{cases} \varepsilon'_{ij}\left(\omega, \vec{k}\right) = \delta_{ij} + 4\pi^2\sigma'_{ij}\left(\omega, \vec{k}\right)\delta(\omega) - \dfrac{4\pi}{\omega}\sigma''_{ij}\left(\omega, \vec{k}\right), \\ \varepsilon''_{ij}\left(\omega, \vec{k}\right) = \dfrac{4\pi}{\omega}\sigma'_{ij}\left(\omega, \vec{k}\right) + 4\pi^2\sigma''_{ij}\left(\omega, \vec{k}\right)\delta(\omega). \end{cases} \tag{1.47}$$

For isotropic and non-gyrotropic media, analogically to Eq. (1.24), one can write

$$\sigma_{ij}\left(\omega, \vec{k}\right) = \left(\delta_{ij} - \frac{k_ik_j}{k^2}\right)\sigma^{\mathrm{tr}}(\omega, k) + \frac{k_ik_j}{k^2}\sigma^l(\omega, k), \tag{1.48}$$

where $\sigma^{\mathrm{tr}}(\omega, k)$ and $\sigma^l(\omega, k)$ are transverse and longitudinal conductivities, correspondingly.

At last, from Eqs. (1.24) and (1.48), it follows that

$$\begin{cases} \varepsilon^{\text{tr}}(\omega, k) = 1 + 8\pi^2\sigma^{\text{tr}}(\omega, k)\delta_+(\omega), \\ \varepsilon^l(\omega, k) = 1 + 8\pi^2\sigma^l(\omega, k)\delta_+(\omega). \end{cases} \tag{1.49}$$

1.3 Dispersion of Dielectric Permittivity

In Sect. 1.2 of the present chapter, we introduced the concept of the dielectric permittivity $\varepsilon_{ij}\left(\omega, \vec{k}\right)$, which considers frequency and spatial dispersions. Now, let us consider the behavior of this function in the regions of small ω and $\vec{k}$, and its connection to the other quantities characterizing the electromagnetic properties of media.

In general, the electromagnetic field varying in time varies in space as well. But, in the case of sharply varying fields in space, it is necessary to consider the field influence of distant points on the electromagnetic properties of the medium at a given point. This means that it is inevitable to consider spatial dispersion. At the same time, if the spatial variation of the electromagnetic field is sufficiently smooth, then one can neglect the spatial correlation and only considers frequency dispersion. On the other hand, if the inhomogeneous field is static, then frequency dispersion can be neglected too. The electromagnetic properties of the medium in these two limiting cases can be described by the limiting expressions of $\varepsilon_{ij}\left(\omega, \vec{k}\right)$ when $k/\omega \to 0$ and $\omega/k \to 0$, correspondingly. We will discuss these limiting cases for isotropic and non-gyrotropic media, when relation (1.24) with two independent functions $\varepsilon^l(\omega, k)$ and $\varepsilon^{\text{tr}}(\omega, k)$ holds.

The external field sources can produce an inhomogeneous potential electric field in the medium in the static limit ($\omega = 0$):

$$\vec{E} = -\nabla\phi, \tag{1.50}$$

where ϕ is the scalar field potential. Expanding $\vec{E}\left(\vec{r}\right)$ and $\phi\left(\vec{r}\right)$ in terms of the Fourier integrals

$$\vec{E}\left(\vec{r}\right) = \int d\vec{k} \exp\left(\imath\vec{k}\cdot\vec{r}\right)\vec{E}\left(\vec{k}\right), \quad \phi\left(\vec{r}\right) = \int d\vec{k} \exp\left(\imath\vec{k}\cdot\vec{r}\right)\phi\left(\vec{k}\right), \tag{1.51}$$

from Eq. (1.50), we find

$$\vec{E}\left(\vec{k}\right) = -\imath\vec{k}\phi\left(\vec{k}\right).$$

Substituting this relation into the field equations (1.32), we obtain the following equation for the scalar potential $\phi\left(\vec{k}\right)$:

$$k^2\varepsilon^l(0,k)\phi\left(\vec{k}\right) = 4\pi\rho_0\left(0,\vec{k}\right). \tag{1.52}$$

If a rest point charge e located at $r = r_0$ represents the field source in the medium, then

$$\rho_0\left(\vec{r}\right) = e\delta\left(\vec{r} - \vec{r}_0\right), \qquad \rho_0\left(0,\vec{k}\right) = \frac{e}{(2\pi)^3}\exp\left(-\imath\vec{k}\cdot\vec{r}\right).$$

As a result, the electrostatic potential of a point charge in an isotropic medium is obtained from Eqs. (1.51) and (1.52):

$$\phi\left(\vec{r}\right) = \frac{e}{2\pi^2}\int d\vec{k}\,\frac{\exp\left[\imath\vec{k}\cdot\left(\vec{r} - \vec{r}_0\right)\right]}{k^2\varepsilon^l(0,k)}. \tag{1.53}$$

When $\varepsilon^l(0,k) = 1$, which takes place in vacuum, this potential coincides with Coulomb potential. But, if $\varepsilon^l(0,k) \neq 1$, then it shows that the field of a point charge in the medium is different from the Coulomb field. For example, if,

$$\varepsilon^l(0,k) = 1 + \frac{1}{k^2 r_{\rm scr}^2}, \tag{1.54}$$

then from Eq. (1.53), we obtain

$$\phi\left(\vec{r}\right) = \frac{e}{\left|\vec{r} - \vec{r}_0\right|}\exp\left(-\frac{\left|\vec{r} - \vec{r}_0\right|}{r_{scr}}\right). \tag{1.55}$$

This potential corresponds to the Debye screening of the field of a point charge in the medium [5]. Debye screening leads to the weakening of field strength on the large distances from the charge. This weakening is the result of this fact that the integrand in Eq. (1.53), according to the specific form of relation (1.54), remains finite when $k \to 0$. As an important point it must be noted that $\varepsilon^l(0,k)$ is positive and has a singularity of the type $1/k^2$ when $k \to 0$. The quantity defined as

$$r_{scr}^{-2} = \lim_{k\to 0}\lim_{\frac{\omega}{k}\to 0} k^2\left[\varepsilon^l(\omega,k) - 1\right], \tag{1.56}$$

characterizes the distance at which the static electric field of the charge is screened in the medium.

In the opposite limit, when $k/\omega \to 0$, the quantity

$$\varepsilon^l(\omega, 0) = \lim_{\frac{k}{\omega}\to 0} \varepsilon^l(\omega, k), \tag{1.57}$$

is the usual dielectric permittivity of the medium when only frequency dispersion is accounted. It must be noted that in this limit, in view of the neglect of spatial dispersion, the dielectric tensor of the isotropic media looks like

$$\varepsilon(\omega)\delta_{ij}.$$

From this fact, it follows that in the isotropic medium when spatial dispersion is negligible $\left(\vec{k} \to 0\right)$ one can build only one tensor of rank 2, which is δ_{ij}. From this, it must be concluded that[8]

$$\varepsilon^l(\omega, 0) = \varepsilon^{\text{tr}}(\omega, 0) = \varepsilon(\omega). \tag{1.58}$$

It is obvious that the quantity $\varepsilon(\omega)$ in the limit $\omega = 0$ does not describe the static field screening in the medium because not only it does not have any singularity of the type $1/k^2$ but also, what is more important, it does not depend on $\vec{k}$ at all.

From the above discussion, it follows that two different limits of the longitudinal permittivity $\varepsilon^l(\omega, k)$ can exist at $\omega = 0$ and $k = 0$:

$$\begin{cases} \varepsilon^l_\omega(0,0) = \lim\limits_{\omega\to 0} \lim\limits_{\frac{k}{\omega}\to 0} \varepsilon^l(\omega, k), \\ \varepsilon^l_k(0,0) = \lim\limits_{k\to 0} \lim\limits_{\frac{\omega}{k}\to 0} \varepsilon^l(\omega, k). \end{cases} \tag{1.59}$$

Of course, these relations are valid only when these limits exist.

In general, the dielectric permittivity $\varepsilon(\omega)$ may have a singularity when ω is small. For conducting media, for example, at small ω [see Eq. (1.47)]

$$\varepsilon(\omega) = \frac{4\pi\imath}{\omega}\sigma_0, \tag{1.60}$$

where σ_0 is the electrostatic conductivity of the conducting medium. At the same time, for dielectrics in the region of small ω, $\varepsilon(\omega)$ has no singularity,

---

[8] Spatial dispersion for isotropic media was first discussed by Vlasov [6]. But, he thought that the equality (1.58) is valid even when $\vec{k} \neq 0$.

$$\varepsilon(\omega) = \varepsilon_0, \tag{1.61}$$

where ε_0 is the static dielectric constant.

Now, let us consider the limiting cases $\omega \to 0$ and $k \to 0$ of the transverse dielectric permittivity $\varepsilon^{tr}(\omega, k)$ of the medium. It must be noted that the electromagnetic field equations (1.4), apart from the electrostatic field, may have a solution, which corresponds to a constant magnetic field produced by external sources in the medium. In this case, field equations (1.4) are more convenient than Eqs. (1.9) for the description of the constant magnetic field in the medium. Thus, suppose the electric field $\vec{E} = 0$ and at the same time the magnetic induction $\vec{B} \neq 0$ in the medium. Of course, the material equation (1.36), in this case, clearly is not acceptable because it does not contain the magnetic induction $\vec{B}$ in an explicit form. On the other hand, the form of the material equation (1.37) is more convenient and in this case looks as

$$\vec{D}' = -\frac{c}{\omega}\frac{4\pi\chi(\omega, k)}{\mu(\omega, k)}\left(\vec{k} \times \vec{B}\right). \tag{1.62}$$

However, this equation is not suitable for description of constant fields in the medium, because it is adopted to the field equations (1.9). To describe the constant magnetic field, it is more suitable to use field equations (1.4) with the material equation (1.28).

Now, let us clarify the conditions at which the magnetic field in the medium can be considered constant. For a slowly varying (in time) magnetic field in the medium, a weak varying electric field is produced through the field equation

$$\nabla \times \vec{E} = -\frac{1}{c}\frac{\partial \vec{B}}{\partial t}.$$

From this equation, the magnitude of varying electric field can be easily evaluated.[9] If ω is the characteristic frequency and $1/k$ is the characteristic size of the inhomogeneity of the electromagnetic field in the medium, then $E \sim (\omega/ck)B$. On the other hand, from Eqs. (1.1) and (1.5), it follows that

$$\nabla \times \vec{B} = \frac{1}{c}\frac{\partial}{\partial t}\left(\vec{E} + 4\pi\vec{P}'\right) + 4\pi\nabla \times \vec{M} + \frac{4\pi}{c}\vec{j}_{\text{cond}}. \tag{1.63}$$

Thus, the magnetic field in the medium can be considered constant if one can neglect the term $(1/c)\partial/\partial t\left(\vec{E} + 4\pi\vec{P}'\right)$ in Eq. (1.63). In dielectrics, $\vec{j}_{\text{cond}} = 0$ at

[9] Note that, at the same time, a constant electric field with arbitrary magnitude can exist in the medium.

small ω and, hence, $\vec{E} + 4\pi\vec{P}' = \vec{D}$. This means that these terms are of the order $\sim\varepsilon_0(\omega/c)E$ where ε_0 is the static dielectric constant. This evaluation also holds for conducting media, since the singularities of $\varepsilon(\omega)$ (see Eq. (1.60)) at small ω are included in $\vec{j}_{\text{cond}}$ and $\nabla \times \vec{M}$ terms in Eq. (1.63). For conducting media, $\vec{E} + 4\pi\vec{P}'$ is different from induction $\vec{D}$ due to the conduction current term. Since the singularity of the dielectric permittivity of conducting media in small frequency regions is caused by finite static conductivity (see expression (1.60)), then

$$\frac{1}{c}\frac{\partial\left(\vec{E} + 4\pi\vec{P}'\right)}{\partial t} \sim \imath\frac{\omega}{c}\varepsilon_0\vec{E},$$

where the quantity ε_0 remains finite when $\omega \to 0$.[10]

Thus, for such media, these terms of Eq. (1.63) can be neglected if

$$\frac{\omega}{ck}\sqrt{\varepsilon_0} \ll 1.$$

In other words, the magnetic field in such media can be considered constant if one passes to the limit $\omega/k \to 0$ in the material equation of media. Then, the static magnetic permeability of the media, in view of Eq. (1.35), is determined by

$$\mu_k(0,0) = \lim_{k\to 0}\lim_{\frac{\omega}{k}\to 0}\left\{1 - \frac{\omega^2}{c^2k^2}\left[\varepsilon^{\text{tr}}(\omega,k) - \varepsilon^{l}(\omega,k)\right]\right\}^{-1}. \tag{1.64}$$

We see that the static magnetic permeability differs from unity (or we have non-zero static susceptibility) if the right side of Eq. (1.35) is non-zero when $\omega/k = 0$, i.e., in the limit of the constant field.

In the opposite limiting case, when $k/\omega \to 0$, one can talk about weak spatial dispersion. In this case, we can expand the quantities $\varepsilon^{l}(\omega,k)$ and $\varepsilon^{\text{tr}}(\omega,k)$ in terms of k/ω

---

[10] The quantity ε_0 is not the static dielectric permittivity of the conducting medium because $E + 4\pi p'$ differs from induction $\vec{D}$. In the region of small frequencies, we have

$$\varepsilon(\omega) \approx \varepsilon_0 + 4\pi\imath\frac{\sigma_0}{\omega},$$

for conducting media. Here, the quantity ε_0 is often called the dielectric constant of the conducting medium. The second term in the aforementioned expression is related to the conduction current.

$$\begin{cases} \varepsilon^{l}(\omega, k) \approx \varepsilon(\omega) + \alpha(\omega)\dfrac{k^2c^2}{\omega^2}, \\ \varepsilon^{\mathrm{tr}}(\omega, k) \approx \varepsilon(\omega) + \beta(\omega)\dfrac{k^2c^2}{\omega^2}. \end{cases} \tag{1.65}$$

In this case, from Eq. (1.35) we obtain

$$\mu_{\omega}(\omega, 0) = \lim_{\frac{k}{\omega} \to 0} \mu(\omega, k) = \frac{1}{1 + \alpha(\omega) - \beta(\omega)}. \tag{1.66}$$

This quantity is independent of the wave vector $\vec{k}$. Moreover, in the static limit ($\omega = 0$) this quantity does not coincide with the static magnetic permeability (1.64).

Specially, it must be noted that for the frequency dispersion of the magnetic permeability corresponding to the material equation (1.10), one should pay attention to the quantity $\mu(\omega, k)$ in the limit $\omega/k \to 0$, instead of the quantity (1.66). It turns out that $\mu(\omega, k)$ in the limit $\omega/k \to 0$ under certain conditions is independent of wave vector $\vec{k}$ but depends on frequency ω. It can be noted that in order to have a medium with the static magnetic permeability different from unity or, in other words, having magnetic properties, the expression $[\varepsilon^{\mathrm{tr}}(\omega, k) - \varepsilon^{\mathrm{tr}}(\omega, k)]$ must have a singularity of the type $(k/\omega)^2$ near the point $\omega/k = 0$. For non-magnetic media, this expression does not have such a singularity near the point $\omega/k = 0$ and, hence, in this medium, $\mu_k(0, 0) = 1$. At the same time, in the expansion of $[\varepsilon^{\mathrm{tr}}(\omega, k) - \varepsilon^{\mathrm{tr}}(\omega, k)]$ in terms of small (k/ω), if the coefficient in front of the term $(k/\omega)^2$ depends on ω, and not k, then $\mu(\omega, k)$, given by Eq. (1.35), will be a function of ω near the point $\omega/k = 0$. Namely, this quantity, in contrast to $\mu_{\omega}(\omega, 0)$ given by Eq. (1.66), represents the magnetic permeability of the medium with account of frequency dispersion. Later on, we denote it as $\mu_k(\omega)$ in order not to confuse it with $\mu_{\omega}(\omega, 0)$. Of course, it has a physical meaning only near the point of $\omega/k \to 0$.

In conclusion, it is noticeable that the frequency dispersion of the magnetic permeability $\mu_k(\omega)$ of the medium is valid only in the limited region of frequency near the point $\omega/k = 0$ where $1/k$ is the characteristic scale of field inhomogeneity in the medium. In the limit of $\omega = 0$, this quantity naturally coincides with the static magnetic permeability.

Now, let us discuss the limiting case of $\omega = 0$ and $k = 0$ for the dielectric permittivity tensor of anisotropic media. In an anisotropic medium, as in the isotropic medium, at $\omega = 0$, i.e., in the static limit, a potential electrostatic field may exist. For example, the scalar field potential produced by a point charge in an anisotropic medium is given by (compare to Eq. (1.53)):

$$\phi\left(\vec{r}\right) = \frac{e}{2\pi^2} \int d\vec{k} \frac{\exp\left[\imath\vec{k} \cdot \left(\vec{r} - \vec{r}_0\right)\right]}{k_i k_j \varepsilon_{ij}\left(0, \vec{k}\right)}. \tag{1.67}$$

The difference of tensor $\varepsilon_{ij}\left(0, \vec{k}\right)$ from a constant leads to the difference in the electric field of a point charge in such a medium from the field of this charge in the anisotropic medium without account of spatial dispersion. Moreover, if the quantity $k_i k_j \varepsilon_{ij}\left(0, \vec{k}\right)$ remains finite when $\vec{k} \to 0$, then in the anisotropic media, as in the isotropic medium, the screening of the electrostatic field of the charge takes place on large distances from it. It must be noted that the field screening can be anisotropic in the anisotropic medium.

In the opposite limit, when $k/\omega \to 0$, the quantity $\varepsilon_{ij}(\omega, 0)$ represents the usual dielectric permittivity of the anisotropic medium, which has only frequency dispersion (spatial dispersion has been completely neglected). For non-conducting media in the static limit, in general, all components of the tensor $\varepsilon_{ij}(\omega, 0)$ are finite. However, some of these components for conducting media may have singularities in the region of small ω, namely

$$\varepsilon_{ij}(\omega, 0) = \frac{4\pi\imath}{\omega}\sigma_{ij}^{0}, \tag{1.68}$$

where σ_{ij}^{0} is the static anisotropic conductivity of the conducting medium. Of course, the tensor (1.66) does not describe static field screening in the medium. Thus, in the anisotropic media, two different limits of the dielectric tensor $\varepsilon_{ij}\left(\omega, \vec{k}\right)$ exist at $\omega = 0$ and $\vec{k} = 0$.

Finally, let us note that field equations (1.4) are more appropriate for describing the static magnetic field in both anisotropic and isotropic media. Therefore, all of the above discussions about the criteria at which the magnetic field can be considered a constant hold in anisotropic media as well. This means that the magnetic field in the medium can be considered constant only near the point $\omega/k = 0$. Namely, the inequality $(\omega/ck)\sqrt{|\,\varepsilon_0\,|} \ll 1$ should hold where ω is the characteristic frequency, $1/k$ is the characteristic size of inhomogeneity, and ε_0 is the static dielectric constant. Then, near the point $\omega/k = 0$, the dielectric permittivity tensor $\varepsilon_{ij}\left(\omega, \vec{k}\right)$ has a singularity of the type k^2/ω^2, which corresponds to the existence of the static magnetic susceptibility. Thus, the medium possesses static magnetic properties.

1.4 Energy of the Electromagnetic Field in a Medium

External sources producing electromagnetic fields naturally change the energy of a medium due to the interaction of electromagnetic fields with sources of fields. The energy of this interaction is determined by the work, which is performed by the field against the external sources. Therefore, the work done during dt in volume $d\vec{r}$ is equal to

$$\vec{E}\left(\vec{r},t\right) \cdot \vec{j}_0\left(\vec{r},t\right) d\vec{r}\, dt, \tag{1.69}$$

where $\vec{E}\left(\vec{r},t\right)$ is the electric field, and $\vec{j}_0\left(\vec{r},t\right)$ is the current density of external sources. To obtain the total work done in the whole space during the action time of external sources till the moment t, we write

$$A(t) = \int\limits_{-\infty}^{t} dt' \int d\vec{r}\, \vec{E}\left(\vec{r},t'\right) \cdot \vec{j}_0\left(\vec{r},t'\right). \tag{1.70}$$

According to the energy conservation law, this work must be balanced by a change in the electromagnetic field energy W. In this case, the rate of charge in energy is determined by

$$\frac{dW}{dt} = -\frac{dA}{dt} = -\int d\vec{r}\, \vec{E}\left(\vec{r},t\right) \cdot \vec{j}_0\left(\vec{r},t\right). \tag{1.71}$$

Using field equations (1.4) to eliminate the current density of external sources in Eq. (1.71), we obtain

$$\frac{dW}{dt} = \int d\vec{r} \left\{ \frac{1}{4\pi} \left(\vec{H} \cdot \frac{\partial \vec{B}}{\partial t} + \vec{E} \cdot \frac{\partial \vec{D}}{\partial t} \right) + \frac{c}{4\pi} \nabla \cdot \left(\vec{E} \times \vec{H} \right) \right\}. \tag{1.72}$$

In a quite similar way, making use of field equations (1.9) leads to

$$\frac{dW}{dt} = \int d\vec{r} \left\{ \frac{1}{4\pi} \left(\vec{B} \cdot \frac{\partial \vec{B}}{\partial t} + \vec{E} \cdot \frac{\partial \vec{D}'}{\partial t} \right) + \frac{c}{4\pi} \nabla \cdot \left(\vec{E} \times \vec{B} \right) \right\}. \tag{1.73}$$

For an unbounded medium (only such a case will be considered), we may suppose that fields $\vec{E}$, $\vec{B}$, and $\vec{H}$ vanish at infinity. Therefore, the surface integrals obtained from the last terms of Eqs. (1.72) and (1.73) can be neglected. Besides, if the magnetic induction $\vec{B}$ and the magnetic field $\vec{H}$ vary continuously in the medium, then the rate of change in field energy in the unbound medium is determined by expressions (1.72) and (1.73) without the last terms:

$$\frac{dW}{dt} = \frac{1}{4\pi} \int d\vec{r} \left\{ \vec{B}\left(\vec{r},t\right) \cdot \frac{\partial \vec{B}\left(\vec{r},t\right)}{\partial t} + \vec{E}\left(\vec{r},t\right) \cdot \frac{\partial \vec{D}'\left(\vec{r},t\right)}{\partial t} \right\}. \tag{1.74}$$

An expression for the amount of the heat released in the medium can be derived from Eq. (1.74). To do this, let us consider the monochromatic field, which depends on time as $\sim e^{\imath\omega t}$ and take into account the reality of functions $\vec{E}\left(\vec{r},t\right)$, $\vec{B}\left(\vec{r},t\right)$, and $\vec{D}\left(\vec{r},t\right)$. So, we can write them as

$$\vec{E}(\vec{r},t) = \{\vec{E}(\vec{r},\omega)\exp(-\imath\omega t) + \vec{E}^{*}(\vec{r},\omega)\exp(\imath\omega t)\}. \quad (1.75)$$

Functions $\vec{B}\left(\vec{r},t\right)$ and $\vec{D}\left(\vec{r},t\right)$ can be written analogically. Substituting these expressions into Eq. (1.74) and averaging over time, we obtain the average amount of energy accumulated in the medium or equivalently the amount of the heat released in the medium per unit time (or energy given up by the medium[11])

$$Q = \frac{\overline{dW}}{dt} = \frac{\imath\omega}{4\pi}\int d\vec{r}\left\{\vec{E}\left(\vec{r},\omega\right)\cdot\vec{D}'^{*}\left(\vec{r},\omega\right) - \vec{E}^{*}\left(\vec{r},\omega\right)\cdot\vec{D}'\left(\vec{r},\omega\right)\right\}. \quad (1.76)$$

For monochromatic electromagnetic field, the material equation (1.12) looks as

$$D_i'\left(\vec{r},\omega\right) = \int d\vec{r}'\,\varepsilon_{ij}\left(\omega,\vec{r},\vec{r}'\right)\vec{E}_j\left(\vec{r}',\omega\right),$$

where according to Eq. (1.16)

$$\varepsilon_{ij}\left(-\omega,\vec{r},\vec{r}'\right) = \varepsilon_{ij}^{*}\left(\omega,\vec{r},\vec{r}'\right).$$

Considering these relations, we can rewrite Eq. (1.76) as

$$Q = \frac{\imath\omega}{4\pi}\int\int d\vec{r}\,d\vec{r}'\left\{\varepsilon_{ij}^{*}\left(\omega,\vec{r},\vec{r}'\right) - \varepsilon_{ji}\left(\omega,\vec{r}',\vec{r}\right)\right\}E_i\left(\vec{r},\omega\right)E_j^{*}\left(\vec{r}',\omega\right). \quad (1.77)$$

In a homogeneous weakly absorbing medium, for the plane monochromatic waves of the type $\exp\left(\imath\vec{k}\cdot\vec{r} - \imath\omega t\right)$, wave vector $\vec{k}$ is approximately a real quantity. Then, making use of the formula (1.77) for such waves, we obtain an expression for the amount of heat delivered per unit time per unit volume of the medium

$$\frac{Q}{V} = \frac{\imath\omega}{4\pi}\left\{\varepsilon_{ij}^{*}(\omega,\vec{k}) - \varepsilon_{ji}(\omega,\vec{k})\right\}E_iE_j^{*}, \quad (1.78)$$

[11]This case is possible when the medium is not in the thermodynamic equilibrium state.

where V is the volume of the medium. Relations (1.77) and (1.78) lead to the following important conclusions: If absorption is very weak and the quantity Q is practically negligible, then from Eq. (1.77) we find:

$$\varepsilon_{ij}^{*}\left(\omega, \vec{r}, \vec{r'}\right) = \varepsilon_{ji}\left(\omega, \vec{r'}, \vec{r}\right). \tag{1.79}$$

Quite similarly, from Eq. (1.78) we have

$$\varepsilon_{ij}^{*}\left(\omega, \vec{k}\right) = \varepsilon_{ji}\left(\omega, \vec{k}\right). \tag{1.80}$$

Thus, for non-dissipative media, the dielectric permittivity tensor is Hermitian. In the case of isotropic and non-gyrotropic media, when the dielectric permittivity tensor is of the form of Eq. (1.24), formula (1.78) can be significantly simplified

$$\frac{Q}{V} = \frac{\omega}{2\pi k^2}\left\{\varepsilon^{l''}(\omega, k)\left|\vec{k}.\vec{E}\right|^2 + \varepsilon^{\text{tr}''}(\omega, k)\left|\left(\vec{k}\times\vec{E}\right)\right|^2\right\}. \tag{1.81}$$

The first term of this expression determines the absorption of the longitudinal field $\left(\vec{E} \parallel \vec{k}\right)$ by the medium, and the second term gives the absorption of the transverse field $\left(\vec{E}\perp\vec{k}\right)$. In this sense, it is possible to talk about the longitudinal and transverse losses in the medium. This highlights an important property of the dielectric tensor of the isotropic medium in thermodynamic equilibrium. It is clear that the entropy of such a medium only increases and heat is released. In this case, $Q \geq 0$ and from Eq. (1.81) it follows that, for $\omega > 0$,

$$\varepsilon^{l''}(\omega, k) \geq 0, \quad \varepsilon^{\text{tr}''}(\omega, k) \geq 0. \tag{1.82}$$

It should be noted that by making use of Eqs. (1.28) and (1.33), Eq. (1.81) can be written as [2]

$$\frac{Q}{V} = \frac{\omega}{2\pi}\left\{\varepsilon''(\omega, k)\left|\vec{E}\right|^2 + \mu''(\omega, k)\left|\vec{H}\right|^2\right\}. \tag{1.83}$$

Moreover, one can obtain this expression directly from Eq. (1.72) by averaging over time.

According to Eq. (1.33) and inequality (1.82), we find $\varepsilon''(\omega, k) \geq 0$. However, we cannot get any analogous result from Eq. (1.83) about $\mu''(\omega, k)$ because the transverse electric field $\vec{E}^{\text{tr}}$ and the magnetic field $\vec{H}$ are related to each other. This means that inequalities $Q \geq 0$ and $\varepsilon''(\omega, k) \geq 0$ in Eq. (1.83) do not result in $Q \geq 0$ and $\mu''(\omega, k) \geq 0$.

The electromagnetic field considered above was assumed to be completely monochromatic with the frequency ω. Now, we consider the almost monochromatic field to determine its rate of change in energy. Actually, the field in a medium consists of a superposition of monochromatic components with frequencies near ω. This means that in the Fourier expansion

$$\vec{E}\left(\vec{r},t\right) = \int_{-\infty}^{+\infty} d\omega' \exp\left(-\imath\omega'\right)\vec{E}\left(\vec{r},\omega'\right),$$

the quantity $\vec{E}\left(\vec{r},\omega'\right)$ has sharp maximums in the vicinity of points $\omega' = \pm\,\omega$. These two frequency values, in view of the reality of the field $\vec{E}\left(\vec{r},t\right)$, are resulted from $\vec{E}^*\left(\vec{r},\omega\right) = \vec{E}\left(\vec{r},-\omega\right)$. Therefore, instead of Eq. (1.75), $\vec{E}\left(\vec{r},t\right)$ must be written as

$$\vec{E}\left(\vec{r},t\right) = E_0\left(\vec{r},t\right)\exp\left(-\imath\omega t\right) + E_0^*\left(\vec{r},t\right)\exp\left(\imath\omega t\right), \tag{1.84}$$

where $\vec{E}_0\left(\vec{r},t\right)$ is a slowly varying function on the period $2\pi/\omega$. It is obvious that

$$\vec{E}_0\left(\vec{r},t\right) = \int_0^{\infty} d\omega' \exp\left[\imath(\omega-\omega')t\right]\vec{E}\left(\vec{r},\omega'\right),$$

$$\vec{E}_0^*\left(\vec{r},t\right) = \int_{-\infty}^{0} d\omega' \exp\left[\imath(\omega+\omega')t\right]\vec{E}\left(\vec{r},\omega'\right).$$

Since $\vec{E}\left(\vec{r},\omega'\right)$ in above integrands is a function with sharp maximums, it is possible to expand it in terms of $(\omega \pm \omega')$. As a result, we can write the following approximate relations

$$\frac{\partial}{\partial t}\vec{E}_0\left(\vec{r},t\right) \approx \imath\int_0^{\infty} d\omega'(\omega-\omega')\vec{E}\left(\vec{r},\omega'\right),$$

$$\frac{\partial}{\partial t}\vec{E}_0^*\left(\vec{r},t\right) \approx -\imath\int_{-\infty}^{0} d\omega'(\omega+\omega')\vec{E}^*\left(\vec{r},\omega'\right).$$

The existence of the quantities $(\omega \pm \omega')$ on the right side of the above relations leads to the fact that $\vec{E}_0$ actually turns out to be a slowly varying function for a sufficiently sharp frequency distribution of the field. Clearly, quite analogous forms can be written for the electric induction $\vec{D}\left(\vec{r},t\right)$ and the magnetic induction $\vec{B}\left(\vec{r},t\right)$.

Now, let us obtain an approximate expression for the time derivative of the electric induction of the almost monochromatic field in terms of the slowly varying function $\vec{E}_0\left(\vec{r}, t\right)$. According to Eq. (1.14), we have

$$\frac{\partial D_i'\left(\vec{r}, t\right)}{\partial t} = -\imath \int_{-\infty}^{+\infty} d\omega' \int d\vec{r'}\, \omega' \varepsilon_{ij}\left(\omega', \vec{r}, \vec{r'}\right) E_j\left(\vec{r}, \omega'\right).$$

Since the integrand has sharp maximums near $\omega' = \pm\,\omega$, and consequently only the frequency regions near $(\omega' \pm \omega)$ have substantial contributions in the integration of almost monochromatic fields over ω', one can expand the integrand in terms of $\omega' \pm \omega$. Considering only the first two terms of this expansion and taking into account the above expressions for $\partial \vec{E}_0/\partial t$ and $\partial \vec{E}_0^*/\partial t$, we obtain

$$\begin{aligned}
&\frac{\partial D_i'\left(\vec{r}, t\right)}{\partial t} \approx -\imath\omega \exp\left(-\imath\omega t\right)\\
&\int d\vec{r'}\, \varepsilon_{ij}\left(\omega, \vec{r}, \vec{r'}\right) E_{0j}\left(\vec{r'}, t\right) + \imath\omega \exp\left(\imath\omega t\right)\\
&\int d\vec{r'}\, \varepsilon_{ij}^*\left(\omega, \vec{r}, \vec{r'}\right) E_{0j}^*\left(\vec{r'}, t\right)\\
&+\exp\left(-\imath\omega t\right) \int d\vec{r'} \frac{\partial}{\partial\omega}\left[\omega\varepsilon_{ij}\left(\omega, \vec{r}, \vec{r'}\right)\right] \frac{\partial E_{0j}\left(\vec{r'}, t\right)}{\partial t} + \exp\left(\imath\omega t\right)\\
&\int d\vec{r'} \frac{\partial}{\partial\omega}\left[\omega\varepsilon_{ij}^*\left(\omega, \vec{r}, \vec{r'}\right)\right] \frac{\partial E_{0j}^*\left(\vec{r'}, t\right)}{\partial t}.
\end{aligned}$$

Similar expressions can be obtained for $\partial \vec{B}\left(\vec{r}, t\right)/\partial t$. From these two expressions, in view of Eq. (1.84), we can obtain the rate of change in electromagnetic field energy in a medium. Actually, averaging Eq. (1.74) over period $2\pi/\omega$ results in:

$$\begin{aligned}
\overline{\frac{dW}{dt}} = &\frac{1}{4\pi} \int d\vec{r}\, d\vec{r'} \left\{ E_{0i}^*\left(\vec{r}, t\right) \frac{\partial E_{0j}\left(\vec{r'}, t\right)}{\partial t} \frac{\partial}{\partial\omega}\left[\omega\varepsilon_{ij}\left(\omega, \vec{r}, \vec{r'}\right)\right]\right.\\
&\left.+E_{0j}\left(\vec{r'}, t\right) \frac{\partial E_{0i}^*\left(\vec{r}, t\right)}{\partial t} \frac{\partial}{\partial\omega}\left[\omega\varepsilon_{ji}^*\left(\omega, \vec{r'}, \vec{r}\right)\right]\right\}\\
&+\frac{1}{4\pi} \int d\vec{r} \frac{\partial}{\partial t}\left[\vec{B}^*\left(\vec{r}, t\right) \cdot \vec{B}\left(\vec{r}, t\right)\right] + Q,
\end{aligned} \tag{1.85}$$

where Q is the heat released per unit time in the medium (see Eq. (1.77)).

Now, we can determine the quantitative criteria at which the medium is non-absorbing. Namely, it is the case when the last term in the right-hand side of Eq. (1.85) is negligible. Then, supposing $\varepsilon_{ji}^{*}\left(\omega, \vec{r}', \vec{r}\right) \approx \varepsilon_{ij}\left(\omega, \vec{r}, \vec{r}'\right)$, from Eq. (1.85), we find

$$\begin{aligned} \frac{\overline{dW}}{dt} = \frac{dU}{dt} = \frac{d}{dt}\Bigg\{ & \frac{1}{4\pi}\int d\vec{r}\left(\vec{B}_0^{*}\left(\vec{r}, t\right).\vec{B}_0\left(\vec{r}, t\right)\right) + \\ & + \frac{1}{4\pi}\int\int d\vec{r}\, d\vec{r}'\, E_{0i}^{*}\left(\vec{r}, t\right).E_{0j}\left(\vec{r}', t\right)\frac{\partial}{\partial\omega}\left[\omega\varepsilon_{ij}\left(\omega, \vec{r}, \vec{r}'\right)\right]\Bigg\}. \end{aligned} \tag{1.86}$$

Thus, in the case of non-absorbing media, the rate of change in electromagnetic field energy is the time derivative of the quantity U. Therefore, the quantity U can be regarded as the average energy of the electromagnetic field in the medium. For plane waves of the type $\exp\left(\imath\vec{k}\cdot\vec{r}\right)$, from Eq. (1.86), it follows that

$$U = \frac{1}{4\pi}\int d\vec{r}\left\{\vec{B}_0^{*}\cdot\vec{B}_0 + E_{0i}^{*}E_{0j}\frac{\partial}{\partial\omega}\left[\omega\varepsilon_{ij}\left(\omega, \vec{k}\right)\right]\right\}. \tag{1.87}$$

In the case of isotropic and non-gyrotropic media, in view of Eq. (1.24), Eq. (1.87) is simplified even further

$$\begin{aligned} U &= \frac{1}{4\pi}\int d\vec{r}\left\{\vec{B}_0^{*}\cdot\vec{B}_0 + \left|\vec{E}_0^{l}\right|^2\frac{\partial}{\partial\omega}\left(\omega\varepsilon^{l}\right) + \left|\vec{E}_0^{\mathrm{tr}}\right|^2\frac{\partial}{\partial\omega}\left(\omega\varepsilon^{\mathrm{tr}}\right)\right\} = \\ &= \frac{1}{4\pi}\int d\vec{r}\left\{\left|\vec{E}_0^{l}\right|^2\frac{\partial}{\partial\omega}\left(\omega\varepsilon^{l}\right) + \left|\vec{E}_0^{\mathrm{tr}}\right|^2\frac{\partial}{\partial\omega}\left(\omega\left[\varepsilon^{tr} - \frac{c^2k^2}{\omega^2}\right]\right)\right\}. \end{aligned} \tag{1.88}$$

Here, $\vec{E}_0^{l}$ and $\vec{E}_0^{\mathrm{tr}}$ are longitudinal (parallel to $\vec{k}$) and transverse ($\nabla\cdot\vec{E}_0 = 0$) components of the electric field. Specially, for media in thermodynamic equilibrium, from this equation, we obtain[12]

$$\frac{\partial}{\partial\omega}\left[\omega\varepsilon^{l}(\omega, k)\right] \geq 0, \qquad \frac{\partial}{\partial\omega}\left[\omega\left(\varepsilon^{\mathrm{tr}}(\omega, k) - \frac{k^2c^2}{\omega^2}\right)\right] \geq 0. \tag{1.89}$$

In the limit $k \to 0$, these inequalities coincide and turn into the usual form [2]

[12]Relations (1.89) are also the result of Kramers–Kronig dispersion relations (see Sect. 1.9).

$$\frac{d}{d\omega}[\omega\varepsilon(\omega)] \geq 0.$$

At the same time, it must be noted that from Eq. (1.88) by making use of Eqs. (1.33), (1.35), and (1.28), we obtain the usual form [2]

$$U = \frac{1}{4\pi}\int d\vec{r}\left\{\left|\vec{E}_0\right|^2 \frac{\partial}{\partial\omega}[\omega\varepsilon(\omega,k)] + \left|\vec{H}_0\right|^2 \frac{\partial}{\partial\omega}[\omega\mu(\omega,k)]\right\}, \tag{1.90}$$

for the isotropic media. In addition, this relation is usually used for non-absorbing isotropic media when only the frequency dispersion of the dielectric permittivity is considered [2]. In conclusion, it must be underlined once more that the quantity U given by Eqs. (1.87), (1.88), and (1.90) is the averaged field energy only when the absorption is negligible.

1.5 Electromagnetic Wave in a Medium

From the vacuum electrodynamics it is well-understood that the plane monochromatic electromagnetic waves $\exp\left(-\imath\omega t + \imath\vec{k}\cdot\vec{r}\right)$ can exist in the absence of external sources. In vacuum, ω and $\vec{k}$ are real values. Such a wave can exist in non-absorbing media as well. However, in absorbing media the situation is more complicated. If, for example, an electromagnetic field in such a medium at the initial time $t = 0$ originates from external sources, then at the subsequent times ($t > 0$) when the external sources are switched off the field will be damped due to the action of dissipative processes. At the same time, the electromagnetic waves damping in time takes place in the medium.

Below, we will consider the variation of electromagnetic fields produced by external sources at the initial time ($t = 0$) in a homogeneous and infinite medium, regarding that external sources are switched off at subsequent times ($t > 0$).

To solve this initial value problem, it is not sufficient to know the quantities $\vec{E}\left(0,\vec{r}\right)$, $\vec{B}\left(0,\vec{r}\right)$, and $\vec{D}'\left(0,\vec{r}\right)$ at the initial moment $t = 0$, but one must also have knowledge of the previous history of the field $\vec{E}\left(t,\vec{r}\right)$ in the medium. Actually, according to the material equation (1.12) of the homogenous infinite medium

$$D_i'\left(t,\vec{r}\right) = \int_{-\infty}^{t} dt' \int d\vec{r'}\,\widehat{\varepsilon}_{ij}\left(t-t',\vec{r}-\vec{r'}\right)E_j\left(\vec{r'},t'\right),$$

the quantity $D'\left(t, \vec{r}\right)$ is determined by the field not only at $t \geq 0$ but also at all of the prior time moments, $t < 0$. Thus, one can represent the electric induction as the sum of two terms $\vec{D'} = \vec{D}^{(0)} + \vec{D}^{(1)}$, where

$$D_i^{(0)}\left(\vec{r},t\right) = \int_{-\infty}^{0} dt' \int d\vec{r'}\,\widehat{\varepsilon}_{ij}\left(t-t',\vec{r}-\vec{r'}\right)E_j\left(\vec{r'},t'\right), \tag{1.91}$$

$$D_i^{(1)}\left(\vec{r},t\right) = \int_{0}^{t} dt' \int d\vec{r'}\,\widehat{\varepsilon}_{ij}\left(t-t',\vec{r}-\vec{r'}\right)E_j\left(\vec{r'},t'\right). \tag{1.92}$$

The quantity $\vec{D}^{(1)}$ depends only on time moments $t > 0$, whereas the quantity $\vec{D}^{(0)}$ depends on the previous history of the field. Therefore, to solve the initial value problem this quantity must be known. The physical reason of this necessity deals with frequency dispersion or, in other words, with the inertia and relaxation processes of the particles in the medium until the moment $t = 0$.[13]

Thus, for the initial value problem of our interest, quantities $\vec{B}\left(\vec{r},0\right)$ and $\vec{D}^{(0)}\left(\vec{r},t\right)$ have to be known. Furthermore, $\vec{D'}\left(\vec{r},0\right) = \vec{D}^{(0)}\left(\vec{r},0\right)$. Then, to solve the field equations we will use the Fourier transformations[14] [7]

$$\vec{E}\left(\vec{r},t\right) = \frac{1}{(2\pi)^4}\int d\vec{k}\exp\left(\imath\vec{k}\cdot\vec{r}\right)\int_{-\infty+\imath\sigma}^{+\infty+\imath\sigma} d\omega\exp\left(-\imath\omega t\right)\vec{E}\left(\vec{k},\omega\right), \qquad (t \geq 0),$$

[13]If one solve the initial problem not only for the field equations but also for the equations of particles' motion, the knowledge of the quantity $\vec{D}^{(0)}$ is not necessary. But, in this case, beside of the fields initial values, the initial conditions of the particles' motion must be given.

[14]From Eqs. (1.93) it follows that the quantity $\vec{E}\left(\vec{k},\omega\right)$ as a function of complex variable ω has no singularity in the complex plane ω above the line $\mathrm{Im}\omega = \sigma$ if the electric field $\vec{E}\left(\vec{r},t\right)$ increases in time not faster than $\exp(\sigma t)$. The same situation happens for

$$\varepsilon_{ij}(\omega,k) = \int_0^{\infty} dt \int d\vec{r}\exp\left(-\imath\vec{k}\cdot\vec{r}+\imath\omega t\right)\widehat{\varepsilon}_{ij}(t,r).$$

$$\vec{E}\left(\vec{k},\omega\right)=\int d\vec{r}\exp\left(-\imath\vec{k}\cdot\vec{r}\right)\times\int\limits_0^\infty dt\exp\left(\imath\omega t\right)\vec{E}\left(\vec{r},t\right),\qquad (\mathrm{Im}\omega=\sigma\geq 0).\tag{1.93}$$

Here, we have used the one-sided Fourier transformation in time because source free field equations for the initial value problem are valid only when $t > 0$.

Quite similar formulas must be considered for the electric and magnetic inductions. Then, from the field equations we obtain

$$\frac{\omega}{c}\vec{B}\left(\vec{k},\omega\right)-\left[\vec{k}\times\vec{E}\left(\vec{k},\omega\right)\right]=\frac{\imath}{c}\vec{B}\left(\vec{k},t=0\right),\qquad \vec{k}.\vec{B}=0,$$
$$\frac{\omega}{c}\vec{D'}\left(\vec{k},\omega\right)+\left[\vec{k}\times\vec{B}\left(\vec{k},\omega\right)\right]=\frac{\imath}{c}\vec{D'}\left(\vec{k},t=0\right),\qquad \vec{k}.\vec{D'}=0.$$

Here, $\vec{D'}\left(\vec{k},t=0\right)=\int d\vec{r}\exp\left(-\imath\vec{k}\cdot\vec{r}\right)\vec{D'}\left(\vec{r},t=0\right)$. An analogous relation holds for $\vec{B}\left(\vec{k},t=0\right)$. Taking Eqs. (1.91) and (1.92) into account for the electric induction $\vec{D'}=\vec{D}^{(0)}+\vec{D}^{(1)}$ and eliminating the magnetic induction $\vec{B}\left(\vec{k},\omega\right)$ from the above system of equations, we find

$$\begin{aligned}&\omega^2\vec{D}^{(1)}\left(\vec{k},\omega\right)+c^2\left[\vec{k}\times\left[\vec{k}\times\vec{E}\left(\vec{k},\omega\right)\right]\right]\\&=-\imath\omega\vec{D'}\left(\vec{k},t=0\right)+\imath c\left[\vec{k}\times\vec{B}\left(\vec{k},t=0\right)\right]-\omega^2\vec{D}^{0}\left(\vec{k},\omega\right),\end{aligned}\tag{1.94}$$

$$\vec{k}\cdot\vec{D}^{1}\left(\vec{k},\omega\right)=-\vec{k}\cdot\vec{D}^{0}\left(\vec{k},\omega\right).\tag{1.95}$$

By adding this system of equations to the material equation

$$D_i^{(1)}\left(\vec{k},\omega\right)=\varepsilon_{ij}\left(\omega,\vec{k}\right)E_j\left(\vec{k},\omega\right),\tag{1.96}$$

which follows from Eq. (1.92), we obtain a system of linear algebraic equations for $\vec{E}\left(\vec{k},\omega\right)$:

$$k_i\varepsilon_{ij}\left(\omega,\vec{k}\right)E_j\left(\vec{k},\omega\right)=-\vec{k}\cdot\vec{D}^{(0)}(k,\omega),\tag{1.97}$$

$$
\begin{aligned}
&\left\{\omega^2\varepsilon_{ij}\left(\omega,\vec{k}\right)-c^2k^2\left(\delta_{ij}-\frac{k_ik_j}{k^2}\right)\right\}E_j\left(\vec{k},\omega\right)\\
&=\imath\omega D_i'\left(\vec{k},t=0\right)+\imath c\left[\vec{k}\times\vec{B}\left(\vec{k},t=0\right)\right]_i-\omega^2D_i^{(0)}\left(\vec{k},\omega\right).
\end{aligned}
\tag{1.98}
$$

Now, let us consider an isotropic and non-gyrotropic medium with the dielectric permittivity tensor (1.24). In this case, the system of Eqs. (1.97) and (1.98) decomposes into two independent equations for the longitudinal $\vec{E}^{l}$ (parallel to wave vector $\vec{k}$) and transverse $\vec{E}^{tr}$ (perpendicular to $\vec{k}$) components of the electric field

$$
\begin{aligned}
&\varepsilon^l(\omega,k)\vec{E}^l\left(\vec{k},\omega\right)=-\vec{D}^{(0)l}\left(\vec{k},\omega\right),\\
&\left[\omega^2\varepsilon^{\rm tr}(\omega,k)-k^2c^2\right]\vec{E}^{\rm tr}\left(\vec{k},\omega\right)=-\omega^2\vec{D}^{(0){\rm tr}}\left(\vec{k},\omega\right)\\
&+\imath\omega\vec{D}'\left(\vec{k},t=0\right)+\imath c\left[\vec{k}\times\vec{B}\left(\vec{k},t=0\right)\right].
\end{aligned}
$$

Here, $\vec{D}^{(0)l}$ and $\vec{D}^{(0)\rm tr}$ are the longitudinal and transverse components of $\vec{D}^{(0)}$. Thus,

$$
\vec{E}^l\left(\vec{k},\omega\right)=-\frac{\vec{D}^{(0)l}\left(\vec{k},\omega\right)}{\varepsilon^l(\omega,k)},
\tag{1.99}
$$

$$
\vec{E}^{\rm tr}\left(\vec{k},\omega\right)=\frac{-\omega^2D^{(0)\rm tr}\left(\vec{k},\omega\right)+\imath\omega\vec{D'}\left(\vec{k},t=0\right)+\imath c\left[\vec{k}\times\vec{B}\left(\vec{k},t=0\right)\right]}{\omega^2\varepsilon^{\rm tr}(\omega,k)-c^2k^2}.
\tag{1.100}
$$

According to these expressions, one can represent the longitudinal $\left(\nabla\times\vec{E}^l\left(\vec{r},t\right)=0\right)$ and transverse components of $\left(\nabla.\vec{E}^{\rm tr}\left(\vec{r},t\right)\right)$ fields as:

$$
\vec{E}^l\left(\vec{r},t\right)=\int_0^\infty dt'\int d\vec{r'}\,G_+^l\left(\vec{r}-\vec{r'},t-t'\right)\nabla^2_{\vec{r'}}\vec{D}^{(0)l}\left(\vec{r'},t'\right),
\tag{1.101}
$$

$$\vec{E}^{\mathrm{tr}}\left(\vec{r},t\right)=\int\limits_{0}^{\infty}dt'\int d\vec{r}'\left\{\vec{D}'\left(\vec{r}',t=0\right)\frac{\partial}{\partial t}G_{+}^{\mathrm{tr}}\left(\vec{r}-\vec{r}',t-t'\right)+\right.$$
$$\left.+G_{+}^{\mathrm{tr}}\left(\vec{r}-\vec{r}',t-t'\right)\left[\frac{\partial^{2}\vec{D}^{(0)}\left(\vec{r}',t'\right)}{\partial t'^{2}}-c\nabla\times\vec{B}\left(\vec{r}',t=0\right)\right]\right\},\tag{1.102}$$

where

$$G_{+}^{l}\left(\vec{r},t\right)=\frac{1}{(2\pi)^{4}}\int\limits_{-\infty+\imath\sigma}^{+\infty+\imath\sigma}d\omega\exp\left(-\imath\omega t\right)\int d\vec{k}\frac{\exp\left(\imath\vec{k}\cdot\vec{r}\right)}{k^{2}\varepsilon^{l}(\omega,k)},\tag{1.103}$$

$$G_{+}^{\mathrm{tr}}\left(\vec{r},t\right)=\frac{1}{(2\pi)^{4}}\int\limits_{-\infty+\imath\sigma}^{+\infty+\imath\sigma}d\omega\exp\left(-\imath\omega t\right)\int d\vec{k}\frac{\exp\left(\imath\vec{k}\cdot\vec{r}\right)}{\omega^{2}\varepsilon^{\mathrm{tr}}(\omega,k)-k^{2}c^{2}},\tag{1.104}$$

are the longitudinal and transverse retarded Green's functions, correspondingly.

It is obvious that for examination of the time dependence of the electromagnetic field the retarded Green's functions form must be studied. To do this, it is appropriate to shift the integration contour in Eqs. (1.103) and (1.104) into the lower half-plane of the complex variable ω.[15] In this case, the integrals vanish on the line parallel to the real axes and infinitely far from it. In shifting the integration contour one must bypass the poles of the integrands and branch cuts caused by the presence of the branch points in the complex plane.[16]

Let us consider the contribution of the pole of the integrand of relation (1.103), which is related to the zeros of the longitudinal dielectric permittivity

$$\varepsilon^{l}(\omega,k)=0.\tag{1.105}$$

In this case, the integral over the closed contour around such a pole, which corresponds to the residue of the integrand of relation (1.103), results in the following form of time dependency

[15] At the same time, in shifting the integration contour in Eq. (1.103), analytical continuation of the integrands from upper half-plane into the lower half-plane of the complex variable ω is necessary. Then, it must be noticed that $\varepsilon^{l}(\omega,k)$ and $\vec{D}^{(0)l}\left(\omega,\vec{k}\right)$ as the functions of ω determined by the one-sided Fourier transformations (1.19) and (1.93) are analytic in the upper half-plane of the complex variable $\omega(\mathrm{Im}\ \omega=\sigma\geq 0)$. A band of finite width σ around the real axis can be an exception.

[16] For more details see Chap. 2.

$$\exp\left(-\imath\omega' t + \omega'' t\right),$$

where $\omega = \omega' + \imath\omega''$ is the solution of Eq. (1.105). This solution determines the dependency of frequency ω' and damping decrement $\gamma = -\omega''$ on the wave vector $\vec{k}$.[17] Note that, in general, different roots of Eq. (1.105) may correspond to one wave vector $\vec{k}$. But, after sufficiently long time only the most slowly damped solutions would become essential. These solutions with minimum decrement correspond to the nearest roots of Eq. (1.105) to the real axis.

The contribution from the branches of the longitudinal dielectric permittivity in integral (1.103) over the bank of these branch cuts of the plane of the complex variable ω does not result in a purely exponential time dependency [8]. This means that a continuous spectrum of frequencies corresponds to one wave vector $\vec{k}$. Specially, an interesting case takes place when near the branch cut the analytic continuation of the function $\varepsilon^l(\omega, k)$ is zero on the adjacent sheets of the complex variable ω. Then, at long time intervals, the main time dependence, which is related to the integration over the bank of the branch cuts of the plane of the complex variable ω, is exponential with a complex frequency, which is determined from Eq. (1.105) for the analytic continuation of the function $\varepsilon^l(\omega, k)$ on the adjacent sheets.

Quite analogous analysis based on expression (1.104) for transverse waves leads to

$$\omega^2 \varepsilon^{\mathrm{tr}}(\omega, k) - k^2 c^2 = 0, \tag{1.106}$$

which determines the frequency spectrum and damping decrement of the transverse field.

In the case of an anisotropic homogenous medium, the solution of the linear system of Eqs. (1.97) and (1.98) is proportional to $\Lambda^{-1}(\omega, k)$ where

$$\Lambda\left(\omega, \vec{k}\right) = \left|\delta_{ij}k^2 - k_i k_j - \frac{\omega^2}{c^2}\varepsilon_{ij}\left(\omega, \vec{k}\right)\right|, \tag{1.107}$$

is the determinant of the system of these linear equations. Therefore, the branch points and zeros of $\Lambda\left(\omega, \vec{k}\right)$ determine the time dependence of the field, which is caused by the properties of the medium and not by preparation of the initial state. Specially, in the case of the anisotropic medium, instead of the dispersion equations (1.105) and (1.106), the relations of oscillation frequency and damping decrement with wave vector are determined by

[17] The quantity ω'' is negative when the pole lies in the lower half-plane. But, if the pole lies in the upper half-plane ($\sigma \geq \omega'' > 0$), then the quantity $\gamma = -\omega''$ must be called as growth increment, which is possible only when the medium is non-equilibrium state (see Sect. 3.7).

$$\left| k^2\delta_{ij} - k_ik_j - \frac{\omega^2}{c^2}\varepsilon_{ij}\left(\omega, \vec{k}\right) \right| = 0. \tag{1.108}$$

It is easy to note that for the isotropic medium when the dielectric permittivity tensor $\varepsilon_{ij}\left(\omega, \vec{k}\right)$ is given by Eq. (1.24) the determinant $\Lambda\left(\omega, \vec{k}\right)$ decomposes into two factors as a result of which Eq. (1.108) splits into Eqs. (1.105) and (1.106).

1.6 Plane Monochromatic Waves in a Medium

As it was noted above, in the material medium and in the absence of absorption, similar to vacuum, the electromagnetic waves of the type

$$\exp\left[-\imath\omega t + \imath\vec{k}\cdot\vec{r}\right] \tag{1.109}$$

can exist. In vacuum the frequency ω and the wave vector $\vec{k}$, which are related to each other by

$$\vec{k} = \frac{\omega}{c}\vec{n}, \tag{1.110}$$

have real values. Here, $\vec{n}$ is the unit vector in the direction of wave's propagation. It is obvious that such a wave is a plane wave with the wavelength

$$\lambda = \frac{2\pi}{k} = 2\pi\frac{c}{\omega}. \tag{1.111}$$

To generally describe these electromagnetic wave's propagation in the material medium, it is necessary to introduce the complex quantities of ω and $\vec{k}$. In the previous section, an initial value problem for electromagnetic wave's propagation in the medium was considered. In other words, we considered the wave excitation by the action of an arbitrary initial perturbation. We obtained Eqs. (1.105), (1.106), and (1.108), which allow us to find out the complex frequency $\omega = \omega' + \imath\omega''$ as a function of the real wave vector $\vec{k}$. The real part ω' is the oscillation frequency, whereas the imaginary part ω'' is the damping decrement (or growth increment) of the wave amplitude in time. However, another approach is possible for the problem when the real frequency is given. In this case, one can investigate the propagation of a monochromatic electromagnetic wave of a given frequency ω in the medium. Such an approach of the problem is known as the boundary value problem. In vacuum, both approaches of the problem are equivalent because ω and $\vec{k}$ are real quantities. In the material medium this problem is more complicated.

Let us consider the propagation of the electromagnetic wave of the type (1.109) through a homogenous unbounded medium. Field equations (1.9) and (1.12), describing the electromagnetic field in the medium, reduce to the following system of linear algebraic equations

$$\begin{cases} k_i \varepsilon_{ij}\left(\omega, \vec{k}\right) E_j = 0, & \vec{k} \cdot \vec{B} = 0, \\ \vec{k} \times \vec{E} = \frac{\omega}{c} \vec{B}, & \left[\vec{k} \times \vec{B}\right]_i = -\frac{\omega}{c} \varepsilon_{ij}\left(\omega, \vec{k}\right) E_j. \end{cases} \tag{1.112}$$

It is noticeable that the second pair of the system of Eqs. (1.112) is a closed system itself and the first pair follows automatically from it. By eliminating the magnetic induction $\vec{B}$, the system of homogeneous equations for the components of the electric field $\vec{E}$ is easily derived,

$$\left\{k^2 \delta_{ij} - k_i k_j - \frac{\omega^2}{c^2} \varepsilon_{ij}\left(\omega, \vec{k}\right)\right\} E_j = 0. \tag{1.113}$$

Non-trivial solutions of this system of equations exist only if

$$\Lambda\left(\omega, \vec{k}\right) = \left| k^2 \delta_{ij} - k_i k_j - \frac{\omega^2}{c^2} \varepsilon_{ij}\left(\omega, \vec{k}\right)\right| = 0. \tag{1.114}$$

This is the dispersion equation of electromagnetic fields in the medium, relating the wave frequency ω to the wave vector $\vec{k}$ in an implicit form.

In vacuum, $\varepsilon_{ij}\left(\omega, \vec{k}\right) = \delta_{ij}$. Therefore, from Eq. (1.114) it follows that $k^2 = \omega^2/c^2$. But, in the material medium, the solution of Eq. (1.114) may be complex

$$\vec{k} = \vec{k'} + \imath \vec{k''},$$

even when ω is real. Here, $\vec{k'}$ and $\vec{k''}$ are real vectors. It must be noted that complex solutions for $\vec{k}$ are not necessarily related to being complex of the dielectric permittivity tensor; the imaginary part of $\vec{k}$ may be non-zero even for a real dielectric permittivity tensor. Actually, for the isotropic and non-gyrotropic medium in the absence of spatial dispersion we can write $\varepsilon_{ij}\left(\omega, \vec{k}\right) = \varepsilon(\omega)\delta_{ij}$. Therefore, from Eq. (1.114) it follows that $k^2 = (\omega^2/c^2)\varepsilon(\omega)$. When $\varepsilon(\omega)$ is real and negative, the roots of this equation are purely imaginary. In such a case, one can assert that the electromagnetic field does not penetrate into the medium.

In the general case of complex $\vec{k}(\omega)$, the wave of the type of Eq. (1.109) with spatial dependence $e^{\imath \vec{k} \cdot \vec{r}} = e^{\imath \vec{k'} \cdot \vec{r} - \vec{k''} \cdot \vec{r}}$ can be called "planar" only conditionally. It should be noted that the planes of constant wave phase, which are perpendicular to vector $\vec{k'}(\omega)$, do not coincide with the planes of constant wave amplitude, which are perpendicular to vector $\vec{k''}(\omega)$; in the direction parallel to $\vec{k''}(\omega)$ damping or increasing of the wave amplitude occurs. Therefore, such waves are called inhomogeneous plane waves in contrast to homogeneous plane waves for them the planes, mentioned above, coincide. The plane waves can be homogeneous when $\vec{k}(\omega)$ is real, or when $\vec{k''}(\omega)$ and $\vec{k'}(\omega)$ are parallel. The media in which a real ω corresponds to a real $\vec{k}(\omega)$ (more exactly, the imaginary part of $\vec{k}$ is negligibly small) are called transparent at given frequencies.

In the boundary value problem, one can determine only one component of $\vec{k}_n(\omega)$ from the dispersion equation (1.114) in a given direction by assuming that ω and two other components are real. For example, we meet such conditions in the problem of reflection and refraction of the plane monochromatic wave from the surface of a medium. The tangential components of the wave vectors of the incident, reflected, and refracted waves are given and are real, whereas the normal components of these waves must be determined from the dispersion equations.

Now, let us consider some simple cases of the plane monochromatic electromagnetic wave's propagation through a specified material medium. A particular simple case happens for transparent media. If, in this case, the medium is isotropic and non-gyrotropic, then the dispersion equation (1.114) splits into two independent equations

$$\varepsilon^{l}(\omega, k) = 0 \tag{1.115}$$

and

$$k^2 - \frac{\omega^2}{c^2}\varepsilon^{\mathrm{tr}}(\omega, k) = 0. \tag{1.116}$$

The first equation describes the longitudinal electromagnetic waves in the isotropic medium. These waves are called longitudinal because for them $\vec{k} \cdot \vec{E} \neq 0$ and $(\delta_{ij} - k_i k_j / k^2)E_j = 0$ (see Eqs. (1.112)) meaning that the electromagnetic field of the wave is parallel to the wave vector $\vec{k}$. The second equation is the dispersion equation of the transverse electromagnetic waves. These waves are called transverse because the electric and magnetic fields of these waves are perpendicular to the wave vector $\vec{k}$, i.e., $\vec{k} \cdot \vec{E} = 0$ and $(\delta_{ij} - k_i k_j / k^2)E_j \neq 0$.

By making use of relations (1.33) and (1.35), these equations can be written in the following forms

$$\varepsilon(\omega, k) = 0, \tag{1.117}$$

$$k^2 - \frac{\omega^2}{c^2}\varepsilon(\omega, k)\mu(\omega, k) = 0. \tag{1.118}$$

This form of the dispersion equations for longitudinal and transverse waves corresponds to the description of the propagation of electromagnetic fields in the isotropic and non-gyrotropic media by making use of field equations (1.4) and material equation (1.11).

As in vacuum [see Eq. (1.110)], in the transparent medium one can introduce the vector $\vec{n}$ by

$$\vec{k} = \frac{\omega}{c}\vec{n}. \tag{1.119}$$

The value of this vector in each direction of wave's propagation determines the wave's phase velocity. Actually, if this direction coincides with ox axis, then we can write

$$\exp\left(\imath\vec{k}\cdot\vec{r} - \imath\omega t\right) = \exp\left(\imath kx - \imath\omega t\right) = \exp\left[\imath\omega\left(\frac{n}{c}x - t\right)\right].$$

We see that the propagation velocity of the constant phase surface ($kx - \omega t = \text{const}$) is equal to

$$v_{\text{ph}} = \frac{\omega}{k} = \frac{c}{n(\omega)}. \tag{1.120}$$

This quantity is called the phase velocity of the wave. Of course, it is obvious that $v_{\text{ph}} \| \vec{n}$. The quantity $n(\omega)$ characterizes the difference of the phase velocity of the wave propagating in a given direction in the medium from the light velocity c, which is the propagation velocity of the electromagnetic waves in vacuum. Besides, the quantity $n(\omega)$ is called the refractive index of the wave in the medium.

In the transparent medium, one can also introduce wave's group velocity. To do this let us consider the propagation of a wave packet, which consists of a group of waves in a narrow frequency and wave vector intervals in the vicinity of ω and $\vec{k}$ in the medium. For simplicity, let us assume that this group consists only of two monochromatic waves

$$\begin{aligned}&\exp\left(\imath\vec{k}\cdot\vec{r} - \imath\omega t\right) + \exp\left[\imath\left(\vec{k} + \Delta\vec{k}\right)\cdot\vec{r} - \imath\left(\omega + \Delta\vec{\omega}\right)t\right]\\ &= \exp\left(\imath\vec{k}\cdot\vec{r} - \imath\omega t\right)\left[1 + \exp\left(\imath\Delta\vec{k}\cdot\vec{r} - \imath\Delta\omega t\right)\right],\end{aligned}$$

where $\Delta\omega$ and $\Delta\vec{k}$ are small quantities in comparison with ω and $\vec{k}$. It is clear that one can treat such a wave packet as a monochromatic wave with a slowly varying amplitude. Then, the propagation velocity of surfaces of constant wave amplitude is equal to

$$\vec{v}_g = \lim_{\Delta \vec{k} \to 0} \frac{\Delta \omega}{\Delta \vec{k}} = \frac{d\omega}{d\vec{k}} = \frac{c}{\frac{d}{d\omega}\left(\vec{n}\omega\right)}. \tag{1.121}$$

This quantity called the group velocity of the wave is the displacement velocity of wave amplitude.

Sometimes, the dispersion equations of longitudinal and transverse waves are written by using the quantities $\vec{n}$ and ω instead of $\vec{k}$ and ω. Thus, for isotropic and non-gyrotropic media the alternative forms of these equations can be written as

$$\varepsilon^l\left(\omega, n\frac{\omega}{c}\right) = 0, \qquad or \qquad \varepsilon\left(\omega, n\frac{\omega}{c}\right) = 0, \tag{1.122}$$

for longitudinal waves and

$$n^2 = \varepsilon^{\mathrm{tr}}\left(\omega, n\frac{\omega}{c}\right), \qquad or \qquad n^2 = \varepsilon\left(\omega, n\frac{\omega}{c}\right)\mu\left(\omega, n\frac{\omega}{c}\right), \tag{1.123}$$

for transverse waves. The wave refractive index does not depend on the wave's propagation direction in an isotropic medium, resulting in this fact that both wave phase and group velocities are parallel to the wave's propagation direction. It should be noted that, in this case, the group velocity may have the opposite direction to that of wave's propagation. In general, the group and phase velocities do not necessarily have the same direction. The angle between vectors $\vec{v}_g$ and $\vec{v}_{\mathrm{ph}}$ can be acute or blunt. In the case of an acute angle, one can speak of a wave with positive dispersion or a forward wave. In the case of a blunt angle, the wave is called backward with negative dispersion. In the latter case, the direction of energy propagation can be opposite to the phase velocity of the wave. It can be easily shown that for the isotropic media Eq. (1.121) becomes

$$\vec{v}_g = \frac{\vec{n}}{n}\frac{c}{\frac{d(n\omega)}{d\omega}}. \tag{1.124}$$

Therefore, if $d(n\omega)/d\omega > 0$, the group velocity is parallel to the phase velocity, but if $d(n\omega)/d\omega < 0$, these velocities are in opposite directions. In this case, one can talk about a wave with the negative group velocity.

In the limit $k/\omega \to 0$, spatial dispersion can be neglected and, according to Eq. (1.58), we have

$$\varepsilon^l(\omega, 0) = \varepsilon^{\mathrm{tr}}(\omega, 0) = \varepsilon(\omega).$$

In this case, from the dispersion equation of longitudinal waves in the isotropic medium

$$\varepsilon(\omega) = 0, \tag{1.125}$$

one can determine the discrete frequencies ω_m of electromagnetic oscillations of the medium. The group velocity of such waves is zero, whereas the phase velocity remains arbitrary. Only when one takes spatial dispersion into account, as it appears in Eqs. (1.115) and (1.122), the frequency of longitudinal waves will become a function of the wave vector and consequently the group velocity turns out to be non-zero. The phase velocity, in this case, is determined by Eq. (1.120). Therefore, longitudinal waves, taking spatial dispersion into account in the medium, become equivalent to a branch of normal modes.

When spatial dispersion is neglected, the dispersion equation of transverse waves given by Eq. (1.123) looks as

$$n^2 = \varepsilon(\omega) \tag{1.126}$$

in the isotropic transparent medium. Since the dielectric permittivity $\varepsilon(\omega)$ is a single-valued function of frequency, one can deduce that in the isotropic medium, under the conditions of neglect of spatial dispersion, only one branch of transverse waves can propagate (of course, in this case, two different polarizations of the electric field are possible). However, if spatial dispersion is taken into account, then, in general, the dispersion equation (1.123) of transverse waves has several (infinite number is possible as well) solutions for $n_i^2(\omega)$. This means that several transverse waves with the same frequency but different refractive indices can propagate in the isotropic medium.

For anisotropic transparent media, the dispersion equation (1.114) of electromagnetic waves in terms of ω and $\vec{n}$ looks as

$$\left| n^2\delta_{ij} - n_i n_j - \varepsilon_{ij}\left(\omega, \frac{\omega}{c}\vec{n}\right) \right| = 0. \tag{1.127}$$

Separation of electromagnetic waves into the longitudinal and transverse waves in anisotropic media, generally, is not possible. In this case, the wave refractive index $n(\omega)$ in the anisotropic medium depends on the direction of wave's propagation. As a result, in the anisotropic medium, in contrast to the isotropic medium, the direction of the group velocity $\vec{v}_g$ is not the same as the direction of wave's propagation as well.

Without spatial dispersion, $\varepsilon_{ij}(\omega, 0) = \varepsilon_{ij}(\omega)$, the dispersion equation of electromagnetic waves in a transparent anisotropic medium

$$\left| n^2\delta_{ij} - n_i n_j - \varepsilon_{ij}(\omega) \right| = 0, \tag{1.128}$$

in the space of (n_x, n_y, n_z) determines a fourth order surface, "the surface of wave vectors." This equation is quadratic relative to n^2 for all directions. This means that two waves with the same frequency ω can propagate in each direction in the anisotropic medium. When spatial dispersion is taken into account, the situation becomes more complicated. In this case, the dispersion equation (1.127) generally represents a higher order surface. Therefore, more than two waves can propagate in each direction in the medium.

As discussed above, we considered the problem of the propagation of plane monochromatic waves of the type of Eq. (1.109) in transparent media when the wave vector $\vec{k}$ is a real quantity. But, as it was noted, in general, it is necessary to introduce the complex wave vector $\vec{k} = \vec{k'} + \imath\vec{k''}$. Then, one can separate the class of homogeneous plane waves for which the vectors $\vec{k'}$ and $\vec{k''}$ are parallel. Such electromagnetic waves may exist, for example, in isotropic dissipative media.

The analysis of homogenous plane wave's propagation in a dissipative medium is quite similar to wave's propagation in a transparent medium. The same dispersion equations as Eq. (1.114) for anisotropic media and Eqs. (1.115), (1.116) for isotropic media determine the complex wave vector $\vec{k}$ in each direction of wave's propagation. At the same time, if the imaginary part $\vec{k''}$ is large, then the "wave" conception is no longer applicable, because the wave amplitude changes considerably on a distance of the order of $\lambda = 2\pi/k'$ and, in fact, the electromagnetic field is exponentially damped in space. In addition, in this case, the conception of the wave's propagation direction fails to serve. However, one can consider this direction conventionally parallel to $\vec{k}$. In an absorbing non-transparent medium, beside the wave vector $\vec{k}$, the refractive index n defined as

$$k = \frac{\omega}{c} n, \tag{1.129}$$

is a complex quantity as well; $n = n' + \imath n''$ where n' is the refractive index, and n'' is the absorption coefficient of the medium.

As an example, let us consider the transverse electromagnetic waves in an isotropic absorbing medium when spatial dispersion is neglected. Then, from Eqs. (1.116) and (1.129) it follows that

$$\begin{cases} n' = \dfrac{1}{\sqrt{2}}\sqrt{\varepsilon'(\omega) + \sqrt{\varepsilon'^2(\omega) + \varepsilon''^2(\omega)}}, \\ n'' = \dfrac{1}{\sqrt{2}}\sqrt{-\varepsilon'(\omega) + \sqrt{\varepsilon'^2(\omega) + \varepsilon''^2(\omega)}}. \end{cases} \tag{1.130}$$

We see that the quantity $n(\omega)$ may be complex even when $\varepsilon(\omega)$ is real, or, when the absorption is absent. For example, if $\varepsilon'(\omega) < 0$ and $\varepsilon''(\omega) = 0$, from Eqs. (1.130), we obtain

$$n' = 0, \qquad n'' = \sqrt{|\varepsilon'(\omega)|}.$$

For conductors, in the low-frequency region when expression (1.60) is valid, from Eqs. (1.130) we find

$$n' = n'' = \sqrt{\frac{2\pi\sigma_0}{\omega}}.$$

Of course, the most general type of plane waves in material media is the inhomogeneous plane waves for which the directions of the vectors $\vec{k}'$ and $\vec{k}''$ are arbitrary (not parallel). Such waves arise, for example, in the problem of reflection and refraction of plane waves from the interface of two homogeneous media. To solve such problems, two real components of the wave vector $\vec{k}$ are usually given and from the dispersion equations (1.114)–(1.116), the third complex component is determined as a function of wave frequency and two given components of the wave vector. Finally, it is worth mentioned that for inhomogeneous waves, the concept of wave's propagation direction and refractive index given by Eq. (1.129) loses meaning.

1.7 Electromagnetic Wave's Propagation in a Weakly Spatially Dispersive Medium

In the study of electromagnetic waves in unbounded spatially homogenous media the material equation

$$D_i' = \varepsilon_{ij}\left(\omega, \vec{k}\right)E_j \tag{1.131}$$

was used. In this equation, there is no restriction whatsoever on the functional dependency of the dielectric permittivity tensor $\varepsilon_{ij}\left(\omega, \vec{k}\right)$ on the wave vector $\vec{k}$. If electromagnetic fields slowly vary in space, then $\varepsilon_{ij}\left(\omega, \vec{k}\right)$ can be expanded in powers of $\vec{k}$. Restricting only on first three terms of this expansion, we can write

$$\varepsilon_{ij}\left(\omega, \vec{k}\right) = \varepsilon_{ij}(\omega) + \imath\gamma_{ijl}(\omega)n_l + \alpha_{ijlm}(\omega)n_l n_m, \tag{1.132}$$

where $\vec{n} = c/\omega\vec{k}$. In the considered case of slowly spatially varying fields, the coefficients $\gamma_{ijl}(\omega)$ and $\alpha_{ijlm}(\omega)$ are small. As a result, expansion (1.132) is a power series of a small parameter. In such cases, one can talk about weak spatial dispersion.[18] Plane electromagnetic wave's propagation in material media with account of weak spatial dispersion can be investigated in more details compared to what was done in Sect. 1.6. Furthermore, when spatial dispersion is neglected, some specific effects arise during electromagnetic wave's propagation in such media.

From Eq. (1.132) it follows that the effects of weak spatial dispersion can become essential when the components of $\varepsilon_{ij}(\omega)$ taking only frequency dispersion into account are small. In this case, the second and third terms, which are related to spatial dispersion, are essential in Eq. (1.132). However, expansion (1.132) cannot always be used for the description of the effects of weak spatial dispersion in the media. If all components of $\varepsilon_{ij}(\omega)$ are large, then it is possible to consider only the first term of Eq. (1.132). At the same time, in this case, the components of $\varepsilon_{ij}^{-1}(\omega)$ are small and in expansion

$$\varepsilon_{ij}^{-1}\left(\omega, \vec{k}\right) = \varepsilon_{ij}^{-1}(\omega) + \imath g_{ijl}(\omega)n_l + \beta_{ijlm}(\omega)n_l n_m, \tag{1.133}$$

the second and third terms taking spatial dispersion into account play an essential role. Therefore, to describe the effects of weak spatial dispersion in the media, regardless of expansion (1.132), we will make use of Eq. (1.133).

Before going to the consideration of electromagnetic waves in media with weak spatial dispersion, it is better to know the symmetry of the coefficients $\gamma_{ijl}(\omega)$ and $\alpha_{ijlm}(\omega)$. The symmetric property of dielectric permittivity tensor (see Sect. 1.9)

$$\varepsilon_{ij}\left(\omega, \vec{k}\right) = \varepsilon_{ji}\left(\omega, -\vec{k}\right) \tag{1.134}$$

directly results in

$$\gamma_{ijl}(\omega) = -\gamma_{jil}(\omega), \qquad \alpha_{ijlm}(\omega) = \alpha_{jilm}(\omega).$$

The tensor $\alpha_{ijlm}(\omega)$ is also symmetric with respect to the indices l and m. Clearly, the coefficients $g_{ijl}(\omega)$ and $\beta_{ijlm}(\omega)$ have the same symmetric property as $\gamma_{ijl}(\omega)$ and $\alpha_{ijlm}(\omega)$. In the absence of absorption in the medium, $\varepsilon_{ij}^{*}\left(\omega, \vec{k}\right) = \varepsilon_{ji}\left(\omega, \vec{k}\right)$ and as a result the tensors $\gamma_{ijl}(\omega)$ and $\alpha_{ijlm}(\omega)$ are real in such a medium. Further simplification

[18]This parameter depends on the electromagnetic properties of the medium. In Sect. 2.3, it is shown that the ratio r_D/λ works as the small parameter of spatial dispersion for longitudinal waves in plasma where r_D is Debye radius and λ is the wavelength of the longitudinal field. This parameter is the order of v/c for transverse waves in plasma where v is the thermal velocity of particles; For crystal media and neutral gases, this parameter is a/λ where a is the lattice constant or the size of the gas molecule. Some effects related to weak spatial dispersion such as natural optical activity are widely known in physics [2, 9]. Weak spatial dispersion was studied in plasmas in [10–12]. Moreover, the phenomenological theory of electromagnetic waves in crystals was investigated in [13, 14]. Furthermore, corresponding microscopic theory was developed in [15–27].

of such quantities is related to the specific symmetry of the medium. Below, we restrict our study only on weakly absorptive media. Therefore, expansion coefficients in Eqs. (1.132) and (1.133) are always considered real quantities in what follows.

In expansions (1.132) and (1.133), regardless of linear terms of $\vec{k}$, we deliberately keep the quadratic terms. In most cases, the expansion of the tensor $\varepsilon_{ij}\left(\omega, \vec{k}\right)$ in terms of $\vec{k}$ does not have any term in odd powers of $\vec{k}$. In fact, if individual molecules of the medium and the crystal unit cell in crystalline media have a center of symmetry, then for such media,

$$\varepsilon_{ij}\left(\omega, \vec{k}\right) = \varepsilon_{ij}\left(\omega, -\vec{k}\right).$$

In this case, the expansion of $\varepsilon_{ij}\left(\omega, \vec{k}\right)$ in powers of $\vec{k}$ clearly contains only even powers of $\vec{k}$. Such media are called non-gyrotropic or optically inactive. In contrast, the medium without the symmetric property mentioned above is called gyrotropic. A gyrotropic medium, especially, can be an isotropic medium as well, for example, the sugarcane solution. For gyrotropic media, it is possible to consider the first two terms in Eqs. (1.132) and (1.133).

Now, let us consider the propagation of electromagnetic waves in media where weak spatial dispersion is taken into account. In isotropic gyrotropic media (also in crystals with cubic symmetry), the symmetric tensor $\varepsilon_{ij}(\omega)$ and the anti-symmetric tensor $\gamma_{ijl}(\omega)n_l$ of rank two become scalar and pseudo scalar quantities, respectively. Introducing $\gamma_{ijl}(\omega) = \gamma(\omega)e_{ijl}$ and $g_{ijl}(\omega) = -g(\omega)e_{ijl}$ where e_{ijl} is the completely anti-symmetric unit tensor of rank 3, we can write expansions (1.132) and (1.133) as follows[19]:

$$\varepsilon_{ij}\left(\omega, \vec{k}\right) = \varepsilon(\omega)\delta_{ij} + \imath\gamma(\omega)e_{ijl}n_l, \tag{1.135}$$

$$\varepsilon_{ij}^{-1}\left(\omega, \vec{k}\right) = \varepsilon^{-1}(\omega)\delta_{ij} - \imath g(\omega)e_{ijl}n_l. \tag{1.136}$$

In this case, the field material equations corresponding to expressions (1.135) and (1.136) are

[19] For isotropic gyrotropic media, the dielectric permittivity tensor taking arbitrary spatial dispersion into account can be expressed as

$$\varepsilon_{ij}\left(\omega, \vec{k}\right) = \left(\delta_{ij} - \frac{k_i k_j}{k^2}\right)\varepsilon^{\text{tr}}(\omega, k) + \frac{k_i k_j}{k^2}\varepsilon^{l}(\omega, k) + \imath f(\omega, k)e_{ijl}k_l.$$

For weak dispersion, this expression reduces to expansion (1.135) if $f(\omega, 0) = \gamma(\omega)(c/\omega)$ and to expansion (1.136) if $f(\omega, 0) = \varepsilon^2(\omega)g(\omega)(c/\omega)$.

$$\vec{D} = \varepsilon(\omega)\vec{E} + \imath\gamma(\omega)\vec{E} \times \vec{n}, \tag{1.137}$$

$$\vec{E} = \frac{\vec{D}}{\varepsilon(\omega)} - \imath g(\omega)\vec{D} \times \vec{n}. \tag{1.138}$$

From these equations, it is obvious that such spatial dispersion affects only the transverse components of electromagnetic fields in the medium. Moreover, only the next terms of expansions (1.132) and (1.133) have a sort of impact on the longitudinal component of the field. As mentioned before, to take weak spatial dispersion into account one should use the expression (1.135) for small values of $\varepsilon(\omega)$ and the expression (1.136) for large values of $\varepsilon(\omega)$. The dielectric permittivity $\varepsilon(\omega)$ with account of frequency dispersion is a non-monotonic function of frequency. For non-conducting media, the interpolated formula [28]

$$\varepsilon(\omega) = 1 - \sum_j \frac{\alpha_j}{\omega^2 - \omega_j^2} \tag{1.139}$$

is often used where α_j and ω_j are the quantities characterizing the medium properties. From relation (1.139), it is clear that $\varepsilon(\omega)$ grows unlimitedly at frequencies near ω_j, and, at the same time, far from these frequencies, $\varepsilon(\omega)$ can be a very small quantity and even becomes zero. It should be noted that in the frequency region close to ω_j, absorption is often large in the medium and, roughly speaking, it is not permitted to use the expression (1.139) in this region. Such a frequency region is called the absorption band and the frequency ω_j is named as the eigen frequency of the medium. Below, in consideration of electromagnetic waves near the absorption band, we use [28]

$$\varepsilon(\omega) = \varepsilon'(\omega) + \imath\varepsilon''(\omega) = \varepsilon_0 - \frac{\omega_0^2}{\omega^2 - \omega_j^2 - \imath\omega\nu}, \tag{1.140}$$

taking absorption in the medium into account. When $\nu = 0$ (ν is related to the relaxation processes), $\varepsilon(\omega)$ in Eq. (1.140) will be a real quantity (absorption is absent).

As expected, from material equations (1.137) and (1.138) we find

$$\varepsilon(\omega) = 0, \tag{1.141}$$

for longitudinal waves, which coincides with the dispersion equation (1.125) for longitudinal waves in isotropic non-gyrotropic media in the absence of spatial dispersion.[20]

Expressions (1.135) and (1.136) for transverse electromagnetic waves result in different dispersion equations. Since, according to the Maxwell's equations, for transverse wave we have $\vec{D} = n^2\vec{E}$, from the expression (1.135) it is possible to find the following dispersion equation:

$$\left[n^2 - \varepsilon(\omega)\right]^2 = \gamma^2(\omega)n^2. \tag{1.142}$$

Furthermore, as the quantity $\gamma^2(\omega)$ is small, the approximate solutions of Eq. (1.142) can be written as

$$n_\pm^2 \approx \varepsilon(\omega) \pm \gamma(\omega)\sqrt{\varepsilon(\omega)}. \tag{1.143}$$

Two solutions of Eq. (1.142) correspond to two ratios between the components of $\vec{E}$ (or $\vec{D}$)

$$E_x = \pm \imath E_y$$

(vector $\vec{k} = (\omega/c)\vec{n}$ is assumed along the oz axis). This means that two transverse waves, which correspond to the solutions of Eq. (1.142), have different polarizations. Waves with $E_x = \imath E_y$ and $E_x = -\imath E_y$ have right-handed and left-handed circular polarization, respectively.

From the expression (1.143) it follows that isotropic gyrotropic media at frequencies far from absorption band, where material Eq. (1.137) is applicable, have the birefractive property (two transverse waves propagate in each direction). In such a medium, the polarization plane of electromagnetic waves rotates. In fact, if a plane-polarized wave propagates in the medium, it can always be represented as the sum of two left-hand and right-hand circularly polarized waves.

As these waves travel a distance l through the medium, they see different refractive indices and consequently acquire different phases $\varphi_+ = n_+ l\omega/c$ and $\varphi_- = n_- l\omega/c$. In this case, the rotation angle of the polarization plane of the wave per unit path length in the medium is[21]

---

[20]The dispersion equation for longitudinal electromagnetic waves in isotropic gyrotropic media with spatial dispersion, $\varepsilon^l\left(\omega, \vec{k}\right) = 0$, is not different from the corresponding equation in isotropic non-gyrotropic media.

[21]A great number of optically active materials with different rotational capacities have been already known. In the optical range of spectra, $\gamma \sim 10^{-5}$ for sugar solution in water, $\gamma \sim 10^{-4}$ for quartz, $\gamma \sim 10^{-3}$ for cinnabar [28]. Rotational capacities of active materials depend on the frequency of the electromagnetic waves [14]. However, far from the absorption band, the quantity $\gamma(\omega)$ can be assumed constant.

$$\psi = \frac{\varphi_+ - \varphi_-}{2l} = \frac{\omega}{2c}(n_+ - n_-) \approx \frac{\omega}{2c}\gamma(\omega).$$

If the frequency of the transverse wave is near the eigen frequencies of the medium, then based on Eq. (1.138) we find [19]

$$g^2(\omega)n^6 - \left(\frac{n^2}{\varepsilon(\omega)} - 1\right)^2 = 0. \tag{1.144}$$

Taking the smallness of the quantity $g^2(\omega)$ into account, we obtain the following solutions[22]

$$n_{1,2}^2 \approx \varepsilon(\omega)\left[1 \pm g(\omega)\varepsilon^{3/2}(\omega)\right], \qquad n_3^2 \approx \frac{1}{\varepsilon^2(\omega)g^2(\omega)}, \tag{1.145}$$

which correspond to three transverse waves. It is simple to show that the waves with refractive indices n_1 and n_2 have right-hand and left-hand circular polarizations, respectively.

Hence, two transverse waves propagate in an isotropic gyrotropic medium far from the absorption band. Besides, near this band (see Eq. (1.145)), there are three transverse waves with the same frequency and different refractive indices.

Three curves $n_{1,2,3}^2(\omega)$ near the absorption band are illustrated in Fig. 1.1. Here, Eq. (1.140) with $\nu = 0$ has been used for the function $\varepsilon(\omega)$. Moreover, in this figure, $g^2 = 10^{-5}$, $\varepsilon_0 = 1$, $\omega_0/\omega_j = 1$. The dotted curve is related to the case of total neglect of spatial dispersion. It is noticeable that the multiple roots of Eq. (1.144) are

$$\varepsilon_m = \frac{2^{2/3}}{3g^{2/3}}, \qquad n_m^2 = \left(\frac{2}{g}\right)^{2/3}, \qquad n^2 = \frac{1}{4}\left(\frac{2}{g}\right)^{2/3}.$$

This means that $\omega^2/\omega_j^2 \approx 0.96$, $n_m^2 \approx 70$ and $n_3^2 \approx 18$. In the optical frequency range, this corresponds to $\Delta\omega \sim 2 \times 10^{-2}\omega_j \sim (6 - 12) \times 10^{13}\,\mathrm{s}^{-1}$ or $\Delta\lambda \sim 80 - 150\,\text{Å}$. This estimation implies that the range of existence of three transverse waves lies sufficiently far from the center of the absorption line (eigen frequency of the medium). In this case, absorption is still negligibly small, making the experimental observation of such waves possible.

As mentioned before, for non-gyrotropic media, the expansion of the tensor $\varepsilon_{ij}\left(\omega, \vec{k}\right)$ begins from the quadratic term with respect to $\vec{k}$. In this case, if the medium is isotropic, expansions (1.132) and (1.133) take the form of

[22] It is assumed that the order of the quantity $g(\omega)$ for different materials is of the order of $\gamma(\omega)$.

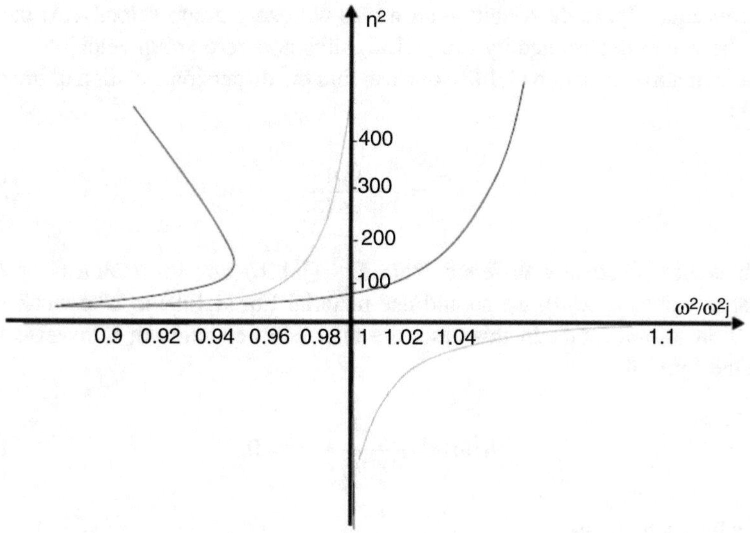

Fig. 1.1 Three curves $n_{1,2,3}^2(\omega)$ near the absorption band

$$\varepsilon_{ij}\left(\omega, \vec{k}\right) = \left[\varepsilon(\omega) - \alpha_1(\omega)n^2\right]\delta_{ij} - \alpha_2(\omega)n_i n_j, \tag{1.146}$$

$$\varepsilon_{ij}^{-1}\left(\omega, \vec{k}\right) = \left[\varepsilon^{-1}(\omega) + \beta_1(\omega)n^2\right]\delta_{ij} + \beta_2(\omega)n_i n_j. \tag{1.147}$$

In derivation of these relations, we made use of this fact that in isotropic media, the tensors $\alpha_{ijlm}(\omega)$ and $\beta_{ijlm}(\omega)$ reduce to the tensors of rank 2 with two independent components. Moreover, expressions (1.146) and (1.147) correspond to field material equations:

$$\vec{D} = \left[\varepsilon(\omega) - \alpha_1(\omega)n^2\right]\vec{E} - \alpha_2(\omega)\,\vec{n}\left(\vec{n}\cdot\vec{E}\right), \tag{1.148}$$

$$\vec{E} = \left[\varepsilon^{-1}(\omega) + \beta_1(\omega)n^2\right]\vec{D} + \beta_2(\omega)\,\vec{n}\left(\vec{n}\cdot\vec{D}\right). \tag{1.149}$$

The material equation (1.148) reduces to the dispersion equation of longitudinal waves in the medium:

$$n^2 = \frac{\varepsilon(\omega)}{\alpha_1(\omega) + \alpha_2(\omega)}. \tag{1.150}$$

This equation qualitatively differs from the longitudinal wave dispersion equation (1.125) obtained by neglecting spatial dispersion. This difference is related to this fact that Eq. (1.125) corresponds only to the oscillations with discrete frequencies and, consequently, to the longitudinal waves with zero group velocity. At the same time, the waves determined by Eq. (1.150) have non-zero group velocity.

From material equation (1.148), one can find the dispersion equation of transverse waves:

$$n^2 = \frac{\varepsilon(\omega)}{1 + \alpha_1(\omega)}, \tag{1.151}$$

which is not practically different from Eq. (1.126) due to smallness of $\alpha_1(\omega)$. For large values of $\varepsilon(\omega)$, we should use material Eq. (1.149) to take weak spatial dispersion into account. In this case, the dispersion equation of transverse waves takes the form of

$$\beta_1(\omega)n^4 + \frac{n^2}{\varepsilon(\omega)} - 1 = 0, \tag{1.152}$$

from which we obtain

$$n_{1,2}^2 = -\frac{1}{2\varepsilon(\omega)\beta_1(\omega)} \pm \sqrt{\left(\frac{1}{2\varepsilon(\omega)\beta_1(\omega)}\right)^2 + \frac{1}{\beta_1(\omega)}}. \tag{1.153}$$

Hence, taking spatial dispersion into account in isotropic media results in the new phenomena near the absorption band, i.e., new transverse waves. Curves $n_{1,2}^2(\omega)$ near the absorption band for real function $\varepsilon(\omega)$, i.e., when $\nu = 0$ in Eq. (1.140), are illustrated in Figs. 1.2 and 1.3. In these figures, $\varepsilon_0 = 1$, $\omega_0/\omega_j = 1$, and $|\beta_1| = 10^{-5}$. The dotted curves in both cases correspond to the expression (1.140).

When $\beta_1 > 0$ (see Fig. 1.2), one of the roots of Eq. (1.152) is negative. Therefore, the corresponding wave cannot propagate in the medium. When $\beta_1 < 0$ (see Fig. 1.3), two waves can propagate. Multiple roots, in this case, are

$$\varepsilon_m = \frac{1}{4\,|\,\beta_1\,|}, \qquad n_m^2 = \frac{1}{\sqrt{|\,\beta_1\,|}},$$

where $n_m^2 \approx 300$ and $\omega^2/\omega_j^2 \approx 0.994$. In the optical range of frequency, this corresponds to $\Delta\omega \sim 3 \times 10^{-3}\omega_j \sim (1-2) \times 10^{13}\mathrm{s}^{-1}$ or $(\Delta\lambda/2\pi)\tilde{1}0 - 20\,\text{Å}$. This is very close to the eigen frequency of the medium so that absorption is very high. As a result, it is difficult to observe these waves. In fact, when $\nu \neq 0$ and $n^2 = (n' + \imath n'')^2$, we find absorption coefficient n'' at the frequencies corresponding to the multiple roots

$$n'' \approx \varepsilon''|\beta_1|^{1/4} \approx 0.5 \times 10^4\ \nu/\omega_j.$$

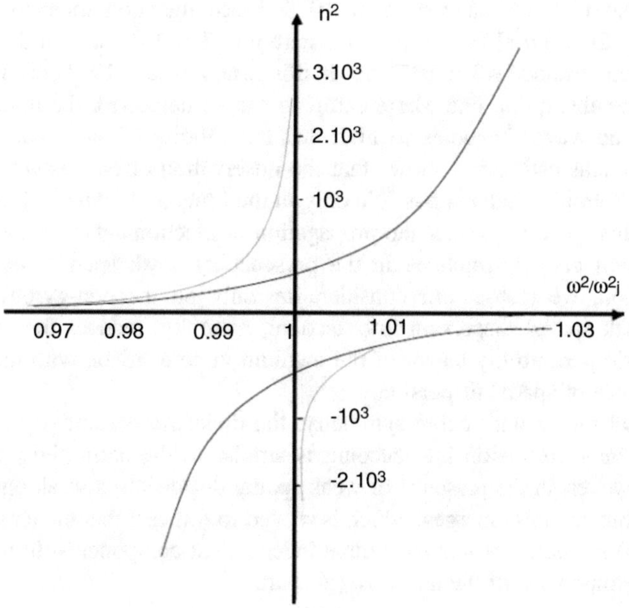

Fig. 1.2 Curves $n_{1,2}^2(\omega)$ near the absorption band when $\beta_1 > 0$

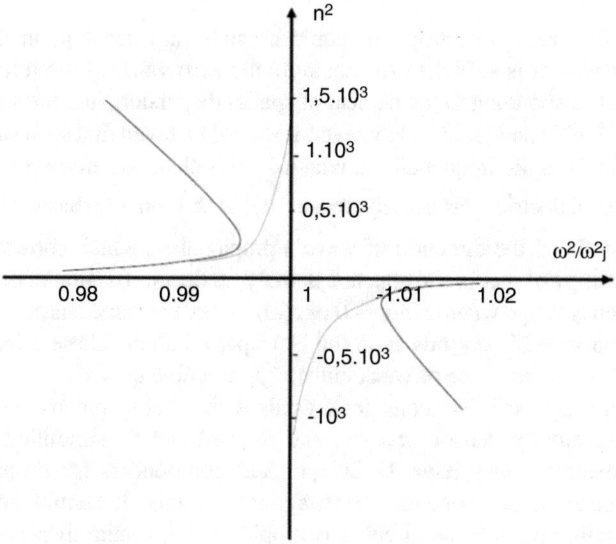

Fig. 1.3 Curves $n_{1,2}^2(\omega)$ near the absorption band when $\beta_1 < 0$

For $\nu/\omega_j \sim 10^{-6}$, we have $n'' \sim 5 \times 10^{-3}$. Since radiation intensity is damped as $\sim \exp[-2(\omega/c)\, n'' z] = \exp(-\mu z)$, where $\mu \approx 3 \times 10^3\ \text{cm}^{-1}$, it decreases by a factor e at the distance $\sim 3 \times 10^{-4}$ cm (in this case, $\lambda\!\!\!^{-} \sim 2 \times 10^{-6}$ cm). Far from the center of the absorption line where damping can be neglected, the refractive index of one of the waves becomes so high that the validity of the expansion (1.147) is violated. This estimation shows that the observation of two waves in isotropic and non-gyrotropic media is possible only in the films of thickness $\leq 10^{-4}$ cm.

In conclusion, we consider the propagation of electromagnetic waves in media with different crystal structures in the presence of weak spatial dispersion. For simplification, we restrict our consideration only on the non-gyrotropic media. Taking weak spatial dispersion into account, naturally, reduces the symmetry of the dielectric permittivity tensor of the medium in comparison with the symmetry in the absence of spatial dispersion.

In crystal media with cubic symmetry, the dielectric permittivity $\varepsilon_{ij}(\omega)$ taking only frequency dispersion into account is similar to the permittivity of isotropic media. However, in the presence of weak spatial dispersion, a weak optical anisotropy of cubic crystals emerges, which is related to this fact that the tensors $\alpha_{ijlm}(\omega)$ and $\beta_{ijlm}(\omega)$ in cubic crystals have three independent components. In this case, the non-zero components of the tensor $\alpha_{ijlm}(\omega)$ are

$$\begin{aligned}
&\alpha_1 = \alpha_{xxxx} = \alpha_{yyyy} = \alpha_{zzzz};\\
&\alpha_2 = \alpha_{xyxy} = \alpha_{xzxz} = \alpha_{yzyz};\\
&\alpha_3 = \alpha_{xxyy} = \alpha_{yyxx} = \alpha_{zzxx} = \alpha_{xxzz} = \alpha_{zzyy} = \alpha_{yyzz}.
\end{aligned} \tag{1.154}$$

To take the weak anisotropy of cubic crystals into account, in the study of transverse waves, it is sufficient to substitute the zero value of the refractive index $n_0^2 = \varepsilon(\omega)$, corresponding to the neglect of spatial dispersion, into the small terms of expansions (1.132) and (1.133). However, it should be noted that such substitution is valid only far from the frequencies at which $\varepsilon(\omega)$ is close to zero or goes to infinity. In this case, dielectric permittivity tensor $\varepsilon_{ij}\left(\omega, \vec{k}\right)$ on the basis of expression (1.132) depends on the direction of wave's propagation, which corresponds to the optical anisotropy of media and appears, namely, in the birefraction of cubic crystals. In the frequency range where $\varepsilon(\omega) \to 0$ or $\varepsilon(\omega) \to \infty$, the same characteristic effects should appear in cubic crystals as in the isotropic medium. These effects are more complicated in the presence of weak anisotropy in cubic crystals.

Similarly, it is possible to consider crystals with other symmetry. Depending on the crystal symmetry, tensors $\alpha_{ijlm}(\omega)$ and $\beta_{ijlm}(\omega)$ can be simplified in different ways. For example, they have 12 independent components for rhombic crystals and 7 independent components for tetragonal crystals. It should be mentioned that in non-cubic crystals, as in cubic isotropic media, spatial dispersion becomes substantially stronger in the presence of gyrotropy.

1.8 Energy Loss of Fast Moving Electrons in the Medium

Fast moving charged particles in the medium excite electromagnetic waves in it. In absorptive media, these waves are damped fast, corresponding to the energy transfer of fast particles to the medium through the excitation of electromagnetic waves in it. Therefore, a fast charged particle loses a part of its energy in the medium.[23] The theory of Tamm–Frank–Fermi for a spatially dispersive medium was generalized in [4, 34–51]. Here, we assume that the energy of electromagnetic wave excitation is small compared to the particle energy and as a result the change of particle's velocity in the medium can be neglected. Energy loss of the moving particle is obviously determined by the work done by the drag force, acting on the particle, from the electromagnetic field produced in the medium by the particle. The work of this force, determined by Eq. (1.3), per unit path length in the medium is

$$W = \frac{\vec{v} \cdot \vec{F}}{v} = \frac{e\vec{v} \cdot \vec{E}}{v}. \tag{1.155}$$

In this relation, one should substitute the electric field $\vec{E}\left(\vec{r}, t\right)$ at the place of the charged particle. Hence, it is necessary to determine the electric field $\vec{E}\left(\vec{r}, t\right)$ produced by the charged particles in order to calculate the energy loss of the charged particles in the medium.

Let us now consider the motion of the fast moving charged particles in a spatially homogeneous and unbounded medium. Using Fourier expansion, we represent the electric field, produced by the particles, as a sum of plane waves $\exp\left(\imath\vec{k} \cdot \vec{r} - \imath\omega t\right)$. Passing on to the Fourier components of the field equations (1.9) and omitting the magnetic induction $\vec{B}$, we obtain an equation for the Fourier component of the electric field $\vec{E}$

$$\left\{k^2\delta_{ij} - k_i k_j - \frac{\omega^2}{c^2}\varepsilon_{ij}\left(\omega, \vec{k}\right)\right\}E_j = \frac{4\pi\imath\omega}{c^2}j_{0i}\left(\omega, \vec{k}\right), \tag{1.156}$$

where $j_0\left(\omega, \vec{k}\right)$ is the Fourier component of the current density of the external field source. The current density produced by a particle with charge e and velocity $\vec{v}$ is

$$\vec{j}_0\left(\vec{r}, t\right) = e\vec{v}\delta\left(\vec{r} - \vec{v}t\right). \tag{1.157}$$

In this case, its Fourier component is equal to

[23] The theory of the energy loss of the fast charged particles in the medium was developed by Tamm, Frank, and Fermi [29, 30]. See also [31–33].

$$\vec{j}_0\left(\omega,\vec{k}\right)=\frac{e\vec{v}}{(2\pi)^3}\delta\left(\omega-\vec{k}\cdot\vec{v}\right). \tag{1.158}$$

In the isotropic non-gyrotropic medium, Eq. (1.156) reduces to

$$\frac{\omega^2}{c^2}\varepsilon^l\left(\omega,\vec{k}\right)\frac{\vec{k}\left(\vec{k}\cdot\vec{E}\right)}{k^2}-\left(k^2-\frac{\omega^2}{c^2}\varepsilon^{\mathrm{tr}}\left(\omega,\vec{k}\right)\right)\left(\vec{E}-\frac{\vec{k}\left(\vec{k}\cdot\vec{E}\right)}{k^2}\right)$$
$$=-\frac{4\pi\imath\omega}{c}\vec{j}_0\left(\omega,\vec{k}\right). \tag{1.159}$$

Scalar multiplying Eq. (1.159) by the vector $\vec{k}$, we find

$$\vec{k}\cdot\vec{E}=-\frac{4\pi\imath}{\omega\varepsilon^l\left(\omega,\vec{k}\right)}\vec{k}\cdot\vec{j}\left(\omega,\vec{k}\right). \tag{1.160}$$

Substituting the latter equation into Eq. (1.159), we obtain

$$E_i=-\frac{4\pi\imath\omega}{k^2}\left\{\frac{k_ik_j}{\omega^2\varepsilon^l\left(\omega,\vec{k}\right)}-\frac{k^2\left(\delta_{ij}-\frac{k_ik_j}{k^2}\right)}{c^2\left[k^2-\frac{\omega^2}{c^2}\varepsilon^{\mathrm{tr}}\left(\omega,\vec{k}\right)\right]}\right\}j_{0j}\left(\omega,\vec{k}\right). \tag{1.161}$$

The electric field in the medium at arbitrary point $\vec{r}$ at the moment t is determined by Fourier transformation

$$\vec{E}\left(\vec{r},t\right)=\int_{-\infty}^{+\infty}d\omega\int d\vec{k}\,\exp\left(\imath\vec{k}\cdot\vec{r}-\imath\omega t\right)\vec{E}\left(\omega,\vec{k}\right). \tag{1.162}$$

Expressions (1.161) and (1.162) allow us to determine the electromagnetic field produced by the arbitrary field source with current density $\vec{j}_0\left(\vec{r},t\right)$ in isotropic non-gyrotropic media. When the field source is a moving charged particle, the electric field is determined by Eqs. (1.158) and (1.162) as follows:

$$\vec{E}\left(\vec{r},t\right) = -\frac{4\pi\imath e}{(2\pi)^3}\int d\vec{k}\,\frac{\exp\left(\imath\vec{k}\cdot\vec{r} - \imath\vec{k}\cdot\vec{v}t\right)}{k^2}$$
$$\times\left\{\frac{\vec{k}}{\varepsilon^l\left(\vec{k}\cdot\vec{v},\vec{k}\right)} - \frac{k^2\left(\vec{k}\cdot\vec{v}\right)\left(\vec{v} - \frac{\vec{k}\left(\vec{k}\cdot\vec{v}\right)}{k^2}\right)}{c^2\left[k^2 - \frac{\left(\vec{k}\cdot\vec{v}\right)^2}{c^2}\varepsilon^{\mathrm{tr}}\left(\vec{k}\cdot\vec{v},\vec{k}\right)\right]}\right\}. \qquad (1.163)$$

Taking the value of the electric field at the place of the charged particle, i.e., at the point $\vec{r} = \vec{v}t$ and making use of Eq. (1.155), we find an expression for the energy loss of the particle per unit length of its path through the medium

$$W = \frac{\imath e^2}{2\pi^2 v}\int d\vec{k}\,\frac{\vec{k}\cdot\vec{v}}{k^2}\left\{\frac{1}{\varepsilon^l\left(\vec{k}\cdot\vec{v},\vec{k}\right)} - \frac{k^2\left(v^2 - \frac{\left(\vec{k}\cdot\vec{v}\right)^2}{k^2}\right)}{c^2\left[k^2 - \frac{\left(\vec{k}\cdot\vec{v}\right)^2}{c^2}\varepsilon^{\mathrm{tr}}\left(\vec{k}\cdot\vec{v},\vec{k}\right)\right]}\right\}. \qquad (1.164)$$

Introducing $\vec{k}\cdot\vec{v} = \omega$ and $q^2 = k^2 - \omega^2/c^2$, from Eq. (1.164) we obtain

$$W = \frac{\imath e^2}{\pi v^2}\int_{-\infty}^{+\infty}\omega\,d\omega\int_0^{\infty}\frac{q\,dq}{q^2 + \frac{\omega^2}{v^2}}\left\{\frac{1}{\varepsilon^l\left(\omega,\sqrt{q^2 + \frac{\omega^2}{v^2}}\right)} - \right.$$
$$\left.\times\frac{v^2}{c^2}\,\frac{q^2}{q^2 + \frac{\omega^2}{v^2}\left[1 - \frac{v^2}{c^2}\varepsilon^{\mathrm{tr}}\left(\omega,\sqrt{q^2 + \frac{\omega^2}{v^2}}\right)\right]}\right\}. \qquad (1.165)$$

Since the real and imaginary parts of the longitudinal and transverse dielectric permittivities are even and odd functions of frequency, respectively, from Eq. (1.165) we find

$$W = W^l + W^{\mathrm{tr}}, \qquad (1.166)$$

where

$$W^l = -\frac{2e^2}{\pi v^2}\int_0^\infty \omega\, d\omega \int_0^\infty \frac{q\, dq}{q^2+\frac{\omega^2}{v^2}} \operatorname{Im} \frac{1}{\varepsilon^l\left(\omega, \sqrt{q^2+\frac{\omega^2}{v^2}}\right)},$$

$$W^{\text{tr}} = -\frac{2e^2}{\pi c^2}\int_0^\infty \omega\, d\omega \int_0^\infty \frac{q^3\, dq}{q^2+\frac{\omega^2}{v^2}} \operatorname{Im} \frac{1}{q^2+\frac{\omega^2}{v^2}\left[1-\frac{v^2}{c^2}\varepsilon^{\text{tr}}\left(\omega, \sqrt{q^2+\frac{\omega^2}{v^2}}\right)\right]}. \tag{1.167}$$

From these relations it is clear that the contribution of the energy loss of the charged particle in the medium is essential only in those ranges of ω and k in which absorption is substantial. However, this is not the case. Those ranges of integration in the right-hand side of Eq. (1.165) in which the imaginary parts of ε^l and ε^{tr} are negligibly small play important role as well. In such regions, the denominator of the first and second terms of the bracket in Eq. (1.165) can pass through zero and, in this case, the integrand can have some poles. In Sect. 1.4, it was shown that for the medium in the thermodynamic equilibrium state, $\operatorname{Im} \varepsilon^l \geq 0$ and $\operatorname{Im} \varepsilon^{\text{tr}} \geq 0$. Taking the latter fact into account and making use of

$$\lim_{\delta\to+0} \frac{1}{x+\imath\delta} = \mathcal{P}\frac{1}{x} - \imath\pi\delta(x),$$

we find expressions for that part of W^l and W^{tr} in the range of ω and $\vec{k}$ where absorption is absent:

$$\Delta W^l = \frac{2e^2}{v^2}\int \omega\, d\omega \int \frac{q\, dq}{q^2+\frac{\omega^2}{v^2}}\, \delta\left[\varepsilon^l\left(\omega, \sqrt{q^2+\frac{\omega^2}{v^2}}\right)\right]; \tag{1.168}$$

$$\Delta W^{tr} = \frac{2e^2}{c^2}\int \omega\, d\omega \int \frac{q^3\, dq}{q^2+\frac{\omega^2}{v^2}}\, \delta\left\{q^2+\frac{\omega^2}{v^2}\left[1-\frac{v^2}{c^2}\varepsilon^{\text{tr}}\left(\omega, \sqrt{q^2+\frac{\omega^2}{v^2}}\right)\right]\right\}. \tag{1.169}$$

Integration of ω and q in these expressions is taken in the region of weak absorption in the medium, i.e., where the imaginary parts of ε^l and ε^{tr} are negligibly small. From Eqs. (1.168) and (1.169), it is clear that in this region, the energy loss of particles is determined by those values of ω, $\vec{k}$ for which the argument of δ-function in these formulas is equal to zero. Regarding Eqs. (1.115) and (1.116), the values

[24] The radiation of transverse electromagnetic waves by fast moving electrons in the medium was first discovered by Cherenkov [50] in the laboratory of S.I.Vavilov who represented the first qualitative description of this phenomenon [51]. In literatures, such a phenomenon is called Vavilov–Cherenkov.

of ω and $\vec{k}$ correspond to the longitudinal and transverse electromagnetic waves in isotropic non-gyrotropic media.

As mentioned, the energy loss of particles in the medium is written as the sum of two terms W^l and W^{tr}. The first term of relation (1.166), W^l, characterizes the energy loss of a non-relativistic electron in the medium, which is stipulated by longitudinal electromagnetic waves. Besides, the second term of relation (1.166), W^{tr}, represents the energy loss of an electron due to the excitation of transverse electromagnetic waves in the medium.[24] The parts of energy loss of the particle, which correspond to the quantities W^l and W^{tr}, are often called polarization and Cherenkov loss, respectively. It should be remarked that such subdivision is somehow related to this fact that the first and second terms of relation (1.166) correspond to the energy loss of the charged particles due to the excitations of longitudinal and transverse electromagnetic waves in the medium, respectively. In the anisotropic medium where it is generally not possible to subdivide electromagnetic waves into longitudinal and transverse waves, partitioning of energy loss into polarization and Cherenkov loss is meaningless.

For a charged particle moving in an anisotropic medium, from Eqs. (1.155), (1.156), and (1.158) we find the energy loss of the particle per unit length of its path as follows:

$$W = \frac{\imath e^2}{2\pi^2 c^2 v} \int \vec{k} \cdot \vec{v} d\vec{k} \left[\left(v_i \alpha_{ij}^{-1} v_j \right) + \frac{\left(v_i \alpha_{ij}^{-1} k_j \right) \left(k_i \alpha_{ij}^{-1} v_j \right)}{1 - k_i \alpha_{ij}^{-1} k_j} \right], \tag{1.170}$$

where

$$\alpha_{ij} = k^2 \delta_{ij} - \frac{\left(\vec{k} \cdot \vec{v} \right)^2}{c^2} \varepsilon_{ij} \left(\vec{k} \cdot \vec{v}, \vec{k} \right).$$

For non-absorptive media, the energy loss of the particle in the medium is determined by the poles of the integrand of Eq. (1.170). These poles coincide with the roots of [see Eq. (1.156)]

$$\left| k^2 \delta_{ij} - k_i k_j - \frac{\omega^2}{c^2} \varepsilon_{ij} \left(\omega, \vec{k} \right) \right| = 0,$$

representing the dispersion equation of electromagnetic waves in anisotropic media.

---

[24] The radiation of transverse electromagnetic waves by fast moving electrons in the medium was first discovered by Cherenkov [52] in the laboratory of S.I.Vavilov who represented the first qualitative description of this phenomenon [53]. In literatures, such a phenomenon is called Vavilov–Cherenkov.

When in the medium $(v^2/c^2)\varepsilon_{ij}\left(\vec{k}\cdot\vec{v},\vec{k}\right) \ll 1$ for all values of ω and $\vec{k}$, Eq. (1.170) is substantially simplified. In this case, from Eq. (1.170) we find

$$W = -\frac{\imath e^2}{2\pi^2 v}\int d\vec{k}\,\frac{\vec{k}\cdot\vec{v}}{k_i\varepsilon_{ij}\left(\vec{k}\cdot\vec{v},\vec{k}\right)k_j}. \tag{1.171}$$

Since Eq. (1.171) is obtained from Eq. (1.170) by passing to the limit $c \to \infty$, as stated, it determines the total non-relativistic energy loss of the charged particle in the anisotropic medium. Although the partitioning of the energy loss into longitudinal and transverse ones in anisotropic media is meaningless, in this case, it is possible to talk about the non-relativistic energy loss of fast moving particles given by Eq. (1.171) and the total energy loss of the particle given by Eq. (1.170).

Now, let us answer to this question: how does taking spatial dispersion into account change the characteristic features of the loss in the isotropic non-gyrotropic media? Furthermore, each term of Eq. (1.166) is considered, separately.

As mentioned before, the quantity W^l, the energy loss of the charged particle, is stipulated by the radiation of longitudinal waves in the medium. Let us assume that a particle with the momentum $\vec{p}$ radiates a longitudinal electromagnetic wave of the frequency ω and wave vector $\vec{k}$ as a result of the interaction with the medium and is scattered at angles $\vartheta \ll 1$. In quantum mechanical language, such a wave can be called a longitudinal quantum with the energy $\hbar\omega$ and the momentum $\hbar\vec{k}$. From the energy-momentum conservation law (Fig. 1.4), making use of the relation between the energy ϵ and the momentum $\vec{p}$ of the particle

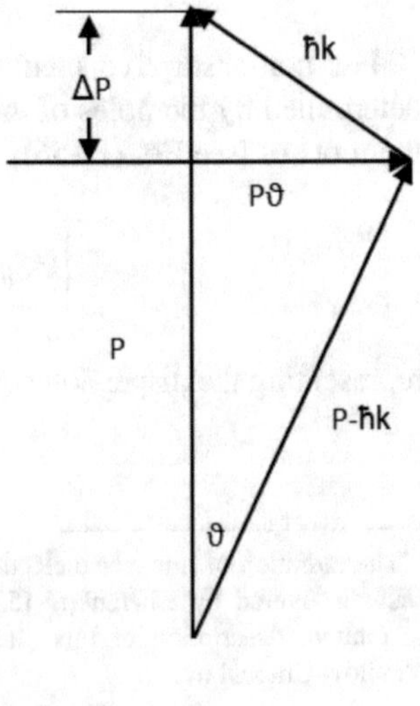

Fig. 1.4 Energy-momentum conservation law

$$\epsilon^2 = c^2p^2 + m^2c^4, \qquad \vec{p} = \frac{\epsilon\,\vec{v}}{c^2},$$

we find

$$\hbar^2k^2 = p^2v^2 + \frac{\hbar^2\omega^2}{v^2} = \left(\Delta\vec{p}\right)^2 + p^2\vartheta^2. \tag{1.172}$$

Then, considering $k^2 = q^2 + \omega^2/v^2$, we find $q = p\vartheta/\hbar$. It should be noted that such a quantum mechanical treatment is valid only in the transparency region of the medium where the imaginary parts of ω and $\vec{k}$ are small. From Eq. (1.167), we find an expression for the scattering probability of a fast particle at angles $\vartheta \ll 1$ alongside with the emission of a longitudinal quantum of the frequency ω during its motion in the medium per unit time as follows:

$$\frac{v dW^l}{\hbar\omega\, d\omega\, \vartheta\, d\vartheta} = -\frac{2e^2}{\pi\hbar v}\frac{1}{\vartheta^2 + \left(\frac{\hbar\omega}{pv}\right)^2}\mathrm{Im}\frac{1}{\varepsilon^l\left(\omega, \sqrt{\left(\frac{p\vartheta}{\hbar}\right)^2 + \frac{\omega^2}{v^2}}\right)}. \tag{1.173}$$

Using relations $\omega = \vec{k}\cdot\vec{v}$ and Eqs. (1.172)–(1.173), we simply find the emission probability of a longitudinal quantum with the wave vector $\vec{k}$ per unit time by a fast electron moving with the velocity $\vec{v}$ in the medium.

$$\frac{v dW^l}{\hbar\omega\, d\vec{k}} = \frac{e^2}{2\pi\hbar}\frac{\delta\left[\varepsilon^l\left(\vec{k}\cdot\vec{v}, \vec{k}\right)\right]}{k^2}. \tag{1.174}$$

If spatial dispersion is negligible, then Eq. (1.173) reduces to

$$\frac{v dW^l}{\hbar\omega\, d\omega\, \vartheta d\vartheta} = -\frac{2e^2}{\pi\hbar v}\frac{1}{\vartheta^2 + \left(\frac{\hbar\omega}{pv}\right)^2}\mathrm{Im}\frac{1}{\varepsilon(\omega)}, \tag{1.175}$$

where $\varepsilon(\omega) = \varepsilon^l(\omega, 0)$. Comparing Eqs. (1.173) and (1.175), we find out that taking spatial dispersion into account changes the angular dependence of the scattering probability of the fast particle in the medium along with the emission of a longitudinal quantum. When the scattering angle is not very small, the difference between expressions (1.172) and (1.175) can be significant.

The probability of the emission of the longitudinal quantum given by Eq. (1.174) in the absence of spatial dispersion takes the form of

$$\frac{vdW^l}{\hbar\omega d\vec{k}} = \frac{e^2}{2\pi\hbar}\frac{\delta\left[\varepsilon''\left(\vec{k}\cdot\vec{v}\right)\right]}{k^2}. \tag{1.176}$$

Finally, we point out another form of the longitudinal polarization loss used in the absence of spatial dispersion. Writing the dielectric permittivity $\varepsilon(\omega)$ in terms of the refractive index n' and absorption coefficient n'',

$$\varepsilon(\omega) = n^2(\omega) = [n'(\omega) + \imath n''(\omega)]^2,$$

and integrating this expression over q, from Eq. (1.167) we find[25]

$$W^l = \frac{2e^2}{\pi v^2}\int_0^\infty \omega\, d\omega \frac{2n'(\omega)n''(\omega)}{\left[n'^2(\omega) + n''^2(\omega)\right]} \ln\frac{q_0 v}{\omega}. \tag{1.177}$$

The quantity q_0 in the latter equation is determined by the condition of the possibility of neglecting the spatial dispersion of the longitudinal dielectric permittivity in Eq. (1.167).

In the same way, as done to obtain Eq. (1.173) from Eq. (1.167), it is possible to find an expression for the scattering probability of a fast particle per unit time along with the emission of a transverse quantum of the frequency ω from Eq. (1.168) as follows:

$$\frac{vdW^{\text{tr}}}{\hbar\omega\, d\omega\, \vartheta d\vartheta} = \frac{2e^2p^2v}{\pi\hbar^3c^2} \times \frac{\vartheta^2}{\vartheta^2 + \left(\frac{\hbar\omega}{pv}\right)^2} \text{Im}\frac{1}{\left(\frac{p\vartheta}{\hbar}\right)^2 + \frac{\omega^2}{v^2}\left[1 - \frac{v^2}{c^2}\varepsilon^{\text{tr}}\left(\omega, \sqrt{\left(\frac{p\vartheta}{\hbar}\right)^2 + \frac{\omega^2}{v^2}}\right)\right]}. \tag{1.178}$$

The angular dependence of Eqs. (1.178) and (1.173) is different from the angular dependence of the corresponding expression obtained by neglecting spatial dispersion. However, taking spatial dispersion into account in Eq. (1.178) for a non-relativistic particle leads a weak effect because changes are related only to the small terms of the order of $(v^2/c^2)\varepsilon^{\text{tr}}$. However, the contribution of Eq. (1.178) in the total probability of particle's scattering in the medium at angles $\vartheta \ll 1$ during the radiation of electromagnetic waves is small, i.e., of the order of v^2/c^2. For relativistic particles, the effect of spatial dispersion can be significant. Taking spatial dispersion into account manifests itself substantially in the spectrum of the angular distribution

[25]This expression for longitudinal loss was found in [39].

of the transverse radiation by particles. To prove this, for a non-absorptive medium, we find

$$W^{\mathrm{tr}} = \frac{2e^2}{c^2} \int_0^\infty \omega\, d\omega \int_0^\infty \times \frac{q^3 dq}{q^2 + \frac{\omega^2}{v^2}}\, \delta\left[q^2 + \frac{\omega^2}{v^2}\left(1 - \frac{v^2}{c^2}\varepsilon^{\mathrm{tr}}\left(\omega, \sqrt{q^2 + \frac{\omega^2}{v^2}}\right)\right)\right]. \tag{1.179}$$

Neglecting spatial dispersion in the latter expression, we obtain

$$W^{\mathrm{tr}} = \frac{2e^2}{c^2} \int_0^\infty \omega\, d\omega \int_0^{q_0} \frac{q^3 dq}{q^2 + \frac{\omega^2}{v^2}}\, \delta\left[q^2 + \frac{\omega^2}{v^2}\left(1 - \frac{v^2}{c^2}\varepsilon(\omega)\right)\right], \tag{1.180}$$

where $\varepsilon(\omega) = \varepsilon^{\mathrm{tr}}(\omega, 0)$. From Eq. (1.180), it follows that Cherenkov radiation with the frequency ω happens only when

$$v \geq \frac{c}{\sqrt{\varepsilon(\omega)}} = \frac{c}{n(\omega)}, \tag{1.181}$$

where $n(\omega)$ is the refractive index for the radiated transverse waves. Integrating Eq. (1.180) over q, we obtain

$$W^{\mathrm{tr}} = \frac{e^2}{c^2} \int_0^{q_0 v} \omega\, d\omega \left(1 - \frac{c^2}{v^2 n^2(\omega)}\right), \tag{1.182}$$

which determines the total intensity of Cherenkov radiation. Introducing the angle between the directions of the charged particle's motion and Cherenkov radiation propagation by θ and noticing that for Cherenkov waves

$$\omega = \vec{k} \cdot \vec{v} = kv \cos\theta = n\frac{\omega}{c} v \cos\theta, \tag{1.183}$$

we find that this radiation is distributed over the surface of a cone with opening angle

$$\cos\theta = \frac{c}{vn(\omega)}. \tag{1.184}$$

In the presence of spatial dispersion, from Eq. (1.179) we find the existence condition of the Cherenkov radiation wave of the frequency ω

$$v \geq \frac{c}{n_i(\omega)}, \tag{1.185}$$

where $n_i(\omega)$ is the one of the roots of the equation

$$n^2 = \varepsilon^{\mathrm{tr}}\left(\omega, \frac{\omega}{c} n\right), \tag{1.186}$$

representing the dispersion equation of transverse electromagnetic waves in the medium [compared to Eq. (1.123)]. In this case, radiation is distributed over the surface of a cone with opening angle θ_i

$$\cos\theta_i = \frac{c}{v n_i(\omega)}. \tag{1.187}$$

Since Eq. (1.186) has, generally speaking, several roots, then the Cherenkov radiation of the frequency ω can be distributed over the surfaces of several cones determined by Eq. (1.187) [13, 44]. At the same time, in the absence of spatial dispersion, all radiations are distributed over the surface of one cone.

1.9 Electromagnetic Field Fluctuations

Using microscopic consideration, the expression for a spectral distribution correlator of current density and electromagnetic field fluctuations is derived for the medium in the general case. Then the analysis concerns electromagnetic field fluctuations in a homogenous isotropic non-gyrotropic medium. Finally, the causality principle and generalization of Kramers–Kronig relations are considered.

1.9.1 Correlation Functions and General Analysis

As a result of fluctuating oscillations, local spontaneous currents or, as said, random currents $\vec{j}_{\mathrm{st}}$ creating fluctuating electromagnetic fields arise in material media.[26] Instead of random currents, it is better to use random inductions $\vec{K}$, which is related to $\vec{j}_{\mathrm{st}}$ by

[26] The theory of electromagnetic fluctuations in material media, taking only frequency dispersion into dielectric permittivity, was investigated in [54–60]. The same issue in the presence of spatial dispersion was studied in [61–64]. In the present book, we follow the general description of the theory of the quantum fluctuations presented in [2].

$$\vec{j}_{\mathrm{st}} = \frac{1}{4\pi}\frac{\partial \vec{K}}{\partial t}. \tag{1.188}$$

These quantities are called extraneous quantities, which emphasize this fact that they play the role of external field sources in the Maxwell's equations of fluctuating fields.

Let us assume that a random current of frequency ω arises in an unbounded material medium. This current, considered as an external source, produces a fluctuating electromagnetic field in the medium. Excluding the magnetic induction $\vec{B}$ and using material equation (1.9), we write Maxwell's equations of the fluctuating field:

$$\begin{gathered}\left[\nabla \times \nabla \times \vec{E}\left(\vec{r},\omega\right)\right]_i = \frac{\omega^2}{c^2}\int d\vec{r'}\,\varepsilon_{ij}\left(\omega,\vec{r},\vec{r'}\right)E_j\left(\vec{r'},\omega\right)\\ +\frac{4\pi\iota\omega}{c}j_{st_i}\left(\vec{r},\omega\right) = \frac{\omega^2}{c^2}\int d\vec{r'}\,\varepsilon_{ij}\left(\omega,\vec{r},\vec{r'}\right)E_j\left(\vec{r'},\omega\right) + \frac{\omega^2}{c^2}K_i\left(\vec{r},\omega\right).\end{gathered} \tag{1.189}$$

This equation makes a connection between the random induction and the strength of the fluctuating electric field by operator $L_{ij}\left(\omega,\vec{r},\vec{r'}\right)$ as follows:

$$\begin{gathered}K_i\left(\vec{r},\omega\right) = \int d\vec{r'}\,L_{ij}\left(\omega,\vec{r},\vec{r'}\right)E_j\left(\vec{r'},\omega\right)\\ \equiv -\int d\vec{r'}\,\varepsilon_{ij}\left(\omega,\vec{r},\vec{r'}\right)E_j\left(\vec{r'},\omega\right) + \frac{c^2}{\omega^2}\left[\nabla\times\nabla\times\vec{E}\left(\vec{r},\omega\right)\right]_i.\end{gathered} \tag{1.190}$$

Moreover, Eq. (1.189) connects the strength of the fluctuating electric field with the random induction. We denote this connection by

$$E_i\left(\vec{r},\omega\right) = \int d\vec{r'}\,L_{ij}^{-1}\left(\omega,\vec{r},\vec{r'}\right)K_j\left(\vec{r'},\omega\right). \tag{1.191}$$

Based on Eq. (1.71), energy change in the medium, which is related to the considered external interaction is determined by

$$\frac{dW}{dt} = \int d\vec{r}\,\vec{E}\left(\vec{r},t\right)\cdot\vec{j}_{\mathrm{st}}\left(\vec{r},t\right) = \frac{1}{4\pi}\int d\vec{r}\,\vec{E}\left(\vec{r},t\right)\cdot\frac{\partial\vec{K}\left(\vec{r},t\right)}{\partial t}. \tag{1.192}$$

Since $\vec{E}\left(\vec{r},t\right)$ and $\vec{K}\left(\vec{r},t\right)$ are real quantities, we represent them as

$$\begin{cases} \vec{E}\left(\vec{r},t\right) = \vec{E}\left(\vec{r},\omega\right)\exp\left(-\imath\omega t\right) + \vec{E}^*\left(\vec{r},\omega\right)\exp\left(\imath\omega t\right), \\ \vec{K}\left(\vec{r},t\right) = \vec{K}\left(\vec{r},\omega\right)\exp\left(-\imath\omega t\right) + \vec{K}^*\left(\vec{r},\omega\right)\exp\left(\imath\omega t\right). \end{cases} \tag{1.193}$$

Substituting these expressions into Eq. (1.192) and averaging over time, we find the averaged energy released in the medium per unit time:

$$Q = \frac{\imath\omega}{4\pi}\int d\vec{r}\left\{\vec{E}\left(\vec{r},\omega\right)\cdot\vec{K}^*\left(\vec{r},\omega\right) - \vec{E}^*\left(\vec{r},\omega\right)\cdot\vec{K}\left(\vec{r},\omega\right)\right\}. \tag{1.194}$$

Making use of Eqs. (1.190) and (1.191), we can rewrite Eq. (1.194) as

$$\begin{aligned} Q &= \frac{\imath\omega}{4\pi}\int d\vec{r}\,d\vec{r'}\left\{L_{ij}^{-1}\left(\omega,\vec{r},\vec{r'}\right) - L_{ji}^{-1*}\left(\omega,\vec{r'},\vec{r}\right)\right\}K_j\left(\vec{r'},\omega\right)K_i^*\left(\vec{r},\omega\right) = \\ &= \frac{\imath\omega}{4\pi}\int d\vec{r}\,d\vec{r'}\left\{L_{ij}^*\left(\omega,\vec{r},\vec{r'}\right) - L_{ji}\left(\omega,\vec{r'},\vec{r}\right)\right\}E_i\left(\vec{r},\omega\right)E_j^*\left(\vec{r'},\omega\right). \end{aligned} \tag{1.195}$$

According to the quantum mechanics rules, we introduce the operator $\widehat{E}_i\left(\vec{r},t\right)$ as the strength of the electric field $\vec{E}\left(\vec{r},t\right)$. The Fourier components of this operator are

$$\begin{cases} \widehat{E}_i\left(\vec{r},\omega\right) = \dfrac{1}{2\pi}\displaystyle\int_{-\infty}^{+\infty} dt\exp\left(\imath\omega t\right)\widehat{E}_i\left(\vec{r},t\right); \\ \widehat{E}_i\left(\vec{r},t\right) = \displaystyle\int_{-\infty}^{+\infty} d\omega\exp\left(-\imath\omega t\right)\widehat{E}_i\left(\vec{r},\omega\right). \end{cases} \tag{1.196}$$

The correlation between the fluctuations of the field $\vec{E}\left(\vec{r},t\right)$ in different spatial and temporal coordinates is characterized by

$$\varphi_{ij}\left(\vec{r},\vec{r'},t,t'\right) = \frac{1}{2}\overline{\left[\widehat{E}_i\left(\vec{r},t\right)\widehat{E}_j\left(\vec{r'},t'\right) + \widehat{E}_j\left(\vec{r'},t'\right)\widehat{E}_i\left(\vec{r},t\right)\right]}, \tag{1.197}$$

where the bar above the quantities means averaging by making use of the exact wave functions of the system. It is assumed that our system (the medium) is in the stationary state. In this case, correlation function (1.197) must depend only on the difference $t' - t = \tau$. Then, using relation (1.196), we obtain

$$\varphi_{ij}\left(\vec{r},\vec{r}',\tau\right)=\frac{1}{2}\int_{-\infty}^{+\infty}d\omega\int_{-\infty}^{+\infty}d\omega'\overline{\left[\hat{E}_i\left(\vec{r},\omega\right)\hat{E}_j\left(\vec{r}',\omega'\right)+\hat{E}_j\left(\vec{r}',\omega'\right)\hat{E}_i\left(\vec{r},\omega\right)\right]}$$
$$\times\exp\left(\imath\omega t+\imath\omega' t\right).\tag{1.198}$$

The integral on the right-hand side of Eq. (1.198) will be a function of $t'-t$, if the integrand contains a δ-function of $\omega+\omega'$, i.e.,

$$\frac{1}{2}\overline{\left[\hat{E}_i\left(\vec{r},\omega\right)\hat{E}_j\left(\vec{r}',\omega'\right)+\hat{E}_j\left(\vec{r}',\omega'\right)\hat{E}_i\left(\vec{r},\omega\right)\right]}=\left(E_i\left(\vec{r}\right)E_j\left(\vec{r}'\right)\right)_\omega\delta(\omega+\omega').\tag{1.199}$$

This relation should be considered as the definition of $\left(E_i\left(\vec{r}\right)E_j\left(\vec{r}'\right)\right)_\omega$, which obviously is a real quantity. Substituting Eq. (1.199) into Eq. (1.198), we obtain

$$\varphi_{ij}\left(\vec{r},\vec{r}',\tau\right)=\int_{-\infty}^{+\infty}d\omega\exp\left(-\imath\omega\tau\right)\left(E_i\left(\vec{r}\right)E_j\left(\vec{r}'\right)\right)_\omega.\tag{1.200}$$

The quantity $\varphi_{ij}\left(\vec{r},\vec{r}',0\right)$ at $\vec{r}=\vec{r}'$ and $i=j$ is the mean square of the fluctuations of the ith components of the electric field $E_i\left(\vec{r},t\right)$.

Assuming that the system is in the specific (nth) stationary state, the averaged value determined by Eq. (1.199) is equal to

$$\frac{1}{2}\left[\hat{E}_i\left(\vec{r},\omega\right)\hat{E}_j\left(\vec{r}',\omega'\right)+\hat{E}_j\left(\vec{r}',\omega'\right)\hat{E}_i\left(\vec{r},\omega\right)\right]_{nn}=$$
$$=\frac{1}{2}\sum_m\left[\left(E_i\left(\vec{r},\omega\right)\right)_{nm}\left(E_j\left(\vec{r}',\omega'\right)\right)_{mn}+\left(E_j\left(\vec{r}',\omega'\right)\right)_{nm}\left(E_i\left(\vec{r},\omega\right)\right)_{mn}\right],$$

where the sum is done over the whole energy spectrum. Calculating the matrix element of the operator $\vec{E}\left(\vec{r},t\right)$ by making use of the wave functions of the stationary states, we find

$$\left(E_i\left(\vec{r},\omega\right)\right)_{nm} = \frac{1}{2\pi}\int\limits_{-\infty}^{+\infty} dt \exp\left[\imath(\omega_{nm}+\omega)t\right]\left(E_i\left(\vec{r}\right)\right)_{nm}$$
$$= \left(E_i\left(\vec{r}\right)\right)_{nm}\delta(\omega_{nm}+\omega). \quad (1.201)$$

Here, $\left(E_i\left(\vec{r}\right)\right)_{nm}$ being independent of time is the matrix element of the operator $\widehat{E}_i\left(\vec{r},t\right)$; $\omega_{nm} = (\epsilon_n - \epsilon_m)/\hbar$ is the transition frequency between states n and m of the system. Hence,

$$\frac{1}{2}\left[\widehat{E}_i\left(\vec{r},\omega\right)\widehat{E}_j\left(\vec{r}',\omega'\right) + \widehat{E}_j\left(\vec{r}',\omega'\right)\widehat{E}_i\left(\vec{r},\omega\right)\right]_{nn} =$$
$$= \frac{1}{2}\sum_m\left\{\left(E_i\left(\vec{r}\right)\right)_{nm}\left(E_j\left(\vec{r}'\right)\right)_{mn}\delta(\omega_{nm}+\omega)\delta(\omega_{mn}+\omega')\right. \quad (1.202)$$
$$\left.+\left(E_j\left(\vec{r}'\right)\right)_{nm}\left(E_i\left(\vec{r}\right)\right)_{mn}\delta(\omega_{nm}+\omega')\delta(\omega_{mn}+\omega)\right\}.$$

Comparing Eqs. (1.199) and (1.202) and noting that $\omega_{nm} = -\omega_{mn}$, we find

$$\left(E_i\left(\vec{r}\right)E_j\left(\vec{r}'\right)\right)_{\omega} = \frac{1}{2}\sum_m\left\{\left(E_i\left(\vec{r}\right)\right)_{nm}\left(E_j\left(\vec{r}'\right)\right)_{mn}\delta(\omega+\omega_{nm})+\right.$$
$$\left.+\left(E_j\left(\vec{r}'\right)\right)_{nm}\left(E_i\left(\vec{r}\right)\right)_{mn}\delta(\omega+\omega_{mn})\right\} \quad (1.203)$$

Now, let us assume that an external periodic (with the frequency ω) perturbation, with energy proportional to $\vec{E}\left(\vec{r},t\right)$, acts on the system. As an example of this perturbation, we can suppose a random induction $\vec{K}\left(\vec{r},t\right)$, which results in the change of medium energy determined by Eq. (1.192). In this case, the operator of the perturbation of the system is

$$\widehat{V} = -\frac{1}{4\pi}\int d\vec{r}\, K_i\left(\vec{r},t\right)\widehat{E}_i\left(\vec{r},t\right). \quad (1.204)$$

In fact, the change of system energy is equal to the mean value of the partial derivative of the Hamiltonian operator of the system with respect to time. However, since only the Hamiltonian of perturbation $\widehat{V}$ depends on time, from Eq. (1.204) we find Eq. (1.192), which characterizes the change of medium energy. Making use of Eq. (1.193), we can write operator $\widehat{V}$ as

$$\widehat{V} = -\frac{1}{4\pi}\int d\vec{r}\,\widehat{E}_i\left(\vec{r},t\right)\left[K_i\left(\vec{r},\omega\right)\exp\left(-\imath\omega t\right) + K_i^*\left(\vec{r},\omega\right)\exp\left(\imath\omega t\right)\right]. \tag{1.205}$$

Transitions between different energy states take place due to the action of perturbation $\widehat{V}$ in the medium. Moreover, the transition probability $n \rightarrow m$ (per unit time) is determined by

$$W_{nm} = \frac{1}{8\pi\hbar^2}\int d\vec{r}\,d\vec{r'}\,K_i\left(\vec{r},\omega\right)K_j^*\left(\vec{r'},\omega\right)$$
$$\times\left\{\left(E_i\left(\vec{r}\right)\right)_{mn}\left(E_i\left(\vec{r'}\right)\right)_{nm}\delta(\omega+\omega_{nm}) + \left(E_i\left(\vec{r}\right)\right)_{nm}\left(E_j\left(\vec{r'}\right)\right)_{mn}\delta(\omega+\omega_{mn})\right\}. \tag{1.206}$$

In each transition, the system absorbs quantum energy $\hbar\omega_{nm}$ from the source of the external perturbation. The averaged energy released in the medium per unit time as the result of transitions is equal to

$$Q = \sum_m W_{nm}\hbar\omega_{nm} = \frac{\omega}{8\pi\hbar}\int d\vec{r}\,d\vec{r'}\sum_m K_i\left(\vec{r},\omega\right)K_j^*\left(\vec{r'},\omega\right)$$
$$\times\left\{\left(E_i\left(\vec{r}\right)\right)_{mn}\left(E_j\left(\vec{r'}\right)\right)_{nm}\delta(\omega+\omega_{nm}) - \left(E_i\left(\vec{r}\right)\right)_{nm}\left(E_j\left(\vec{r'}\right)\right)_{mn}\delta(\omega+\omega_{mn})\right\}. \tag{1.207}$$

Comparing Eqs. (1.195) and (1.207), we find

$$L_{ij}^{*-1}\left(\omega,\vec{r},\vec{r'}\right) - L_{ji}^{-1}\left(\omega,\vec{r'},\vec{r}\right)$$
$$= \frac{\imath}{2\hbar}\sum_m\left\{\left(E_i\left(\vec{r}\right)\right)_{mn}\left(E_j\left(\vec{r'}\right)\right)_{nm}\delta(\omega+\omega_{nm}) - \left(E_i\left(\vec{r}\right)\right)_{nm}\left(E_j\left(\vec{r'}\right)\right)_{mn}\delta(\omega+\omega_{mn})\right\}. \tag{1.208}$$

Now, we average relations (1.203) and (1.208) over Gibbs distribution of the different stationary states of the system.[27] From Eq. (1.203) we obtain[28]

---

[27]Gibbs function

$$W_n = \exp\left(\frac{F - \epsilon_n}{\kappa T}\right)$$

[28]is the energy distribution of every macroscopic body, which is a relatively small part of a large system and is in equilibrium with it. From the normalization condition of the distribution $\sum_m W_n = 1$, it is found that the free energy of the body, F, is equal to

$$\left(E_i\left(\vec{r}\right)E_j\left(\vec{r}'\right)\right)_\omega = \frac{1}{2}\sum_n\sum_m \exp\left(\frac{F-\epsilon_n}{\kappa T}\right)$$
$$\times\left\{\left(E_i\left(\vec{r}\right)\right)_{nm}\left(E_j\left(\vec{r}'\right)\right)_{mn}\delta(\omega+\omega_{nm})+\left(E_j\left(\vec{r}'\right)\right)_{nm}\left(E_i\left(\vec{r}\right)\right)_{mn}\delta(\omega+\omega_{mn})\right\},$$

where ϵ_n is the energy levels of the system, F is free energy, T is temperature, and κ is Boltzmann's constant. After some algebraic calculations, we find

$$\left(E_i\left(\vec{r}\right)E_j\left(\vec{r}'\right)\right)_\omega = \frac{1}{2}\left(1+\exp\left(-\frac{\hbar\omega}{\kappa T}\right)\right)$$
$$\sum_m\sum_n \exp\left(\frac{F-\epsilon_n}{\kappa T}\right)\left(E_i\left(\vec{r}\right)\right)_{nm}\left(E_j\left(\vec{r}'\right)\right)_{mn}\delta(\omega+\omega_{nm}). \tag{1.209}$$

Averaging expression (1.209) over Gibbs distribution is performed in the same way. In this case, we find

$$L_{ji}^{-1*}\left(\omega,\vec{r}',\vec{r}\right)-L_{ij}^{-1}\left(\omega,\vec{r},\vec{r}'\right)=\frac{\imath}{2\hbar}\left(1-\exp\left(-\frac{\hbar\omega}{\kappa T}\right)\right)$$
$$\times\sum_m\sum_n \exp\left(-\frac{F-\epsilon_n}{\kappa T}\right)\left(E_j\left(\vec{r}'\right)\right)_{mn}\left(E_i\left(\vec{r}\right)\right)_{nm}\delta(\omega+\omega_{nm}). \tag{1.210}$$

From Eqs. (1.209) and (1.210), it follows that

$$\left(E_i\left(\vec{r}\right)E_j\left(\vec{r}'\right)\right)_\omega = \imath\hbar\coth\left(\frac{\hbar\omega}{2\kappa T}\right)\left\{L_{ji}^{-1*}\left(\omega,\vec{r}',\vec{r}\right)-L_{ij}^{-1}\left(\omega,\vec{r},\vec{r}'\right)\right\}. \tag{1.211}$$

As mentioned before [compare to Eq. (1.200)], the quantity $\left(E_i\left(\vec{r}\right)E_j\left(\vec{r}\right)\right)_\omega$ characterizes the fluctuation of the electric field in the medium. From Eq. (1.211) an expression is obtained for the correlation of the random inductions $\vec{K}$ whose actions result in the equivalent spontaneous fluctuation of $\vec{E}$. For this aim, we notice that from Eqs. (1.190) and (1.191), the following equality is obtained:

$$F=-\kappa T\ln\sum_n\exp\left(-\frac{\epsilon_n}{\kappa T}\right).$$

This formula allows us to calculate the thermodynamic functions of the body if its energy spectrum ϵ_n is known [5].

In what follows, all quantities are averaged by Gibbs distribution. Besides, any specific notation will not be used for them. We hope it does not make any misunderstanding.

$$\int d\vec{r}' L_{ij}\left(\omega, \vec{r}'', \vec{r}'\right) L_{jk}^{-1}\left(\omega, \vec{r}', \vec{r}\right) = \delta\left(\vec{r} - \vec{r}''\right)\delta_{ik}. \tag{1.212}$$

Making use of this equality along with Eqs. (1.190) and (1.211), we find

$$\left(K_i\left(\vec{r}\right)K_j\left(\vec{r}'\right)\right)_\omega = \imath\hbar \coth\left(\frac{\hbar\omega}{2\kappa T}\right)\left[L_{ij}\left(\omega, \vec{r}, \vec{r}'\right) - L_{ji}^*\left(\omega, \vec{r}', \vec{r}\right)\right]. \tag{1.213}$$

From Eq. (1.190), it follows that

$$L_{ij}\left(\omega, \vec{r}, \vec{r}'\right) - L_{ji}^*\left(\omega, \vec{r}', \vec{r}\right) = \varepsilon_{ji}^*\left(\omega, \vec{r}', \vec{r}\right) - \varepsilon_{ij}\left(\omega, \vec{r}, \vec{r}'\right).$$

Therefore, relation (1.213) can be written as

$$\left(K_i\left(\vec{r}\right)K_j\left(\vec{r}'\right)\right)_\omega = \imath\hbar \coth\left(\frac{\hbar\omega}{2\kappa T}\right)\left[\varepsilon_{ji}^*\left(\omega, \vec{r}', \vec{r}\right) - \varepsilon_{ij}\left(\omega, \vec{r}, \vec{r}'\right)\right]. \tag{1.214}$$

At the same time, from relations (1.188) and (1.42), we find

$$\left(j_{st_i}\left(\vec{r}\right)j_{st_j}\left(\vec{r}'\right)\right)_\omega = -\frac{\hbar\omega}{4\pi}\coth\left(\frac{\hbar\omega}{2\kappa T}\right)\left[\sigma_{ij}\left(\omega, \vec{r}, \vec{r}'\right) + \sigma_{ji}^*\left(\omega, \vec{r}', \vec{r}\right)\right], \tag{1.215}$$

for the correlation of random currents $\vec{j}_{st}$. It should be remarked that the non-locality of the correlation of random inductions (1.214) and random currents (1.215) is stipulated by spatial dispersion. In the dispersion free case, it follows that

$$\varepsilon_{ij}\left(\omega, \vec{r}, \vec{r}'\right) = \varepsilon(\omega)\delta\left(\vec{r} - \vec{r}'\right), \qquad \sigma_{ij}\left(\omega, \vec{r}, \vec{r}'\right) = \sigma_{ij}(\omega)\delta\left(\vec{r} - \vec{r}'\right).$$

Relations (1.214) and (1.215), in this case, take the form of

$$\left(K_i\left(\vec{r}\right)K_j\left(\vec{r}'\right)\right)_\omega = -\imath\hbar \coth\left(\frac{\hbar\omega}{2\kappa T}\right)\left[\varepsilon_{ij}(\omega) - \varepsilon_{ji}^*(\omega)\right]\delta\left(\vec{r} - \vec{r}'\right), \tag{1.216}$$

$$\left(j_{st_i}\left(\vec{r}\right)j_{st_j}\left(\vec{r}'\right)\right)_\omega = -\frac{\hbar\omega}{4\pi}\coth\left(\frac{\hbar\omega}{2\kappa T}\right)\left[\sigma_{ij}(\omega) + \sigma_{ji}^*(\omega)\right]\delta\left(\vec{r} - \vec{r}'\right). \tag{1.217}$$

1.9.2 Electromagnetic Field Fluctuations in Homogenous Isotropic Non-gyrotropic Media

In the above formulations, the general formulas of the fluctuations of the electric field in the medium were obtained by taking spatial dispersion into account. Now, we consider a spatially homogeneous isotropic non-gyrotropic medium. In this case, it is better to use the Fourier transformation in space

$$F_i\left(\vec{r}\right) = \int d\vec{k}\, \exp\left(\imath\vec{k}\cdot\vec{r}\right)F_i\left(\vec{k}\right),\quad F_i\left(\vec{k}\right) = \frac{1}{(2\pi)^3}\int d\vec{r}\, \exp\left(-\imath\vec{k}\cdot\vec{r}\right)F_i\left(\vec{r}\right).$$

Then, Eqs. (1.214) and (1.215) based on relations (1.19) and (1.24) take the form of

$$\left(K_i\left(\vec{k}\right)K_j\left(\vec{k'}\right)\right)_\omega = -\frac{\imath\hbar}{(2\pi)^3}\coth\left(\frac{\hbar\omega}{2\kappa T}\right)\left[\varepsilon_{ij}\left(\omega,\vec{k}\right) - \varepsilon_{ji}^*\left(\omega,\vec{k}\right)\right]\delta\left(\vec{k}+\vec{k'}\right) =$$
$$= \frac{2\hbar}{(2\pi)^3}\coth\left(\frac{\hbar\omega}{2\kappa T}\right)\left\{\left(\delta_{ij}-\frac{k_ik_j}{k^2}\right)\mathrm{Im}\varepsilon^{\mathrm{tr}}(\omega,k)+\frac{k_ik_j}{k^2}\mathrm{Im}\varepsilon^l(\omega,k)\right\}\delta\left(\vec{k}+\vec{k'}\right);\tag{1.218}$$

$$\left(j_{\mathrm{st}\,i}\left(\vec{k}\right)j_{\mathrm{st}\,j}\left(\vec{k'}\right)\right)_\omega = \frac{\hbar\omega}{4\pi(2\pi)^3}\coth\left(\frac{\hbar\omega}{2\kappa T}\right)\left[\sigma_{ij}\left(\omega,\vec{k}\right) + +\sigma_{ji}^*\left(\omega,\vec{k}\right)\right]\delta\left(\vec{k}+\vec{k'}\right)$$
$$= \frac{\hbar\omega}{(2\pi)^4}\coth\left(\frac{\hbar\omega}{2\kappa T}\right)\left\{\left(\delta_{ij}-\frac{k_ik_j}{k^2}\right)\mathrm{Re}\,\sigma^{\mathrm{tr}}(\omega,k)+\frac{k_ik_j}{k^2}\mathrm{Re}\,\sigma^l(\omega,k)\right\}\delta\left(\vec{k}+\vec{k'}\right).\tag{1.219}$$

To determine the correlation of the electric field fluctuations given by Eq. (1.211), we have to find the explicit form of the operator $L_{jk}^{-1}\left(\omega,\vec{r'},\vec{r}\right)$. To do this, we find the solution of Eq. (1.189) for the Fourier components of the field, which for an isotropic non-gyrotropic medium takes the form

$$\left\{k^2\delta_{ij} - k_ik_j - \frac{\omega^2}{c^2}\left[\left(\delta_{ij}-\frac{k_ik_j}{k^2}\right)\varepsilon^{\mathrm{tr}}(\omega,k) + \frac{k_ik_j}{k^2}\varepsilon^l(\omega,k)\right]\right\}E_j = \frac{\omega^2}{c^2}K_i.\tag{1.220}$$

The solution of this equation (see Sect. 1.8) can be written as

$$E_i\left(\vec{k}\right) = -\frac{\omega^2}{k^2}\left\{\frac{k_i k_j}{\omega^2 \varepsilon^l(\omega, k)} - \frac{k^2\left(\delta_{ij} - k_i k_j/k^2\right)}{c^2\left[k^2 - \frac{\omega^2}{c^2}\varepsilon^{\mathrm{tr}}(\omega, k)\right]}\right\} K_j\left(\vec{k}\right). \tag{1.221}$$

From this, taking Eq. (1.218) into account, we find

$$\left(E_i\left(\vec{k}\right)E_j\left(\vec{k'}\right)\right)_\omega = \frac{2\hbar}{(2\pi)^3}\coth\left(\frac{\hbar\omega}{2\kappa T}\right)\psi_{ij}\left(\omega, \vec{k}\right)\delta\left(\vec{k} + \vec{k'}\right), \tag{1.222}$$

where[29]

$$\psi_{ij}\left(\omega, \vec{k}\right) = \frac{k_i k_j}{k^2}\frac{\mathrm{Im}\,\varepsilon^l(\omega, k)}{\left|\varepsilon^l(\omega, k)\right|^2} + \frac{\omega^4}{c^4}\left(\delta_{ij} - \frac{k_i k_j}{k^2}\right)\frac{\mathrm{Im}\,\varepsilon^{\mathrm{tr}}(\omega, k)}{\left|k^2 - \frac{\omega^2}{c^2}\varepsilon^{\mathrm{tr}}(\omega, k)\right|^2}. \tag{1.223}$$

Finally, making use of the inverse Fourier transformation, we find

$$\left(E_i\left(\vec{r}\right)E_j\left(\vec{r'}\right)\right)_\omega = 2\hbar\coth\left(\frac{\hbar\omega}{2\kappa T}\right)\phi_{ij}\left(\omega, \vec{r} - \vec{r'}\right), \tag{1.224}$$

where

$$\phi_{ij}\left(\omega, \vec{R}\right) = \frac{1}{(2\pi)^3}\int d\vec{k}\,\exp\left(\imath\vec{k}.\vec{R}\right)\psi_{ij}\left(\omega, \vec{k}\right). \tag{1.225}$$

We have to mark that the first term on the right-hand side of Eq. (1.223) corresponds to the correlation of the longitudinal field. At the same time, the second term is stipulated by the transverse field. In this connection, the correlation formula for longitudinal fields can be transformed into

$$\left(\vec{E}^l\left(\vec{r}\right).\vec{E}^l\left(\vec{r'}\right)\right)_\omega = 2\hbar\coth\left(\frac{\hbar\omega}{2\kappa T}\right)\int\frac{d\vec{k}}{(2\pi)^3}\exp\left[\imath\vec{k}\cdot\left(\vec{r} - \vec{r'}\right)\right] \times \frac{\mathrm{Im}\varepsilon^l\left(\omega, \vec{k}\right)}{\left|\varepsilon^l\left(\omega, \vec{k}\right)\right|^2}. \tag{1.226}$$

Equation (1.224) can be written as

[29] It should be noted that inequality (1.82), $\mathrm{Im}\varepsilon^l(\omega, k) \geq 0$ and $\mathrm{Im}\varepsilon^{\mathrm{tr}}(\omega, k) \geq 0$, follows from expressions (1.222) and (1.223).

$$\left(\vec{E}\left(\vec{r}\right)\cdot\vec{E}\left(\vec{r'}\right)\right)_{\omega}=2\hbar\coth\left(\frac{\hbar\omega}{2\kappa T}\right)\int\frac{d\vec{k}}{(2\pi)^{3}}\exp\left(\imath\vec{k}\cdot\vec{R}\right)$$
$$\times\left\{\frac{\mathrm{Im}\varepsilon^{l}\left(\omega,\vec{k}\right)}{\left|\varepsilon^{l}\left(\omega,\vec{k}\right)\right|^{2}}+2\frac{\omega^{4}}{c^{4}}\frac{\mathrm{Im}\,\varepsilon^{\mathrm{tr}}\left(\omega,\vec{k}\right)}{\left|k^{2}-\frac{\omega^{2}}{c^{2}}\varepsilon^{\mathrm{tr}}\left(\omega,\vec{k}\right)\right|^{2}}\right\},\tag{1.227}$$

where $\vec{R}=\vec{r}-\vec{r'}$. Neglecting spatial dispersion in this equation and taking into account that

$$\varepsilon^{l}(\omega,0)=\varepsilon^{\mathrm{tr}}(\omega,0)=\varepsilon(\omega),$$

we find another form of Eq. (1.227) as

$$\left(\vec{E}\left(\vec{r}\right).\vec{E}\left(\vec{r'}\right)\right)_{\omega}=2\hbar\coth\left(\frac{\hbar\omega}{2\kappa T}\right)\left\{\frac{\mathrm{Im}\,\varepsilon(\omega)}{|\varepsilon(\omega)|^{2}}\delta\left(\vec{R}\right)-\right.$$
$$\left.-\frac{1}{4\pi}\frac{\imath\omega^{2}}{Rc^{2}}\left[\exp\left(-\frac{\omega}{c}\sqrt{-\varepsilon(\omega)}R\right)-\exp\left(-\frac{\omega}{c}\sqrt{-\varepsilon^{*}(\omega)}R\right)\right]\right\}.\tag{1.228}$$

The presence of the δ-function in front of the term being proportional to the imaginary part of the dielectric permittivity $\varepsilon(\omega)$ and leading to infinitely large fluctuations of the longitudinal field in absorptive media is one of the singularities of this formula. Moreover, expression (1.227) does not contain the singularity of the type of $\delta\left(\vec{R}\right)$, which was mentioned. The second singularity of Eq. (1.228) is the divergence of the fluctuations of the longitudinal field at $\varepsilon(\omega)=0$, i.e., when both the imaginary and real parts of the dielectric permittivity are equal to zero. The physical reason of such divergence is rather simple. The point is that the condition $\varepsilon(\omega)=0$ is the existence condition of longitudinal oscillations, and the oscillation frequency is independent of the wave vector as well. Consequently, an infinite number of waves with arbitrary wave vectors correspond to one frequency of longitudinal oscillations. The latter means that the fluctuating longitudinal field with the frequency of longitudinal oscillations corresponds to the thermal excitation of an infinite number of the degrees of freedom, which leads to a singularity in Eq. (1.228) at the frequency of longitudinal oscillations. It is simply understood that by taking spatial dispersion into account when the frequency of longitudinal waves is a function of the wave vector, such a singularity is not possible to happen. In fact, from Eq. (1.227) it is clear that the correlations of the fluctuations of the longitudinal fields do not contain such a singularity when $\varepsilon^{l}(\omega,k)=0$. This is stipulated by the fact that when spatial dispersion is taken into account, both longitudinal and transverse waves become the equivalent branch of normal waves in the medium and result in analogous effects, which arise from the transverse electromagnetic waves [64].

In conclusion, we consider the symmetry of the dielectric permittivity tensor of the medium. Here, we use the property of the time symmetry of the electric field fluctuations. Since the strength of the electric field is time-invariant, then the left-hand side of Eq. (1.199) does not change by replacement $\omega' \to -\omega'$ and $\omega \to -\omega$. Making use of this fact that expression (1.199) is practically different from zero only at $\omega' = -\omega$, we find a relation, which characterizes the property of time symmetry of the electric field fluctuations

$$\left(E_i\left(\vec{r}\right)E_j\left(\vec{r}'\right)\right)_\omega = \left(E_j\left(\vec{r}'\right)E_i\left(\vec{r}\right)\right)_\omega. \tag{1.229}$$

The fluctuations of the random inductions $\vec{K}$ possess the analogous property as a result of which the following identity holds [compare to Eq. (1.213)]:

$$\varepsilon_{ij}\left(\omega, \vec{r}, \vec{r}'\right) - \varepsilon^*_{ji}\left(\omega, \vec{r}', \vec{r}\right) = \varepsilon_{ji}\left(\omega, \vec{r}', \vec{r}\right) - \varepsilon^*_{ij}\left(\omega, \vec{r}, \vec{r}'\right). \tag{1.230}$$

From the latter relation, we find

$$\varepsilon'_{ij}\left(\omega, \vec{r}, \vec{r}'\right) = \varepsilon'_{ji}\left(\omega, \vec{r}', \vec{r}\right), \tag{1.231}$$

where $\varepsilon'_{ij}\left(\omega, \vec{r}, \vec{r}'\right)$ is the real part of the tensor $\varepsilon_{ij}\left(\omega, \vec{r}, \vec{r}'\right)$. But, the real and imaginary parts of the dielectric permittivity tensor of the medium in the equilibrium state are related to each other by linear integral relations, i.e., the formulas of Kramers–Kronig [2]. In fact, $\varepsilon_{ij}\left(\omega, \vec{r}, \vec{r}'\right)$, as a function of ω, which is determined with the help of one-sided Fourier transformation (1.15), is an analytic function everywhere on the upper half-plane of the complex variable ω with exception of possible poles of finite width Im $\omega \geq \sigma \geq 0$. For a medium in the equilibrium state, we have $\sigma \to +0$ (see Sect. 1.5). Therefore, for such media, we find [22]

$$\varepsilon_{ij}\left(\omega, \vec{r}, \vec{r}'\right) - \delta_{ij}\delta\left(\vec{r} - \vec{r}'\right) = \frac{1}{\pi i}\int_{-\infty}^{+\infty} d\omega' \mathcal{P}\frac{\varepsilon_{ij}\left(\omega', \vec{r}, \vec{r}'\right) - \delta_{ij}\delta\left(\vec{r} - \vec{r}'\right)}{\omega' - \omega}, \tag{1.232}$$

where $\mathcal{P}$ denotes the principal part of the integrand. Separating real and imaginary parts in Eq. (1.232), we find the well-known Kramers–Kronig formulas[30]

$$\begin{cases} \varepsilon'_{ij}\left(\omega,\vec{r},\vec{r}'\right)-\delta_{ij}\delta\left(\vec{r}-\vec{r}'\right)=\dfrac{1}{\pi}\displaystyle\int\limits_{-\infty}^{+\infty}d\omega'\,\mathcal{P}\dfrac{\varepsilon''_{ij}\left(\omega',\vec{r},\vec{r}'\right)}{\omega'-\omega}, \\ \varepsilon'_{ij}\left(\omega,\vec{r},\vec{r}'\right)=-\dfrac{1}{\pi}\displaystyle\int\limits_{-\infty}^{+\infty}d\omega'\,\mathcal{P}\dfrac{\varepsilon'_{ij}\left(\omega,\vec{r},\vec{r}'\right)-\delta_{ij}\delta\left(\vec{r}-\vec{r}'\right)}{\omega'-\omega}. \end{cases} \tag{1.233}$$

From these relations, it follows that the real and imaginary parts of the dielectric permittivity tensor, denoted by $\varepsilon'_{ij}\left(\omega,\vec{r},\vec{r}'\right)$ and $\varepsilon''_{ij}\left(\omega,\vec{r},\vec{r}'\right)$, respectively, have the symmetric property, which is described by relation (1.231).

Hence,

$$\varepsilon_{ij}\left(\omega,\vec{r},\vec{r}'\right)=\varepsilon_{ji}\left(\omega,\vec{r}',\vec{r}\right). \tag{1.234}$$

For unbounded spatially homogeneous media, we have

[30]For isotropic media, the Kramers–Kronig formulas (1.233) are valid for both longitudinal $\varepsilon^{l}\left(\omega,\vec{k}\right)$ and transverse $\varepsilon^{\mathrm{tr}}\left(\omega,\vec{k}\right)$ dielectric permittivities. Therefore, making use of Eq. (1.209) for the magnetic permeability $\mu\left(\omega,\vec{k}\right)$, we obtain

$$1-\mathrm{Re}\left(\frac{1}{\mu(\omega,k)}\right)=-\frac{1}{\pi}\int\limits_{-\infty}^{+\infty}d\omega'\,\mathcal{P}\frac{1}{\omega'-\omega}\mathrm{Im}\left(\frac{1}{\mu(\omega',k)}\right),$$

$$\mathrm{Im}\left(\frac{1}{\mu(\omega,k)}\right)=\frac{1}{\pi}\int\limits_{-\infty}^{+\infty}d\omega'\,\mathcal{P}\frac{1}{\omega'-\omega}\left\{1-\mathrm{Re}\left(\frac{1}{\mu(\omega',k)}\right)\right\}.$$

If the function $\mathrm{Im}\mu\left(\omega,\vec{k}\right)$ at $\omega=0$ does not have any singularity, then from these formulas, we can find a relation for the static magnetic permeability of the isotropic medium:

$$\mathrm{Re}\left(\frac{1}{\mu(0,k)}\right)=1-\frac{2}{\pi}\int\limits_{0}^{\infty}\frac{d\omega'}{\omega'}\frac{\mathrm{Im}\mu(\omega',k)}{|\mu(\omega',k)|^2}.$$

In Sect. 1.4, it was mentioned that $\mathrm{Im}\mu\left(\omega,\vec{k}\right)$ in contrast to the quantities $\mathrm{Im}\varepsilon^{l}\left(\omega,\vec{k}\right)$ and $\mathrm{Im}\varepsilon^{tr}\left(\omega,\vec{k}\right)$ can be positive or negative. In fact, inequality $\mathrm{Im}\mu\left(\omega,\vec{k}\right)\geq 0$ (or $\mathrm{Im}\mu\left(\omega,\vec{k}\right)\leq 0$) would result in $\mu\left(0,\vec{k}\right)\geq 1$ (or $\mu\left(0,\vec{k}\right)\leq 1$). This would mean the impossibility of the existence of diamagnetic (or paramagnetic) media.

$$\varepsilon_{ij}\left(\omega, \vec{k}\right) = \varepsilon_{ji}\left(\omega, -\vec{k}\right). \tag{1.235}$$

Relations (1.234) and (1.235) are changed if an external magnetic field $\vec{B}_0$, produced by external sources, exists in the medium. In this case, for the sign change of time, we should replace $\vec{B}_0 \to -\vec{B}_0$. Thus, instead of Eq. (1.234), we find

$$\varepsilon_{ij}\left(\omega, \vec{r}, \vec{r}', \vec{B}_0\right) = \varepsilon_{ji}\left(\omega, \vec{r}', \vec{r}, -\vec{B}_0\right). \tag{1.236}$$

Moreover, for unbounded spatially homogenous media, we have

$$\varepsilon_{ij}\left(\omega, \vec{k}, \vec{B}_0\right) = \varepsilon_{ji}\left(\omega, -\vec{k}, -\vec{B}_0\right). \tag{1.237}$$

1.9.3 Causality Principle and Generalization of Kramers–Kronig Relations

Now let us consider the problem of the field of the external sources in media. This problem is directly related to the energy loss of fast moving charged particles in the medium. Suppose that the field sources are $\rho_0\left(\vec{r}, t\right)$ and $\vec{j}_0\left(\vec{r}, t\right)$. In particular, these sources for fast moving charged particles are written as

$$\rho_0\left(\vec{r}, t\right) = q\delta\left(\vec{r} - \vec{v}t\right), \qquad \vec{j}_0\left(\vec{r}, t\right) = q\delta\left(\vec{r} - \vec{v}t\right), \tag{1.238}$$

where v is the particle velocity. Making use of the Fourier transformation

$$A\left(\vec{r}, t\right) = \int d\vec{k} e^{-\imath\omega t + \imath\vec{k}\cdot\vec{r}} A\left(\omega, \vec{k}\right), \tag{1.239}$$

we simply reduce the system of Maxwell's equations (1.9) to the following algebraic equations [see Eqs. (1.156)]

$$\left\{k^2\delta_{ij} - k_i k_j - \frac{\omega^2}{c^2}\varepsilon_{ij}\left(\omega, \vec{k}\right)\right\}E_j = \frac{4\pi\imath\omega}{c^2} j_{0i}\left(\omega, \vec{k}\right), \tag{1.240}$$

where $\vec{j}_{0i}\left(\omega, \vec{k}\right)$ is the Fourier transformed of the current density of the external sources $\vec{j}_{0i}\left(\vec{r}, t\right)$.

Equations (1.240) are simply solved for an isotropic medium as

$$E_i\left(\omega,\vec{k}\right)=-\frac{4\pi\iota\omega}{k^2}\left\{\frac{k_ik_j}{\omega^2\varepsilon^l\left(\omega,\vec{k}\right)}-\frac{k^2\left(\delta_{ij}-\frac{k_ik_j}{k^2}\right)}{c^2k^2-\omega^2\varepsilon^{\text{tr}}\left(\omega,\vec{k}\right)}\right\}j_{0j}\left(\omega,\vec{k}\right). \quad (1.241)$$

The first term describes the longitudinal field produced by the source in the medium, while the second term is the transverse electromagnetic field. Substituting this expression in Fourier transformation (1.239), we find the spatial-temporal distribution of the field produced by the external source.

Now we consider the general properties of the dielectric permittivity tensor of the isotropic medium which follow from the above relations. First of all, we note that according to the Eq. (1.241) the electromagnetic field in the medium is completely determined by the field source $\vec{j}_0\left(\omega,\vec{k}\right)$ which we can specify from the outside. In this sense, the source is the cause and the field is the effect. The quantities

$$\frac{1}{\varepsilon^l(\omega,k)},\qquad \frac{1}{k^2c^2-\omega^2\varepsilon^{\text{tr}}(\omega,k)}, \quad (1.242)$$

connecting the cause to the effect are called the response function of the medium to the external influences. In fact, these quantities, being causal functions in time, must possess the properties of frequency analyticity and do not have any poles in the upper half-plane of complex ω. Therefore, these functions should satisfy Cauchy's integral or, in other words, the Kramers–Kronig relations:

$$\begin{aligned}\frac{1}{\varepsilon^l(\omega,k)}-1&=\frac{1}{\pi\iota}\int_{-\infty}^{\infty}d\omega'\frac{\frac{1}{\varepsilon^l(\omega',k)}-1}{\omega'-\omega},\\ \frac{1}{k^2c^2-\omega^2\varepsilon^{\text{tr}}(\omega,k)}&=\frac{1}{\pi\iota}\int_{-\infty}^{\infty}\frac{d\omega'}{(\omega'-\omega)}\frac{1}{\left[k^2c^2-\omega^2\varepsilon^{\text{tr}}(\omega',k)\right]}.\end{aligned} \quad (1.243)$$

These relations connect the real parts of dielectric permittivities $\varepsilon^l(\omega,k)$ and $\varepsilon^{\text{tr}}(\omega,k)$ to their imaginary parts and thus reflect this fact that these quantities are not completely independent. In other words, knowing the imaginary parts of $\varepsilon^l(\omega,k)$ and $\varepsilon^{\text{tr}}(\omega,k)$, we can find their real parts and vice versa.

Now we consider the corollaries that follow from relation (1.243) in the static limit when $\omega\to 0$. From the first relation, we can find

$$\frac{1}{\text{Re}\,\varepsilon^l(\omega,k)}-1=\frac{2}{\pi}\int_0^{\infty}\frac{d\omega'}{\omega'}\text{Im}\frac{1}{\varepsilon^l(\omega',k)}. \quad (1.244)$$

In a thermodynamically equilibrium medium, inequality $\mathrm{Im}\varepsilon^l(\omega,k) > 0$ always holds. Therefore, from (1.244) we find

$$\frac{1}{\mathrm{Re}\,\varepsilon^l(0,k)} < 1. \tag{1.245}$$

This inequality says that in the static limit, longitudinal dielectric permittivity of thermodynamically equilibrium media can be either negative (as it happens in superconductors where the same charges attract each other in this circumstance) or larger than unity (as in the classical model of gaseous and solid-state plasmas with zero imaginary part of dielectric permittivity).

Analogously, from the second relation (1.243), we find a restriction on the static value of $\omega^2\varepsilon^{\mathrm{tr}}(\omega,k)$. In this case, in accordance with definition (1.35), it is convenient to introduce quantity $\mu(\omega,k)$ in the following form:

$$\frac{\mu(0,k)}{c^2k^2} = -\frac{2}{\pi}\int\limits_0^\infty \frac{d\omega'}{\omega'}\mathrm{Im}\left(\frac{1}{\omega'^2\varepsilon^l(\omega',k) - c^2k^2/\mu(\omega',k)}\right). \tag{1.246}$$

In thermodynamically equilibrium media, we have $\mathrm{Im}\varepsilon^l(\omega,k) > 0$ and $\mathrm{Im}\mu(\omega,k) > 0$. Therefore, from Eq. (1.246) we find that

$$\mu(0,k) > 0. \tag{1.247}$$

This means that the static magnetic permeability of thermodynamically equilibrium media, in contrast to their static dielectric permittivity (which coincides with longitudinal dielectric permittivity), is always positive. In this case, inequality $\mu(0,k) > 1$ corresponds to the paramagnetic, ferromagnetic, anti-ferromagnetic materials (in the limit $\mu(0,k) \to \infty$), while inequality $\mu(0,k) < 1$ corresponds to the diamagnetic materials.

1.10 Electromagnetic Properties of Inhomogeneous Media

The method of geometrical optics is described for spatially inhomogeneous media, with spatial dispersion, the eikonal equation, and quasi-classical quantization rules being derived.

1.10.1 *Inhomogeneous Media Without Spatial Dispersion. Approximation of Geometrical Optics*

In the preceding sections we considered only homogeneous media with parameters which do not depend on the coordinate. Real media, being bounded, are however inhomogeneous. In this section, we adopt the model of a spatially unbounded inhomogeneous medium. The method to describe spatially bounded media (plasmas) will be studied in the next chapter. The characteristic length scale of the inhomogeneity of some material media such as laboratory plasmas usually is the dimension of the experimental set-up. For example, in controlled thermonuclear fusion devices or in gas discharges the charged particle density varies significantly at distances of the order of the dimension of a plasma filament (the radius of the discharge tube). The characteristic length of the density homogeneity L_N is of the order of 1–10 cm. The charged particle temperature can be independent of the coordinates. In ionospheric plasma the characteristic lengths of the regular inhomogeneity are $L_N \approx 10^7$ cm for the charged particle density, $L_T \approx 5 \times 10^7$cm for the temperature and $L_B \approx 10^8$cm to 10^9 cm for the inhomogeneity of the earth's magnetic field. In solid state plasmas the characteristic length for the inhomogeneity is often determined by the method of creating the charge carriers. It is of the order of 10^{-1} to 1 cm.

We begin our study of the electromagnetic properties of an inhomogeneous medium with the formulation of its dielectric permittivity $\varepsilon_{ij}\left(\omega, \vec{k}\right)$. Since in the case of an inhomogeneous medium the kernels of the material equations (1.17) and (1.39) are not functions of the difference of the coordinates $\vec{r}$ and $\vec{r'}$, they depend both on $\vec{r} - \vec{r'}$ and on $\vec{r}$ and $\vec{r'}$ separately. Thus, one cannot describe the general electromagnetic properties within the concept of the tensor of the dielectric permittivity (or conductivity), as it has been defined in Eqs. (1.18)–(1.19) and (1.44)–(1.45).

1.10.1.1 Field Equation for an Inhomogeneous Medium Without Spatial Dispersion

We obtain the analysis of the electromagnetic properties of inhomogeneous media with the simplest model where spatial dispersion can be ignored. In this case, for homogeneous media the operator $\varepsilon_{ij}(t - t')$ does not depend on the difference $\vec{r} - \vec{r'}$. For inhomogeneous media the operator $\varepsilon_{ij}\left(t - t', \vec{r}\right)$ can depend on $\vec{r}$ only. We can apply the decomposition

$$\varepsilon_{ij}\left(\omega, \vec{r}\right) = \int_{0}^{\infty} dt_1 \varepsilon_{ij}\left(t_1, \vec{r}\right) e^{\imath\omega t_1}, \tag{1.248}$$

i.e., actually the expression for the dielectric tensor $\varepsilon_{ij}(\omega, 0)$ of the medium without spatial dispersion in the limit $k/\omega \rightarrow 0$. However, the corresponding parameters (density, temperature, etc.) should be regarded as functions of the coordinates. Even in this simple model the formulation of the propagation theory of electromagnetic waves presents a complex problem since it is necessary to solve the field equation

$$\nabla^2 \vec{E} - \nabla\nabla \cdot \vec{E} + \frac{\omega^2}{c^2} \vec{D} = 0, \tag{1.249}$$

where

$$D_i = \varepsilon_{ij}\left(\omega, \vec{r}\right) E_j.$$

Equation (1.249) is the basic equation of the theory of electromagnetic wave propagation in inhomogeneous media. For the most common case of normal incidence of a wave on a plane stratified medium, it has the same form as the stationary Schrodinger equation. When the medium is isotropic, $\varepsilon_{ij}\left(\omega, \vec{r}\right) = \varepsilon\left(\omega, \vec{r}\right)\delta_{ij}$, and inhomogeneous along the x-axis only, we write $\varepsilon\left(\omega, \vec{r}\right) = \varepsilon(\omega, x)$ and $\vec{E}\left(\omega, \vec{r}\right) = \vec{E}(\omega, x)$. Then, for the transverse waves $\left(\vec{E} \perp ox\right)$ Eq. (1.249) simplifies to

$$\frac{d^2 \vec{E}}{dx^2} + \frac{\omega^2}{c^2} \varepsilon(\omega, x) \vec{E} = 0. \tag{1.250}$$

Comparing with the one-dimensional Schrodinger equation

$$\frac{d^2 \Psi}{dx^2} + \frac{2m}{\hbar^2} [W - V(x)] \Psi = 0, \tag{1.251}$$

where Ψ is the wave function, W the total energy of a particle, and $V(x)$ its potential energy, we see that Eqs. (1.250) and (1.251) are closely related.

1.10.1.2 The Method of Geometrical Optics and the WKB Method

A thorough investigation of possible ways to solve equations of the type of Eqs. (1.250) and (1.251) and, particularly, to derive exact solutions for specific functions $\varepsilon(\omega, x)$ shows that they are fundamental in physics. Exact solutions are known for linear, parabolic, and some other dependences of $\varepsilon(\omega, x)$ on x. Approximate methods of solving the wave equation for an arbitrary dependence of $\varepsilon(\omega, x)$ on x are elaborated in the theory of electromagnetic wave propagation and quantum mechanics. The main methods are the method of *geometrical optics* in electrodynamics and the WKB (*Wentzel–Kramers–Brillouin*) method in quantum mechanics. We summarize the essentials of this method, which is of great importance for our further analysis.

The method of geometrical optics can be applied when the medium is weakly inhomogeneous on the scale of the wavelength of the electromagnetic oscillation, i.e., when the wavelength λ is smaller than the characteristic length L_0 of the inhomogeneity

$$\lambda / L_0 \ll 1.$$

In this case, the wave propagation is similar to that in a homogeneous medium with the corresponding parameters. For example, plane waves are eigen solutions of the wave equation in a homogeneous unbounded medium. In an inhomogeneous medium this is not true, if the properties of the medium do not vary much on the scale λ, the wave behaves as nearly plane.

In the approximation of geometrical optics any quantity characterizing the wave is of the form

$$\vec{E} = \vec{E}_0 \exp\left[-\imath\omega t + \imath\Psi(r)\right], \tag{1.252}$$

$\Psi(r)$ being called the *eikonal*. Physically it is the phase of the wave dependent on the coordinates. The numerical value of the eikonal is large since it must take the value 2π at $r = \lambda$, and since the approximation of geometrical optics corresponds to the limit $\lambda \to 0$.

In homogeneous media, we have

$$\Psi\left(\vec{r}\right) = \vec{k} \cdot \vec{r} = \frac{\omega}{c} \vec{n} \cdot \vec{r}. \tag{1.253}$$

For an inhomogeneous medium, we assume quite analogously

$$\nabla\Psi\left(\vec{r}\right) = \vec{k}\left(\vec{r}\right) = \frac{\omega}{c} \vec{n}\left(\vec{r}\right). \tag{1.254}$$

In weakly inhomogeneous media $\vec{k}\left(\vec{r}\right)$ is a slowly varying function of $\vec{r}$ which is determined by the variations of the properties of the medium in space. We can suppose that the scale of inhomogeneity of $\vec{k}\left(\vec{r}\right)$ coincides with L_0, i.e.,

$$\frac{\partial}{\partial \vec{r}} \vec{k}\left(\vec{r}\right) \sim \frac{\vec{k}\left(\vec{r}\right)}{L_0}.$$

We call $\vec{k}\left(\vec{r}\right)$ a wave vector in a weakly inhomogeneous medium, $\lambda = 2\pi/k$ a wavelength and $\vec{n}\left(\vec{r}\right)$ a refractive index. Since the wave vector depends weakly on the coordinates, we can still construct the solutions of electrodynamic problems for weakly inhomogeneous media in the form of expansion in the parameter λ/L_0. In the zero-order approximation the wave is considered plane, i.e., all terms of the order λ/L_0 and of higher order are fully neglected, in the first-order approximation only the terms of the first order of λ/L_0 are accounted for, etc. In other words, we neglect all derivatives of $\vec{k}\left(\vec{r}\right)$ in the zero-order approximation, the first-order derivatives are accounted for in the first-order approximation, etc. When calculating the terms of higher order one can simultaneously solve the field equations with any desired degree of accuracy.

We apply the described method to solve Eq. (1.250), assuming the field E in the form of

$$E(x) = E_0 \exp\left[\imath\Psi(x)\right]. \tag{1.255}$$

Substituting this expression into Eq. (1.250), we obtain an equation for $\Psi(x)$ which is called the *eikonal* equation:

$$\Psi'^2 - \imath\Psi'' = \varepsilon(\omega, x)\frac{\omega^2}{c^2}. \tag{1.256}$$

The prime denotes differentiation with respect to the coordinate. Since the condition $\lambda/L_0 \ll 1$ was assumed, one can expand the solution $\Psi(x)$ in the powers of this small parameter

$$\Psi = \Psi_0 + \Psi_1 + \Psi_2 + \ldots \tag{1.257}$$

Here, Ψ_0 is the value of Ψ in the zero-order approximation of geometrical optics, following from Eq. (1.256) when the second term on the left-hand side is ignored:

$$\Psi_0'^2 = \frac{\omega^2}{c^2}\varepsilon(\omega, x), \tag{1.258}$$

or

$$\Psi_0(x) = \pm\frac{\omega}{c}\int_0^x \sqrt{\varepsilon(\omega,x')}\,dx', \tag{1.259}$$

In the theory of wave propagation the domain where $\varepsilon(\omega,x) > 0$ holds is of special interest, since $\Psi_0(x)$ is a real function in this domain. The field $E(x)$ has a wave (oscillatory) character with the wavelength

$$\lambda \sim 1/\Psi_0' \sim c/\left[\omega\sqrt{\varepsilon(\omega,x)}\right].$$

At first sight one could think that the dependence of $\varepsilon(\omega,x)$ on the coordinate x should be completely neglected in the zero-order approximation of geometrical optics. However, this is not true, since [in spite of the weak inhomogeneity of $\varepsilon(\omega,x)$] in integration range in Eq. (1.259) can be very large and even significantly exceed the characteristic scale L_0 of the inhomogeneity. Therefore, the function $\varepsilon(\omega,x)$ cannot be regarded constant.

From the condition which allows to neglect the second term of Eq. (1.256) we obtain an inequality which defines the range of applicability of the zero-order approximation of Eq. (1.259):

$$\frac{\Psi_0''}{\left(\Psi_0'\right)^2} \sim \frac{c}{\omega}\frac{d}{dx}\frac{1}{\sqrt{\varepsilon(\omega,x)}} \sim \frac{\lambda}{L_0} \ll 1. \tag{1.260}$$

Expectedly, this equation gives us the condition of applicability for the approximation of geometrical optics. Equation (1.260) is violated when $\lambda/L_0 \sim 1$. This is possible for finite values of the dielectric $\varepsilon(\omega,x) \sim 1$ and not too small values of $c/(\omega L_0) \equiv \lambda_0/L_0 \geq 1$ or for any value of $\lambda_0/L_0 \ll 1$ near the points where $\varepsilon(\omega,x) \approx 0$.

Thus, according to Eq. (1.259), the field can be written in the zero-order approximation as

$$E = C_+\exp\left[\imath\frac{\omega}{c}\int^x \sqrt{\varepsilon(\omega,x')}\,dx'\right] + C_-\exp\left[-\imath\frac{\omega}{c}\int^x \sqrt{\varepsilon(\omega,x')}\,dx'\right], \tag{1.261}$$

or, using the wave vector which is defined in a weakly inhomogeneous medium by Eq. (1.254)

$$E = C_+\exp\left[\imath\int^x dx'\,k(x')\right] + C_-\exp\left[-\imath\int^x dx'\,k(x')\right]. \tag{1.262}$$

In order to calculate the next term of the expansion (1.257) we substitute this expansion into Eq. (1.256), take account of Eq. (1.258), and keep all the first-order terms. As a result, we obtain

$$2\Psi_0'\Psi_1' - \imath\Psi_0'' = 0, \tag{1.263}$$

hence

$$\Psi_1 = \frac{\imath}{2}\ln\Psi_0'. \tag{1.264}$$

Accounting for this correction to expression (1.255) we obtain, by a simple calculation, the expression for the field which is correct up to the terms of the first-order approximation of geometrical optics:

$$E(x) = \frac{C_1}{\sqrt[4]{\varepsilon(\omega,x)}}\exp\left[\imath\frac{\omega}{c}\int^{x}\sqrt{\varepsilon(\omega,x')}\,dx'\right] + \frac{C_2}{\sqrt[4]{\varepsilon(\omega,x)}}\exp\left[-\imath\frac{\omega}{c}\int^{x}\sqrt{\varepsilon(\omega,x')}dx'\right]. \tag{1.265}$$

As stated above, the ranges where the function $\varepsilon(\omega,x)$ is real and positive are of special importance in the theory of wave propagation in inhomogeneous media. The electromagnetic field (1.265) has an oscillatory character. In other words, propagation of waves is possible. These ranges are called the ranges of transparency in geometrical optics. In contrast to them, the ranges where the function $\varepsilon(\omega,x)$ is negative are called the ranges of opacity. In these domains the field E varies exponentially with the coordinate x, either increasingly or decreasingly. In quantum mechanics the range of transparency corresponds to the range where the function $U(x) = W - V(x)$ is positive, i.e., where the motion of the particles can be described classically [see Eq. (1.251)]. The ranges where $U(x) < 0$ holds are inaccessible for a classical particle, however. Therefore, the point $U(x) = 0$ is called a turning point. By analogy, the points which separate the ranges of transparency and opacity, i.e., the points where $\varepsilon(\omega,x) = 0$, are called turning points in geometrical optics, too. An electromagnetic wave propagating the transparency range is reflected at these points.

Since the approximation of geometrical optics is inapplicable near the turning points, the solution (1.265) loses its sense. However, near the turning point the function $\varepsilon(\omega,x)$ can be expanded in a series which represents the exact solution of Eq. (1.250) when it tends asymptotically to expression (1.265) far from this point. We assume that the range of classically accessible solutions, or the range of transparency, lies between the turning points a and b, which are the solutions of equation $\varepsilon(\omega,x) = 0$, i.e., $\varepsilon(\omega,x) \geq 0$ for $a \leq x \leq b$. Then the asymptotic solution of Eq. (1.250) at the left of the point b which goes over into the damped solution (1.265) for $x > b$ is of the form ($x < b$)

$$E = \frac{C}{\sqrt[4]{\varepsilon(\omega, x)}} \sin\left[\frac{\omega}{c}\int_x^b dx' \sqrt{\varepsilon(\omega, x')} + \frac{\pi}{4}\right]. \tag{1.266}$$

The asymptotic solution of Eq. (1.250) at the right of the point b, passing over into the damped solution (1.265) for $x < a$, is written analogously in the form

$$E = \frac{C'}{\sqrt[4]{\varepsilon(\omega, x)}} \sin\left[\frac{\omega}{c}\int_a^x dx' \sqrt{\varepsilon(\omega, x')} + \frac{\pi}{4}\right]. \tag{1.267}$$

1.10.1.3 The Bohr–Sommerfeld Quasi-Classical Quantization Rules

Naturally, expressions (1.266) and (1.267) must coincide in the entire range $a \leq x \leq b$. To ensure this, the sum of their phases must be an integer multiple of π. This is in fact given if

$$\frac{\omega}{c}\int_a^b dx \sqrt{\varepsilon(\omega, x)} = \pi\left(n + \frac{1}{2}\right), \tag{1.268}$$

n being an arbitrary integer: $n = 0, \pm 1, \pm 2, \ldots$ The integration constants C and C' are connected by the relation $C = (-1)^n C'$. In the approximation of geometrical optics, due to relation (1.260), the eikonal has a large value, i.e., $|n| \gg 1$. Therefore, Eq. (1.268) can be written approximately as

$$\frac{\omega}{c}\int_a^b dx \sqrt{\varepsilon(\omega, x)} = \int_a^b dx\, k(\omega, x) = \pi n, \tag{1.269}$$

with $n \gg 1$. Note that Eq. (1.269) can also be obtained when an additional circular integral along the closed contour around the turning points in the complex k-plane is considered in the solution (1.262). The phase of the solution (1.262) grows, the additional phase being $\imath \oint k(\omega, x)\, dx$, and since the field $E(x)$ must remain unchanged the following condition (the condition of a single-valued solution)

$$\oint k(\omega, x)\, dx = 2\pi n$$

should be satisfied. In the case of two turning points, Eq. (1.269) follows.

In geometrical optics, Eq. (1.269) is the dispersion equation defining the frequency spectrum ω, or the wave vector spectrum $\vec{k}$ of electromagnetic waves which are trapped in the transparency range of the medium. Thus, in the approximation of geometrical optics, when there are two turning points present, the eigenvalue spectrum of the wave equation is discrete in the inhomogeneous medium. This is the important qualitative difference of the electromagnetic properties of inhomogeneous and homogeneous media.

Equation (1.269) has a simple physical meaning: there must be a place for an integer number of half waves between the turning points in the transparency range of the medium (Fig. 1.5). Oscillations of a string with fixed end points are analogous to this situation, since Eq. (1.250) is identical to the equation of string oscillations of a medium with a weakly inhomogeneous modulus of elasticity. The fixed end points of the string may be identified with the turning points, since the solutions of the wave equation decrease exponentially in the complementary domain outside these points. An analogous situation occurs in quantum mechanics in the quasi-classical limit. Therefore, Eqs. (1.268) and (1.269) are called the quasi-classical quantization rules and the integrals on the left-hand sides of these relations are known as Bohr–Sommerfeld phase integrals.

In the above conclusions, we considered the function $\varepsilon(\omega, x)$ real. However, it becomes complex when the medium is dissipative, e.g., due to particle collisions. As stated before, weakly damped electromagnetic waves can exist in the medium only when the dissipation is weak. Then $\varepsilon(\omega, x)$ is an almost real function and it is not difficult to generalize Eq. (1.269) for this case of weakly dissipative inhomogeneous media. Due to $\mathrm{Re}\{k(\omega, x)\} \gg \mathrm{Im}\,\{k(\omega, x)\}$ and because the electromagnetic oscillations are thus weakly damped ($\omega \to \omega + \imath\delta, \omega \gg \delta$), the integrand takes the form

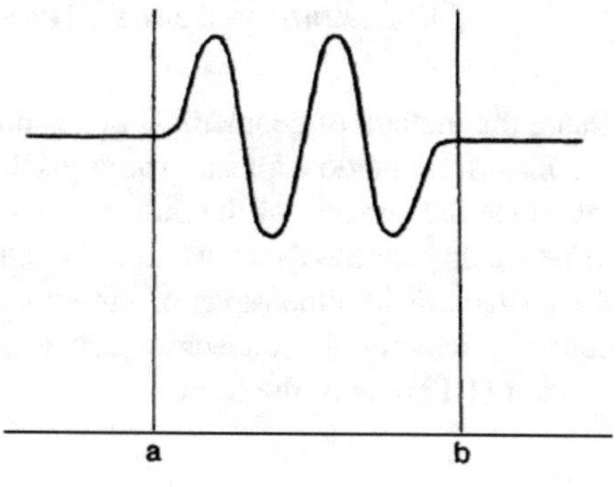

Fig. 1.5 An integer number of half waves between the turning points in the transparency range

$$k(\omega + \imath\delta, x) \approx \mathrm{Re}\,\{k(\omega,x)\} + \imath \mathrm{Im}\{k(\omega,x)\} + \imath\delta \frac{\partial\, \mathrm{Re}\,\{k(\omega,x)\}}{\partial\omega}.$$

Substituting this expansion into Eq. (1.269) and separating its real and imaginary parts yields

$$\int_a^b \mathrm{Re}\,\{k(\omega,x)\}dx = \pi n, \qquad \delta = \frac{\int_a^b \mathrm{Im}\{k(\omega,x)\}dx}{\int_a^b \frac{\partial}{\partial\omega}\mathrm{Re}\,\{k(\omega,x)\}dx}. \tag{1.270}$$

Here, the turning points a and b are the points where $\mathrm{Re}\{k^2(\omega,x)\} = 0$.

In homogeneous media without spatial dispersion, Eqs. (1.270) determines the frequency spectrum and the damping decrement of electromagnetic waves in the approximation of geometrical optics. The second equation of Eqs. (1.270) should be used only when the frequency ω, defined by the first equation, is real.

To summarize the above, we have considered the simplest case of an inhomogeneous medium with two turning points, between which the range of transparency lies. Electromagnetic waves can propagate in the medium between these turning points. Therefore, they are often referred to as trapped oscillations and the corresponding solutions of the field equation as finite solutions. Under real conditions other situations are possible in inhomogeneous media, for example, when there exists only one turning point in the medium or, conversely, when there are several ranges of transparency, separated from each other and included between pairs of corresponding turning points, or when there is no turning point at all and the medium is transparent in the entire space. From this variety of cases we shall treat only those which are most frequently realized in material media.

1.10.2 *Approximation of Geometrical Optics for Inhomogeneous Media with Spatial Dispersion*

Using the method of geometrical optics presented in Sect. 1.10.1, we can formulate the material equation for an arbitrary inhomogeneous medium and, in particular, introduce the concept of the dielectric tensor for such media. We assume medium to be weakly inhomogeneous, i.e., the wavelength to be significantly smaller than the characteristic dimensions of inhomogeneity of the medium. Here, the material equation relating the dielectric displacement to the electric field strength, [see relation (1.12)], is of the form

$$D_i\left(t,\vec{r}\right) = \int_{-\infty}^{t} dt' \int d\vec{r'}\,\varepsilon_{ij}\left(t-t',\vec{r}-\vec{r'},\vec{r}\right)E_j\left(t',\vec{r'}\right). \tag{1.271}$$

With respect to a weak inhomogeneity of the medium, it is allowed to keep the dependence of the kernel of Eq. (1.271) on $\vec{r}-\vec{r'}$, only, and to neglect that on $\vec{r}$. The dependence on $\vec{r}-\vec{r'}$ is related to the wavelength of the oscillation and in a homogeneous medium, thus the dependence of $\varepsilon_{ij}\left(\omega,\vec{k}\right)$ on $\vec{k}$ arises after the Fourier transformation. The dependence on $\vec{r}$ is determined by the inhomogeneity of the medium and related to the characteristic dimension L_0 of the inhomogeneity. Since the parameter λ/L_0 is small, we can again apply the approximation of geometrical optics, considering the wave vector $\vec{k}$ as a weak function of the coordinates in the zero-order approximation. Then, Eq. (1.271) can be written as

$$D_i\left(\omega,\vec{k}\right) = \varepsilon_{ij}\left(\omega,\vec{k},\vec{r}\right)E_j\left(\omega,\vec{k}\right), \tag{1.272}$$

where

$$\varepsilon_{ij}\left(\omega,\vec{k},\vec{r}\right) = \int_{0}^{\infty} dt_1 \int d\vec{R}\,\varepsilon_{ij}\left(t_1,\vec{R},\vec{r}\right)\exp\left(\imath\omega t_1 - \imath\vec{R}\cdot\vec{k}\right) \tag{1.273}$$

is the dielectric tensor of the weakly inhomogeneous medium, taking account of spatial dispersion in the zero-order approximation of geometrical optics.

1.10.2.1 Eikonal Equation for Inhomogeneous Medium with Spatial Dispersion

Writing the Maxwell's equations in the zero-order approximation of geometrical optics:

$$\begin{aligned} \left(\vec{k}\times\vec{B}\right) &= -\frac{\omega}{c}\vec{D}, \qquad \vec{k}\cdot\vec{B}=0, \\ \left(\vec{k}\times\vec{E}\right) &= \frac{\omega}{c}\vec{B}, \qquad \vec{k}\cdot\vec{D}=0, \end{aligned} \tag{1.274}$$

yields together with Eq. (1.272) the condition for the existence of solutions of this system of equations

$$\left| k^2\delta_{ij} - k_i k_j - \frac{\omega^2}{c^2}\varepsilon_{ij}\left(\omega, \vec{k}, \vec{r}\right) \right| = 0. \tag{1.275}$$

This relation can be obtained from Eq. (1.249) for media without spatial dispersion, too, when the method of geometrical optics in the zero-order approximation is applied.

Note that in the homogeneous medium an equation of the type Eq. (1.275) is a dispersion equation, [see Eq. (1.114)], defining the frequency spectrum $\omega\left(\vec{k}\right)$ of the natural electromagnetic oscillations. If Eq. (1.275) is solved for ω, the "frequency of the natural oscillations of the medium," the eigen modes depend on the coordinates, i.e., $\omega = \omega\left(\vec{k}, \vec{r}\right)$. This is not possible physically. The principle difference between Eqs. (1.275) and (1.114) is the following. The former does not constitute a dispersion relation, but simply defines $\vec{k}\left(\vec{r}\right)$ or the eikonal $\Psi\left(\vec{r}\right)$. Therefore, it is called the eikonal equation. It generalizes Eq. (1.258) in the zero-order approximation of geometrical optics for weakly inhomogeneous media with spatial dispersion.

When the electric field of a wave in the inhomogeneous medium is longitudinal, Eq. (1.275) simplifies to

$$\varepsilon\left(\omega, \vec{k}, \vec{r}\right) = \frac{k_i k_j}{k^2}\varepsilon_{ij}\left(\omega, \vec{k}, \vec{r}\right) = 0. \tag{1.276}$$

Here, quantity $\varepsilon\left(\omega, \vec{k}, \vec{r}\right)$ is known as the longitudinal dielectric permittivity of a weakly inhomogeneous medium and Eq. (1.276) constitutes the eikonal equation for longitudinal waves in the zero-order approximation of geometrical optics.

It is convenient to calculate the electric field as a function of the coordinates with the aid of the eikonal equation, i.e., to solve the boundary value problem. In order to solve the initial value problem, i.e., to calculate the frequency spectra of the electromagnetic oscillations of the weakly inhomogeneous medium, the eikonal equation is not sufficient. Dispersion equations of the type of Eqs. (1.270), derived for the media without spatial dispersion, must be derived in this case.

1.10.2.2 Quantization Rules

In general, when the inhomogeneous medium is treated three-dimensionally, one must know the spectrum of eigenvalues of partial integro-differential equations of higher order in order to derive the dispersion equation. This problem is more or less solved for media with a one-dimensional inhomogeneity. In this case, we can determine the projection of the wave vector in the direction of the inhomogeneity $k_x(\omega, k_y, k_z, x)$. In the most interesting frequency ranges, Eqs. (1.275) and (1.276) have the pair of non-degenerate roots $\pm k_{x,s}(\omega, x)$. In the zero-order approximation of geometrical optics, the differential equation

$$\frac{d^2y}{dx^2} + k_{xs}^2(\omega, x)y = 0 \tag{1.277}$$

can describe each pair. The general theory, given in Sect. 1.10.1, can be applied here. To determine the frequency spectrum, we can write relations defining the frequency and the damping decrement in analogy with (1.270) ($\omega \to \omega + \imath\delta$):

$$\int\limits_{x_\mu}^{x_\nu} \mathrm{Re}\,\{k_{xs}(\omega, x)\}dx = \pi n, \qquad \delta = -\frac{\int\limits_{x_\mu}^{x_\nu} \mathrm{Im}\{k_{xs}(\omega, x)\}dx}{\int\limits_{x_\mu}^{x_\nu} \frac{\partial}{\partial\omega} \mathrm{Re}\,\{k_{xs}(\omega, x)\}\, dx}. \tag{1.278}$$

In these relations, the domain of integration is the range where the medium is transparent with respect to these oscillations and the turning points x_μ and x_ν are determined from $\mathrm{Re}\left\{k_{xs}^2(\omega, x)\right\} = 0$. Thus, the basic idea for calculating the electromagnetic wave propagation in a weakly inhomogeneous dispersive medium is similar to that applied to obtain the dispersion of the medium. The method of solution is to calculate from the eikonal equation (1.275), or (1.276), the wave vectors $k_{xs}^2(\omega, x)$ corresponding to the S-branches (azimuthally symmetric modes) of the oscillations, their frequency spectra, and damping decrements being determined by the quantization rules (1.278).

Finally, note that the transition to unbounded homogeneous media is included formally. In this case, there are no turning points, but we can introduce the arbitrary points a and b. With the aid of them, Eq. (1.269) is written as

$$\int\limits_a^b k_{xs}(\omega, x)\, dx = k_{xs}(\omega)(b - a) = \pi n. \tag{1.279}$$

Assuming $\pi n/(b - a) = \text{const}$, we obtain the equation determining the mode spectrum of homogeneous media

$$k_{xs}(\omega) = k_{xs} = \text{const}. \tag{1.280}$$

Naturally, the roots of this equation $\omega\left(\vec{k}\right)$ are identical with those of Eq. (1.275), which is the dispersion equation for homogeneous media.

1.11 Problems

1.11.1 Show the regions of charge carrier degeneracy and applicability of the gas approximation in the diagram $N(T)$ of electron plasma.

Solution The degeneracy condition for electron plasma reads

$$\epsilon_F > kT$$

In the diagram of ln$N$ versus ln$T$ (Fig. 1.6) the condition

$$\epsilon_F = kT = \frac{(3\pi^2)^{2/3}\hbar^2 N^{2/3}}{2m} \tag{1.281}$$

defines the straight line 1 separating the region of degenerate plasma from the non-degenerate one. The applicability condition of the gas approximation in the non-degenerate state is

$$\eta_{cl} = \frac{e^2 N^{1/3}}{kT} < 1. \tag{1.282}$$

In the same diagram $\eta_{cl} = 1$ provides the line 2. In the degenerate state for applicability of the gas approximation one must satisfy the condition

$$\eta_{qu} = \frac{e^2 N^{1/3}}{\epsilon_F} < 1. \tag{1.283}$$

As the Fermi energy depends only on N and is independent of T the condition $\eta_{qu} = 1$ gives the straight line 3 passing through the point A, where the three lines intersect, defined by $\epsilon_F = kT = e^2 N^{1/3}$. In region I we have non-degenerate plasma with weak interaction for interaction for which the gas approximation is valid. In region II there is non-degenerate plasma with strong interaction, i.e., a classical fluid. In region III we have degenerate plasma with strong interaction, i.e., a quantum fluid. In regions II and III the gas approximation is not valid. Finally, region IV belongs to degenerate plasma with weak interaction for which the gas approximation is valid.

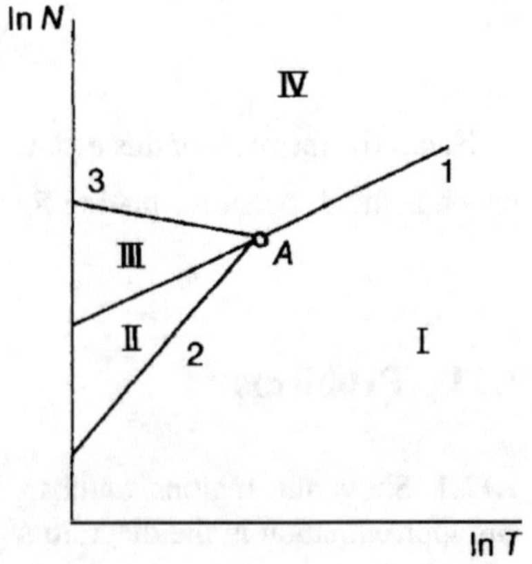

Fig. 1.6 $N(T)$ for electron plasma

1.11.2 Find the velocity of the coordinate system in which $\vec{E}$ and $\vec{B}$ fields are parallel.

Solution Without loss of generality, we can orient one of the components of $\vec{E}'$ and $\vec{B}'$ fields along the velocity of the coordinate system. Besides we choose ox axis along this velocity. Thus, $v_x = v$, and $E'_x = E_x = B'_x = B_x = 0$. According to the Lorentz transformation formulas, for fields, we have

$$E'_y = \frac{E_y - B_z v/c}{\sqrt{1 - v^2/c^2}}, \quad E'_z = \frac{E_z + B_y v/c}{\sqrt{1 - v^2/c^2}},$$
$$B'_y = \frac{B_y + E_z v/c}{\sqrt{1 - v^2/c^2}}, \quad B'_z = \frac{B_z - E_y v/c}{\sqrt{1 - v^2/c^2}}. \tag{1.284}$$

Since $\vec{E}'$ and $\vec{B}'$ are parallel, then their vector product is zero. Consequently,

$$\left(\vec{E}' \times \vec{B}'\right) = E'_y B'_z - E'_z B'_y = 0. \tag{1.285}$$

Substituting expression (1.284) in these relations, we find an equation to determine the velocity v as follows:

$$\frac{v/c}{1 - v^2/c^2} = \frac{E_y B_z - E_z B_y}{E^2 + B^2} = \frac{\left(\vec{E} \times \vec{B}\right)_x}{E^2 + B^2}. \tag{1.286}$$

Hence, it follows that

$$a^2 = \frac{\left(\vec{E} \times \vec{B}\right)}{E^2 + B^2} < 1, \tag{1.287}$$

and

$$\frac{v}{c} = \frac{1}{2a} \pm \sqrt{\frac{1}{4a^2 - 1}}. \tag{1.288}$$

Of course, it is necessary to select the signs before the radical so that $v < c$. This means that

$$\frac{v}{c} = \begin{cases} \dfrac{1}{2a} - \sqrt{\dfrac{1}{4a^2} - 1}, & \text{at } a > 0, \\ -\dfrac{1}{2|a|} + \sqrt{\dfrac{1}{4a^2} - 1}, & \text{at } a < 0. \end{cases} \tag{1.289}$$

1.11.3 Study potential cylindrical waves which propagate with light speed in the radial direction.

Solution The solution of D'Alembert's equation

$$\frac{1}{r}\frac{\partial}{\partial r}\left(r\frac{\partial A}{\partial r}\right) + \frac{1}{r^2}\frac{\partial^2 A}{\partial \vartheta^2} + \frac{\partial^2 A}{\partial z^2} - \frac{1}{c^2}\frac{\partial^2 A}{\partial t^2} = 0 \tag{1.290}$$

is sought in the form of $A = R(r)f(r \pm ct)$. Substituting this solution into Eq. (1.290), we find

$$\left(R'' + \frac{R'}{r}\right)f(\tau) + \left(2R' + \frac{R}{r}\right)f'(\tau) = 0, \tag{1.291}$$

where $\tau = r \pm ct$. This equation is solved by separating variables method:

$$\frac{R'' + R'/r}{2R + R/r} = -\frac{f'}{f} = k = \text{const.} \tag{1.292}$$

In this case, function $f(\tau)$ turns to be non-arbitrary and is given by

$$f(\tau) = C\exp(-k\tau) = C\exp(x\tau/2r), \tag{1.293}$$

where $x = 2rk$. In this notation, equation for function $R(r) = Y(x)$ reduces to the equation for hypergeometric functions

$$xY'' + (b - x)Y' - aY = 0. \tag{1.294}$$

For the case, when $a = 1/2$ and $b = 1$, we have

$$Y(1/2, x) = \frac{1}{\Gamma(1/2)}\int_0^1 \frac{\exp(-xt)}{\sqrt{t(1-t)}}dt, \quad \Gamma(1/2) = \int_0^\infty \frac{\exp(-t)}{\sqrt{t}}dt. \tag{1.295}$$

At large r, we have $R \sim 1/r$.

Above, we assumed wave amplitude C constant. In this case, this wave is a homogeneous wave. In the cylindrical case, an inhomogeneous wave can be created on the surface with the given radius $r = \text{const}$. In this case, for amplitude $C(z, \vartheta) = \Phi$, we find

$$\frac{\partial^2 \Phi}{\partial z^2} + \frac{\partial^2 \Phi}{\partial x^2} = 0. \tag{1.296}$$

Here, $C(z, \vartheta) = \Phi(z, x)$, where $x = r\vartheta$. Further analysis of Eq. (1.296) is simple.

1.11.4 Analyze the oscillations occurring in homogenous gaseous plasma due to a small displacement of the electrons with respect to the ions.

Solution Denote the displacement vector of the electrons with respect to the ions by $\vec{S}$. The density of the uncompensated electron charge due to this displacement is equal to

$$\rho = \nabla \cdot en\vec{S} = en\nabla \cdot \vec{S}, \tag{1.297}$$

where n is the constant electron density. This charge produces the electric field $\vec{E}$ which is determined by

$$\nabla \cdot \vec{E} = 4\pi\rho = 4\pi en\nabla \cdot \vec{S}. \tag{1.298}$$

Since in the equilibrium state, $\vec{E} = 0$ for $\vec{S} = 0$, then it follows

$$\vec{E} = 4\pi en\vec{S}. \tag{1.299}$$

Thus, the field $\vec{E}$ is parallel to the electron displacement $\vec{S}$ and acts on each electron with the force

$$\vec{F} = -e\vec{E} = -4\pi e^2 n\vec{S}. \tag{1.300}$$

This force tends to return electrons into the initial equilibrium position $\vec{S}$. As a result, we obtain the equation of electron motion in the form

$$m\frac{d^2\vec{S}}{dt^2} = -e\vec{E} = -4\pi e^2 n\vec{S}. \tag{1.301}$$

It describes plasma oscillations near equilibrium ($\vec{S} = 0$) with the Langmuir (plasma) frequency

$$\omega = \sqrt{\frac{4\pi e^2 n}{m}} = \omega_{\text{pe}}. \tag{1.302}$$

1.11.5 Find the force of mutual interaction of two moving point charges and show that the Newton third law about the equality and oppositeness of these forces is violated.

Solution First, let us solve an auxiliary problem: we find the electric and magnetic fields created by a point charge q_i which moves with velocity $\vec{v}_i$. In the rest frame of the charge, the field is purely potential

$$\varphi' = \frac{q_i}{R'}, \quad \vec{A}' = 0, \tag{1.303}$$

where R' is the radius vector with components (x', y', z'). In the laboratory frame, we have

$$\varphi = \frac{\varphi}{\sqrt{1 - v_i^2/c^2}}, \quad \vec{A} = \varphi\frac{\vec{v}_i}{c}. \tag{1.304}$$

From these two formulas, we find

$$\vec{E} = \left(1 - \frac{v_i^2}{c^2}\right)\frac{q_i\vec{R}}{R^{*3}}, \quad \vec{B} = \frac{1}{c}\left(\vec{v}_i \times \vec{E}\right), \tag{1.305}$$

where $\vec{R} = \vec{r}_i - \vec{r}$, $\vec{r}_i$ is the position of the charge q_i, and $\vec{r}$ is the observation point, and

$$R^{*2} = R^2\left(1 - \sin^2\vartheta v_i^2/c^2\right), \quad \cos\vartheta = \left(\vec{v}_i \cdot \vec{R}\right)/v_i R. \tag{1.306}$$

If the charge q_2, moving with velocity $\vec{v}_2$, is placed at the point $\vec{r} = \vec{r}_2$ in the field created by the charge q_1, then the force

$$\vec{F}_{1,2} = q_2\left\{\vec{E}_1\left(\vec{r}_2\right) + \frac{1}{c}\left[\vec{v}_2 \times \vec{B}_1\left(\vec{r}_2\right)\right]\right\} \tag{1.307}$$

acts on it. Simple calculations give

$$\vec{F}_{1,2} = \frac{q_1 q_2}{R_{12}^3} \frac{\left(1 - v_1^2/c^2\right)\left\{\vec{R}_{12} + \left[\vec{v}_2 \times \left(\vec{v}_1 \times \vec{R}_{12}\right)\right]/c^2\right\}}{1 - v_1^2/c^2 + \left(\vec{v}_1 \cdot \vec{R}_{12}\right)^2/c^2 R_{12}^2} \tag{1.308}$$

where $\vec{R}_{12} = \vec{r}_1 - \vec{r}_2$. Analogously, we find

$$\vec{F}_{2,1} = \frac{q_1 q_2}{R_{21}^3} \frac{\left(1 - v_2^2/c^2\right)\left\{\vec{R}_{21} + \left[\vec{v}_1 \times \left(\vec{v}_2 \times \vec{R}_1\right)\right]/c^2\right\}}{1 - v_2^2/c^2 + \left(\vec{v}_2 \cdot \vec{R}_{21}\right)^2/c^2 R_{21}^2}. \tag{1.309}$$

It is easy to see that $\vec{F}_{1,2} \neq \vec{F}_{2,1}$. This violation of Newton third law is stipulated by the relativistic second-order terms (magnetic forces).

1.11.6 Find the average transverse force acting on the plasma electrons in a weakly inhomogeneous stationary magnetic field.

Solution We first consider electrons without longitudinal (parallel to the magnetic field) velocity component which rotate around the magnetic field lines with the angular velocity $\Omega = eB(r)/mc$. Their position is $r = r_0(r) + \xi(t)$, where $r_0(t)$ denotes the coordinate of the center of the Larmor rotation and $\xi(t)$ is the position of the electron in the orbit; $r_0(t)$ is in general large and slowly varying, $\xi(t)$ is small and quickly varying. In the model of non-interacting particles, we have

$$\vec{\dot{\xi}} = \Omega\left(\vec{\xi} \times \frac{\vec{B}}{B}\right), \quad \vec{\xi} = \frac{1}{\Omega}\left(\frac{\vec{B}}{B} \times \vec{v}_\perp\right), \tag{1.310}$$

where $v_\perp$ is the unperturbed velocity of the electron rotation. Expanding $B(r)$ in the powers of ξ and averaging the Lorentz force over the time, we find the average force perpendicular to the magnetic field acting on the plasma electrons:

$$\begin{aligned} F_{1av} &= \frac{e}{c}\left[\vec{\dot{\xi}} \times \left(\vec{\xi} \cdot \nabla\right)\vec{B}\right] = -\frac{mv_\perp^2}{2B}\left[\left(\frac{\vec{B}}{B} \times \nabla\right)\vec{B}\right] \\ &= -\frac{mv_\perp^2}{2}\left(\frac{\vec{B} \cdot \nabla}{B}\right)\frac{\vec{B}}{B} = \frac{mv_\perp^2}{2R}\vec{n}. \end{aligned} \tag{1.311}$$

Here, R is the radius of curvature of the magnetic field lines and $\vec{n}$ is the unit vector directed from the center of curvature to the position of the electron. In the derivation of relation (1.311), Eq. (1.310) and the Maxwell's equation for the stationary magnetic field $B(r)$ have been used:

$$\nabla \times \vec{B} = 0, \qquad \nabla \cdot \vec{B} = 0. \tag{1.312}$$

The averages over the gyro phase involved are

$$\overline{\xi_i \xi_i} = \frac{1}{2}\xi^2 \delta_{ij} = \frac{v_\perp^2}{2\Omega^2}\delta_{ij}. \tag{1.313}$$

We now admit, along with the Larmor rotation, that the plasma electrons can move longitudinally with the velocity $v_\parallel$. Passing over to the coordinate system which rotates with the angular velocity $v_\parallel/R$ around the momentary center of curvature of the magnetic field lines, we again have the case of electrons without a longitudinal velocity. In this system, however, there emerges an additional transverse inertial force, i.e., the centrifugal force equal to

$$\vec{F}_{2\text{av}} = \frac{m v_\parallel^2}{R}\vec{n}. \tag{1.314}$$

The sum of expressions (1.311) and (1.314) gives the total average force

$$\vec{F}_{\text{av}} = \vec{F}_{1\text{av}} + \vec{F}_{2\text{av}} = \vec{n}\frac{m}{R}\left(v_\parallel^2 + \frac{v_\perp^2}{2}\right). \tag{1.315}$$

This force on the electrons is equivalent to a gravitational acceleration

$$\vec{g} = \frac{\vec{n}}{R}\left(v_\parallel^2 + \frac{v_\perp^2}{2}\right). \tag{1.316}$$

Such a force evidently affects the plasma ions as well and since it is independent of the sign of the charge, both components experience forces in the same direction.

The particles drift in toroidal installations for plasma magnetic confinement is described in this way. Since the curvature radius R of the magnetic field in such installations greatly exceeds the localized region of fusion plasma r_0 (R is larger radius of torus and r_0 is the radius of the toroidal plasma column). Then, with a good accuracy, the magnetic field $\vec{B}$ lines can be considered rectilinear and parallel to the oz axis and curvature can be taken into account in the form of the gravitational field along the torus radius. Such a gravitational force causes an azimuthal drift with velocity

$$v_g = \frac{mcg}{eB} = \frac{g}{\Omega}. \tag{1.317}$$

1.11.7 From the expression of energy variation in a medium, find the average force acting on the charge species on a plasma medium in a high-frequency electromagnetic field.

Solution The average (per volume) variation of the energy density of the electromagnetic field in a non-absorbing medium due to the small variation of the dielectric permittivity is equal to

$$\frac{\delta W}{V}=\frac{1}{4\pi}\vec{E}\cdot\delta\vec{D}=\frac{1}{4\pi}\left(\vec{E}\cdot\delta\vec{D}^{*}+\vec{E}^{*}\cdot\delta\vec{D}\right)=\frac{1}{4\pi}\left(E_iE_j^*\delta\varepsilon_{ij}^*+E_i^*E_j\delta\varepsilon_{ij}\right). \quad (1.318)$$

Quantity $\varepsilon_{ij}\left(\omega,\vec{k}\right)$ evidently depends on the carriers' density. Therefore, the variation of the density δn_α of α charge species causes the variation of the dielectric permittivity

$$\delta\varepsilon_{ij}\left(\omega,\vec{k}\right)=\frac{\partial\varepsilon_{ij}}{\partial n_\alpha}\delta n_\alpha. \quad (1.319)$$

In this case, for a non-absorbing medium (i.e., tensor ε_{ij} is Hermitian), from Eq. (1.318) we find

$$\frac{\delta W}{V}=\frac{1}{4\pi}E_iE_j^*\delta\varepsilon_{ij}^*=\frac{E_i^*E_j}{4\pi}\frac{\partial\varepsilon_{ij}}{\partial n_\alpha}\delta n_\alpha. \quad (1.320)$$

For $\delta n_\alpha\to n_\alpha$, this quantity is the average potential energy (the sign inversed) of all α particle species in the field of the electromagnetic wave. Therefore, the gradient of this quantity, which is non-zero for the spatially inhomogeneous electromagnetic field $\vec{E}\left(\omega,\vec{k},\vec{r}\right)$, gives the average force acting on the α particle species in the field of an inhomogeneous electromagnetic wave:

$$\vec{F}_{\text{av}}=\frac{1}{4\pi}n_\alpha\frac{\partial\varepsilon_{ij}\left(\omega,\vec{k}\right)}{\partial n_\alpha}\nabla E_i^*E_j. \quad (1.321)$$

In the case of an isotropic medium, we obtain

$$\vec{F}_{\text{av}}=\frac{1}{4\pi}n_\alpha\left\{\frac{\partial\varepsilon^l\left(\omega,\vec{k}\right)}{\partial n_\alpha}\nabla\left|\vec{E}^l\right|^2+\frac{\partial\varepsilon^{\text{tr}}\left(\omega,\vec{k}\right)}{\partial n_\alpha}\nabla\left|\vec{E}^{\text{tr}}\right|^2\right\}. \quad (1.322)$$

For $\frac{\partial\varepsilon^{l,\text{tr}}\left(\omega,\vec{k}\right)}{\partial n_\alpha}<0$, particles species are pushed out from the region of the strong field, while for $\frac{\partial\varepsilon^{l,\text{tr}}\left(\omega,\vec{k}\right)}{\partial n_\alpha}>0$, in the opposite, they are attracted to this region. In particular, for $\varepsilon^l=\varepsilon^{\text{tr}}=\varepsilon=1-\frac{\omega_{\text{pe}}^2}{\omega^2}$, from Eq. (1.322) we find the average force, called Miller's force, acting on the electrons of a plasma medium in the inhomogeneous high-frequency field:

$$\vec{F}_{\text{av}} = -\frac{n_\alpha e^2}{m\omega^2}\nabla\left|\vec{E}\right|^2. \tag{1.323}$$

This force pushes out the electrons of the medium from the strong field regions.

1.11.8 Let us generalize the material Equation (1.28) for uniform isotropic equilibrium media and introduce the different dielectric permittivities for the longitudinal and transverse electric fields:

$$D^{\|}\left(\vec{k},\omega\right) = \varepsilon_{\|}(k,\omega)E^{\|}\left(\vec{k},\omega\right), \quad \vec{D}^{\perp}\left(\vec{k},\omega\right) = \varepsilon_{\perp}(k,\omega)\vec{E}^{\perp}\left(\vec{k},\omega\right),$$
$$\vec{B}\left(\vec{k},\omega\right) = \mu_{\perp}(k,\omega)\vec{H}\left(\vec{k},\omega\right).$$

Both magnetic vectors are transverse and a single permeability has been introduced to them. Show that, the permittivity and permeability $\varepsilon_\perp$ and $\mu_\perp$ can be arbitrarily changed, but the quantity

$$\eta(k,\omega) = \frac{c^2k^2/\mu_\perp - \omega^2\varepsilon_\perp}{c^2k^2 - \omega^2} = \frac{c^2k^2 - \omega^2\varepsilon^{tr}}{c^2k^2 - \omega^2}$$

is unchanged. It is just the real physical characteristic of the medium for transverse electromagnetic field [65].

Solution From Eqs. (1.5)–(1.7), and the Maxwell's Eqs. (1.1) we find

$$\rho_{\text{int}} = -\frac{\imath\left(\varepsilon_{\|} - 1\right)}{4\pi}\vec{k}\cdot\vec{E}, \qquad \vec{j}^{\perp}_{\text{int}} = \frac{\imath\left(\omega^2 - c^2k^2\right)}{4\pi\omega}(1-\eta)\vec{E}^{\perp},$$

the longitudinal current is expressed in terms of ρ_{int}. In every medium ρ_{int}, $\vec{j}^{\perp}_{\text{int}}$, and $\vec{E}$ are well-defined physical quantities, and the coefficients of proportionality between them are the electromagnetic characteristics (response functions) of the given medium. These are $\varepsilon_{\|} = \varepsilon$ and η, but not $\varepsilon_\perp$ and $\mu_\perp$, which are combined in η.

1.11.9 Calculate the electromagnetic field of a point charge q situated at $\vec{r} = \vec{r}_0$ in an anisotropic homogeneous medium.

Solution The static charge density

$$\rho_0\left(\vec{r}\right) = q\delta\left(\vec{r} - \vec{r}_0\right) \tag{1.324}$$

produces an electric field $\left(\nabla\times\vec{E} = 0, \vec{E} = -\nabla\phi\right)$ which is subject to

$$\nabla \cdot \vec{D} = -4\pi\rho_0 = 4\pi q\delta\left(\vec{r} - \vec{r}_0\right). \tag{1.325}$$

Expanding all the quantities into Fourier components

$$A\left(\vec{r}\right) = \int d\vec{k}\, e^{\imath\vec{k}\cdot\vec{r}} A\left(\vec{k}\right) \tag{1.326}$$

and taking the static limit as $\omega \to 0$, we obtain

$$\begin{aligned} \imath k_i D_i\left(0, \vec{k}\right) &= \imath k_i \varepsilon_{ij}\left(0, \vec{k}\right) E_j\left(0, \vec{k}\right) = k_i k_j \varepsilon_{ij}\left(0, \vec{k}\right)\phi\left(\vec{k}\right) = 4\pi\rho_0\left(\vec{k}\right) \\ &= \frac{4\pi q}{(2\pi)^3} e^{-\imath\vec{k}\cdot\vec{r}_0}. \end{aligned} \tag{1.327}$$

It follows

$$\phi\left(\vec{k}\right) = \frac{4\pi q}{(2\pi)^3} \frac{e^{-\imath\vec{k}\cdot\vec{r}}}{k_i k_j \varepsilon_{ij}\left(0, \vec{k}\right)}, \tag{1.328}$$

and finally

$$\phi\left(\vec{r}\right) = \frac{q}{2\pi^2} \int d\vec{k} \frac{\exp\left[\imath\vec{k}\cdot\left(\vec{r} - \vec{r}_0\right)\right]}{k_i k_j \varepsilon_{ij}\left(0, \vec{k}\right)}. \tag{1.329}$$

For an isotropic medium the denominator has the form $k_i k_j \varepsilon_{ij}\left(0, \vec{k}\right) = k^2 \varepsilon^l(0, k)$, i.e., the static field of the charge is given by the longitudinal dielectric permittivity.

Note that Eqs. (1.328), (1.329) can be generalized to the case of an oscillating charge $q \sim \exp(-\imath\omega t)$, if the frequency is sufficiently small. Then the field can still be derived from a potential

$$\begin{aligned} \phi\left(\omega, \vec{k}\right) &= \frac{4\pi q}{(2\pi)^3} \frac{e^{\imath\vec{k}\cdot\vec{r}_0}}{k_i k_j \varepsilon_{ij}\left(\omega, \vec{k}\right)}, \\ \phi\left(\omega, \vec{r}\right) &= \frac{q}{2\pi^2} \int d\vec{k} \frac{\exp\left[\imath\vec{k}\cdot\left(\vec{r} - \vec{r}_0\right)\right]}{k_i k_j \varepsilon_{ij}\left(\omega, \vec{k}\right)}. \end{aligned} \tag{1.330}$$

For the isotropic medium when $k_i k_j \varepsilon_{ij}\left(\omega, \vec{k}\right) = k^2 \varepsilon^l(\omega, k)$ holds, these formulas are valid for arbitrary ω.

In the vacuum limit $\varepsilon_{ij}\left(0, \vec{k}\right) \to \delta_{ij}$ the potential from Eq. (1.329) reduces to the well-known Coulomb potential

$$\phi\left(\vec{r}\right) = \frac{q}{2\pi^2} \int d\vec{k} \frac{e^{\imath \vec{k} \cdot \vec{R}}}{k^2} = \frac{q}{R}, \qquad \vec{R} = \vec{r} - \vec{r}_0. \tag{1.331}$$

If the relation $k_i k_j \varepsilon_{ij}\left(0, \vec{k}\right) = k^2 + 1/r_{\text{scr}}^2$ holds, we obtain the screened potential

$$\phi\left(\vec{r}\right) = \frac{q}{R} \exp\left(-R/r_{\text{scr}}\right), \tag{1.332}$$

where r_{scr} is the screening radius. For electron-ion plasma the Debye length gives the distance of screening.

If $k_i k_j \varepsilon_{ij}(0, k) = k^2 - k_0^2$ holds we get

$$\phi\left(\vec{r}\right) = \frac{q}{R} \cos k_0 R. \tag{1.333}$$

The field of a test charge in such a medium has a periodic character which indicates instability.

1.11.10 Calculate the magnetic field produced by a straight stationary current filament in an isotropic and homogeneous medium.

Solution By orienting the z-axis along the direction of the current, we have

$$j_0(r) = \vec{e}_z j_0 \delta(x) \delta(y), \qquad \rho_0 = 0, \tag{1.334}$$

where $\vec{e}_z$ is the unit vector along the z-axis. Expanding all the quantities in Fourier components

$$A\left(\vec{r}\right) = \int e^{\imath \vec{k} \cdot \vec{r}} A\left(\vec{k}\right) d\vec{k},$$

and assuming for convenience $\vec{j}_0\left(\vec{r}\right) = \lim_{\omega \to 0} \vec{j}_0\left(\vec{r}\right) \exp\left(-\imath \omega t\right)$, we obtain from the field equations

$$\left[\vec{k} \times \vec{B}\left(\omega, \vec{k}\right)\right]_i + \frac{\omega}{c}\left[\left(\delta_{ij} - \frac{k_i k_j}{k^2}\right)\varepsilon^{\text{tr}} + \frac{k_i k_j}{k^2}\varepsilon^l\right] E_j\left(\omega, \vec{k}\right) = -\frac{4\pi \imath}{c} j_{0i}\left(\omega, \vec{k}\right),$$
$$\vec{k} \cdot \vec{B}\left(\omega, \vec{k}\right) = 0, \quad \left[\vec{k} \times \vec{E}\left(\omega, \vec{k}\right)\right] = \frac{\omega}{c} \vec{B}\left(\omega, \vec{k}\right), \quad \varepsilon^l\left(\omega, \vec{k}\right) \vec{k} \cdot \vec{E}\left(\omega, \vec{k}\right) = 0, \tag{1.335}$$

the magnetic field in the form

$$\vec{B}\left(\vec{r}\right) = \frac{4\pi\imath}{c} \lim_{\omega\to 0} \int d\vec{k} \frac{e^{\imath \vec{k}.\vec{r}} \left[\vec{k} \times \vec{j}_0\left(\vec{k}\right)\right]}{k^2 - \frac{\omega^2}{c^2}\varepsilon^{tr}\left(\omega, \vec{k}\right)}. \tag{1.336}$$

Here, $\vec{j}_0\left(\vec{k}\right)$ is the Fourier transform of the current density $\vec{j}_0\left(\vec{r}\right)$, i.e.,

$$\vec{j}_0\left(\vec{k}\right) = \frac{j_0 \vec{e}_z}{(2\pi)^2}\delta(k_z).$$

In vacuum, $\varepsilon^{\text{tr}} = 1$ holds and (1.336) reduces to

$$\vec{B}\left(\vec{r}\right) = \frac{2j_0}{cr}\vec{e}_\phi, \tag{1.337}$$

where $\vec{e}_\phi$ is the unit vector in the azimuthal direction. Equation (1.337) is

$$\lim_{\omega\to 0} \frac{\omega^2}{c^2}\varepsilon^{\text{tr}}\left(\omega, \vec{k}\right) = 0. \tag{1.338}$$

This assumption holds true for classical media in the thermodynamic equilibrium (Sect. 1.8). Violation of condition (1.338) indicates that the classical medium is in a state of non-equilibrium.

Thus, the magnetic field produced by a static current within a classical medium in equilibrium does not differ from that produced in vacuum. This result no longer holds for a time-varying current, of course, valid for arbitrary frequencies ω of the current.

1.11.11 Find the dispersion equation for arbitrarily polarized electromagnetic waves in the system comprising two moving media.

Solution The analysis of arbitrary wave fields in multi-component moving medium is rather complicated. Here, therefore, we restrict our analysis on the system consisting of two media: medium 1 resting in the laboratory frame, and medium 2 moving with velocity $\vec{u}$ in the laboratory frame. Indeed, each medium is isotropic and non-gyrotropic in its intrinsic frame [see Eqs. (3.169) and (3.170)].

Avoiding writing the dielectric permittivity tensor of this system (it is very cumbersome), we directly write the whole dispersion equation which splits into two equations:

$$\left\{k^2 - \frac{\omega^2}{c^2}\left[\varepsilon_1^{\text{tr}}(\omega, k) + \frac{\omega'^2}{\omega^2}\left(\varepsilon_2^{\text{tr}}(\omega', k') - 1\right)\right]\right\} = 0, \tag{1.339}$$

$$\left\{k^2 - \frac{\omega^2}{c^2}\left[\varepsilon_1^{tr}(\omega,k) + \frac{\omega'^2}{\omega^2}\left(\varepsilon_2^{tr}(\omega',k') - 1\right)\right]\right\}\left[\varepsilon_1^l(\omega,k) + \varepsilon_2^l(\omega',k') - 1\right] -$$
$$-\frac{k^2u^2 - \left(\vec{k}\cdot\vec{u}\right)^2}{c^2(1-u^2/c^2)}\left\{\varepsilon_2^l(\omega',k') - 1 + \frac{\omega'^2}{c^2k'^2}\left[\varepsilon_2^{tr}(\omega',k') - \varepsilon_2^l(\omega',k')\right]\right\}\times$$
$$\times\left\{\varepsilon_1^l(\omega,k) - 1 + \frac{\omega^2}{k^2c^2}\left[\varepsilon_1^{tr}(\omega,k) - \varepsilon_2^l(\omega,k)\right]\right\} = 0. \tag{1.340}$$

Here, u is the relative velocity of two media. Equation (1.339) describes the purely transverse wave with $\vec{E}\perp\vec{k}$, while Eq. (1.340) describes the propagation of mixed transverse-longitudinal wave, assuming $\vec{E}$ and $\vec{k}$ are arbitrarily oriented. However, for purely longitudinal wave propagation, assuming $\vec{u} \parallel \vec{k}$, we find a purely longitudinal wave with dispersion equation

$$\varepsilon^l(\omega,k) = \varepsilon_1^l(\omega,k) + \varepsilon_2^l(\omega',k') - 1 = 0, \tag{1.341}$$

and a purely transverse two-fold degenerated wave with two mutually perpendicular polarizations of the field $\vec{E}$ with dispersion equation

$$\left\{k^2 - \frac{\omega^2}{c^2}\left\{\varepsilon_1^{tr}(\omega,k) + \frac{\omega'^2}{\omega^2}\left(\varepsilon_2^{tr}(\omega',k') - 1\right)\right\}\right\}^2 = 0. \tag{1.342}$$

The whole effective transverse dielectric permittivity of the system is

$$\varepsilon_{ef}^{tr}(\omega,k) = \varepsilon_1^{tr}(\omega,k) + \frac{\omega'^2}{\omega^2}\left[\varepsilon_2^{tr}(\omega',k') - 1\right], \tag{1.343}$$

which has an imaginary part with odd frequency dependence. Hence, the solution of this equation under the condition

$$\omega - \vec{k}\cdot\vec{u} < 0 \tag{1.344}$$

can be oscillatory. However, such a condition is possible if the wave with frequency ω can exist in the system. This is possible when

$$\frac{\partial}{\partial\omega}\,\mathrm{Re}\left[k^2 - \frac{\omega^2}{c^2}\varepsilon_{ef}^{tr}(\omega,k)\right] > 0, \tag{1.345}$$

and velocity u exceeds the threshold velocity determined by

$$\mathrm{Im}\varepsilon_{\mathrm{ef}}^{\mathrm{tr}}(\omega,k)=\mathrm{Im}\left\{\varepsilon_1^{\mathrm{tr}}(\omega,k)\right\}+\frac{\omega'^2}{\omega^2}\left[\varepsilon_2^{\mathrm{tr}}(\omega',k')-1\right]=0. \tag{1.346}$$

Naturally, the instability here is the stimulated Cherenkov radiation of a moving medium in the system.

1.11.12 Find the transformation law of dielectric and magnetic permittivity tensors in EHDB representation of electrodynamics.

Solution We recall that such a representation is related to the writing of the induced current in terms of the electric dipole and magnetic moment of the medium

$$\vec{j}=\frac{\partial\vec{P}}{\partial t}+c\nabla\times\vec{M}=-\imath\omega\vec{P}+\imath c\left(\vec{k}\times\vec{M}\right), \tag{1.347}$$

where

$$\begin{aligned}P_i&=\alpha_{ij}\left(\omega,\vec{k}\right)E_j=\frac{1}{4\pi}\left(\varepsilon_{ij}-\delta_{ij}\right)E_j,\\M_i&=\chi_{ij}\left(\omega,\vec{k}\right)H_j=\frac{1}{4\pi}\left(\mu_{ij}-\delta_{ij}\right)H_j.\end{aligned} \tag{1.348}$$

Now we can use the transformation of $\vec{P}$ and $\vec{M}$. It is obvious that $\vec{P}$ and $\vec{M}$ should be transformed as electric and magnetic fields, respectively. Thus, writing relations (1.348) in the rest frame and making use of transformation rules for $\vec{P}$ and $\vec{M}$, we finally find

$$\begin{aligned}P_i\left(\omega,\vec{k}\right)&=\gamma_{i\mu}\left(\vec{u}\right)P'_\mu\left(\omega',\vec{k}'\right)=\gamma_{i\mu}\left(\vec{u}\right)\frac{\varepsilon_{\mu\nu}\left(\omega',\vec{k}'\right)-\delta_{\mu\nu}}{4\pi}\beta_{\nu j}\left(\vec{u}\right)E_j\left(\omega,\vec{k}\right),\\M_i\left(\omega,\vec{k}\right)&=\widehat{\delta}_{i\mu}\left(\vec{u}\right)H'_\mu\left(\omega',\vec{k}'\right)=\widehat{\delta}_{i\mu}\left(\vec{u}\right)\frac{\mu_{\mu\nu}\left(\omega',\vec{k}'\right)-\delta_{\mu\nu}}{4\pi}\eta_{\nu j}\left(\vec{u}\right)H_j\left(\omega,\vec{k}\right).\end{aligned} \tag{1.349}$$

We easily can write the transformation formulas of tensors ε_{ij} and μ_{ij}:

$$\begin{aligned}\varepsilon_{ij}\left(\omega,\vec{k}\right)&=\delta_{ij}+\gamma_{i\mu}\left(\varepsilon_{\mu\nu}\left(\omega',\vec{k}'\right)-\delta_{\mu\nu}\right)\beta_{\nu j},\\\mu_{ij}\left(\omega,\vec{k}\right)&=\delta_{ij}+\widehat{\delta}_{i\mu}\left(\mu_{\mu\nu}\left(\omega',\vec{k}'\right)-\delta_{\mu\nu}\right)\eta_{\nu j}.\end{aligned} \tag{1.350}$$

It should be remarked that tensors $\gamma_{ij}, \beta_{ij}, \widehat{\delta}_{ij}, \eta_{ij}$ carry out direct and inverse Lorentz transformations for fields $\vec{E}$ and $\vec{H}$. It is not difficult to generalize these transformation formulas for the multi-components media with independent motion of each component with respect to the laboratory frame.

1.11.13 From the expansion in powers of the wave vector, study the electromagnetic properties of isotropic media with no inversion center.

Solution At the very high-frequency range, or at the optical range of frequency when $\omega \gg kv_0$, i.e., the wave's phase velocity greatly exceeds the characteristic velocity of the chaotic motion of particles, spatial dispersion can be neglected. However, there is an optical frequency range in which, nevertheless, we have to take into account weak spatial dispersion, which leads to the appearance of a whole series of qualitative new effects. In the general case, weak spatial dispersion can be taken into account by simple expansion of dielectric permittivity in terms of the wave vector $\vec{k}$ [see Eq. (1.132)]

$$\varepsilon_{ij}\left(\omega, \vec{k}\right) = \varepsilon(\omega)\delta_{ij} + \imath\gamma_{ijl}k_l + \alpha_{ijlm}k_l k_m. \tag{1.351}$$

The second term in this expansion exists only in the media with no inversion center. This is the consequence of symmetry relation (1.134) which is valid for completely isotropic media.

For isotropic media with no inversion center (such a medium is, for example, a solution of sugar whose crystals do not have a center of inversion; about an order of magnitude more than the effects of quartz) the second term in relation (1.351) is non-zero and we can limit ourselves to its consideration only. In this case, we have

$$\varepsilon_{ij}\left(\omega, \vec{k}\right) = \varepsilon(\omega)\delta_{ij} + \imath\gamma(\omega)e_{ijl}k_l\frac{c}{\omega}. \tag{1.352}$$

This relation is equivalent to the following form of the material equation

$$\vec{D} = \varepsilon(\omega)\vec{E} - \imath\gamma(\omega)\frac{c}{\omega}\left(\vec{k} \times \vec{E}\right). \tag{1.353}$$

It is obvious that the second term in Eq. (1.353) is important at the frequency range of the longitudinal eigen modes, wherein

$$\varepsilon(\omega) \simeq 0. \tag{1.354}$$

Substituting expression (1.352) into the general dispersion equation for the electromagnetic field oscillations in the medium (1.114), we find two independent equations:

$$
\begin{aligned}
&\varepsilon(\omega) = 0, \\
&\left[k^2 - \frac{\omega^2}{c^2}\varepsilon(\omega)\right]^2 = \gamma^2(\omega)\frac{\omega^2}{c^2}k^2.
\end{aligned}
\tag{1.355}
$$

The first equation describes purely longitudinal waves ($\vec{k}\|\vec{E}$), while the second equation describes transverse waves with different polarizations.

In this way, the spectrum of the longitudinal oscillations does not change if we take into account weak spatial dispersion in the isotropic medium with no inversion center. As about the transverse waves, they are two-fold degenerate in the isotropic medium with no weak dispersion. Taking into account the dispersion removes degeneracy and gives two spectra:

$$
k^2 = \frac{\omega^2}{c^2}\varepsilon(\omega) \pm |\gamma(\omega)|\frac{\omega}{c}k = \frac{\omega^2}{c^2}\varepsilon(\omega) \pm |\gamma(\omega)|\frac{\omega^2}{c^2}\sqrt{\varepsilon(\omega)}. \tag{1.356}
$$

In the Fresnel's problem (in the boundary-value problem), this leads to birefraction with polarizations (for $\vec{k}\|oz$)

$$
E_x = \pm E_y, \tag{1.357}
$$

which is used to determine the sugar content in the solution.

A somewhat different situation occurs near the poles of the components of dielectric permittivity which correspond to single-particle oscillations of the medium. In this case, for the isotropic medium with no inversion center, we have

$$
\varepsilon_{ij}^{-1}\left(\omega, \vec{k}\right) = \frac{1}{\varepsilon(\omega)}\delta_{ij} - \imath g(\omega)\frac{c}{\omega}e_{ijl}k_l. \tag{1.358}
$$

This relation is equivalent to the following material equation

$$
\vec{E} = \frac{1}{\varepsilon(\omega)}\vec{D} + \imath g(\omega)\frac{c}{\omega}\left(\vec{k}\times\vec{D}\right), \tag{1.359}
$$

which implies the transversality of both the field $\vec{D}$ and $\vec{E}$. The substitution of this relation into the Maxwell's equations, which in this case is reduced to the form of

$$
\vec{E} = \frac{\omega^2}{c^2k^2}\vec{D}, \tag{1.360}
$$

leads to the following dispersion relation for the transverse fields in plasma

$$\left(\frac{k^2}{\varepsilon(\omega)}-\frac{\omega^2}{c^2}\right)^2-g^2\frac{c^2}{\omega^2}k^6=0. \tag{1.361}$$

Here, also there is the splitting of the transverse field of the waves with two polarizations:

$$k^2=\frac{\omega^2}{c^2}\varepsilon(\omega)\pm|g(\omega)|\frac{c}{\omega}k^3=\frac{\omega^2}{c^2}\varepsilon(\omega)\pm|g(\omega)|\frac{\omega^2}{c^2}\varepsilon^{3/2}(\omega). \tag{1.362}$$

The last term in Eq. (1.362), as well as in Eq. (1.356), indicates the appearance of birefraction in the isotropic medium with weak spatial dispersion of the first order, i.e., taking into account the material equations in the form of Eq. (1.353) or Eq. (1.359) for media with no inversion center. The first of these relations essentially shows spatial dispersion at the range of eigen frequencies of the collective longitudinal oscillations, wherein $\varepsilon(\omega)\simeq 0$. At the same time, the second relation indicates the importance of taking into account the effect of weak spatial dispersion in the range of the eigen frequencies of oscillations of the individual particles of the medium, i.e., near the poles of $\varepsilon(\omega)$, when $1/\varepsilon(\omega)\to 0$.

There is, however, an important difference in Eqs. (1.361) and (1.355). The fact is that the first of them is cubic with respect to k^2, while the second remains bi-quadratic. This means that in the latter case, corresponding to the direct expansion of $\varepsilon_{ij}\left(\omega,\vec{k}\right)$ in the form of relation (1.352), the number of transverse waves does not change. Moreover, near the resonance frequency of collective longitudinal oscillations, the degeneracy of two branches of transverse waves with different polarizations is removed. In the case of the expansion of the inverse tensor $\varepsilon_{ij}^{-1}\left(\omega,\vec{k}\right)$ in relation (1.358), near the eigen frequencies of the individual particles of the medium, in addition to removing the polarization degeneracy, which corresponds to relations (1.362), there is one more branch of oscillations with the following dispersion relation

$$k^2\simeq\frac{\omega^2/c^2}{\varepsilon^2(\omega)g^2(\omega)}. \tag{1.363}$$

Rewriting the latter relation in the form of

$$\omega^2=k^2c^2\varepsilon^2(\omega)g^2(\omega), \tag{1.364}$$

we see that the wave is most strongly manifested in the region of frequencies near the poles of $\varepsilon(\omega)$, i.e., single-frequency absorption bands of the medium.

1.11.14 A particle of charge e moves with a velocity $\vec{v}=\mathrm{const}$ through the homogeneous isotropic medium of permittivity $\varepsilon(\omega)$ and magnetic permeability $\mu=1$. Determine the components of the electromagnetic field produced by this particle [65].

Solution By decomposition the field vectors into the Fourier integral over space and time coordinates

$$\vec{E}\left(\vec{R},t\right)=\int\vec{E}\left(\vec{k},\omega\right)e^{\imath\left(\vec{k}\cdot\vec{R}-\omega t\right)}\frac{d^3k}{(2\pi)^3}\frac{d\omega}{2\pi}\cdots$$

we obtain, from Maxwell's equations, a system of algebraic equations relative to the Fourier amplitudes:

$$\begin{cases}\kappa\vec{n}\times\vec{E}\left(\vec{k},\omega\right)=\vec{B}\left(\vec{k},\omega\right),\\ \kappa\vec{n}\times\vec{B}\left(\vec{k},\omega\right)=-\varepsilon(\omega)\vec{E}\left(\vec{k},\omega\right)-\imath\dfrac{8\pi^2ev}{\omega^2}\delta\left(\dfrac{\kappa}{c}\vec{n}\cdot\vec{v}-1\right),\\ \kappa\varepsilon(\omega)\vec{n}\cdot\vec{E}\left(\vec{k},\omega\right)=-\imath\dfrac{8\pi^2ec}{\omega^2}\delta\left(\dfrac{\kappa}{c}\vec{n}\cdot\vec{v}-1\right),\\ \kappa\vec{n}\cdot\vec{B}\left(\vec{k},\omega\right)=0;\end{cases}\tag{1.365}$$

Here $\vec{B}\left(\vec{k},\omega\right)$ is the Fourier amplitude of the magnetic field, $\vec{k}=\omega\kappa\vec{n}/c$, κ is the parameter which is depends on ω and $\vec{k}$, and $\vec{n}$ is the unit vector. When deriving Eq. (1.365), one should take into account that the amplitude of the Fourier function $\delta\left(\vec{R}-\vec{v}t\right)$ is equal to $2\pi\delta\left(\vec{k}\cdot\vec{v}-\omega\right)$ and $\delta(\alpha x)=(1/|\alpha|)\delta(x)$. From the system of Eq. (1.365), we can determine $\vec{E}$ and $\vec{B}$:

$$\begin{cases}\vec{E}\left(\vec{k},\omega\right)=-\imath\dfrac{8\pi^2ec}{\omega^2}\cdot\dfrac{\kappa\vec{n}-(\vec{v}/c)\varepsilon}{\varepsilon(\kappa^2-\varepsilon)}\delta\left(\dfrac{\kappa}{c}\vec{n}\cdot\vec{v}-1\right),\\ \vec{B}\left(\vec{k},\omega\right)=\imath\dfrac{8\pi^2e\kappa}{\omega^2}\cdot\dfrac{\vec{n}\times\vec{v}}{\kappa^2-\varepsilon}\delta\left(\dfrac{\kappa}{c}\vec{n}\cdot\vec{v}-1\right).\end{cases}\tag{1.366}$$

In order to determine the fields, we have to inverse the Fourier transformation. First, we calculate $E_z\left(\vec{R},t\right)$. It follows from Eq. (1.366) that

$$E_z\left(\vec{k},\omega\right)=-\imath\frac{8\pi^2ec}{\omega^2}\cdot\frac{\kappa\cos\theta-\beta\varepsilon}{\varepsilon(\kappa^2-\varepsilon)}\delta(\beta\kappa\cos\theta-1),$$

and, hence,

$$E_z\left(\vec{R},t\right) = -\frac{\imath e}{2\pi^2 c^2}\int_{-\infty}^{\infty}\omega d\omega e^{-\imath\omega t}\int_0^{\infty}\kappa^2 d\kappa\int\frac{\kappa\cos\theta-\beta\varepsilon}{\varepsilon(\kappa^2-\varepsilon)}$$
$$\times\exp\left\{\imath\frac{\omega}{c}\kappa[r\sin\theta\cos(\Phi-\varphi)-z\cos\theta]\right\}\delta(\beta\kappa\cos\theta-1)\sin\theta d\theta d\Phi. \quad (1.367)$$

Here, $\vec{r}$ is the component of $\vec{R}$ in the *xy*-plane, φ is the angle between $\vec{r}$ and the *ox*-axis, $\beta = v/c$; θ and Φ are the polar angles of the unit vector $\vec{n}$. The integral over Φ is expressed in terms of the Bessel function J_0 of the argument $(\omega/c)\kappa r\sin\theta$. The integral over θ has the form:

$$\int_0^{\pi} f(\theta)\delta(\beta\kappa\cos\theta-1)\sin\theta d\theta = \frac{1}{\beta\kappa}\int_{-\beta\kappa}^{\beta\kappa}\varphi(y)\delta(y-1)\,dy. \quad (1.368)$$

It has non-zero value only at $\beta\kappa \geq 1$, and hence the lower limit of κ is equal to $1/\beta$. In Eq. (1.367), this is automatically allowed for because of the presence of the delta function. However, after integration over *dy* the delta function will vanish and it will be necessary to take into account explicitly the lower limit of integration. The integration of Eq. (1.368) over *dy* gives

$$\frac{1}{\beta\kappa}\varphi(1) = \frac{1}{\beta\kappa}f(\theta)\bigg|_{\cos\theta=1/\beta\kappa}. \quad (1.369)$$

Substitute Eq. (1.369) in Eq. (1.367) and change the variable κ by $x = \sqrt{\kappa^2 - 1/\beta^2}$; because κ varies in the limits from $1/\beta$ to ∞, x will vary between 0 and ∞. Then $E_z\left(\vec{R},t\right)$ will be of the form:

$$E_z\left(\vec{R},t\right) = \frac{\imath e}{\pi c^2}\int_{-\infty}^{\infty}\omega d\omega\exp\left[\imath\omega\left(\frac{z}{v}-t\right)\right]\left(1-\frac{1}{\beta^2\varepsilon}\right)\int_0^{\infty}\frac{J_0(\omega rx/c)x dx}{x^2+1/\beta^2-\varepsilon}.$$

Integration over x with the aid of formula

$$\int_0^{\infty}\frac{J_0(xr)x dx}{x^2+k^2} = K_0(kr), \quad (1.370)$$

gives

$$E_z\left(\vec{R},t\right) = \frac{\imath e}{\pi c^2}\int\limits_{-\infty}^{\infty}\left(1-\frac{1}{\beta^2\varepsilon}\right)K_0(sr)\exp\left[\imath\omega\left(\frac{z}{v}-t\right)\right]\omega\, d\omega, \qquad (1.371)$$

where $K_n(x)$ is the MacDonald function of the order of n and $s^2 = \omega^2/v^2 - (\omega^2/c^2)\varepsilon(\omega)$. The sign of s should be chosen so that Re $s > 0$; otherwise the integral over ω will diverge. The integration over ω in Eq. (1.370) can only be completed by specifying the precise form of the function $\varepsilon(\omega)$. In order to evaluate $E_x\left(\vec{R},t\right)$, we again begin with integration over Φ. Integration over θ can be carried out with the aid of the delta function. Subsequent integration over $x = \sqrt{\kappa^2 - 1/\beta^2}$ can be done with the formula

$$\int\limits_0^{\infty}\frac{J_1(xr)x^2dx}{x^2+k^2} = kK_1(kr),$$

which is obtained from Eq. (1.370) by differentiating with respect to r and taking into account that $J_0' = -J_1$ and $K_0' = -K_1$. The result is

$$E_x\left(\vec{R},t\right) = \cos\varphi\frac{e}{\pi v}\int\limits_{-\infty}^{\infty}\frac{s}{\varepsilon}K_1(sr)\exp\left[\imath\omega\left(\frac{z}{v}-t\right)\right]d\omega.$$

The components $E_y\left(\vec{R},t\right)$ and $\vec{B}\left(\vec{R},t\right)$ can be determined in a similar way. E_y differs from E_x in that cosφ is replaced by sinφ; hence, in the cylindrical coordinates we have

$$E_r\left(\vec{R},t\right) = \frac{e}{\pi v}\int\limits_{-\infty}^{\infty}\frac{s}{\varepsilon}K_1(sr)\exp\left[\imath\omega\left(\frac{z}{v}-t\right)\right]d\omega, \qquad E_\varphi = 0. \qquad (1.372)$$

For $\vec{B}$, we obtain

$$B_\varphi\left(\vec{R},t\right) = \frac{e}{\pi c}\int\limits_{-\infty}^{\infty}sK_1(sr)\exp\left[\imath\omega\left(\frac{z}{v}-t\right)\right]d\omega, \qquad B_z = B_r = 0. \qquad (1.373)$$

According to Eqs. (1.371)–(1.373), the electromagnetic field is axially symmetric. The above formulas will hold only in the region $r \gg a$, where a is of the order of the interatomic distance. In the region $r \le a$, it is necessary to take into account spatial dispersion of the permittivity.

1.11.15 Derive the expression for magnetic fluctuations in equilibrium isotropic plasma.

Solution From relation (1.227) we can write the correlator of fluctuations of the self-consistent electric fields of isotropic non-degenerate plasma:

$$\langle E_i E_j \rangle_{\omega,\vec{k}} = \frac{8\pi T}{\omega}\left[\frac{k_i k_j}{k^2}\frac{\mathrm{Im}\{\varepsilon^l(\omega,k)\}}{|\varepsilon^l(\omega,k)|^2} + \left(\delta_{ij} - \frac{k_i k_j}{k^2}\right)\frac{\mathrm{Im}\{\varepsilon^{\mathrm{tr}}(\omega,k)\}}{\left|\varepsilon^{\mathrm{tr}}(\omega,k) - k^2c^2/\omega^2\right|^2}\right]. \quad (1.374)$$

Let us determine from Eq. (1.374) the fluctuation correlator of the transverse electric field in isotropic plasma:

$$\langle E^{\mathrm{tr}2} \rangle_{\omega,\vec{k}} = \frac{8\pi T}{\omega}\frac{\mathrm{Im}\{\varepsilon^{\mathrm{tr}}(\omega,k)\}}{\left|\varepsilon^{\mathrm{tr}}(\omega,k) - k^2c^2/\omega^2\right|^2}. \quad (1.375)$$

Using the relationship $\vec{B} = (c/\omega)\left(\vec{k}\times\vec{E}\right)$, we obtain

$$\langle B^2 \rangle_{\omega,\vec{k}} = \frac{k^2c^2}{\omega^2}\langle E^{\mathrm{tr}2} \rangle_{\omega,\vec{k}} = \frac{8\pi T k^2 c^2 \mathrm{Im}\{\varepsilon^{\mathrm{tr}}(\omega,k)\}}{\omega^3\left|\varepsilon^{\mathrm{tr}}(\omega,k) - k^2c^2/\omega^2\right|^2}. \quad (1.376)$$

In the plasma transparency range, i.e., in the limit $\omega \gg k v_{\mathrm{Te}}$, when $\varepsilon^{\mathrm{tr}}(\omega,k) \approx 1 - \omega_{\mathrm{pe}}^2/\omega^2$, we have

$$\langle B^2 \rangle_{\omega,\vec{k}} = \frac{16\pi^2 T k^2 c^2}{|\omega|}\delta\left(\omega^2 - \omega_{\mathrm{pe}}^2 - k^2c^2\right). \quad (1.377)$$

Integrating this expression over frequencies yields

$$\langle B^2 \rangle_{\vec{k}} = \frac{8\pi^2 T k^2 c^2}{\omega_{\mathrm{pe}}^2 + k^2c^2}. \quad (1.378)$$

Hence, it is easy to derive a space correlation function

$$\langle B^2 \rangle_{\vec{r}} = 8\pi^2 T\left[\delta\left(\vec{r}\right) - \frac{1}{4\pi}\frac{\exp\left(-r/r_{\mathrm{cor}}\right)}{r r_{\mathrm{cor}}^2}\right], \quad (1.379)$$

where $r_{\mathrm{cor}} = c/\omega_{\mathrm{pe}}$ is the correlation length of magnetic fluctuations in isotropic plasma.

For $r \approx r_{\rm cor}$ we have from Eq. (1.379)

$$\frac{\langle B^2\rangle_r}{4\pi N_e T} \sim \frac{\langle E^{\rm tr2}\rangle_r}{4\pi N_e T} \sim \frac{1}{r_{\rm cor}^3 N_e} \sim \frac{v_{Te}^3}{c^3}\frac{1}{r_{De}^3 N_e} \approx \frac{v_{Te}^3}{c^3}\frac{\langle (E^l)\rangle_r^2}{4\pi N_e T} \tag{1.380}$$

Thus, the fluctuation energy of the transverse field is $c^3/v_{\rm Te}^3$ times smaller than that of the longitudinal field.

References

1. I.E. Tamm, *Principles of Theory of Electricity* (Gostekhizdat, Moscow, 1946)
2. L.D. Landau, E.M. Lifshitz, *Electrodynamics of Continuous Media*, 2nd edn. (Pergamon, New York, 1984)
3. M.E. Gertsenshtein, J. Exp. Theor. Phys. **22**(10), 303 (1952)
4. J. Lindhard, Det. Kong. Danske vid. Selskab. Dan. Mat. Fys. Med. **28**, 2 (1954)
5. L.D. Landau, E.M. Lifshitz, *Course of Theoretical Physics, Vol. 5: Statistical Physics*, 3rd edn. (Nauka, Moscow, 1976).; Pergamon Press, Oxford, 1980), Part 1
6. A.A. Vlasov, *Macroscopic Electrodynamics* (Gostkhizdat, Moscow, 1955)
7. J. Doetsch, *Theorie und Wending der Laplace-Transformation* (Dover Publication, New York, 1948)
8. V.P. Silin, Fiz. Met. Metalloved. **10**, 942 (1960)
9. P. Drude, *The Theory of Optics* (Dover Publications, New York, 1959)
10. V.L. Ginzburg, *Theory of Radio Wave Propagation in Ionosphere* (Gostechizdat, Moscow, 1949)
11. Y.A. Alpert, V.L. Ginzburg, E.L. Feinberg, *Radio Wave Propagation* (Gostechizdat, Moscow, 1953)
12. V.L. Ginzburg, *Electromagnetic Waves in Plasma* (Fizmatgiz, Moscow, 1960)
13. V.L. Ginzburg, J. Exp. Theor. Phys. **34**, 1594 (1958)
14. V.M. Agranovich, A.A. Rukhadze, J. Exp. Theor. Phys. **35**, 982 (1958)
15. A.S. Davydov, Theory of light absorption in molecular crystals, Transactions of the Institute of Physics, Academy of Sciences of the Ukrainian SSR [in Russian], Izd. AN UkrSSR, Kiev (1951)
16. A.S. Davydov, J. Exp. Theor. Phys. **19**, 930 (1949)
17. U. Fano, Phys. Rev. **103**, 1202 (1956)
18. S. Pekar, J. Exp. Theor. Phys. **33**, 1022 (1957).; J. Exp. Theor. Phys. **34**, 1176 (1958); J. Exp. Theor. Phys. **35**, 522 (1958); J. Exp. Theor. Phys. **36**, 451 (1959)
19. J.J. Hopfield, Phys. Rev. **112**, 1555 (1958)
20. V.M. Agranovich, J. Exp. Theor. Phys. **35**, 430 (1959)
21. A.S. Davydov, A.F. Lubchenko, J. Exp. Theor. Phys. **35**, 1499 (1958)
22. O.V. Konstantinov, V.I. Perel, J. Exp. Theor. Phys. **37**, 786 (1959)
23. V.S. Mashkvich, J. Exp. Theor. Phys. **38**, 906 (1960)
24. V.L. Strizhevskii, Russ. Solid State Phys. **2**, 1806 (1960)
25. S.I. Pekar, B.E. Tsekvava, Russ. Solid State Phys. **2**, 211 (1960)
26. B.E. Tsekvava, Russ. Solid State Phys. **2**, 482 (1960)
27. M.S. Brodin, S.I. Pekar, J. Exp. Theor. Phys. **38**, 74 (1960).; J. Exp. Theor. Phys. **38**, 1910 (1960)
28. G.S. Landsberg, *Optics* (Gostechizdat, Moscow, 1957)
29. I.E. Tamm, I.M. Frank, Dokl. Acad. Sci. USSR **14**, 107 (1937)

30. E. Fermi, Phys. Rev. **57**, 485 (1940)
31. N. Bohr, *The Penetration of Atomic Particles Through Matter* (Hafner Publishing Company, New York, 1948)
32. B.M. Bolotovskii, Usp. Fiz. Nauk **62**, 201 (1957)
33. J. Jelly, *Cherenkov Radiation* (Pergamon Press, New York, 1958)
34. A.A. Vlasov, *Many Particles Theory* (Gostechizdat, Moscow, 1950)
35. A.I. Akhiezer, A.G. Sitenko, J. Exp. Theor. Phys. **23**, 161 (1952)
36. D. Pines, D. Bohm, Phys. Rev. **82**, 625 (1951).; Phys. Rev. **85**, 338 (1952)
37. J. Hubbard, Proc. Phys. Soc. **A62**, 441 (1955)., 977
38. J. Neufeld, R.H. Ritchie, Phys. Rev. **98**, 1632 (1955)
39. H. Froehlich, H. Pelzer, Proc. Phys. Soc. **A68**, 525 (1955)
40. D. Pines, Rev. Mod. Phys. **28**, 184 (1956).; D. Pincs, D. Nozieres, Nuovo Cimento, 9, 470 (1958)
41. R.H. Ritchie, Phys. Rev. **106**, 874 (1957)
42. A.A. Rukhadze, V.P. Silin, Fiz. Met. Metalloved. **12**(2), 287 (1961)
43. E.L. Feinberg, J. Exp. Theor. Phys. **34**, 1125 (1958)
44. V.M. Agranovich, A.A. Rukhadze, J. Exp. Theor. Phys. **35**, 1171 (1958)
45. A.G. Sitenko, K.N. Stepanov, Bull. Univ. A.M. Gorki, Kharkov **7**, 5 (1958)
46. V.D. Shafranov, J. Exp. Theor. Phys. **34**, 1475 (1958)
47. V.P. Silin, J. Exp. Theor. Phys. **37**, 873 (1959)
48. Y.L. Klimontovich, V.P. Silin, Usp. Fiz. Nauk **70**, 247 (1960)
49. A.L. Larkin, J. Exp. Theor. Phys. **37**, 264 (1959)
50. B.L. Zhilnov, J. Exp. Theor. Phys. **40**, 527 (1961)
51. J.E. Drummond, *Plasma Physics* (McGraw Hill, New York, 1961)
52. P.A. Cherenkov, Dokl. Acad. Sci. USSR **2**, 451 (1934)
53. S.I. Vavilov, Dokl. Acad. Sc. USSR **2**, 457 (1934)
54. V.L. Granovskii, *Electric Fluctuations* (Gostechizdat, Moscow, 1936)
55. V.L. Ginzburg, Usp. Fiz. Nauk **46**, 348 (1952)
56. M.A. Leontovich, S.M. Rytov, J. Exp. Theor. Phys. **23**, 246 (1952)
57. M.E. Gertsenshtein, J. Exp. Theor. Phys. **25**, 827 (1955)
58. S.M. Rytov, *Theory of Electric Fluctuations And Thermal Radiation* (U.S. Air Force, 1959)
59. M.L. Levin, Dokl. Acad. Sci. USSR **102**, 53 (1955)
60. F.V. Bunkin, Dissertation, FIAN USSR, 1955
61. Y.L. Klimontovich, J. Exp. Theor. Phys. **34**, 173 (1958)
62. F.G. Bass, M.I. Kaganov, J. Exp. Theor. Phys. **34**, 1154 (1958)
63. V.D. Shafranov, *Plasma Physics and Problem of Controlled Thermonuclear Reactions*, vol 4 (Izd-vo Academy Science of USSR, Moscow, 1958)
64. V.P. Silin, Radio Phys. **2**, 198 (1959)
65. I.N. Toptygin, *Electromagnetic Phenomena in Matter: Statistical and Quantum Approaches*, 1st edn. (Wiley-VCH, Weinheim, 2015)

Chapter 2
Isotropic Plasma

2.1 Kinetic Equation with Self-consistent Fields

Plasma can be considered as an ionized gas consisting of a large number of charged particles. The specific properties distinguishing plasma from ordinary gases are stipulated by the motion of charged particles that create substantial electromagnetic fields. Because of these fields, the usual Boltzmann kinetic theory taking in account only the pair collisions of particles is obviously insufficient for a thorough description of plasmas. In this sense, it is necessary to consider the influence of electromagnetic fields on the particles' motion in plasma.

At the same time, it is obvious that when the ionization degree is sufficiently small, one can consider plasma as a gas of neutral particles. On the other hand, there exist a lot of phenomena, which can be described by usual gaseous kinetic method even if plasma is almost completely ionized.[1] Such phenomena are not considered practically in this book. We will concentrate on the phenomena in which the influence of electromagnetic fields on the particles' motion is extremely important.

The specific properties of plasma manifest themselves when the particle distribution under, for example, the action of electromagnetic fields is inhomogeneous. The inhomogeneous distribution of charged particles of plasma leads to the

[1] At first, it seems impossible to use the scheme of gaseous kinetics to the collisions of charged particles. Really, it is well-known that the scattering cross section of particles with Coulomb interaction diverges. This is the consequence of Coulomb interaction at large distances, which leads to the angular dependence of the scattering probability as $1/\vartheta^4$ for small angles ϑ. However, the charged particles interaction in plasma is screened at large distances and as a result the scattering cross section becomes finite [1–7]. Such screening is obvious for the static potential of a charged particle in plasma, which is known as Debye screening ($r_D = (\kappa T/4\pi e^2 N)^{3/2}$ is the Debye radius). For the fast varying processes, the screening radius is equal to the length traversed by the particle during the characteristic time of process change provided that this length is less than Debye length [8, 9]. These circumstances justify the application of usual scattering theory to the scattering of charged particles in plasma.

B. Shokri, A. A. Rukhadze, *Electrodynamics of Conducting Dispersive Media*, Springer Series on Atomic, Optical, and Plasma Physics 111,
https://doi.org/10.1007/978-3-030-28968-3_2

appearance of induced current and charge densities. The latter, in turn, creates electromagnetic fields, which again influence in turn the particles' motion. Thus, to study the electromagnetic properties of plasmas, the self-consistent interaction between the particles and electromagnetic fields must be taken into account. Clearly, an accurate description of the particles system of plasma requires self-consistence. One can write the following classical (non-quantum) equation of motion for each particle interacting with other particles through the electromagnetic field created by other particles:

$$\frac{d\vec{p}_s}{dt} = e_s\left\{\vec{E}\left(\vec{r}_s\right) + \frac{1}{c}\left[\vec{v}_s \times \vec{B}\left(\vec{r}_s\right)\right]\right\}, \tag{2.1}$$

where $\vec{r}_s$ and $\vec{p}_s$ are the coordinate and momentum of the s-th particle, $\vec{v}_s = d\vec{r}_s/dt$ is the velocity, and e_s is the charge of the s-th particle. The electric field and magnetic induction in Lorentz force Eq. (2.1) are determined themselves by the current and charge densities of other particles. Equation (2.1) must be solved simultaneously with field equations (1.4), in which the induced charge and current densities are

$$\rho = \sum_s e_s \delta\left(\vec{r} - \vec{r}_s\right), \quad \vec{j} = \sum_s e_s \vec{v}_s \delta\left(\vec{r} - \vec{r}_s\right).$$

The summation extends over all particles of plasma. Thus the complete problem is a self-consistent one.

It is obvious that the solution of a system of infinite number of equations (2.1) together with field equations (1.4) is practically impossible. Therefore, it is advisable to use, as it was done in kinetic theory of gases, the distribution function of all particles in plasma. This distribution is defined as the probability density for a particle with the definite momentum p_s at a given time t to be at the given space point $\vec{r}_s$. Such a function depends on an extremely large number of variables and, therefore, shows no advantage in comparison with the system of Eq. (2.1). But by this way, it is easier to obtain the appropriate kinetic description of plasma.

According to the above definition, plasma is regarded as a gaseous system containing a large number of charged particles. But, we cannot consider any arbitrary system of particles as a gas. A set of charged particles constitutes a gas or, in other words, the gas approximation will be valid if the average energy of the particle's interaction is smaller than their average kinetic energy. The average energy of charged particles interaction in plasma is equal to $e^2N^{1/3}$, where e is the particle charge and N is the number of particles per unit volume (or density of particles) and $N^{-1/3}$ is the average distance between the particles in plasma. The average kinetic energy is of the order of κT, where T is the temperature and κ is Boltzmann constant.[2] Thus, for validity of the gas approximation, it is necessary that

[2]This concerns to non-degenerate plasma. For degenerate gaseous plasma of electrons, the average kinetic energy is equal to the Fermi energy (see Sect. 4.1).

$$e^2 N^{1/3} \ll \kappa T. \tag{2.2}$$

This inequality is usually satisfied for all real plasmas.

Under condition (2.2), plasma particles are almost free and interact weakly with each other. In this case, one can talk about the independent motion of individual particles. Therefore, the probability distribution of momenta and coordinates of each independent particle can be described by its own single particle distribution function of coordinates and momenta $f_\alpha\left(\vec{p}, \vec{r}, t\right)$, which determines the probability of finding the particle of the type α at time t and at the space point $\vec{r}$ with the momentum $\vec{p}$.

From conservation of probability, it follows that[3]

$$\frac{df_\alpha}{dt} = \frac{\partial f_\alpha}{\partial t} + \frac{d\vec{r}}{dt} \cdot \frac{\partial f_\alpha}{\partial \vec{r}} + \frac{d\vec{p}}{dt} \cdot \frac{\partial f_\alpha}{\partial \vec{p}} = 0,$$

where $\vec{v} = d\vec{r}/dt$ is the particle velocity, and $d\vec{p}/dt$ for a charged particle coincides with Lorentz force (1.3). Thus, the equation of the distribution function has the form

$$\frac{\partial f_\alpha}{\partial t} + \vec{v} \cdot \frac{\partial f_\alpha}{\partial \vec{r}} + e_\alpha \left\{ \vec{E} + \frac{1}{c} \left[\vec{v} \times \vec{B} \right] \right\} \cdot \frac{\partial f_\alpha}{\partial \vec{p}} = 0. \tag{2.3}$$

Such an equation must be satisfied for each type of particles in plasma. The charge and current densities ρ and $\vec{j}$ are determined by the distribution function as below:

$$\rho = \sum_\alpha e_\alpha \int f_\alpha d\vec{p}, \quad \vec{j} = \sum_\alpha \rho_\alpha \int \vec{v} f_\alpha d\vec{p}, \tag{2.4}$$

where the summation extends over all types of charged particles in plasma. Field equations contain the electric field $\vec{E}$ and the magnetic induction $\vec{B}$. In addition, charge and current densities themselves in the field equations are determined by f_α, which is the solution of Eq. (2.3). Therefore, the self-consistent motion of charged particles takes place. This is the reason why Eq. (2.3) by taking into account the field equations

[3] About the continuity equation in the phase space, see [10].

$$\nabla \cdot \vec{E} = 4\pi(\rho + \rho_0), \quad \nabla \times \vec{E} = -\frac{1}{c}\frac{\partial \vec{B}}{\partial t},$$

$$\nabla \times \vec{B} = \frac{1}{c}\frac{\partial \vec{E}}{\partial t} + \frac{4\pi}{c}\left(\vec{j} + \vec{j}_0\right), \quad \nabla \cdot \vec{B} = 0,$$

are called the kinetic equations with self-consistent fields or Vlasov equation (ρ_0 and $\vec{j}_0$ are the charge and current densities of external sources). The advantage of these equations for the description of plasma properties was first demonstrated by Vlasov [11]. In present times, it is applied for studying collisionless plasma.

It must be noted that Eq. (2.3) is an approximation, because it is valid only in collisionless limit. Actually, if we take into account the correlations in particles' motions due to their collisions, then a non-zero term will be derived in the right-hand side of Eq. (2.3), which we denote it as

$$\left(\frac{\partial f}{\partial t}\right)_c .$$

The explicit form of collision integral $(\partial f/\partial t)_c$ depends on the concrete conditions of plasma. Some of the expressions of this integral will be discussed in Sect. 2.7.[4]

Below, kinetic equation with self-consistent fields without any collision integral will be used. This means that there is no collision in plasma. However, in some cases, we consider the effects resulted from collisions.

In conclusion, let us consider the energy conservation law for collisionless plasma. Denoting the kinetic energy of a particle by $\epsilon\left(\vec{p}\right)$ and considering that $\vec{v} = \partial \epsilon_\alpha / \partial \vec{p}$, from Eq. (2.3), we obtain[5]

$$\frac{\partial}{\partial t}\sum_\alpha \int d\vec{p} f_\alpha \epsilon_\alpha + \nabla \cdot \sum_\alpha \int d\vec{p}\, \vec{v} f_\alpha \epsilon_\alpha = \vec{j} \cdot \vec{E}. \tag{2.5}$$

Here, $\Sigma_\alpha \int d\vec{p} f_\alpha \epsilon_\alpha$ is the density of the kinetic energy of particles, $\Sigma_\alpha \int d\vec{p}\, \vec{v} f_\alpha \epsilon_\alpha$ is the density of kinetic energy flux, and $\vec{j} \cdot \vec{E}$ is the work done per unit time by the electric field $\vec{E}$ interacting with the current densities of plasma particles.

On the other hand, from the field equations, it follows that

[4]For relatively slow processes, in the absence of constant magnetic fields, the collision integral in plasma was first obtained by Landau, and was generalized for magnetized plasma in [12, 13]. For fast varying processes, it was obtained in [9].

[5]Equation (2.5) is also valid in collisional plasma if one considers only elastic collisions in plasma because in the presence of such particle collisions, the energy conservation law should be satisfied, $\sum_\alpha \int d\vec{p} \epsilon_\alpha (\partial f_\alpha / \partial t)_c = 0$.

$$\frac{\partial}{\partial t}\frac{1}{8\pi}\left(\vec{E}^2+\vec{B}^2\right)+\nabla\cdot\frac{c}{4\pi}\left[\vec{E}\times\vec{B}\right]=-\vec{j}\cdot\vec{E}-\vec{j}_0\cdot\vec{E}\cdot \tag{2.6}$$

By taking the sum of Eqs. (2.5) and (2.6), we obtain the equation, which represents the energy conservation law in plasma

$$\frac{\partial}{\partial t}\left\{\frac{1}{8\pi}\left(\vec{E}^2+\vec{B}^2\right)+\sum_\alpha\int d\vec{p}\epsilon_\alpha f_\alpha\right\}+\nabla\cdot\left\{\frac{c}{4\pi}\left[\vec{E}\times\vec{B}\right]+\sum_\alpha\int d\vec{p}\,\vec{v}\epsilon_\alpha f_\alpha\right\}$$
$$=-\vec{j}_0\cdot\vec{E}, \tag{2.7}$$

or

$$\frac{\partial}{\partial t}\int_V d\vec{r}\left\{\frac{1}{8\pi}\left(\vec{E}^2+\vec{B}^2\right)+\sum_\alpha\int d\vec{p}\epsilon_\alpha f_\alpha\right\}$$
$$=-\oint_s d\vec{\Sigma}\left\{\frac{c}{4\pi}\left[\vec{E}\times\vec{B}\right]+\sum_\alpha\int d\vec{p}\,\vec{v}f_\alpha\epsilon_\alpha\right\}-\int_V d\vec{r}\,\vec{j}_0\cdot\vec{E}. \tag{2.8}$$

Thus, the rate of change in energy in a volume V of plasma is equal to the total energy flux through the surrounding surface of this volume and the work done by the field $\vec{E}$ interacting with the current density $\vec{j}_0$ of external sources. But, according to relation (1.8), we have

$$\vec{j}=\frac{1}{4\pi}\left(\frac{\partial\vec{D}'}{\partial t}-\frac{\partial\vec{E}}{\partial t}\right).$$

Therefore, relation (2.6) can be written as

$$\frac{1}{4\pi}\left(\vec{B}\cdot\frac{\partial\vec{B}}{\partial t}+\vec{E}\cdot\frac{\partial\vec{D}'}{\partial t}\right)+\nabla\cdot\frac{c}{4\pi}\left[\vec{E}\times\vec{B}\right]=-\vec{j}_0\cdot\vec{E}. \tag{2.9}$$

Comparison of relations (2.8) and (2.9) shows that for the unbounded medium when the field and energy flux vanish at infinity, the following equality is valid:

$$\frac{\partial}{\partial t}\int d\vec{r}\left[\frac{1}{8\pi}\left(\vec{E}^2+\vec{B}^2\right)+\sum_\alpha\int d\vec{p}\epsilon_\alpha f_\alpha\right]$$
$$=\int d\vec{r}\,\frac{1}{4\pi}\left(\vec{B}\cdot\frac{\partial\vec{B}}{\partial t}+\vec{E}\cdot\frac{\partial\vec{D}'}{\partial t}\right). \tag{2.10}$$

Thus, the right-hand side of this relation determines the rate of change in energy in the unbounded medium, as it was shown in Chap. 1 [see Eq. (1.74)].

2.2 Dielectric Permittivity of Collisionless Isotropic Plasma

Let us consider homogeneous unbounded isotropic and electrically neutral plasma. Actually, for isotropic plasma, there is no constant field (electric and magnetic fields), which could result in the anisotropic properties. In addition, we assume that there is no regular motion in plasma, which could lead to the appearance of currents and non-uniform charge distributions. Then, in such a basic equilibrium state of plasma, the distribution functions of particles can be considered homogeneous in space, i.e., they are independent of spatial coordinates. Moreover, for isotropic plasma, the equilibrium distribution functions of particles can be functions of only the absolute value of momentum $|\vec{p}|=p$, i.e., $f_{0\alpha}=f_{0\alpha}(p)$.

In linear electrodynamics, the plasma state with induced charges and currents can be considered weakly different from the basic equilibrium state. In this case, one must consider a small perturbation of the equilibrium distribution function $f_{0\alpha}(p)$. Such deviation of the basic equilibrium state occurs due to the small electric and magnetic fields $\vec{E}\left(\vec{r},t\right)$ and $\vec{B}\left(\vec{r},t\right)$, which in turn are caused by the perturbation of the equilibrium state. Let us write the perturbed distribution function as

$$f_\alpha\left(\vec{p},\vec{r},t\right)=f_{0\alpha}\left(\vec{p}\right)+\delta f_\alpha\left(\vec{p},\vec{r},t\right). \tag{2.11}$$

Here, $f_{0\alpha}\left(\vec{p}\right)$ is the distribution function of the particles of the type α in the equilibrium state of plasma and $\delta f_\alpha\left(\vec{p},\vec{r},t\right)$ is the corresponding small perturbation determined by the induced currents and charges.

For sufficiently dilute plasma, when particle collisions are negligible, the distribution function $f_\alpha\left(\vec{p},\vec{r},t\right)$ satisfies Vlasov equation (2.3). In the absence of external fields, this equation results in the following equation for δf_α:

$$\frac{\partial \delta f_\alpha}{\partial t} + \vec{v} \cdot \frac{\partial \delta f_\alpha}{\partial \vec{r}} + e_\alpha \vec{E} \cdot \frac{\partial f_{0\alpha}}{\partial \vec{p}} = 0. \tag{2.12}$$

The self-consistent electric field $\vec{E}$ in this equation is determined by the induced currents and charges, determined by δf_α, through solving the Maxwell's equations (1.1).

In this section, we will try to solve Eq. (2.12) to obtain the complex tensor of dielectric permittivity. For this aim, it is sufficient to express the induced current $\vec{j}$ in plasma as a linear function of the electric field $\vec{E}$.

The general solution of Eq. (2.12) with initial value $\delta f_\alpha\left(\vec{p}, \vec{r}, t_0\right)$ at $t = t_0$ can be written in the following form:

$$\begin{aligned} \delta f_\alpha\left(\vec{p}, \vec{r}, t\right) &= \delta f_\alpha\left(\vec{p}, \vec{r} - \vec{v}(t - t_0), t_0\right) \\ &\quad - e_\alpha \frac{\partial f_{o\alpha}}{\partial \vec{p}} . \int_{t_0}^{t} dt' \vec{E}\left(\vec{r} - \vec{v}(t - t'), t'\right). \end{aligned} \tag{2.13}$$

If $\delta f_\alpha\left(\vec{p}, \vec{r}, t_0\right) \to 0$ when $t_0 \to -\infty$, then this solution can be presented as[6]

$$\delta f_\alpha\left(\vec{p}, \vec{r}, t\right) = -e_\alpha \frac{\partial f_{o\alpha}}{\partial \vec{p}} . \int_{-\infty}^{t} dt' \vec{E}(r - \vec{v}(t - t'), t'). \tag{2.14}$$

This expression allows us to represent the induced current density in plasma as

$$\begin{aligned} j_i\left(\vec{r}, t\right) &= \sum_\alpha e_\alpha \int d\vec{p} v_i \delta f_\alpha \\ &= -\sum_\alpha e_\alpha^2 \int_{-\infty}^{t} dt' \int d\vec{p} \frac{\partial f_{0\alpha}}{\partial p_j} v_i E_j\left(\vec{r} - \vec{v}(t - t'), t'\right). \end{aligned} \tag{2.15}$$

This formula can be easily transformed into

$$j_i\left(\vec{r}, t\right) = \int_{-\infty}^{t} dt' \int d\vec{r}' \hat{\sigma}_{ij}\left(\vec{r} - \vec{r}', t - t'\right) E_j\left(\vec{r}', t'\right),$$

where

[6]Then, one can talk about the adiabatic switching of the field in the infinite past.

$$\hat{\sigma}_{ij}\left(\vec{r},t\right) = -\sum_{\alpha} e_{\alpha}^{2} \int d\vec{p} v_i \frac{\partial f_0}{\partial p_j} \delta\left(\vec{r} - \vec{v}t\right). \tag{2.16}$$

Using Eq. (2.15) and considering the definition of the electric induction $\overrightarrow{D'}$ [see Eq. (1.8)], we obtain

$$D_i'\left(\vec{r},t\right) = E_i\left(\vec{r},t\right) - \sum_{\alpha} 4\pi e_{\alpha}^{2} \int_{-\infty}^{t} dt' \int d\vec{r'} E_j\left(\vec{r'},t'\right) \times \int d\vec{p} v_i \frac{\partial f_{o\alpha}}{\partial p_j} \int_0^{t-t'} d\tau \delta\left(\vec{r} - \vec{r'} - \vec{v}\tau\right). \tag{2.17}$$

This relation looks similar to material equation (1.12). Actually, it can be written as

$$D_i'\left(\vec{r},t\right) = \int_{-\infty}^{t} dt' \int d\vec{r'} \hat{\varepsilon}_{ij}\left(\vec{r} - \vec{r'}, t - t'\right) E_j\left(\vec{r'},t'\right),$$

where [see Eq. (1.40)]

$$\hat{\varepsilon}_{ij}\left(\vec{r},t\right) = \delta_{ij}\delta(t)\delta\left(\vec{r}\right) - 4\pi \sum_{\alpha} e_{\alpha}^{2} \int d\vec{p} v_i \frac{\partial f_{0\alpha}}{\partial p_j} \int_0^{t} d\tau \delta\left(\vec{r} - \vec{v}\tau\right). \tag{2.18}$$

Relations (2.16) and (2.18) enable us to obtain the complex tensors of conductivity and permittivity of plasma. Thus, from Eq. (2.16) by making use of definition (1.44), one can obtain

$$\begin{aligned} \sigma_{ij}\left(\omega, \vec{k}\right) &= -2\pi \sum_{\alpha} e_{\alpha}^{2} \int d\vec{p} v_i \frac{\partial f_{o\alpha}}{\partial p_j} \delta_{+}\left(\omega - \vec{k}\cdot\vec{v}\right) \\ &= -\imath \sum_{\alpha} e_{\alpha}^{2} \int d\vec{p} \frac{\partial f_{0\alpha}}{\partial p_j} v_i \left[\mathcal{P} \frac{1}{\omega - \vec{k}\cdot\vec{v}} - \imath\pi\delta\left(\omega - \vec{k}\cdot\vec{v}\right)\right], \end{aligned} \tag{2.19}$$

where

$$\delta_{+}(z) = \frac{1}{2\pi} \int_0^{\infty} dt \exp\left(\imath z t\right) = \frac{1}{2}\delta(z) + \frac{\imath}{2\pi}\mathcal{P}\frac{1}{z},$$

and $\mathcal{P}$ denotes the prescription that the principal value is to be taken at the singularity at $r = 0$. Note that the singular δ_{+} function can be considered as

$$\frac{\imath}{2\pi}\lim_{\Delta\to+0}\frac{1}{z+\imath\Delta}.$$

Therefore,

$$\int d\omega' F(\omega')\left[\mathcal{P}\frac{1}{\omega-\omega'}-\imath\pi\delta(\omega-\omega')\right]=\lim_{\Delta\to+0}\int d\omega'\frac{F(\omega')}{\omega-\omega'+\imath\Delta}.$$

If the function $F(\omega')$ does not have any singularity on the real axis ω', then

$$\lim_{\Delta\to+0}\int d\omega'\frac{F(\omega')}{\omega-\omega'+\imath\Delta}=\lim\int_{C(\Delta)}d\omega'\frac{F(\omega')}{\omega-\omega'},$$

where $C(\Delta)$, the contour in the plane of complex variable ω', is shifted down by an infinitely small distance from the real axes. As a result, relation (2.19) can be rewritten in the form of

$$\sigma_{ij}\left(\omega,\vec{k}\right)=-\imath\sum_{\alpha}e_{\alpha}^{2}\int_{C}d\vec{p}\,v_{i}\frac{\partial f_{0\alpha}}{\partial p_{j}}\frac{1}{\omega-\vec{k}\cdot\vec{v}}. \tag{2.20}$$

This means that one must bypass the singularity at $\vec{k}\cdot\vec{v}=\omega$ by integrating not over the real axes, but over contour C, as it is shown in Fig. 2.1.

From expression (2.18) or by using relation (1.46), we find the complex tensor of dielectric permittivity

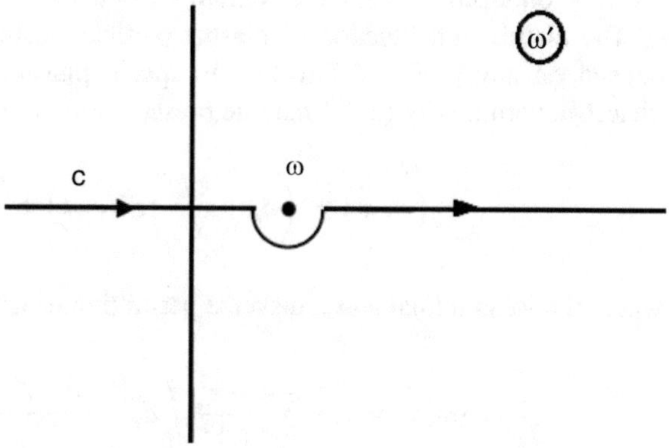

Fig. 2.1 Landau prescription

$$\varepsilon_{ij}\left(\omega,\vec{k}\right) = \delta_{ij} + 8\pi^2\delta_+(\omega)\sigma_{ij}\left(\omega,\vec{k}\right)$$
$$= \delta_{ij} + 4\pi\left(\frac{1}{\omega} - \imath\pi\delta(\omega)\right)\sum_\alpha e_\alpha^2 \int d\vec{p}\, v_i \frac{\partial f_0}{\partial p_j}\left[\frac{1}{\omega - \vec{k}\cdot\vec{v}} - \imath\pi\delta\left(\omega - \vec{k}\cdot\vec{v}\right)\right].$$

This relation may be presented in a much simpler way if we suppose that ω has an infinitely small positive imaginary part Δ:

$$\varepsilon_{ij}\left(\omega,\vec{k}\right) = \delta_{ij} + \sum_\alpha \frac{4\pi e_\alpha^2}{\omega} \int d\vec{p}\, v_i \frac{\partial f_{0\alpha}}{\partial p_j} \frac{1}{\omega - \vec{k}\cdot\vec{v}}. \tag{2.21}$$

Above, we regarded the dielectric permittivity as a complex function of real ω and $\vec{k}$. However, Eq. (2.21) indicates that this function can be taken as the limit of the function of the complex variable ω. Since in relation (2.21) ω has a small positive imaginary part, this limit is understood when the complex variable ω tends to the real axis from the upper half-plane. The integrals defining the complex tensors of conductivity (2.20) and dielectric permittivity (2.21) are the Cauchy-type integrals [14]. Such integrals determine the functions, which are analytical everywhere with the exception of the points of the integration contour. Specially, the functions defined by the Cauchy integrals undergo a jump when the integration contour is intersected. Therefore, to determine the value of such an integral at the points of the integration contour, it is necessary to take the limit from both sides of the contour.

Below, we will use expression (2.21) for the complex tensor of dielectric permittivity $\varepsilon_{ij}\left(\omega,\vec{k}\right)$ for real ω in the limit of $\Delta \to +0$. Specially, in explicit calculations, it can be obtained by using the contour C, as it was done in relation (2.20).

The distribution function of plasma particles in the equilibrium state does not depend on any preferred direction in space (plasma is isotropic) and, therefore, dielectric permittivity (2.21) may be presented in the form [see Eq. (1.24)][7]:

$$\varepsilon_{ij}\left(\omega,\vec{k}\right) = \left(\delta_{ij} - \frac{k_ik_j}{k^2}\right)\varepsilon^{\mathrm{tr}}(\omega,k) + \frac{k_ik_j}{k^2}\varepsilon^l(\omega,k),$$

where the longitudinal and transverse permittivities are defined as below:

$$\varepsilon^l(\omega,k) = 1 + \sum_\alpha \frac{4\pi e_\alpha^2}{\omega k^2}\int_C d\vec{p}\, \frac{\vec{k}\cdot\vec{v}}{\omega - \vec{k}\cdot\vec{v}}\left(\vec{k}\cdot\frac{\partial f_{0\alpha}}{\partial \vec{p}}\right), \tag{2.22}$$

[7] It must be noted that the longitudinal and transverse permittivities of electron plasma were firstly introduced in [15].

$$\varepsilon^{\mathrm{tr}}(\omega, k) = 1 + \sum_{\alpha} \frac{2\pi e_{\alpha}^{2}}{\omega} \int_{c} d\vec{p} \frac{1}{\omega - \vec{k} \cdot \vec{v}} \times \left[\left(\vec{v} \cdot \frac{\partial f_{0\alpha}}{\partial \vec{p}} \right) - \frac{\vec{k} \cdot \vec{v}}{k^{2}} \left(\vec{k} \cdot \frac{\partial f_{0\alpha}}{\partial \vec{p}} \right) \right]. \tag{2.23}$$

By denoting $f'_{0\alpha} \equiv \partial f_{0\alpha}/\partial \epsilon_{\alpha}$ and remembering that $\vec{v} = \partial \epsilon_{\alpha}/\partial \vec{p}$, these expressions may be represented in the form

$$\varepsilon^{l}(\omega, k) = 1 + \sum_{\alpha} \frac{4\pi e_{\alpha}^{2}}{\omega k^{2}} \int_{C} d\vec{p} \frac{\left(\vec{k} \cdot \vec{v} \right)^{2}}{\omega - \vec{k} \cdot \vec{v}} f'_{0\alpha}, \tag{2.24}$$

$$\varepsilon^{\mathrm{tr}}(\omega, k) = 1 + \sum_{\alpha} \frac{2\pi e_{\alpha}^{2}}{\omega k^{2}} \int_{C} d\vec{p} \frac{\left(\vec{k} \times \vec{v} \right)^{2}}{\omega - \vec{k} \cdot \vec{v}} f'_{0\alpha}. \tag{2.25}$$

In the long-wave limit ($\vec{k} \to 0$), from Eqs. (2.24) and (2.25), it follows that

$$\varepsilon^{l}(\omega, 0) = \varepsilon^{\mathrm{tr}}(\omega, 0) = \varepsilon(\omega) = 1 + \sum_{\alpha} \frac{4\pi e_{\alpha}^{2}}{3\omega^{2}} \int d\vec{p} v^{2} f'_{0\alpha}. \tag{2.26}$$

This expression defines the dielectric permittivity by taking into account only frequency dispersion.

In the low-frequency (quasi-static) limit $\omega \ll \vec{k} \cdot \vec{v}$, from Eq. (2.24), we obtain

$$\varepsilon^{l}(0, k) = 1 - \sum \frac{4\pi e_{\alpha}^{2}}{k^{2}} \int d\vec{p} f'_{0\alpha}. \tag{2.27}$$

Substituting this expression into relation (1.56), we find out that the electric field of a charge in plasma will be screened at distance r_{scr}, which is equal to

$$r_{\mathrm{scr}}^{-2} = -\sum_{\alpha} 4\pi e_{\alpha}^{2} \int d\vec{p} f'_{0\alpha}. \tag{2.28}$$

In the thermodynamic equilibrium state, the distribution function $f_{0\alpha}$ depends only on the difference of the particle energy and chemical potential μ_{α}. Therefore,[8]

[8]The temperature is assumed to be constant in this relation.

$$f'_{0\alpha} = -\left(\frac{\partial f_{0\alpha}}{\partial \mu_\alpha}\right)_T.$$

If we consider the normalization condition

$$\int d\vec{p} f_{0\alpha} = N_\alpha, \tag{2.29}$$

where N_α is the number of particles of the type α per unit volume, then relation (2.28) reads

$$r_{\rm scr}^{-2} = \sum_\alpha 4\pi e_\alpha^2 \left(\frac{\partial N_\alpha}{\partial \mu_\alpha}\right)_T. \tag{2.30}$$

This formula corresponds to the expression obtained for the shielding radius of electrostatic fields in the Debye–Huckel's theory of heavy electrolytes.

Now, let us consider the transverse dielectric permittivity in the low-frequency limit, when $\omega \ll \vec{k} \cdot \vec{v}$. From Eq. (2.25) it follows that

$$\varepsilon^{\rm tr}_{\omega\to 0}(\omega, k) = 1 - \imath \sum_\alpha \frac{2\pi^2 e_\alpha^2}{\omega} \int d\vec{p}\, v^2 \delta\left(\vec{k}\cdot\vec{v}\right) f'_{0\alpha}. \tag{2.31}$$

Comparing this expression to formula (1.60), we find plasma conductivity in the low-frequency limit

$$\sigma^{\rm tr}_{\omega\to 0}(\omega, k) = \frac{G}{|k|}, \tag{2.32}$$

where

$$G = -\sum_\alpha \frac{\pi e_\alpha^2}{2} \int d\vec{p}\, v^2 \delta\left(\frac{\vec{k}\cdot\vec{v}}{k}\right) f'_{0\alpha}.$$

The existence of such conductivity of plasma and, more generally, the imaginary parts of the transverse and longitudinal permittivities are consequences of appearance of $\delta\left(\omega - \vec{k}\cdot\vec{v}\right)$ functions in the corresponding integrands. It has a relatively simple physical interpretation. Namely, it means that only particles satisfying the condition of $\omega = \vec{k}\cdot\vec{v}$ are responsible for the imaginary part of the dielectric permittivity and consequently for the electromagnetic wave absorption in plasma. However, this condition is nothing but the stipulation of Cherenkov radiation by a moving charged particle (see Sect. 1.8). The same condition governs the inverse process, i.e., the Cherenkov wave absorption by particles, which leads to the dissipation in the medium. Namely, such Cherenkov dissipation can occur in the absence of particle collisions in plasma.

Let us consider the absorption of weakly damped electromagnetic waves in collisionless plasma in more details. In order to describe this process, it is necessary to know the imaginary parts of the longitudinal and transverse dielectric permittivities (see Sect. 1.4). They can be written easily by making use of expressions (2.24) and (2.25)

$$\varepsilon^{l''}(\omega,k) = -\sum_{\alpha} \frac{4\pi^2 e_{\alpha}^2 \omega}{k^2} \int d\vec{p}\, \delta\left(\vec{k}\cdot\vec{v} - \omega\right) f'_{0\alpha}, \tag{2.33}$$

$$\varepsilon^{\text{tr}''}(\omega,k) = -\sum_{\alpha} \frac{2\pi^2 e_{\alpha}^2}{\omega k^2} \int d\vec{p}\, \delta\left(\vec{k}\cdot\vec{v} - \omega\right) \left[\vec{k}\times\vec{v}\right]^2 f'_{0\alpha}. \tag{2.34}$$

Here, ω and k are the real quantities. Using these expressions with the help of relation (1.80), we can find the amount of heat released per unit volume of plasma

$$\frac{Q}{V} = \frac{\omega}{2\pi}\left\{\varepsilon^{l''}(\omega,k)\left|\vec{E}^{l}\right|^2 + \varepsilon^{\text{tr}''}(\omega,k)\left|\vec{E}^{\text{tr}}\right|^2\right\}.$$

When the absorption is weak, one can introduce the average energy of the electromagnetic field in plasma. Using relation (1.87), we can write the following expressions for the energy of longitudinal and transverse waves, respectively,

$$\frac{U^l}{V} = \frac{1}{4\pi}\frac{\partial}{\partial\omega}\left[\omega\varepsilon^{l'}(\omega,k)\right]\left|\vec{E}^{l}\right|^2,$$

$$\frac{U^{\text{tr}}}{V} = \frac{1}{4\pi}\left\{\frac{c^2k^2}{\omega^2} + \frac{\partial}{\partial\omega}\left[\omega\varepsilon^{\text{tr}'}(\omega,k)\right]\right\}\left|\vec{E}^{\text{tr}}\right|^2.$$

Then, in the case of weak absorption, from Eqs. (1.100) and (1.101), we obtain the following conditions of longitudinal and transverse waves propagation in plasma:[9]

$$\varepsilon^{l'}(\omega,k) = 0, \qquad k^2 - \frac{\omega^2}{c^2}\varepsilon^{\text{tr}'}(\omega,k) = 0.$$

At last, considering the above relations, we can find the wave damping decrements as the ratio of the half of wave energy losses to the wave energy

[9]The real parts of the longitudinal and transverse dielectric permittivities are defined by the relations, which coincide with Eqs. (2.24) and (2.25) but only the principal value in the integrals has to be taken.

$$\gamma^{l} = \frac{Q^{l}}{2U^{l}} = \frac{\varepsilon^{l''}(\omega, k)}{\frac{\partial}{\partial\omega}\left[\omega\varepsilon^{l'}(\omega, k)\right]}, \tag{2.35}$$

$$\gamma^{\mathrm{tr}} = \frac{Q^{\mathrm{tr}}}{2U^{\mathrm{tr}}} = \frac{\omega^{2}\varepsilon^{\mathrm{tr}''}(\omega, k)}{\frac{\partial}{\partial\omega}\left[\omega^{2}\varepsilon^{\mathrm{tr}'}(\omega, k)\right]}. \tag{2.36}$$

Similar results can be obtained by taking into account the quantum consideration of wave absorption in plasma [16–20]. This way is more obvious for treating the electromagnetic wave dissipation in collisionless plasma rather than Cherenkov absorption. The fundamental of this approach is the Einstein's coefficient formalism [8, 21, 22].

The number of photons absorbed in plasma in transition from state $1 \to 2$ is equal to

$$Z_{\mathrm{ab}} = B_{12} N_1 U_{\vec{k}},$$

where B_{12} is the absorption coefficient, N_1 is the number of particles in state 1, and $U_{\vec{k}}$ is the radiation energy density per wave vector interval $d\vec{k}$. The number of photons emitted in transition from state $2 \to 1$ is equal to

$$Z_{\mathrm{em}} = \left(A_{21} + B_{21} U_{\vec{k}}\right) N_2,$$

where A_{21} is the probability (coefficient) of spontaneous emission per unit time and $B_{21} U_{\vec{k}} N_2$ is the number of photons radiated by stimulated emission. In our case of emission and absorption of photons by free particles of ionized plasma, the statistical weights of states 1 and 2 are equal. Therefore, in the equilibrium state when $Z_{\mathrm{em}} = Z_{\mathrm{ab}}$ and $N_2/N_1 = \exp(-\hbar\omega/\kappa T)$ (this takes place for the Boltzmann distribution and ω is the frequency of emitted or absorbed photons), we obtain

$$U_{\vec{k}} = \frac{A_{21}}{B_{12}\exp(\hbar\omega/\kappa T) - B_{21}},$$

for the density of radiation energy. From this fact that this expression should coincide with the Planck distribution, regardless of the system with which the field is in equilibrium, the well-known Einstein's relations are obtained

$$B_{12} = B_{21}, \qquad A_{21} = B_{12}\frac{\hbar\omega}{(2\pi)^3} g,$$

where g is the statistical weight. Here, $g = 1$ for the longitudinal photons and $g = 2$ for the transverse ones. The effective number of the absorbed quanta $Z_{\mathrm{ab\ eff}}$ of the non-equilibrium radiation in the system is equal to

$$Z_{\text{ab eff}} = B_{12} N_1 U_{\vec{k}} - B_{21} U_{\vec{k}} N_2 = (N_1 - N_2) A_{21} U_{\vec{k}} \frac{(2\pi)^3}{g\hbar\omega}.$$

If we consider the limiting case $\hbar\omega \ll \kappa T$ and note that N depends only on particle's energy $\epsilon\left(\vec{p}\right)$, then we obtain

$$N_2 \simeq N_1 + \hbar\omega \frac{\partial N_1}{\partial \epsilon_1}.$$

Therefore,

$$Z_{\text{ab eff}} = -(2\pi)^3 A_{21} \frac{\partial N_1}{\partial \epsilon_1} \frac{U_{\vec{k}}}{g}. \tag{2.37}$$

Absorption is dominant if $N_1 > N_2$ (or $\partial N_1 / \partial \epsilon_1 < 0$). Otherwise, $Z_{\text{ab eff}} < 0$, which corresponds to the case in which the stimulated emission is dominant.

One can obtain the total number of the quanta absorbed in the system by summation of Eq. (2.37) over all states from which the particles during quantum absorption can transit. This leads to the expression

$$\langle Z_{\text{ab eff}} \rangle = -(2\pi)^3 \frac{U_{\vec{k}}}{g} \sum_{\alpha} \int d\vec{p}_1 A_{21} \frac{\partial f_{0\alpha}}{\partial \epsilon_1},$$

from which we can represent the electromagnetic wave damping decrement

$$\gamma = \frac{\hbar\omega \langle Z_{\text{ab eff}} \rangle}{2U_{\vec{k}}} = -(2\pi)^3 \frac{\hbar\omega}{2g} \sum_{\alpha} \int d\vec{p}_1 A_{21} \frac{\partial f_{0\alpha}}{\partial \epsilon_1}. \tag{2.38}$$

In the absorption processes, the conservation laws must be taken into account

$$\vec{p}_2 = \vec{p}_1 + \hbar\vec{k}, \qquad \epsilon\left(\vec{p}_2\right) = \epsilon\left(\vec{p}_1\right) + \hbar\omega.$$

In the classical limit, when $\vec{p} \gg \hbar\vec{k}$, it follows that $\omega = \vec{k} \cdot \vec{v}_1$, which coincides with the condition of Cherenkov radiation [see Eq. (1.184)].

From the relations of Sect. 1.8, one can immediately obtain the expression for the probability of spontaneous emission A_{21}. Actually, in accordance to Eq. (1.175), the probability of emission of a longitudinal quantum with the wave vector $\vec{k}$ and frequency $\omega = \vec{k} \cdot \vec{v}_1$ in the interval $d\vec{k}$ by a particle with the charge e_α and velocity $\vec{v}_1$ per unit time is equal to

$$A_{21}^{l} = \frac{e_{\alpha}^{2}}{2\pi\hbar k^{2}} \delta\left[\varepsilon^{l}\left(\vec{k}\cdot\vec{v}_{1}, k\right)\right]. \tag{2.39}$$

In a quite similar way, we obtain the emission probability of a transverse quantum from Eq. (1.179)

$$A_{21}^{\text{tr}} = \frac{e_{\alpha}^{2}}{2\pi\hbar k^{2}} \left[\vec{k}\times\vec{v}_{1}\right]^{2} \delta\left[\left(\vec{k}\cdot\vec{v}_{1}\right)^{2} \varepsilon^{\text{tr}}\left(\vec{k}\cdot\vec{v}_{1}, k\right) - k^{2}c^{2}\right]. \tag{2.40}$$

Substituting expression (2.39) into formula (2.38), we find

$$\gamma^{l} = \frac{-1}{\frac{\partial}{\partial\omega}\varepsilon^{l}(\omega, k)} \sum_{\alpha} \frac{4\pi^{2}e_{\alpha}^{2}\omega}{k^{2}} \int d\vec{p} f'_{0\alpha} \delta\left(\omega - \vec{k}\cdot\vec{v}\right), \tag{2.41}$$

where ω is determined by the condition of $\varepsilon^{l}(\omega, k) = 0$ ($\omega > 0$). Similarly, from formula (2.40), it follows that

$$\gamma^{\text{tr}} = \frac{-1}{\frac{\partial}{\partial\omega}\left[\omega^{2}\varepsilon^{\text{tr}}(\omega, k)\right]} \sum_{\alpha} \frac{2\pi^{2}e_{\alpha}^{2}\omega}{k^{2}} \int d\vec{p} f'_{0\alpha} \left[\vec{k}\times\vec{v}\right]^{2} \delta\left(\omega - \vec{k}\cdot\vec{v}\right), \tag{2.42}$$

where ω satisfies the condition of $\omega^{2}\varepsilon^{\text{tr}}(\omega, k) - k^{2}c^{2} = 0$. It can be easily shown that expressions (2.41) and (2.42) can be obtained from Eqs. (2.33), (2.35) and (2.34), (2.36), respectively.

2.3 Dielectric Permittivity and Electromagnetic Oscillations of Isotropic Collisionless Non-relativistic Electron Plasma

If one can neglect the ion motion, then plasma resembles a pure electronic medium. Therefore, the ion distribution can be considered homogenous and, as a result, one can neglect the induced currents and spatially inhomogeneous charges stipulated by ions motions. The effect of ions' charges, in this case, appears only in the total charge of plasma, which is zero. Therefore, we can keep only the electron terms in the equations. Of course, the electromagnetic properties of pure electron plasma are simpler than the properties of electron–ion plasma. Therefore, below, we begin our consideration from electron plasma. The actual range of validity of results obtained by neglecting the ions motions can be obtained by taking the ions motions into account.

The theory of electromagnetic oscillations of plasma was developed by many authors.[10] The theory turns out to be very simple in the long-wave limit when the spatial dispersion of the dielectric permittivity can be neglected. In this limit, from Eq. (2.26), it follows that

$$\varepsilon^l(\omega, 0) = \varepsilon^{\mathrm{tr}}(\omega, 0) \equiv \varepsilon(\omega) = 1 - \frac{\omega_0^2}{\omega^2}, \tag{2.43}$$

where

$$\omega_0^2 = -\frac{4\pi e^2}{3} \int d\vec{p} v^2 f'_{0\alpha}, \tag{2.44}$$

From the existence condition of longitudinal oscillations given by Eq. (1.105) ($\varepsilon^l = 0$), we find

$$\omega^2 = \omega_0^2, \tag{2.45}$$

whereas for the transverse oscillations, from Eq. (1.106), it follows that

$$\omega^2 = \omega_0^2 + k^2 c^2. \tag{2.46}$$

Thus, when spatial dispersion is neglected, the frequency of longitudinal oscillations is constant and equals ω_0. This frequency is also known as the plasma frequency. It coincides with the limiting frequency of transverse waves when $k \to 0$.

Now, let us consider the effects arising when spatial dispersion is taken into account. We assume the particle distribution to be Maxwellian with temperature T_e and density N_e:

$$f_{0e}\left(\vec{p}\right) = \frac{N_e}{(2\pi m \kappa T_e)^{3/2}} \exp\left(-p^2/2m\kappa T_e\right). \tag{2.47}$$

Then, integration in Eq. (2.44) leads to

$$\omega_0^2 = \frac{4\pi e^2 N_e}{m} = \omega_{\mathrm{pe}}^2. \tag{2.48}$$

The frequency ω_{pe} is called the electron Langmuir frequency.

---

[10] The first investigations were carried out by Rayleigh [23], Tonks and Langmuir [24, 25]. The investigation of plasma oscillations by using the kinetic equation with self-consistent fields was firstly done by Vlasov [11, 26]. The well-known Landau's significant work dealing with the problem of longitudinal oscillation damping affected the theory of plasma oscillations [27].

In Maxwellian electron plasma, the expressions of longitudinal and transverse dielectric permittivities (2.22) and (2.23) can be represented as

$$\varepsilon^{l}(\omega,k) = 1 - \frac{\omega_{\mathrm{pe}}^2}{\sqrt{2\pi}\omega} \int\limits_C \frac{dx x^2 \exp\left(-x^2/2\right)}{\omega - xk\sqrt{\frac{\kappa T_e}{m}}}, \tag{2.49}$$

$$\varepsilon^{\mathrm{tr}}(\omega,k) = 1 - \frac{\omega_{\mathrm{pe}}^2}{\sqrt{2\pi}\omega} \int\limits_C \frac{dx \exp\left(-x^2/2\right)}{\omega - xk\sqrt{\frac{\kappa T_e}{m}}}. \tag{2.50}$$

By substituting expression (2.47) into formulas (2.22) and (2.23), the integration is performed over the momentum components perpendicular to the wave vector $\vec{k}$. Furthermore, we perform the change of variable $\vec{p} \cdot \vec{k} = xk\sqrt{\frac{\kappa T_e}{m}}(k > 0)$.

In view of expression (2.50), let us begin the analysis of transverse oscillations of plasma. In accordance to Eq. (2.46), the phase velocity of transverse waves exceeds the light velocity and therefore the influence of spatial dispersion is weak. Moreover, the denominator of the integrand in Eq. (2.50), in this case, can become zero only for particles with velocity higher than the light velocity, which does not make sense. Therefore, the imaginary part of the transverse dielectric permittivity appeared in Eq. (2.50) is out of accuracy of this expression. Thus, it must be neglected in the non-relativistic approximation. As a result, it turns out that transverse waves of electron plasma would not be damped in this approximation. This result remains correct in the relativistic theory as well.

The effects of weak spatial dispersion can be obtained by expanding the integrand of Eq. (2.50) in terms of k/ω. Then

$$\varepsilon^{\mathrm{tr}}(\omega,k) \approx 1 - \frac{\omega_{\mathrm{pe}}^2}{\omega^2}\left(1 + \frac{\kappa T_e}{m}\frac{k^2}{\omega^2}\right). \tag{2.51}$$

Substituting this expression into Eq. (1.116), we find the following relation for transverse oscillations of electron plasma:

$$\omega^2 = \omega_{\mathrm{pe}}^2 + k^2\left(c^2 + \frac{\kappa T_e}{m}\right). \tag{2.52}$$

In non-relativistic plasma, when $mc^2 \gg \kappa T_e$, the influence of spatial dispersion is weak.

Now, let us consider the longitudinal oscillations. Its description is relatively simple, in the case of sufficiently long-wave limit, when the spectrum slightly differs from relation (2.45) and the absorption of longitudinal waves is small. One can calculate this correction to relation (2.45) by expanding expression (2.49) in terms of k/ω. Hence,

$$\varepsilon^{l}(\omega,k) = 1 - \frac{\omega_{\mathrm{pe}}^2}{\omega^2}\left(1 + \frac{3\kappa T_e k^2}{m\omega^2}\right). \tag{2.53}$$

Equating this expression to zero, we obtain the following formula for the spectrum of longitudinal oscillations, which was first found out by Vlasov [11]:

$$\omega^2 = \omega_{\mathrm{pe}}^2 + \frac{3\kappa T_e}{m}k^2. \tag{2.54}$$

By making use of the screening radius of the electric field in electron plasma given by Eq. (2.28), which in the case of Maxwellian plasma with the distribution function (2.47) is equal to

$$r_{\mathrm{scr}} = r_D = \sqrt{\frac{\kappa T_e}{4\pi e^2 N_e}} = \sqrt{\frac{\kappa T_e}{m\omega_{\mathrm{pe}}^2}},$$

formula (2.54) can be represented in the following form:

$$\omega^2 = \omega_{\mathrm{pe}}^2\left[1 + 3(kr_D)^2\right].$$

From this relation, one can notice that expressions (2.53) and (2.54) are valid only if $(kr_D)^2 \ll 1$. In the opposite limit, when the wavelength of longitudinal waves is less than the Debye length r_D, these relations will break down. Moreover, in this case, significant damping of longitudinal oscillations occurs in plasma, which we will investigate below.

Therefore, longitudinal oscillations of plasma are possible only for sufficiently long waves. Longitudinal wave absorption in electron plasma can be investigated by using the results of Sect. 2.2. Thus, according to Eqs. (2.33) and (2.47), the imaginary part of the transverse dielectric permittivity of electron plasma is equal to

$$\begin{aligned}\varepsilon^{\mathrm{tr}\prime\prime}(\omega,k) &= \sqrt{\frac{\pi}{2}}\frac{\omega_{\mathrm{pe}}^2\omega}{k^3}\left(\frac{m}{\kappa T_e}\right)^{3/2}\exp\left(-\frac{m\omega^2}{2\kappa T_e k^2}\right)\\ &= \sqrt{\frac{\pi}{2}}\frac{\omega}{\omega_{\mathrm{pe}}}\frac{1}{(kr_{\mathrm{De}})^3}\exp\left[-\frac{1}{2}\frac{\omega^2}{\omega_{\mathrm{pe}}^2}\frac{1}{(kr_{\mathrm{De}})^2}\right].\end{aligned} \tag{2.55}$$

By substituting Eqs. (2.43) and (2.55) into Eq. (2.35), one can obtain an expression for damping decrement of longitudinal waves

$$\gamma^{l} = \sqrt{\frac{\pi}{8}}\frac{\omega^4}{\omega_{\mathrm{pe}}^3}\frac{1}{(kr_D)^3}\exp\left[-\frac{1}{2}\frac{\omega^2}{\omega_{\mathrm{pe}}^2}\frac{1}{(kr_D)^2}\right]. \tag{2.56}$$

Furthermore, considering Eq. (2.54), we find

$$\gamma^l = \sqrt{\frac{\pi}{8}}\frac{\omega_{\mathrm{pe}}}{(kr_D)^3}\exp\left[-\frac{3}{2}-\frac{1}{2(kr_D)^2}\right]. \tag{2.57}$$

This expression was first obtained by Landau [21, 27]. It is obvious from the above formulas that the damping of longitudinal waves (Landau damping) is relatively small only where wavelengths exceed the Debye length.

The strong damping of longitudinal oscillations in region $\lambda \leq r_D$ is the consequence of the following aspect. In this region, the phase velocity of such waves is of the order of the thermal velocity of electrons. Therefore, Cherenkov absorption condition $\omega = \vec{k} \cdot \vec{v}$ is satisfied for a large number of electrons, which move in phase with waves and results in the strong damping of waves.

Instead of expressions (2.49) and (2.50), another form of representation is used very often for the longitudinal and transverse dielectric permittivities of electron plasma, which is more reliable for the investigation of their analytical properties.[11] Before this case, it must be noted that the integrals in expressions (2.49) and (2.50) are Cauchy-type integrals. Since the integration contour C, being parallel to the real axis, intersects the whole plane of the complex variable ω, then such integrals define an analytical function in the upper half-plane of the complex variable ω, above the contour C. On the other hand, when we considered the initial value problem in Sect. 1.5, the necessity of analytical continuation of the dielectric permittivity from the upper half-plane of the complex variable $\omega = \omega' + \imath\omega''$ into the lower half-plane was unavoidable. Such continuation can be realized by shifting the integration contour C in Eqs. (2.49) and (2.50) into the lower half-plane x.[12] Such shifting process is possible because the integrands in Eqs. (2.49) and (2.50) may be continued into the lower half-plane x. Since for such a shifting when $\mathrm{Im}\omega < 0$, the only singularity of integrands is located at the point $x = \omega/k\sqrt{m/\kappa T_e}$, then the integration contour C takes the form shown in Fig. 2.2. However, the integrand has an essential singular point at infinity. This means that the analytical continuation of the integral should have the same singular point as well.

Now, let us apply the result obtained in [30][13] where it was shown that[14]

It must be noted that the analytical properties of the dielectric permittivity are closely connected to the singularity of the solution of the kinetic equation [28, 29].

[12] Just such continuation was proposed in [27].

[13] This result was applied to plasma in [31].

[14] One can be convinced of this result by representing the integral in the right side of Eq. (2.58) for $\mathrm{Im}\beta > 0$ in the following form:

$J_+(\beta) = \imath \int_{-\infty}^{+\infty} dx \exp(-x^2/2) \int_{+\infty}^{0} dt \exp[\imath(\beta - x)t]$.

Using $y = x + \imath t$, we obtain

$J_+(\beta) = \int_{+\infty}^{0} dt \exp(\imath\beta t - t^2/2) \int_{-\infty+\imath t}^{+\infty+\imath t} dy \exp(-y^2/2) = \sqrt{2\pi}\,\imath \int_{\infty}^{0} \exp(\imath\beta t - t^2/2)dt$.

Finally, introducing $\tau = \imath t + \beta$, we find

$J_+(\beta) = \sqrt{2\pi}\exp(-\beta^2/2)\int_{+\imath\infty}^{\beta} d\tau \exp(\tau^2/2)$.

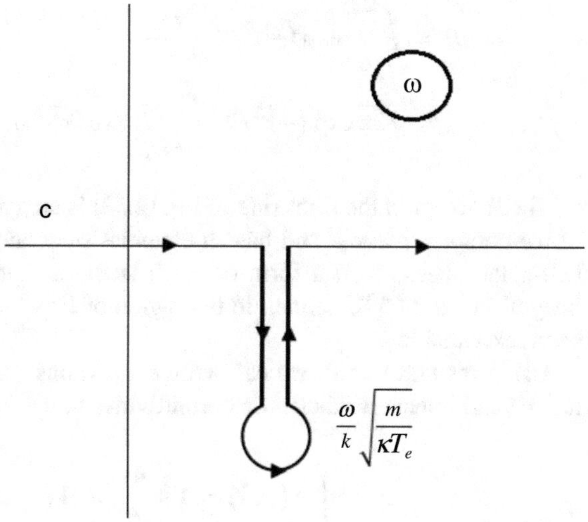

Fig. 2.2 Integration contour

On the other hand, if $\mathrm{Im}\beta < 0$ the integral in the left side of Eq. (2.58) can be represented as:

$$J_-(\beta) = \imath \int_{-\infty}^{+\infty} dx \exp\left(-x^2/2\right) \int_{-\infty}^{0} \exp\left[\imath(\beta - x)t\right]dt = \sqrt{2\pi}\exp\left(-\beta^2/2\right) \int_{-i\infty}^{\beta} \exp\left(\tau^2/2\right)d\tau$$

$$= \sqrt{2\pi}\exp\left(-\beta^2/2\right) \int_{i\infty}^{\beta} \exp\left(\tau^2/2\right)d\tau + \sqrt{2\pi}\exp\left(-\beta^2/2\right) \int_{-i\infty}^{+i\infty} \exp\left(\tau^2/2\right)d\tau.$$

Since $\int_{-i\infty}^{+i\infty} \exp\left(\tau^2/2\right)d\tau = \imath\sqrt{2\pi}$, then function $J_-(\beta)$, which is analytical in the lower half-plane, differs from the analytical continuation of $J_+(\beta)$ into this half-plane by the quantity $2\pi\imath \exp\left(-\beta^2/2\right)$. It follows, from this fact, that the analytical continuation of $J_+(\beta)$ has no singularities except at infinity in the β complex plane. In what follows, we will use the function

$$I_+(\beta) = \beta e^{-\beta^2/2} \int_{i\infty}^{\beta} d\tau \exp\left(\tau^2/2\right) = \frac{\beta}{\sqrt{2\pi}} J_+(\beta).$$

For small $|\beta| \ll 1$, the usual expansion

$$I_+(\beta) \approx \beta \int_{i\infty}^{0} d\tau \exp\left(\tau^2/2\right) = -\imath\sqrt{\frac{\pi}{2}}\beta$$

is valid. One must be more cautious about asymptotic expansion in the region of $|\beta| \gg 1$, because there exists an essential singular point at infinity. The following asymptotic expansions of function $I_+(\beta)$ are useful:

$$I_+(\beta) \approx 1 + \frac{1}{\beta^2} + \frac{3}{\beta^4} + \cdots - \imath\sqrt{\frac{\pi}{2}}\beta \exp\left(-\beta^2/2\right),\ |\beta| \gg 1,\ |\mathrm{Re}\beta| \gg |Im\beta|,$$

$$I_+(\beta) \approx -\imath\sqrt{2\pi}\beta \exp\left(-\beta^2/2\right),\ |\beta| \gg 1,\quad |Im\beta| \gg |\mathrm{Re}\beta|, Im\beta < 0.$$

Finally, it must be noticed that the function

$$W(z) = \frac{2\imath}{\sqrt{\pi}} e^{-z^2} \int_{i\infty}^{z} dt \exp\left(t^2\right)$$

has been tabulated in [32, 33].

$$J_{+}(\beta) = \int_{-\infty}^{+\infty} \exp\left(-x^2/2\right) \frac{dx}{\beta - x}$$
$$= \sqrt{2\pi} \exp\left(-\beta^2/2\right) \int_{+i\infty}^{\beta} d\tau \exp\left(\tau^2/2\right), \qquad (\mathrm{Im}\beta > 0). \tag{2.58}$$

The function in the right side of Eq. (2.58) is analytical in the whole finite range of the complex plane β and has an essential singular point at infinity in the lower half-plane. Thus, such a form of $J_{+}(\beta)$ is the analytical continuation of Cauchy integral in Eq. (2.58), defined in the region of $\mathrm{Im}\beta > 0$, on the whole plane of the complex variable β.

Using relation (2.58), we can write expressions (2.49) and (2.50) for the longitudinal and transverse dielectric permittivities in the following forms:

$$\begin{cases} \varepsilon^{l}(\omega, k) = 1 + \dfrac{\omega_{\mathrm{pe}}^2}{\omega^2} \beta^2 [1 - I_{+}(\beta)], \\ \varepsilon^{\mathrm{tr}}(\omega, k) = 1 - \dfrac{\omega_{\mathrm{pe}}^2}{\omega^2} I_{+}(\beta), \end{cases} \tag{2.59}$$

which are valid not only for $\mathrm{Im}\omega > 0$, but also for $\mathrm{Im}\omega < 0$. Here,

$$\beta = \frac{\omega}{k} \sqrt{\frac{m}{\kappa T_e}}, \quad I_{+}(\beta) = \beta \exp\left(-\beta^2/2\right) \int_{i\infty}^{\beta} d\tau \exp\left(\tau^2/2\right). \tag{2.60}$$

The existence of an essential singular point at infinity in the analytical continuation of the dielectric permittivity into the lower half-plane of ω makes the application of the results of Sect. 1.5 difficult for studying the time dependence of the field. This difficulty is connected to the shifting of the integration contour in Eqs. (1.99) and (1.100) to the lower half-plane of ω. Actually, in Sect. 1.5, we neglected the integral over a contour at infinity, which is correct if the integrands are regular functions. But, in our case, the integrand may become arbitrarily large since there exists an essential singular point at infinity and therefore it is impossible to neglect the integral over the contour at infinity. However, if we are only interested in the asymptotic time behavior of the field, then it can be obtained easily [27, 34]. For example, if in the integral of

$$\int_{C} \frac{\exp\left(-\imath\omega t\right)}{\varepsilon^{l}(\omega, k)} d\omega,$$

one shifts the contour C into the lower half-plane of ω and bypasses the first pole connected to the nearest zero of $\varepsilon^{l}(\omega, k)$, then the integral over the contour, which is shown in Fig. 2.3, will be obtained. The integration around the pole leads to the exponential time dependence of the field as it takes place for damping oscillations.

At the same time, the integral over the line, which is parallel to the real axis and on which $|\,\mathrm{Im}\,\omega|$ is much larger than the imaginary part of the pole of $\varepsilon^l(\omega,k)$ for sufficiently large time t, is exponentially small. Thus, we see that the roots of the analytical continuation of $\varepsilon^l(\omega,k)$ determine the damped oscillations of longitudinal fields. A quite similar statement is valid for the roots of the analytical continuation of $\omega^2\varepsilon^{\mathrm{tr}}(\omega,k) - k^2c^2$.

When the poles of integrands in Eqs. (1.99) and (1.100) are settled sufficiently close together, then one can calculate their contributions by expanding the integrands in the Loran series near such a singular point. It is obvious that the contributions from poles in Eqs. (1.99) and (1.100) result in the exponential dependence, which corresponds to the waves. The spectra of such waves are determined by Eqs. (1.101) and (1.102).

Specially, for longitudinal oscillations[15]

$$1 = \frac{\omega_{\mathrm{pe}}^2}{\omega^2}\frac{1}{\sqrt{2\pi}}\int_{C'}\frac{dx x^2 \exp\left(-x^2/2\right)}{\omega - xk\sqrt{\frac{\kappa T_e}{m}}},$$

or

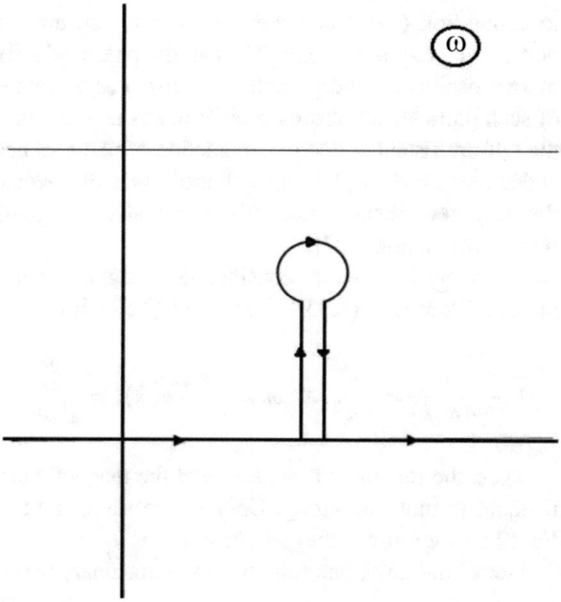

Fig. 2.3 Integration contour

[15]This equation without account of the contributions of poles was obtained first by Vlasov [11]. The contributions of poles were pointed out firstly by Landau [27].

$$1 = \frac{\omega_{\rm pe}^2}{\omega^2}\beta^2[I_+(\beta) - 1]. \tag{2.61}$$

From this equation, in the long-wave limit or $\beta \gg 1$, expressions (2.54) and (2.56) are obtained for the frequency spectra and damping decrement of longitudinal oscillations, respectively. In the opposite short-wave limit, when $\lambda/2\pi = 1/k \ll r_D$, Landau found out that the solution of Eq. (2.60) corresponds to strong damping of plasma oscillations. The frequency of these oscillations is much less than their damping decrement

$$\gamma \approx \sqrt{\frac{\kappa T_e}{m}}k\xi, \tag{2.62}$$

where ξ is the solution of equation $\sqrt{2\pi}\xi e^{\xi^2/2} = (kr_D)^2$ and is of the order of $\sim \sqrt{\ln(kr_D)} \gg 1$.

Langmuir was the first who explained why in the short-wave limit, where wavelength is less than the Debye radius or when the phase velocity of waves is not higher than the thermal velocity of particles, weakly damped plasma oscillations cannot exist [24, 35]. Only in the opposite limit, when the wave phase velocity is much higher than the particles' velocity, the thermal velocity can be neglected and in this condition, the force acting on the particles could be determined. This leads to expression (2.43) and consequently to the undamped oscillations. But, if the particle velocity is much higher than the phase velocity of waves, then in the period of one oscillation the particles traverse many wavelengths. Therefore, the effect of such particles on plasma oscillations is negligible. Hence, in this approximation, the self-consistent collective interaction of particles becomes negligible as well. The collective nature of plasma oscillations was discovered by Langmuir. In this sense, the only new phenomenon discovered after Langmuir is the Landau damping of plasma oscillations [27].

In conclusion, let us consider the magnetic permeability $\mu(\omega, k)$ of isotropic plasma. From Eqs. (1.35), (2.58), and (2.59), it follows that

$$1 - \frac{1}{\mu(\omega, k)} = \frac{\omega^2}{c^2k^2}\left[\varepsilon^{\rm tr}(\omega, k) - \varepsilon^l(\omega, k)\right] = \frac{\omega_{pe}^2}{k^2c^2}\left[I_+(\beta)\left(\beta^2 - 1\right) - \beta^2\right]. \tag{2.63}$$

Thus, the statement often said that the magnetic permeability of classical plasma is equal to unity is wrong. Only in the static limit, when $\omega \to 0$ ($\beta \to 0$), from Eq. (2.62), it follows that $\mu(0, k) = 1$.

From Eq. (2.62), one can obtain the imaginary part of the magnetic permeability[16]

[16]Of course relations (2.63) do not take into account the paramagnetic part of the magnetic permeability which is important only when $\hbar^2k^2/2m > \kappa T_e$, see Chap. 4.

$$\frac{Im\mu(\omega,k)}{|\mu(\omega,k)|^2} = -\sqrt{\frac{\pi}{2}}\frac{\omega_{\rm pe}^2}{c^2k^2}\frac{\omega}{k}\sqrt{\frac{m}{\kappa T_e}}\left(\frac{m\omega^2}{k^2\kappa T_e}-1\right)\exp\left(\frac{-m\omega^2}{2\kappa T_e k^2}\right). \tag{2.64}$$

For $\omega^2 = k^2\kappa T_e/m$, the imaginary part of the magnetic permeability changes sign. In the low-frequency range $\omega^2 < k^2\kappa T_e/m$, we have $\mathrm{Im}\mu(\omega,k) > 0$. In the opposite case, in the high-frequency range, we have $\mathrm{Im}\mu(\omega,k) > 0$.

2.4 Dielectric Permittivity and Electromagnetic Oscillations of Relativistic Collisionless Electron Plasma[17]

The relativistic effects must be taken into account in isotropic plasma in two cases: for high temperatures when κT_e is sufficiently large so that it cannot be neglected in comparison with mc^2; for low temperatures when the phenomenon of our interest is stipulated by those particles in the Maxwellian distribution whose velocities are of the order of the light velocity. To describe such processes, one must use the following distribution function [41]:

$$f_{0e}\left(\vec{p}\right) = \frac{N_e}{4\pi(mc)^3}\frac{\exp\left[-\frac{c\sqrt{p^2+m^2c^2}}{\kappa T_e}\right]}{\frac{\kappa T_e}{mc^2}K_2\left(\frac{mc^2}{\kappa T_e}\right)}, \tag{2.65}$$

where $K_2(x)$ is the MacDonald function. This function characterizes the momentum distribution of an ideal relativistic gas of particles. In contrast to the Maxwell distribution (2.47), the velocity of particles in Eq. (2.65) is limited by the light velocity.

When the temperature is very high ($\kappa T_e \gg mc^2$), one can talk about the ultra-relativistic particle gas. Then, instead of Eq. (2.65), the following distribution must be used:

$$f_{0e} = \frac{N_e}{8\pi}\left(\frac{c}{\kappa T_e}\right)^3\exp\left(-\frac{cp}{\kappa T_e}\right). \tag{2.66}$$

Since $\epsilon = cp$, then the velocity of all particles is equal to the light velocity c.

From the distribution function (2.65), one can find the frequency of long wave ($k \to 0$) plasma oscillations by making use of Eq. (2.44)

[17]This problem was discussed in [16, 36–40].

$$\omega_0^2 = \frac{4\pi e^2 c^2 N_e}{\kappa T_e} K_2^{-1}\left(\frac{mc^2}{\kappa T_e}\right) \int\limits_1^\infty \frac{dz}{z^2} K_2\left(\frac{mc^2}{\kappa T_e} z\right). \tag{2.67}$$

In the non-relativistic limit $\kappa Te \ll mc^2$, this expression coincides with Eq. (2.48), whereas in the opposite limit of ultra-relativistic temperature ($\kappa T_e \gg mc^2$), we have

$$\omega_{0p}^2 = \frac{4\pi e^2 N_e c^2}{3\kappa T_e}. \tag{2.68}$$

Of course, this formula can be obtained from the distribution (2.66) as well.

Since spatial dispersion is stipulated by the thermal motion of particles, one can expect that, in a relativistic gas in which the thermal velocity of electrons is comparable to the light velocity, the influence of spatial dispersion becomes essential. In order to be convinced, let us consider the longitudinal and transverse dielectric permittivities of relativistic electron plasma. These expressions can be easily obtained by substituting distribution (2.65) into Eqs. (2.24) and (2.25). After integrating over the transverse components of velocity perpendicular to $\vec{k}$, these expressions may be represented in the form of Cauchy integrals:

$$\varepsilon^l(\omega,k) = 1 - \frac{2\pi e^2 N_e}{\omega k^3 c\kappa T_e} K_2^{-1}\left(\frac{mc^2}{\kappa T_e}\right) \int\limits_{-kc}^{+kc} d\omega' \frac{\omega'^2}{\omega-\omega'} \exp\left[-\frac{mc^2}{\kappa T_e}\frac{1}{\sqrt{1-\omega'^2/c^2k^2}}\right]$$
$$\times\left[\frac{1}{1-\omega'^2/c^2k^2} + 2\left(\frac{\kappa T_e}{mc^2}\right)\frac{1}{\sqrt{1-\omega'^2/c^2k^2}} + 2\left(\frac{\kappa Te}{mc^2}\right)^2\right], \tag{2.69}$$

$$\varepsilon^{tr}(\omega,k) = 1 - \frac{2\pi e^2 N_e c}{\omega k\kappa T_e} K_2^{-1}\left(\frac{mc^2}{\kappa T_e}\right) \int\limits_{-kc}^{+kc} \frac{d\omega'}{\omega-\omega'}\frac{\kappa T_e}{mc^2} \exp\left[-\frac{mc^2}{\kappa T_e}\frac{1}{\sqrt{1-\omega'^2/c^2k^2}}\right]$$
$$\times\left[\left(1-\omega'^2/c^2k^2\right)\frac{\kappa T_e}{mc^2} + \sqrt{1-\omega'^2/c^2k^2}\right]. \tag{2.70}$$

As above (see Sect. 2.2), one must suppose that the frequency ω in Eqs. (2.69) and (2.70) has a small positive imaginary part. In contrast to non-relativistic formulas (2.49) and (2.50), in the relativistic case, the integrations extend over a finite region $(-kc, +kc)$. The dielectric permittivity as a function of real ω may be considered as the limit of expressions (2.69) and (2.70) from the upper half-plane of the complex variable ω when the integration contour is located on the real axis, as shown in Fig. 2.4. Then, these integrals are of Cauchy type and, therefore, expressions (2.69) and (2.70) determine the dielectric permittivity in the whole

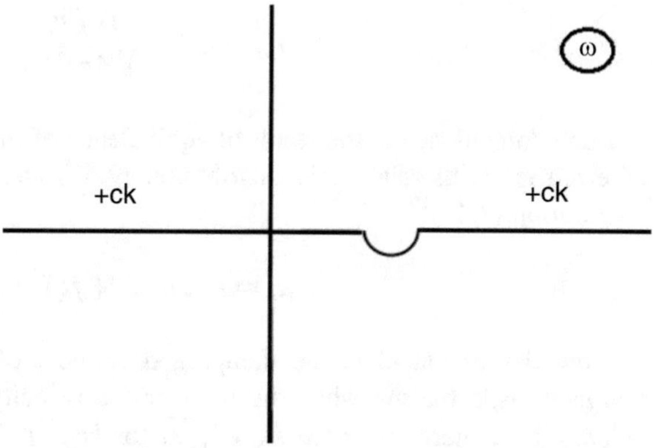

Fig. 2.4 Integration contour

plane of the complex variable ω except the integration contour. When the integration contour is intersected, the Cauchy integral undergoes a jump. Therefore, the integration contour is a cutting line in the complex plane with branching points $\pm ck$ being the ends of the contour. Thus, in relativistic plasma, as a consequence of velocity limitation of plasma particles, the singular points of the dielectric permittivity are located in the finite region of ω, whereas in the non-relativistic case, such points are only at infinity.[18]

In the long-wave limit ($k \to 0$), the longitudinal and transverse dielectric permittivities coincide and are equal to

$$\varepsilon^l(\omega, 0) = \varepsilon^{\mathrm{tr}}(\omega, 0) = 1 - \frac{\omega_{0p}^2}{\omega^2},$$

where ω_{0p} is determined by Eq. (2.68).

In the static limit ($\omega \to 0$), the longitudinal dielectric permittivity looks as

$$\varepsilon^l(0, k) = 1 + \frac{4\pi e^2 N_e}{k^2 \kappa T_e}.$$

This means that the screening length [see Eq. (1.54)] in relativistic plasma, as in the non-relativistic case, coincides with the Debye length

[18]The existence of branching points $\omega = \pm ck$ is the reason why in consideration of the initial value problem of field behavior, in a medium without external sources (see Sect. 1.5), it should not be considered only the plasma oscillations with spectra defined by Eqs. (1.100) and (1.101) but it is necessary to take into account the contributions from the cutting line in the plane of complex variable in the integral of the type of Eq. (1.99) as well.

$$r_{\text{scr}} = r_D = \sqrt{\frac{\kappa T_e}{4\pi e^2 N_e}}. \tag{2.71}$$

Such coincidence is the result of equivalence of the chemical potential of a gas of electrons with relativistic distribution (2.65) and non-relativistic Maxwellian distribution (2.47)[19]

$$\mu_e = \kappa T_e \ln N_e + f(T_e).$$

Now, let us calculate the damping decrement of longitudinal oscillations in non-relativistic plasma when the wave phase velocity is of the order of the light velocity and, therefore, $\omega \gg k\sqrt{\kappa T_e/m}$ (or $kr_D \ll 1$). In this condition, relation (2.54) stays valid. One can obtain this result by using the expansion of the integrand in Eq. (2.69) in powers of ω' for non-relativistic temperatures ($\kappa T_e \ll mc^2$) in the integration over the real axis. However, such an expansion is impossible around the pole of $\omega = \omega'$ for calculation of the damping decrement, because the small values of ω' compared to ck play a major role. In contrast, when the phase velocity ω/k is comparable to the light velocity, the values of ω', which is of the order of ck, have the main contribution.

In non-relativistic plasma, in accordance to Eq. (2.56), the damping decrement of oscillations must be small and this means that the singular point of the integrand in Eq. (2.69) is located near the real axis. It allows us to represent the pole as

$$\frac{1}{\omega - \omega'} = \mathcal{P}\frac{1}{\omega - \omega'} - \imath\pi\delta(\omega - \omega').$$

As a result, from $\varepsilon^l(\omega, k) = 0$, we obtain[20]

$$\gamma = \sqrt{\frac{\pi}{8}} \frac{\omega_{\text{pe}}}{(kr_D)^3} \frac{1}{1 - \omega^2/c^2k^2} \exp\left[-\frac{mc^2}{\kappa T_e}\left(\frac{1}{\sqrt{1 - \omega^2/c^2k^2}} - 1\right)\right]. \tag{2.72}$$

In the limit of non-relativistic phase velocities, when $ck \gg \omega$, this expression coincides with expression (2.56). As $ck \to \omega$, the right-hand side of this expression vanishes and, therefore, $\gamma \to 0$. At last, if $\omega > ck$, this expression is invalid because the pole of $\omega = \omega'$ in the integrand of Eq. (2.69) is not on the real axis. Therefore,

---

[19]In non-relativistic plasma $f(T_e) = (3/2)\kappa T_e \ln(2\pi\hbar^2/m\kappa T_e)$, whereas in relativistic plasma $f(T_e) = \kappa T_e\{\ln(2\pi^2\hbar^3/c\kappa T_e) - \ln(K_2(mc^2/\kappa T_e))\}$.

[20]For relativistic plasma, the damping decrement of longitudinal oscillations is equal to

$$\gamma = \frac{\pi}{2}\frac{\omega_{\text{pe}}^2\omega_0^2 m}{k^3 c\kappa T_e} K_2^{-1}\left(\frac{mc^2}{\kappa T_e}\right) \exp\left[-\frac{mc^2}{\kappa T_e}\frac{1}{\sqrt{1-\omega^2/c^2k^2}}\right]$$
$$\times\left[\frac{1}{1-\omega^2/c^2k^2} + 2\frac{\kappa T_e}{mc^2}\frac{1}{\sqrt{1-\omega^2/c^2k^2}} + 2\left(\frac{\kappa T_e}{mc^2}\right)^2\right] \times \begin{cases} 1, & \textit{for} \quad \omega < kc, \\ 0. & \textit{for} \quad \omega > kc. \end{cases}$$

the whole integral and, as a result, the dielectric permittivity will become real. Thus, absorption in plasma is absent and $\gamma \equiv 0$. The physical interpretation of this result is that if $\omega > ck$ there is no particle in plasma with the velocity equal to the phase velocity $\left(\omega \neq \vec{k}.\vec{v}\right)$ and, therefore, the Cherenkov absorption of oscillations is impossible. For transverse electromagnetic waves in electron plasma, relativistic consideration results in the absence of damping because the phase velocity of such waves is always larger than the light velocity.

Now, let us consider another example, i.e., electron plasma with ultra-relativistic temperature ($\kappa T_e \gg mc^2$). In this case, the distribution function of electrons is given by Eq. (2.66). It leads to the following expressions for longitudinal and transverse dielectric permittivities as functions of real $\omega > 0$ and $k > 0$:

$$\varepsilon^l(\omega, k) = 1 + \frac{4\pi e^2 N_e}{k^2 \kappa T_e}\left\{1 + \frac{\omega}{2ck}\ln\left|\frac{\omega - ck}{\omega + ck}\right| + \frac{\imath\pi\omega}{4ck}\left(\frac{ck - \omega}{| ck - \omega |} + 1\right)\right\}, \tag{2.73}$$

$$\varepsilon^{\text{tr}}(\omega, k) = 1 + \frac{\pi e^2 N_e c}{\omega k \kappa T_e}$$
$$\times\left\{-\frac{2\omega}{ck} + \left(1 - \frac{\omega^2}{c^2k^2}\right)\ln\left|\frac{\omega - ck}{\omega + ck}\right| + \frac{\imath\pi}{2}\left(1 - \frac{\omega^2}{c^2k^2}\right)\left(\frac{ck - \omega}{| ck - \omega |} + 1\right)\right\}. \tag{2.74}$$

Substituting Eq. (2.74) into Eq. (1.106), we obtain the frequency spectra of transverse waves in the long- and short-wave limits:

$$\omega^2 = \frac{4\pi e^2 N_e c^2}{3\kappa T_e} + \frac{6}{5}c^2k^2, \quad \text{for} \quad \omega \gg ck, \tag{2.75}$$

$$\omega^2 = \frac{2\pi e^2 N_e c^2}{\kappa T_e} + c^2k^2, \quad \text{for} \quad \omega \to ck. \tag{2.76}$$

The difference of these expressions with formula (2.52), which corresponds to non-relativistic plasma, is stipulated by the increasing role of spatial dispersion. This effect also causes the difference between the constant terms in Eqs. (2.75) and (2.76). In fact, since the particles' velocity of the ultra-relativistic gas is equal to the light velocity, then the correction of term ~k^2c^2 in formula (2.75), which is caused by thermal motion of electrons, is about 20 percent.

It must be noted that the phase velocity of the transverse waves exceeds the light velocity in accordance to Eqs. (2.75) and (2.76). Therefore, they cannot be absorbed by electrons and this leads to zero damping decrement, $\gamma^{tr} = 0$.

Quite analogically, we obtain the frequency spectra of longitudinal waves of ultra-relativistic electron plasma using Eq. (2.73)

$$\omega^2 = \frac{4\pi e^2 N_e c^2}{3\kappa T_e} + \frac{3}{5}c^2k^2, \qquad for \qquad \omega \gg ck, \tag{2.77}$$

$$\omega = ck\left[1 + 2\exp\left(-\frac{k^2\kappa T_e}{2\pi e^2 N_e} - 2\right)\right], \qquad for \qquad \omega \to ck. \tag{2.78}$$

We see that the phase velocity of longitudinal and transverse waves exceeds the velocity of light and, therefore, $\gamma^l = 0$.

In the range of small phase velocities (low frequencies) $\omega \ll ck$, the field is either strongly absorbed or screened. Indeed, for $\omega \ll ck$, we have

$$\begin{cases} \varepsilon^l(\omega, k) = 1 + \dfrac{4\pi e^2 N_e}{k^2 T_e \kappa}, \\ \varepsilon^{\text{tr}}(\omega, k) = 1 + \imath\pi \dfrac{\pi e^2 N_e c}{\omega k \kappa T_e}. \end{cases} \tag{2.79}$$

The screening radius of the longitudinal field, as it was shown above [see Eq. (2.71)], is equal to the Debye length. For the low-frequency transverse field, the conductivity of both ultra-relativistic and relativistic plasmas has the same structure as Eq. (2.32), where $C = \pi e^2 N_e c / 4\kappa T_e$.

2.5 Oscillations of Isotropic Electron–Ion Plasma

In this section, we will consider the ion's motion and its influence on the electromagnetic field oscillations of plasma. Besides, we will try to determine the range of validity of the obtained results when the ion motion is neglected.

Let us suppose that plasma consists of electrons and only one type of ions (generalization to multi-component electron–ion plasma is obvious). Moreover, in non-relativistic plasma, we assume that the distribution functions of electrons and ions are Maxwellian in the equilibrium state[21]

$$\begin{cases} f_{0e}\left(\vec{p}\right) = \dfrac{N_e}{(2\pi m \kappa T_e)^{3/2}} \exp\left(-\dfrac{p^2}{2m\kappa T_e}\right), \\ f_{0i}\left(\vec{p}\right) = \dfrac{N_i}{(2\pi M \kappa T_i)^{3/2}} \exp\left(-\dfrac{p^2}{2M\kappa T_i}\right). \end{cases} \tag{2.80}$$

Here, N_e and N_i are the concentrations of electrons and ions in plasma at equilibrium, which are related together by the neutrality condition

[21] About relativistic electron–ion plasma, see [40].

$$eN_e + e_i N_i = 0,$$

where e, m and e_i, M are the charge and mass of the electron and ion, respectively. The temperature of electrons T_e and ions T_i in collisionless plasma may be different. The process of temperature relaxation between electrons and ions is determined by energy transfer through collisions, which is a very long-time process.

The longitudinal and transverse dielectric permittivities of such plasma can be obtained by making use of Eqs. (2.24), (2.25), (2.59), and (2.60):

$$\varepsilon^l(\omega, k) = 1 + \frac{\omega_{\mathrm{pe}}^2}{\omega^2}\beta_e^2[1 - I_+(\beta_e)] + \frac{\omega_{\mathrm{pi}}^2}{\omega^2}\beta_i^2[1 - I_+(\beta_i)], \tag{2.81}$$

$$\varepsilon^{\mathrm{tr}}(\omega, k) = 1 - \frac{\omega_{\mathrm{pe}}^2}{\omega^2} I_+(\beta_e) - \frac{\omega_{\mathrm{pi}}^2}{\omega^2} I_+(\beta_i). \tag{2.82}$$

Here, $\omega_{\mathrm{pe}} = (4\pi e^2 N_e/m)^{1/2}$ and $\omega_{\mathrm{pi}} = \left(4\pi e_i^2 N_i/M\right)^{1/2}$ are the Langmuir frequencies of electrons and ions, respectively. Moreover, $\beta_e = (\omega/k)\sqrt{m/\kappa T_e}$, $\beta_i = (\omega/k)\sqrt{M/\kappa T_i}$, and $I_+(\beta) = \beta \exp\left(-\beta^2/2\right) \int\limits_{+i\infty}^{\beta} d\tau \exp\left(\tau^2/2\right)$ (see Sect. 2.3). By substituting these expressions into relations (1.100) and (1.101), we obtain the following dispersion equations for longitudinal and transverse waves for electron–ion plasma:

$$\omega^2 = \omega_{pe}^2 \beta_e^2 [I_+(\beta_e) - 1] + \omega_{pi}^2 \beta_i^2 [I_+(\beta_i) - 1], \tag{2.83}$$

$$\omega^2 - c^2 k^2 = \omega_{\mathrm{pe}}^2 I_+(\beta_e) + \omega_{\mathrm{pi}}^2 I_+(\beta_i). \tag{2.84}$$

Let us begin by analyzing Eq. (2.84), which describes the transverse electromagnetic waves in electron–ion plasma. One can easily find that the generalization of spectrum (2.52), which corresponds to the phase velocity higher than the light velocity, to the case of electron–ion plasma is trivial. Using the asymptotic relations of functions $I_+(\beta_e)$ and $I_+(\beta_i)$, when $\beta_{e,\,i} \gg 1$, from Eq. (2.84), we obtain

$$\omega^2 = \omega_{\mathrm{pe}}^2 + \omega_{\mathrm{pi}}^2 + k^2\left(c^2 + \frac{\kappa T_e}{m} + \frac{m}{M}\frac{\kappa T_i}{M}\right), \tag{2.85}$$

for undamped transverse waves. It is obvious that ion's contribution to the frequency spectrum of transverse waves is very small, i.e., of the order of $\sim m/M$ or $\kappa T_i m/M^2 c^2$ and so it can be neglected with respect to the main terms. Furthermore, the phase velocity of such waves exceeds the light velocity and, therefore, they are undamped (see Sect. 2.3).

On the other hand, longitudinal waves experience a quite different situation. In Sect. 2.3, it was shown that electron plasma supports longitudinal waves with phase velocities less than the light velocity. However, they are weakly damped when their phase velocity is higher than the thermal velocity of particles. Two possibilities may happen in electron–ion plasma. The first possibility is when the phase velocity exceeds the thermal velocities of both electrons and ions, or when $\beta_{e,\,i} \gg 1$. Using the asymptotical representations of functions $I_+(\beta_e)$ and $I_+(\beta_i)$ for $\beta_e \gg 1$ and $\beta_i \gg 1$,

$$I_+(\beta) \approx 1 + \frac{1}{\beta^2} + \ldots - \imath\sqrt{\frac{\pi}{2}}\ \beta \exp\left(-\frac{\beta^2}{2}\right),$$

from Eq. (2.81), one can find the frequency spectra and damping decrement of fast longitudinal waves [compared to Eqs. (2.54) and (2.56)]

$$\omega^2 = \omega_{\rm pe}^2 + \omega_{\rm pi}^2 + k^2\left(\frac{3\kappa T_e}{m} + \frac{m}{M}\frac{3\kappa T_i}{M}\right), \tag{2.86}$$

$$\gamma = +\sqrt{\frac{\pi}{8}}\frac{\omega^2}{k^3} \times \left[\omega_{\rm pe}^2\left(\frac{m}{\kappa T_e}\right)^{3/2} \exp\left(-\frac{\omega^2}{k^2}\frac{m}{2\kappa T_e}\right) + \omega_{\rm pi}^2\left(\frac{M}{\kappa T_i}\right)^{3/2} \exp\left(-\frac{\omega^2}{k^2}\frac{M}{2\kappa T_i}\right)\right], \tag{2.87}$$

where we have replaced ω with $\omega - \imath\gamma$.

From expressions (2.86) and (2.87), it is clear that ion's influence in the frequency dependence of longitudinal waves of the wave vector may be essential only if $T_i/T_e \geq M^2/m^2$. In the opposite case, the ion's contribution can be neglected and the results of Sect. 2.3 are valid.

Concerning the wave damping, when the phase velocity becomes of the order of the thermal velocity of particles, the ion's impact turns out to be more essential. In this case, increasing the wave vector k (i.e., by decreasing the wavelength) increases damping. In particular, quite similar to what happened in electron plasma for sufficiently short waves, when the phase velocity is less than the ion thermal velocity, it is possible to talk about the aperiodic damping of fields. Making use of $I_+(\beta) \approx -\imath\sqrt{2\pi}\beta \exp\left(-\beta^2/2\right)$, $|\,\mathrm{Im}\,\beta\,| \gg |\,\mathrm{Re}\,\beta|$, and $\mathrm{Im}\beta < 0$, Eq. (2.83) gives the following relation:

$$1 = \omega_{\rm pe}^2\left(\frac{m}{\kappa T_e}\right)^{3/2}\frac{\sqrt{2\pi}}{k^3}\gamma \exp\left(\frac{\gamma^2 m}{2k^2\kappa T_e}\right) + \omega_{\rm pi}^2\left(\frac{M}{\kappa T_i}\right)^{3/2}\frac{\sqrt{2\pi}}{k^3}\gamma \exp\left(\frac{\gamma^2 M}{2k^2\kappa T_i}\right), \tag{2.88}$$

where $\gamma \equiv$ Im ω. If $T_e/T_i < m/M$, the ion's contribution, in this relation, can be neglected. In this condition, it coincides with Eq. (2.61). In the opposite case, when

$T_e/T_i > m/M$, the ion's contribution becomes dominant. When the latter inequality holds strongly, $T_e/T_i \gg m/M$, from Eq. (2.88), it follows that[22]

$$\gamma = \sqrt{\frac{\kappa T_i}{M}} k\xi, \tag{2.89}$$

where

$$\sqrt{2\pi}\xi \exp\left(\xi^2/2\right) = k^2 r_{\text{Di}}^2. \tag{2.90}$$

In contrast to Eq. (2.62), in this equation, the ion Debye length $r_{\text{Di}} = \left(\kappa T_i/4\pi e_i^2 N_i\right)^{1/2}$ appears instead of the electron Debye length. It should be noted that $\xi \sim \sqrt{\ln\left(kr_{Di}\right)}$. Thus, in the short-wave region, the ion's influence increases. Furthermore, in this case, when the phase velocity becomes less than the ion thermal velocity, as it is clear from Eq. (2.89), the ion's motion becomes dominant.

In the long-wave range, when the wavelength exceeds the Debye length, in accordance to Eq. (2.86), the influence of ions is negligible. But more importantly, this statement only holds in the case of isothermal plasma. Below, it will be shown that when plasma is non-isothermal, two branches of long-wave longitudinal oscillations appear. The first branch is described by Eqs. (2.86) and (2.87) (the ion terms may be neglected). The second branch corresponds to the acoustic type oscillations of electron–ion plasma. The frequency of these oscillations tends to zero when $k \to 0$ and, therefore, in the left-hand side of Eq. (2.83), one can suppose $\omega^2 = 0$. Then, taking into account the charge neutrality of plasma, $eN_e + e_iN_i = 0$, and denoting $Z = |e_i/e|$, from Eq. (2.83), it follows that

$$\frac{T_i}{ZT_e}[1 - I_+(\beta_e)] + 1 - I_+(\beta_i) = 0. \tag{2.91}$$

Since $v_{T_i} \ll v_{T_e}$, and the velocity of such acoustic wave is much less than the electron velocity, we can expand this equation in powers of $\sqrt{m/M}$, leading to

$$\frac{T_i}{ZT_e}\left[1 + \imath\sqrt{\frac{\pi}{2}\frac{m}{M}\frac{T_i}{T_e}}\beta_i\right] + 1 - I_+(\beta_i) = 0, \tag{2.92}$$

where $\beta_i = [(\omega - \imath\gamma)/k]\sqrt{M/\kappa T_i}$, ω is the frequency of oscillations, and γ is the damping decrement. It can be easily shown that when $T_i \sim ZT_e$, this equation has a solution only if $I_+(\beta_i) \sim 1$, which is possible only for a complex β_i with $\text{Re}\beta_i \sim \text{Im}\,\beta_i$. Thus, in isothermal plasma, the acoustic oscillations are strongly damped [44]. On the other hand, in non-isothermal plasma with $ZT_e \gg T_i$, the acoustic oscillations of plasma are relatively weakly damped. Thus, the phase velocity of acoustic

[22]This result was first obtained in [42, 43].

oscillations is much higher than the ion thermal velocity and, therefore, one can use the asymptotic representation of $I_+(\beta_i)$ for $|\beta_i| \gg 1$. As a result, Eq. (2.92) takes the form

$$\frac{T_i}{ZT_e}\left(1+\imath\sqrt{\frac{\pi}{2}\frac{m}{M}\frac{T_i}{T_e}}\beta_i\right)-\frac{1}{\beta_i^2}-\frac{3}{\beta_i^4}+\imath\sqrt{\frac{\pi}{2}}\beta_i\exp\left(-\beta_i^2/2\right)=0. \tag{2.93}$$

Hence, it follows that [42,45]

$$\omega^2 \approx k^2\left(Z\frac{\kappa T_e}{M}\right)\left(1+3\frac{T_i}{ZT_e}\right), \tag{2.94}$$

$$\gamma \approx \sqrt{\frac{\pi}{8}}\left[\sqrt{Z\frac{m}{M}}+\left(\frac{ZT_e}{T_i}\right)^{3/2}\exp\left(-\frac{ZT_e}{2T_i}\right)\right]\omega. \tag{2.95}$$

Since $ZT_e \gg T_i$, the ion's contribution to the damping decrement is very small and the complete absorption of acoustic waves is practically stipulated by the electrons moving in phase with the wave. Note that one can obtain the expression (2.95) for the damping decrement using the quantum considerations, as it was done in Sect. 1.3 of the present chapter.

The acoustic oscillations with spectrum (2.94), similar to plasma oscillations with spectrum (2.86), represent two branches of longitudinal waves in electron–ion plasma. In plasma oscillations, a non-equilibrium electric charge density arises, whereas the acoustic oscillations are electrically neutral in the first approximation because the non-equilibrium charge densities of electrons and ions neutralize each other. This fact is the physical nature of the difference between these branches of long-wave longitudinal oscillations of plasma. For this reason, plasma oscillations are known as charge oscillations, whereas the acoustic oscillations represent the oscillations of the mass density. In this sense, these two categories of long-wave oscillations of plasma are similar to the well-known optical (Born) and acoustic branches of oscillations in the solid-state physics.

2.6 Hydrodynamics of Collisionless Plasma

The hydrodynamic description of moving liquids and gases is based on three functions, which determine the distributions of flow velocity $\vec{V}\left(\vec{r},t\right)$, mass density $W\left(\vec{r},t\right)$, and pressure $P\left(\vec{r},t\right)$. Hydrodynamic equations have some advantages compared to kinetic equations provided that they can be given in a closed form. They are simpler than kinetic equation due to the fact that the hydrodynamic quantities depend only on four variables $\vec{r}$ and t, whereas the distribution functions depend on the particle's momentum $\vec{p}$ as well, and, as a result, $f\left(\vec{p}, \vec{r}, t\right)$ is a function of seven variables.

Below, we will consider the possibility of derivation of the hydrodynamic equations for collisionless plasmas by the kinetic equation (2.3) with the self-consistent field (Vlasov equation):

$$\frac{\partial f_\alpha}{\partial t} + \vec{v}.\frac{\partial f_\alpha}{\partial \vec{r}} + e_\alpha\left\{\vec{E} + \frac{1}{c}\left[\vec{v} \times \vec{B}\right]\right\}.\frac{\partial f_\alpha}{\partial \vec{p}} = 0. \tag{2.96}$$

It must be emphasized that the particle collisions play a major role in derivation of hydrodynamics from the equations of kinetic theory [46–48]. Moreover, the usual hydrodynamic equations are applicable only when the characteristic sizes of inhomogeneity of the hydrodynamic quantities are large in comparison with the mean free path of liquid of gaseous particles. Therefore, it must be noted that the applicability of usual hydrodynamics to collisionless plasma is hopeless. However, under certain conditions some approximate equations similar to the hydrodynamic equations can be derived, which are quiet suitable for describing such plasmas.

Let us determine the mass density of particles of the type α by the help of the distribution function f_α:

$$W^{(\alpha)}\left(\vec{r}, t\right) = m_\alpha \int d\vec{p} f_\alpha\left(\vec{p}, \vec{r}, t\right). \tag{2.97}$$

The quantity characterizing the velocity distribution of particles is given by

$$W^{(\alpha)}\left(\vec{r}, t\right) \vec{V}^{(\alpha)}\left(\vec{r}, t\right) = \int d\vec{p}\, \vec{p} f_\alpha\left(\vec{p}, \vec{r}, t\right). \tag{2.98}$$

Using Eqs. (2.97) and (2.98), after integrating Eq. (2.96) over the momentum space, we obtain the continuity equation for the particles of the type α,

$$\frac{\partial}{\partial t} W^{(\alpha)}\left(\vec{r}, t\right) + \nabla \cdot \left[W^{(\alpha)}\left(\vec{r}, t\right) \vec{V}^{(\alpha)}\left(\vec{r}, t\right)\right] = 0. \tag{2.99}$$

Let us introduce the momentum flux density tensor for α species particles as

$$\Pi_{ij}^{(\alpha)}\left(\vec{r}, t\right) = \int d\vec{p} p_i v_j f_\alpha\left(\vec{p}, \vec{r}, t\right). \tag{2.100}$$

Multiplying Eq. (2.96) by $\vec{p}$ and integrating over the momentum, we obtain

$$\frac{\partial}{\partial t} W^{(\alpha)} \vec{V}_i^{(\alpha)} = -\frac{\partial}{\partial r_j} \Pi_{ij}^{(\alpha)} + \left\{ \rho^{(\alpha)} E_i + \frac{1}{c} \left(\vec{j}^{(\alpha)} \times \vec{\beta} \right)_i \right\}, \tag{2.101}$$

where

$$\rho^{(\alpha)}\left(\vec{r}, t\right) = e_\alpha \int d\vec{p} f_\alpha \left(\vec{p}, \vec{r}, t\right), \tag{2.102}$$

$$j^{(\alpha)}\left(\vec{r}, t\right) = e_\alpha \int d\vec{p} \;\; \vec{v} f_\alpha \left(\vec{p}, \vec{r}, t\right) \tag{2.103}$$

are the charge and current densities of the particles of the type α, respectively.

Equations (2.99) and (2.101) are similar to the hydrodynamic equations of a charged liquid [49, 50]. But, this system of equations is not closed since the stress tensor

$$W^{(\alpha)} V_i^{(\alpha)} V_j^{(\alpha)} - \Pi_{ij}^{(\alpha)}$$

is determined not only by hydrodynamic quantities $W^{(\alpha)}\left(\vec{r}, t\right)$ and $V^{(\alpha)}\left(\vec{r}, t\right)$ but also by the distribution function $f^{(\alpha)}\left(\vec{p}, \vec{r}, t\right)$.

In general, the calculation of the stress tensor is rather complicated. But, the situation becomes simpler when the stress tensor can be neglected. Hence, we assume

$$\Pi_{ij}^{(\alpha)} = W^{(\alpha)} V_i^{(\alpha)} V_j^{(\alpha)}. \tag{2.104}$$

Furthermore, for non-relativistic plasma, we have

$$\rho^{(\alpha)} = \frac{e_\alpha}{m_\alpha} W^{(\alpha)}, \quad \vec{j}^{(\alpha)} = \frac{e_\alpha}{m_\alpha} W^{(\alpha)} \vec{V}^{(\alpha)}. \tag{2.105}$$

Then, from Eqs. (2.99) and (2.101), it follows that

$$\frac{dV_i^{(\alpha)}}{dt} \equiv \frac{\partial V_i^{(\alpha)}}{\partial t} + \left(\vec{V}^{(\alpha)} \cdot \frac{\partial}{\partial \vec{r}} \right) V_i^{(\alpha)} = \frac{e_\alpha}{m_\alpha} \left\{ E_i + \frac{1}{c} \left[\vec{V}^{(\alpha)} \times \vec{B} \right]_i \right\}. \tag{2.106}$$

Equations (2.99) and (2.106) are often used in the elementary theory of plasma oscillations. By obtaining an expression for the dielectric permittivity of isotropic non-magnetized plasma, one can easily be convinced of the limitation of such a description, which was derived from Eq. (2.106). In the linear approximation, from Eq. (2.106), we can write

$$\vec{V}^{(\alpha)}\left(\vec{r},t\right) = \frac{e_\alpha}{m_\alpha}\int_{-\infty}^{t} dt'\vec{E}\left(\vec{r},t'\right).$$

Substituting this expression into the right-hand side of Eq. (2.105) and using relations (1.39), (1.45), and (1.46), we find the dielectric permittivity of plasma (for $\omega \neq 0$)

$$\varepsilon(\omega) = 1 - \sum_\alpha \frac{4\pi e_\alpha^2 N_\alpha}{m_\alpha \omega^2}.$$

This expression corresponds to Eq. (2.26) when spatial dispersion is completely neglected. This is quite natural because Eq. (2.106) was derived in the absence of the thermal motion of particles, which could lead to the spatial dispersion of the dielectric permittivity.

The influence of thermal motion of particles in plasma, as it was shown in Sects. 2.3 and 2.5, is inessential for transverse oscillations.[23] In addition, the longitudinal oscillations, in the long-wave limit, when their wavelength in this region is much larger than the Debye length, are weakly affected by the thermal effects. Consequently, the effect of thermal motion results only in small corrections. This means that the stress tensor (2.104) characterizing the difference between Eqs. (2.106) and (2.101) is small. Therefore, it is natural to suppose that when the wavelength or the characteristic size of inhomogeneity of plasma is larger than the Debye length (r_D), one can obtain a relatively simple expression for the stress tensor. To do this, let us use the solution of the kinetic equation in the form of Eq. (2.14)

$$\delta f\left(\vec{p},\vec{r},t\right) = -e\frac{\partial f_0}{\partial \vec{p}}\int_{-\infty}^{t} dt'\vec{E}\left[\vec{r} - \vec{v}(t-t'),t'\right].$$

Then, the momentum flux Π_{ij} can be written as

$$\Pi_{ij}\left(\vec{r},t\right) = -e\int_{-\infty}^{t} dt' \int d\vec{p}\; p_i v_j \frac{\partial f_0}{\partial p_l} E_l\left[\vec{r} - \vec{v}(t-t'),t'\right]. \tag{2.107}$$

For a dependency of the type of $\exp\left(-\imath\omega t + \imath\vec{k}\cdot\vec{r}\right)$, from Eq. (2.107), it follows that

[23] The exception is the problem of transverse wave penetration into plasma, see Sects. 2.8 and 2.9.

$$\Pi_{ij}\left(\omega,\vec{k}\right) = -\imath e\int d\vec{p}\; p_i v_j \frac{\partial f_0}{\partial p_l}$$
$$\times\left[\mathcal{P}\frac{1}{\omega-\vec{k}\cdot\vec{v}} - \imath\pi\delta\left(\omega-\vec{k}\cdot\vec{v}\right)\right]E_l\left(\omega,\vec{k}\right). \qquad (2.108)$$

In the long-wave limit, which is in our interest, the first term of the integrand in Eq. (2.108) may be expanded in powers of $\vec{k}$. Besides, if we consider $f_0(p)$ as the distribution function of particles in isotropic plasma and introduce a new notation $f_0' = \partial f_0/\partial\epsilon$ (see Sect. 2.2), then after some algebraic calculations we obtain

$$\Pi_{ij}\left(\omega,\vec{k}\right) = \frac{\imath e}{m\omega^2}P_0\left(k_j\delta_{il} + k_i\delta_{jl} + k_l\delta_{ij}\right)E_l - em\pi\frac{\omega^3}{k^6}$$
$$\times\int d\vec{p}\;\delta\left(\omega-\vec{k}\cdot\vec{v}\right)k_ik_j\left(\vec{k}\cdot\vec{E}\right)f_0',$$

where

$$P_0 = \frac{1}{m}\int d\vec{p}\,\frac{p^2}{3}f_0.$$

Suppose that the electric field is longitudinal. Therefore,

$$\vec{E} = \frac{\vec{k}}{\imath k^2}4\pi\delta\rho,$$

where $\delta\rho$ is the non-equilibrium charge density. For pure electron plasma, $\delta\rho = (e/m)\delta W$ and consequently

$$\Pi_{ij}\left(\omega,\vec{k}\right) = \frac{4\pi P_0 e^2}{m^2\omega^2}\left(\delta_{ij} + \frac{2k_ik_j}{k^2}\right)\delta W\left(\omega,\vec{k}\right)$$
$$+\imath 4\pi^2e^2\int d\vec{p}\delta\left(\omega-\vec{k}\cdot\vec{v}\right)f_0'\frac{\omega^3}{k^6}k_ik_j\delta W\left(\omega,\vec{k}\right). \qquad (2.109)$$

The first term in the right-hand side of Eq. (2.109) characterizes the elastic properties of electron plasma, which are stipulated by the thermal motion of particles, whereas the second term determines dissipation [44, 51]. According to Eq. (2.109), the connection between quantities $\Pi_{ij}\left(\vec{r},t\right)$ and $\delta W\left(\vec{r},t\right)$ is spatially non-local, which leads to the non-locality of the hydrodynamic equations. This non-locality is the result of consideration of spatial dispersion. In this sense, one can talk about hydrodynamics with spatial dispersion and just this is the principal difference between the hydrodynamics of collisionless plasmas and the usual hydrodynamics of liquids and gases. Certainly, such a difference arises only when the thermal motion of particles is taken into account. Therefore, when the thermal motion of particles is absent, Eqs. (2.99) and (2.106) are applicable.

The approximate equations of the hydrodynamic quantities of electron–ion plasma, as well as electron plasma, generally are the integral equations in time and in space coordinates. However, there exists one special case when such equations become local in the absence of dissipation. This case corresponds to non-isothermal plasma in which $T_e \gg T_i$ and the ion temperature may be considered zero. Besides, the characteristic size of inhomogeneity of the hydrodynamic quantities should be considered larger than the Debye screening radius and the characteristic time intervals have to be larger than the period of the ion Langmuir oscillations $2\pi/\omega_{pi}$. Below, we will consider this case.

If we suppose that the ion temperature is zero, then Eqs. (2.99) and (2.106) will become valid for ions as well. But, since the ions compose almost the whole mass of plasma, Eq. (2.99) may be considered as the continuity equation of matter:

$$\frac{\partial W}{\partial t} + \nabla \cdot \left(W\vec{V} \right) = 0. \tag{2.110}$$

Moreover, the ion velocity $\vec{V}$ must be considered as the velocity of matter.

In isotropic plasma, the equation of motion (2.106) becomes simpler for the small perturbations from the equilibrium state:

$$\frac{d\vec{V}}{dt} = \frac{e_i}{M}\vec{E}. \tag{2.111}$$

This form of equation is far from the usual hydrodynamic form because it includes the electric field strength, which must be determined by the Maxwell's equations. In the case of non-isothermal plasma considered below, this quantity can be expressed in terms of the hydrodynamic quantities.

In order to do this, let us consider the electrons' motion. As the thermal velocity of electrons is sufficiently large, the time derivative term in the kinetic equation for electrons can be neglected. Then, in the absence of the constant magnetic field, we have

$$\vec{v} \cdot \frac{\partial f_e}{\partial \vec{r}} + e\vec{E} \cdot \frac{\partial f_e}{\partial \vec{p}} = 0.$$

The solution of this equation may be written as

$$f_e\left(\vec{p}, \vec{r}\right) = f_0\left(\vec{p}\right) \exp\left(-\frac{e\phi}{\kappa T_e} \right),$$

where $f_0\left(\vec{p}\right)$ is Maxwell distribution (2.47) and ϕ is the scalar potential $\left(\vec{E} = -\nabla\phi\right)$, satisfying the Poisson's equation

$$\nabla^2\phi = -4\pi(\rho_i + \rho_e) = -4\pi\left(\rho_i + eN_e \exp\left(-\frac{e\phi}{\kappa T_e}\right)\right).$$

When the characteristic size of variation of $\phi\left(\vec{r}\right)$ is much smaller than the Debye length, the left-hand side of this equation may be omitted. Then,

$$\phi = -\frac{\kappa T_e}{e} \ln\left(-\frac{\rho_i}{eN_e}\right),$$

and as a result

$$\vec{E} = \frac{\kappa T_e}{e} \frac{\vec{\nabla}\rho_i}{\rho_i} = \frac{\kappa T_e}{e} \frac{\vec{\nabla} W}{W}.$$

Substituting this expression into Eq. (2.111), we obtain

$$W\frac{d\vec{V}}{dt} = -\frac{Z\kappa T_e}{M}\vec{\nabla} W, \tag{2.112}$$

where $Z = |e_i/e|$. This equation was applied to the problem of ion oscillation in plasma by Tonks and Langmuir, but they did not consider the restrictions of this equation, namely the requirement that the electron temperature must be much higher than the ion temperature [25, 50]. Let us show the necessity of this requirement.[24]

Equation (2.112) has the hydrodynamic form and the quantity $Z\kappa T_e/M$ plays the role of compressibility of plasma $\partial P/\partial W$ in this equation. In this sense, it must be noted that the sound velocity in accordance to Eq. (2.112) is equal to $v_s = \sqrt{Z\ \kappa T_e/M}$. But, this result corresponds to Eq. (2.94), which is correct only when the ion temperature is much less than the electron temperature. As soon as this condition is broken, the sound velocity will become of the order of the ion thermal velocity. Then, the majority of the ions' population may move in phase with the sound wave, which in turn results in the strong dissipation of sound waves. In other words, the dissipative terms will be comparable to the elastic terms and, as a result, Eq. (2.112) is no longer applicable.

---

[24]The longitudinal dielectric permittivity of plasma in this hydrodynamic limit may be represented as

$\varepsilon^l(\omega, k) = \frac{4\pi e^2 N_e}{\kappa T_e k^2} - \frac{4\pi e_i^2 N_i}{M\omega^2}$.

This expression corresponds to the ion contribution when $k = 0$ and to the electron contribution when $\omega = 0$ in the dielectric permittivity.

2.7 Dielectric Permittivity of Plasma; Taking account of Particle Collisions

So far, we have completely neglected particle collisions in plasma. Taking them into account enables us to determine the limits of applicability of the collisionless plasma approximation used above. For simplicity, we will restrict our consideration to the regions of frequency and wavelength, in which the spatial dispersion of dielectric permittivity is relatively weak.

For transverse waves in non-relativistic plasma, such a situation takes place in the high-frequency region when the frequency is much higher than the Langmuir frequency of electrons ω_{pe}. Longitudinal waves, besides, must satisfy the inequality $\lambda \gg r_D$ (see Sects. 2.3 and 2.5). Under these conditions, the expressions for the transverse and longitudinal dielectric permittivities for collisionless plasma can be represented as

$$\varepsilon^{\text{tr}}(\omega, k) = 1 - \frac{\omega_{\text{pe}}^2}{\omega^2}, \tag{2.113}$$

$$\varepsilon^{l}(\omega, k) = 1 - \frac{\omega_{\text{pe}}^2}{\omega^2}\left(1 + \frac{3\kappa T_e k^2}{m\omega_{\text{pe}}^2}\right) + \imath\sqrt{\frac{\pi}{2}}\frac{\omega_{\text{pe}}^2\omega}{k^3\left(\frac{\kappa T_e}{m}\right)^{3/2}}\exp\left(-\frac{\omega^2 m}{2\kappa T_e k^2}\right). \tag{2.114}$$

The ion's contributions are small and, therefore, we neglect them in these expressions.

For generalizing expressions (2.113) and (2.114) to the case of collisional plasma, one should start from the kinetic equation (2.3)

$$\frac{\partial f_\alpha}{\partial t} + \vec{v}\cdot\frac{\partial f_\alpha}{\partial \vec{r}} + e_\alpha\left\{\vec{E} + \frac{1}{c}\left[\vec{v}\times\vec{B}\right]\right\}\cdot\frac{\partial f_\alpha}{\partial \vec{p}} = \left(\frac{\partial f_\alpha}{\partial t}\right)_c, \tag{2.115}$$

where $(\partial f_\alpha/\partial t)_c$ is the collision integral considering the collisions of particles of the type α with each other and with all types of other particles.

Firstly proposed by Boltzmann, the collision integral

$$\left(\frac{\partial f_\alpha}{\partial t}\right)_c = \sum_\beta \int d\vec{p}_1 d\Omega \frac{d\sigma(u,\vartheta)}{d\Omega} u\left\{f_\alpha\left(\vec{p}\right)f_\beta\left(\vec{p}_1\right) - f_\alpha\left(\vec{p}'\right)f_\beta\left(\vec{p}'_1\right)\right\} \tag{2.116}$$

is often used. Here, $\vec{p}, \vec{p}_1$ and $\vec{p}', \vec{p}'_1$ are the momenta of the colliding particles before and after their collisions, respectively; $\vec{u} = \mid \vec{p}/m_\alpha - \vec{p}_1/m_\beta \mid$, and ϑ is the angle between $\left(\vec{p}/m_\alpha - \vec{p}_1/m_\beta\right)$ and $\left(\vec{p}'/m_\alpha - \vec{p}'_1/m_\beta\right)$ (scattering angle); $d\sigma/d\Omega$ is the differential scattering cross section over solid angle Ω, and $d\Omega = \sin\vartheta d\vartheta d\varphi$; m_α and m_β are the mass of α and β particles species, respectively. We shall consider only elastic collisions and, as a result, the quantities $\vec{p}, \vec{p}_1$ and $\vec{p}', \vec{p}'_1$ satisfy the energy and momentum conservation laws.

We will begin our analysis of the collisional effects by investigating completely ionized plasma, when only collisions of charged particles are important.[25] As the strength of the Coulomb interaction between the charged particles falls very slowly with distance, the collisions with large impact parameters play the major role. Therefore, the energy and momentum exchange in collisions are small. As it was shown by Landau, this allows us to simplify the collision integral (2.116) and represent it in the following form[26] [12]:

$$\left(\frac{\partial f_\alpha}{\partial t}\right)_c = \frac{\partial}{\partial p_i}\int d\vec{p}'\sum_\beta I_{ij}^{\alpha\beta}\left(\frac{\vec{p}}{m_\alpha}-\frac{\vec{p}'}{m_\beta}\right) \times \left\{f_\beta\left(\vec{p}'\right)\frac{\partial f_\alpha\left(\vec{p}\right)}{\partial p_j} - f_\alpha\left(\vec{p}\right)\frac{\partial f_\beta\left(\vec{p}'\right)}{\partial p'_j}\right\}, \tag{2.117}$$

where

$$I_{ij}^{\alpha\beta}(\vec{v}) = 2\pi e_\alpha^2 e_\beta^2 \frac{v^2\delta_{ij} - v_i v_j}{v^3}L, \qquad L = \ln\frac{r_D}{\rho_{\min}}. \tag{2.118}$$

Here, $\rho_{\min}$ is the minimum impact parameter when the deflection of colliding particles is small. In the classic scattering theory $(e^2/\hbar v \gg 1)$, $\rho_{\min}$~$e^2/\kappa T$, whereas in the quantum limit $(e^2/\hbar v \ll 1)$ $\rho_{\min} \sim \lambda_b$, where $\lambda_b = \hbar/mv$ is the de Broglie wavelength of a free particle.

In the high-frequency region, when $\omega \geq \omega_{pe}$, which is in our interest, the analysis of collision effects becomes simple, because the collision term $(\partial f/\partial t)_{st}$ in the kinetic equation (2.115) turns out to be small. Below, it will be shown that, in this case, the collision term results in the corrections in the plasma dielectric permittivity of order of $\sim\nu_{eff}/\omega\sim\nu_{eff}/\omega_{pe} \ll 1$, where ν_{eff} is the effective collision frequency.

---

[25] For more details, see [52, 53].

[26] This expression can be obtained from Eq. (2.116) by expanding it in powers of the momentum change via collisions. For fast varying processes, the collision integral was obtained in [9]

$$\left(\frac{\partial f_\alpha}{\partial t}\right)_c = \frac{\partial}{\partial p_i}\int d\vec{p}'\sum_\beta \int_{\tau_{\min}}^{\tau_{\max}}\frac{d\tau}{\tau}\frac{1}{L}I_{ij}^{\alpha\beta}\left(\frac{\vec{p}}{m_\alpha}-\frac{\vec{p}'}{m_\beta}\right) \left\{f_\beta\left(\vec{p}',t-\tau\right)\frac{\partial f_\alpha\left(\vec{p},t-\tau\right)}{\partial p_j} - f_\alpha\left(\vec{p}',t-\tau\right)\frac{\partial f_\beta\left(\vec{p}',t-\tau\right)}{\partial p'_j}\right\},$$

where $\tau_{\max} = 1/\omega_{pe}$ and $\tau_{\min} = \rho_{\min}/v$.

Then, because of the smallness of the spatial derivative term in Eq. (2.115) in the region of weak spatial dispersion, this term can be omitted. This allows us to write the kinetic equation in a linear approximation in the following form:

$$-\imath\omega\delta f_\alpha + e_\alpha \vec{E}\cdot\frac{\partial f_{0\alpha}}{\partial\vec{p}} = \left(\frac{\partial f_\alpha}{\partial t}\right)_c. \tag{2.119}$$

Here, we supposed that the time dependence of the perturbations is of the form of $\exp(-\imath\omega t)$. Besides, if we consider that the collision integral vanishes for the Maxwellian distribution, we can write it in the linear approximation as follows:

$$\begin{aligned}\left(\frac{\partial f_\alpha}{\partial t}\right)_c = \frac{\partial}{\partial p_i}\int d\vec{p'}\sum_\beta I_{ij}^{\alpha\beta}\left(\frac{\vec{p}}{m_\alpha}-\frac{\vec{p'}}{m_\beta}\right)\Bigg\{ f_{0\beta}\left(\vec{p'}\right)\frac{\partial\delta f_\alpha\left(\vec{p}\right)}{\partial p_j} + \delta f_\beta\left(\vec{p'}\right)\frac{\partial f_{0\alpha}\left(\vec{p}\right)}{\partial p_j} \\ - f_{0\alpha}\left(\vec{p}\right)\frac{\partial\delta f_\beta\left(\vec{p'}\right)}{\partial p'_j} - \delta f_\alpha\left(\vec{p}\right)\frac{\partial f_{0\beta}\left(\vec{p'}\right)}{\partial p'_j}\Bigg\}.\end{aligned} \tag{2.120}$$

Regarding that the collision term in Eq. (2.119) is small, we find the solution of Eq. (2.119) in the first approximation as

$$\delta f_\alpha^{(1)}\left(\vec{p}\right) = -\frac{\imath e_\alpha}{\omega}\vec{E}\cdot\frac{\partial f_{0\alpha}\left(\vec{p}\right)}{\partial\vec{p}}. \tag{2.121}$$

This solution gives the following expression for the current density:

$$\vec{j} = -\frac{\imath}{\omega}\sum_\alpha e_\alpha^2\int\vec{v}\left(\vec{E}\cdot\frac{\partial f_{0\alpha}\left(\vec{p}\right)}{\partial\vec{p}}\right)d\vec{p} = \frac{\imath}{\omega}\sum_\alpha\frac{e_\alpha^2 N_\alpha}{m_\alpha}\vec{E}. \tag{2.122}$$

If we neglect the small ion's contribution, i.e., terms of the order of $\sim m/M < 0.1$, the well-known dielectric permittivity of collisionless electron plasma in the absence of spatial dispersion would appear [compared to Eq. (2.113)]

$$\varepsilon(\omega) = 1 - \frac{\omega_{\text{pe}}^2}{\omega^2}.$$

Now, let us determine the collisional correction for this expression by taking into account the collision integral in Eq. (2.119). Substituting Eq. (2.121) into Eq. (2.120), we find

$$\delta f_\alpha^{(2)}\left(\vec{p}\right) = \frac{\imath}{\omega}\left(\frac{\partial f_\alpha}{\partial t}\right)_c = \frac{1}{\omega^2} E_l \frac{\partial}{\partial p_i} \int d\vec{p'} \sum_\beta I_{ij}^{\alpha\beta}\left(\frac{\vec{p}}{m_\alpha} - \frac{\vec{p'}}{m_\beta}\right) \times$$
$$\times \left\{ f_{0\beta}\left(\vec{p'}\right) e_\alpha \frac{\partial^2 f_{0\alpha}\left(\vec{p}\right)}{\partial p_j \partial p_l} + e_\beta \frac{\partial f_{0\beta}\left(\vec{p'}\right)}{\partial p'_l} \frac{\partial f_{0\alpha}\left(\vec{p}\right)}{\partial p_j} \right. \tag{2.123}$$
$$\left. - e_\beta f_{0\alpha}\left(\vec{p}\right) \frac{\partial^2 f_{0\beta}\left(\vec{p'}\right)}{\partial p'_j \partial p'_l} - e_\alpha \frac{\partial f_{0\alpha}\left(\vec{p}\right)}{\partial p_l} \frac{\partial f_{0\beta}\left(\vec{p'}\right)}{\partial p'_j} \right\}.$$

This expression provides a correction to the induced current density in plasma. Here, only the electron current takes a significant part because the ion current is inversely proportional to their mass. Moreover, it can be easily shown that the contribution from electron–electron collisions vanishes (this is a consequence of the momentum conservation law). Furthermore, due to large mass, the ion distribution can be considered in equilibrium. Then, by neglecting the terms proportional to ion charge with respect to the contribution of the order of ~m/M, we obtain the following relation for the collisional correction of the current density:

$$j_i^{(2)} = \frac{e^2 E_j}{m^2 \omega^2 \kappa T_e} \int d\vec{p}\, d\vec{p'} \sum_\beta I_{ij}^{\alpha\beta}\left(\frac{\vec{p}}{m_\alpha} - \frac{\vec{p'}}{m_\beta}\right) f_{0e}\left(\vec{p}\right) f_{0\beta}\left(\vec{p'}\right) = \sigma_{ij}(\omega) E_j.$$

Here, the summation extends over all types of ions in plasma, and $f_{0\alpha}(p)$ is the Maxwellian distribution function. After integrations, it follows that

$$\sigma_{ij}(\omega) = \sigma(\omega)\delta_{ij}, \qquad \sigma(\omega) = \frac{e^2 N_e}{m\omega^2} \nu_{\text{eff}}, \tag{2.124}$$

where the effective collision frequency is equal to

$$\nu_{\text{eff}} = \frac{4}{3}\sqrt{\frac{2\pi}{m}} \frac{e^2 L}{\left(\kappa T_e\right)^{3/2}} \sum_\beta e_\beta^2 N_\beta. \tag{2.125}$$

Here, the term ~$(m/M)\,(T_i/T_e)$ has been neglected. By considering the collisional correction, the complete dielectric permittivity in the high-frequency region can be written as

$$\varepsilon(\omega) = 1 - \frac{\omega_{\text{pe}}^2}{\omega^2}\left(1 - \imath \frac{\nu_{\text{eff}}}{\omega}\right).$$

Now, we can estimate the collisional correction and justify the above assumption about the collision integral. We have

$$\frac{\nu_{\text{eff}}}{\omega_{\text{pe}}} = \sqrt{\frac{8}{9}} \frac{eL}{(\kappa T_e)^{3/2}} \frac{\sum_{\beta} e_{\beta}^2 N_{\beta}}{\sqrt{N_e}} \approx \left(\frac{e^2 N_e^{1/3}}{\kappa T_e}\right)^{3/2}$$

$$\ln \frac{e^2 N^{1/3}}{\kappa T_e} = \left(\frac{e^2 N_e^{1/3}}{\kappa T_e}\right)^{3/2} \ln \frac{r_D}{\rho_{\min}} \ll 1.$$

Strictly speaking, the last inequality is valid only for classical plasmas, but a quite similar inequality takes place in the quantum limit. It justifies our assumption. In the consideration taken above, we have completely neglected spatial dispersion. Now, let us consider it. In the long-wave limit, when spatial dispersion is weak, it is always negligible for the transverse dielectric permittivity. But, for the longitudinal dielectric permittivity, as it is seen from expression (2.114), the influence of weak spatial dispersion may be essential. Actually, if we add the correction expression (2.125), which takes into account the particle collisions to expression (2.114), then the longitudinal dielectric permittivity takes the following form:

$$\varepsilon^l(\omega, k) = 1 - \frac{\omega_{\text{pe}}^2}{\omega^2}\left(1 + 3\frac{k^2 \kappa T_e}{m\omega_{\text{pe}}^2}\right) + \imath \frac{\nu_{\text{eff}} \omega_{\text{pe}}^2}{\omega^3} + \imath \sqrt{\frac{\pi}{2}} \frac{\omega_{\text{pe}}^2 \omega}{k^3 \left(\frac{\kappa T_e}{m}\right)^{3/2}} \exp\left(-\frac{\omega^2 m}{2k^2 \kappa T_e}\right). \tag{2.126}$$

Now, we can compare the collisional dissipative term to the collisionless Cherenkov dissipation. Considering that the frequency of the long-wave longitudinal oscillation is of the order of ω_{pe}, we conclude that if

$$\nu_{\text{eff}} \gg \frac{\omega_{\text{pe}}}{k^3 r_D^3} \exp\left(-\frac{1}{k^2 r_D^2}\right), \tag{2.127}$$

the collisional dissipation will be dominant. It can be neglected in the opposite limit.

Thus, one can determine the critical wavelength of the high-frequency longitudinal oscillations

$$\lambda_{cr} \sim r_D \sqrt{\ln \frac{\kappa T_e}{e^2 N_e^{1/3}}}. \tag{2.128}$$

When $\lambda > \lambda_{\text{cr}}$, the dissipation is determined by the electron collisions, whereas in the opposite limit, when $\lambda < \lambda_{\text{cr}}$, the Cherenkov dissipation dominates.

It must be noticed that formulas (2.124) and (2.126) can be obtained from the kinetic equation with approximate collision integral of the form[27] [8]

$$\left(\frac{\partial f}{\partial t}\right)_c = -\nu_{\text{eff}}\delta f. \tag{2.129}$$

Of course, such an equation is much simpler than Eqs. (2.115) and (2.117). Here, it should be mentioned that the kinetic equation

$$\frac{df_\alpha}{dt} = \frac{\partial f_\alpha}{\partial t} + \frac{d\vec{r}}{dt} \cdot \frac{\partial f_\alpha}{\partial \vec{r}} + \frac{d\vec{p}}{dt} \cdot \frac{\partial f_\alpha}{\partial \vec{p}} = \left(\frac{\partial f}{\partial t}\right)_c$$

for the distribution function of the particles of type α which takes into account close collisions with all particles of type β through collision integral (2.117) can be written as

$$\frac{df_\alpha}{dt} = \frac{\partial f_\alpha}{\partial t} + \frac{d\vec{r}}{dt} \cdot \frac{\partial f_\alpha}{\partial \vec{r}} + \frac{d\vec{p}}{dt} \cdot \frac{\partial f_\alpha}{\partial \vec{p}} = \sum_\beta \left(\frac{\partial f}{\partial t}\right)_c^{\alpha\beta} = \frac{\partial}{\partial p_{\alpha i}}\left(D_{ij}\frac{\partial f_\alpha}{\partial p_{\alpha j}} - A_i f_\alpha\right),$$

where

$$D_{ij} = \sum_\beta \int d\vec{p}_\beta I_{ij}^{\alpha\beta}\left(\vec{p}_\alpha, \vec{p}_\beta\right) f_\beta\left(\vec{p}_\beta\right),\quad A_i = \sum_\beta \int d\vec{p}_\beta I_{ij}^{\alpha\beta}\left(\vec{p}_\alpha, \vec{p}_\beta\right) \frac{\partial f_\beta\left(\vec{p}_\beta\right)}{\partial p_{\beta j}}$$

are the diffusion and friction coefficients in the momentum space, respectively. This kinetic equation is frequently called the Fokker–Planck equation. The collision integral (2.117) and the Fokker–Planck equation are valid quite generally since they do not specify the interaction law. In particular, they can be applied to plasmas with any degree of ionization if the law of binary interaction is known.

To conclude, let us consider the collisional effects in weakly ionized plasma where elastic collisions of charged particles with neutrals dominate. The description of such collisions by means of Boltzmann integral is mathematically very complicated. Therefore, below we will use a model for the collisional term of charged particles with neutrals, the so-called Bhatnagar–Gross–Krook (BGK) model, which is of great importance [54]. It cannot be derived from the general Boltzmann integral

[27] Collision integral (2.120) is suitable only for frequencies $\omega \ll \omega_{pe}$. As it was noted above, for fast varying processes, i.e., in high frequencies region, the Coulomb logarithm $L = \ln(r_D/\rho_{\min})$ in the collision integral must be replaced by $L' = \ln[v_T/(\omega\rho_{\min})]$. For longitudinal oscillations, $\omega \sim \omega_{pe}$; therefore, such a replacement is inessential. But, for transverse waves in the frequency region $\omega \gg \omega_{pe}$, such a replacement can result in significant effects [8, 9].

by means of any approximation; it can be only constructed by general physical reasoning.

In weakly ionized plasma, the conservation laws of particle number, momentum, and energy are satisfied by the BGK integral

$$\left(\frac{\partial f}{\partial t}\right)_c = -\sum_\beta \nu_{\alpha\beta}\left(f_\alpha - N_\alpha \phi_{\alpha\beta}\right), \tag{2.130}$$

where velocity-independent collision frequency $\nu_{\alpha\beta}$ describes the momentum relaxation of the particles of the type α due to collisions with the particles of the type β; it should be determined in experiments. Here,

$$\phi_{\alpha\beta} = \frac{1}{\left(2\pi m_\alpha \kappa T_{\alpha\beta}\right)^{3/2}} \exp\left[-\frac{m_\alpha\left(\vec{v} - \vec{V}_\alpha\right)^2}{2\kappa T_{\alpha\beta}}\right],$$

$$N_\alpha = \int d\vec{p} f_\alpha, \vec{V}_\alpha = \frac{1}{N_\alpha}\int d\vec{p}\,\vec{v} f_\alpha,\ T_{\alpha\beta} = \frac{m_\alpha T_\beta + m_\beta T_\alpha}{m_\alpha + m_\beta},$$

$$\kappa T_\alpha = \frac{m_\alpha}{3N_\alpha}\int d\vec{p}\left(\vec{v} - \vec{V}_\alpha\right)^2 f_\alpha. \tag{2.131}$$

The conservation of momentum and energy requires $m_\alpha \nu_{\alpha\beta} N_\alpha = m_\beta \nu_{\beta\alpha} N_\beta$.

We can estimate $\nu_{\alpha\beta}$ by using some simple molecular kinetic considerations for the scattering of charged particles from neutrals and writing $\nu_{\alpha n} = v_{T\alpha}\sigma_0 N_n$, where σ_0, the effective scattering cross section, is of the order of $\sigma_0 = \pi a^2$, and $a \approx 10^{-8} cm$ is the radius of the neutral particles with density N_n. Further simplification of the BGK integral can be obtained by assuming that $m_i = M_n = M$ and $T_n = T_i$. This is the case if plasma ions are generated by ionization of neutral particles of the same substance. By neglecting the terms of the order of m/M, as above, we have $T_{\text{en}} = T_e$. Under this condition, function $\phi_{\alpha n}$ coincides with the Maxwellian distribution function normalized to one, or

$$\phi_{\alpha n} = \frac{1}{N_{0\alpha}} f_{0\alpha}(p) = \frac{1}{\left(2\pi m_\alpha \kappa T_\alpha\right)^{3/2}} \exp\left(-\frac{m_\alpha v^2}{2\kappa T_\alpha}\right).$$

For small perturbation δf_α, we can linearize the kinetic equation with the BGK collision integral

$$\frac{\partial \delta f_\alpha}{\partial t} + \vec{v}\cdot\frac{\partial \delta f_\alpha}{\partial \vec{r}} + e_\alpha \vec{E}\cdot\frac{\partial f_{0\alpha}}{\partial \vec{p}} = -\nu_{\alpha n}\left(\delta f_\alpha - \phi_{\alpha n}\int d\vec{p}\,\delta f_\alpha\right). \tag{2.132}$$

The solution of this linear integral equation for the plane monochromatic waves of the type $e^{\left(-\imath\omega t+\imath\vec{k}\cdot\vec{r}\right)}$ without any restriction on ω and $\vec{k}$ can be written as

$$\delta f_{\alpha}=\frac{\imath e_{\alpha}}{\kappa T_{\alpha}}\frac{\left(\vec{v}\cdot\vec{E}\right)f_{0\alpha}(p)}{\omega+\imath\nu_{\alpha n}-\vec{k}\cdot\vec{v}}+\imath\frac{\nu_{\alpha n}\eta_{\alpha}f_{0\alpha}(p)}{\omega+\imath\nu_{\alpha n}-\vec{k}\cdot\vec{v}}, \tag{2.133}$$

where

$$\eta_{\alpha}=\frac{1}{N_{0\alpha}}\int d\vec{p}\,\delta f_{\alpha}.$$

Integrating Eq. (2.133) over momentum and using the continuity equation for the particles of the type α, we obtain

$$\eta_{\alpha}=\frac{\vec{k}\cdot\vec{j}_{\alpha}}{e_{\alpha}\omega N_{0\alpha}},\qquad \vec{j}_{\alpha}=e_{\alpha}\int d\vec{p}\,\vec{v}\,\delta f_{\alpha}. \tag{2.134}$$

From Eqs. (2.133) and (2.134), we can determine current density $\vec{j}_{\alpha}$ in terms of $\vec{E}$ and then the following expressions for the longitudinal and transverse dielectric permittivities:

$$\varepsilon^{l}(\omega,k)=1+\sum_{\alpha}\frac{\omega_{p\alpha}^{2}}{k^{2}v_{T\alpha}^{2}}\frac{1-I_{+}\left(\frac{\omega+\imath\nu_{\alpha n}}{kv_{T\alpha}}\right)}{1-\frac{\imath\nu_{\alpha n}}{\omega+\imath\nu_{\alpha n}}I_{+}\left(\frac{\omega+\imath\nu_{\alpha n}}{kv_{T\alpha}}\right)}, \tag{2.135}$$

$$\varepsilon^{\mathrm{tr}}(\omega,k)=1-\sum_{\alpha}\frac{\omega_{p\alpha}^{2}}{\omega(\omega+\imath\nu_{\alpha n})}I_{+}\left(\frac{\omega+\imath\nu_{\alpha n}}{kv_{T\alpha}}\right). \tag{2.136}$$

Here, the summation extends over the charged particles species only. In the long-wave limit $\left(\vec{k}\to 0\right)$, when spatial dispersion can be neglected, from Eqs. (2.135) and (2.136), it follows that

$$\varepsilon^{l}(\omega,0)=\varepsilon^{\mathrm{tr}}(\omega,0)=\varepsilon(\omega)=1-\frac{\omega_{\mathrm{pe}}^{2}}{\omega(\omega+\imath\nu_{\mathrm{en}})}. \tag{2.137}$$

In this case, the ion's contribution is negligible. Besides, in the high-frequency limit, $\omega\gg\nu_{\mathrm{en}}$, if we change $\nu_{\mathrm{eff}}\to\nu_{\mathrm{en}}$, this expression coincides with Eq. (2.125). Hence, we can determine a condition on the ionization degree under which plasma can be considered weakly ionized. If

$$\frac{\nu_{\text{en}}}{\nu_{\text{eff}}} \sim \frac{a^2(\kappa T_e)^2}{e^4 L} \frac{N_n}{N_e} \gg 1, \tag{2.138}$$

plasma is considered weakly ionized, whereas in the opposite limit, we can talk about completely ionized plasma. It is of interest to notice that plasma can be considered completely ionized even when N_e/N_n is very small. For example, if $T_e \approx 10^4 k^{\circ}$, then the opposite limit of the inequality (2.138) will be satisfied, beginning from $N_e/N_n > 10^{-4}$. In conclusion, it must be noticed that, in the static limit ($\omega \to 0$), from Eq. (2.135), we have

$$\varepsilon^l(0,k) = 1 + \sum_\alpha \frac{\omega_{\text{pe}}^2}{k^2 v_{T_\alpha}^2} = 1 + \frac{1}{k^2 r_D^2}.$$

This means that in both collisional and collisionless plasma, the Debye screening of the electrostatic field happens (see Sect. 2.3).

2.8 Boundary Problem of Fields in Plasma

In this part, we consider the boundary problem for a transverse monochromatic field in semi-bounded electron plasma.[28] This problem is connected to the reflection and absorption of electromagnetic waves normally incident to the plasma surface. Here, the vector of the electric field is parallel to the boundary surface of plasma ($E_x \neq 0, E_y = E_z = 0$).

To set up a closed system of field equations, it is necessary to calculate the current density induced in semi-bounded plasma. For this purpose, we introduce the effective collision frequency ν in the right-hand side of Eq. (2.12) [see Eq. (2.129)]. For non-relativistic plasma, this frequency coincides with ν_{eff} when $\omega \gg \nu_{\text{eff}}$. In the opposite case, we can completely neglect the particles' collisions. In this case, we use the kinetic equation for the perturbation of distribution function δf with collision integral (2.129) with $\nu \equiv \nu_{\text{eff}} \to 0$. Then, for the perturbation $\delta f = \delta f (z) \exp(-\imath\omega t + \imath kx)$, we obtain

$$(-\imath\omega + \nu)\delta f + v_z \frac{\partial \delta f}{\partial z} + e\left(\vec{v} \cdot \vec{E}\right) f'_{0e} = 0, \tag{2.139}$$

[28] For degenerate electron plasma, this problem was considered in [55], and for Maxwellian plasma in [16, 51, 56, 57]. The boundary problem for the longitudinal field was studied in [16, 27]. In all of these works, the question of plasma confinement was not discussed. Such a problem for semi-bounded plasma, confined by the strong magnetic field, was considered in [58]. To read the boundary problem for moving plasmas, see [59].

where f'_{0e} is the energy derivative of the equilibrium distribution function. The general solution of Eq. (2.139) can be written as

$$\delta f(z) = -\int_{C(\vec{p})}^{z} \frac{e\vec{v}\cdot\vec{E}(z')}{v_z} f'_{0e} dz' \exp\left[\frac{z-z'}{v_z}(-\imath\omega + \nu + \imath k_x v_x)\right]. \tag{2.140}$$

Arbitrary function $C\left(\vec{p}\right)$ may be determined from the boundary conditions. By introducing

$$\delta f^{(\mp)} = \begin{cases} \delta f(v_z < 0, z), \\ \delta f(v_z > 0, z), \end{cases}$$

from condition $\delta f^{(-)}(z = +\infty) = 0$, we find

$$\delta f^{(-)} = \int_z^{\infty} dz' \frac{e\vec{v}.\vec{E}(z')}{v_z} f'_{0e} \exp\left[-\frac{z-z'}{v_z}(-\imath\omega + \nu)\right]. \tag{2.141}$$

To determine function $C\left(\vec{p}\right)$ for $v_z > 0$, let us consider the boundary condition for δf on the plasma surface $z = 0$. For mirror reflection, we have $f\left(v_x, v_y, v_z, z = 0\right) = f\left(v_x, v_y, -v_z, z = 0\right)$. In the opposite, for purely diffusive reflection of particles, $\delta f^{(+)}$, the non-equilibrium distribution function of electrons flying from the surface $z = 0$ should be equal to zero. We will assume that a part of electrons (denoted by ρ) undergo a mirror reflection from this surface, whereas the other part $(1 - \rho)$ undergoes a diffusive reflection. This implies that at $z = 0$, the condition

$$f_{0e} + \delta f^{(+)}(z = 0) = \rho\left[f_{0e} + \delta f^{(-)}(z = 0)\right] + (1 - \rho)f_{0e},$$

is satisfied. Then, we find

$$\delta f^{(+)}(z = 0) = \rho\delta f^{(-)}(z = 0). \tag{2.142}$$

Using this relation, we find

$$\delta f^{(+)}(\vec{v},z) = -\int_0^z dz' \frac{e\vec{v}\cdot\vec{E}(z')}{v_z} f'_{0e} \exp\left[-\frac{(z-z')}{v_z}(-\imath\omega+\nu+\imath k_x v_x)\right] +$$

$$+\rho\int_0^\infty dz' \frac{e\vec{v}\cdot\vec{E}(z')}{v_z} f'_{0e} \exp\left[-\frac{(z+z')}{v_z}(-\imath\omega+\nu+\imath k_x v_x)\right]. \tag{2.143}$$

Substituting Eqs. (2.141) and (2.143) into the current density relation, after simple calculations, we obtain

$$j_i(z) = \int_0^\infty dz' \{K_{ij}(\mid z-z'\mid) + \rho K_{ij}(\mid z+z'\mid)\} E_j(z'), \qquad (z \geq 0). \tag{2.144}$$

Here, we have defined

$$K_{ij}(|z|) = -e^2 \int_{v_z \geq 0} d\vec{p} \frac{v_i v_j}{v_z} f'_{0e} \exp\left[\imath \frac{\mid z\mid}{v_z}(\omega+\imath\nu-k_x v_x)\right]. \tag{2.145}$$

To determine the fields in plasma, we have to consider the following equation:

$$\frac{d^2 E_x}{dz^2} + \frac{\omega^2}{c^2} E_x = -\frac{4\pi\imath\omega}{c^2} j \qquad (z \geq 0), \tag{2.146}$$

in which the induced current density is determined by Eq. (2.144).

Let us begin from the purely mirror reflection case when $\rho = 1$. Then, field equation (2.146) can be extended into negative value of z by considering $E_x(-z) = E_x(z)$. In this case, we find

$$j(z) = \int_{-\infty}^{+\infty} dz' K(|z-z'|) E_x(z'), \qquad (-\infty \leq z \leq +\infty). \tag{2.147}$$

It should be noted that under the continuation of field equation (2.146) into region $z < 0$, the field derivative E'_x undergoes a jump at $z = 0$.

Since the kernel of Eq. (2.147) is a difference function, and the integration is performed over infinite limits, the solution can be obtained by Fourier transformation

$$E_x(z) = \frac{1}{2\pi} \int_{-\infty}^{+\infty} dk \exp(\imath kz) E(k), \quad E(k) = \int_{-\infty}^{+\infty} dz \exp(-\imath kz) E_x(z).$$

In this case, from Eq. (2.146), we find

$$\left\{ \frac{\omega^2}{c^2} \varepsilon^{\text{tr}}(\omega, k) - k^2 \right\} E(k) = 2E_x'(0), \tag{2.148}$$

where

$$\varepsilon^{\text{tr}}(\omega, k) = 1 + \frac{4\pi e^2}{\omega} \int d\vec{p} \, \frac{v_x^2 f_{0e}'}{\omega + \imath\nu - kv_z}. \tag{2.149}$$

Then, we find

$$E_x(z) = \frac{E_x'(0)}{\pi} \int_{-\infty}^{+\infty} dk \frac{\exp(\imath kz)}{\frac{\omega^2}{c^2} \varepsilon^{\text{tr}}(\omega, k) - k^2}. \tag{2.150}$$

Relation (2.150) is the solution of the field equation.

It should be noted that the dielectric permittivity in Eqs. (2.148) and (2.149) is different from Eq. (2.25) because of the presence of a finite positive imaginary term in the denominator of the integrand. The latter simplifies the problem because for real ω and k the singular point of the integrand is not located on the real axis.

From relation (2.150), it follows the complex effective depth of field penetration in plasma

$$\lambda_M = \int_0^{\infty} \frac{B_y(z)}{B_y(0)} dz = -\frac{E_x(0)}{E_x'(0)} = -\frac{1}{\pi} \int \frac{dk}{\frac{\omega^2}{c^2} \varepsilon^{\text{tr}}(\omega, k) - k^2}. \tag{2.151}$$

This formula characterizes the field inside plasma under the condition of mirror reflection of particles from the boundary. In the study of plasma reflection and absorption of electromagnetic fields, it is sufficient to make a connection between the quantities E_x and E_x' on the plasma surface. In other words, it is sufficient to know the expression of the effective penetration depth (2.151).

The problem of fields in semi-bounded plasma, in the case of diffusive reflection of electrons from the boundary, is mathematically more complicated than that of mirror reflection. In diffusive reflection ($\rho = 0$), Eq. (2.146) reduces to

$$\frac{d^2E_x}{dz^2} + \frac{\omega^2}{c^2}E_x = -\frac{4\pi\imath\omega}{c^2}\int\limits_0^\infty dz'\, K(|z-z'|)E_x(z'), \qquad (z \geq 0). \tag{2.152}$$

To solve this equation, the Wiener–Hopf method is used [34, 60]. Equation (2.152) determines the electric field for $z \geq 0$. Extending this equation into the left half-axis z and assuming $E_x = 0$ for $z < 0$, from Eq. (2.152), we find

$$\frac{d^2E_x}{dz^2} + \frac{\omega^2}{c^2}E_x + \frac{4\pi\imath\omega}{c^2}\int\limits_{-\infty}^{+\infty} dz'\, K(|z-z'|)E_x(z') = F_x(z), \tag{2.153}$$

where

$$F_x(z) = \begin{cases} 0, & if \quad z \geq 0, \\ \dfrac{4\pi\imath\omega}{c^2}\int\limits_0^\infty dz'\, K(|z-z'|)E_x(z'). & if \quad z < 0. \end{cases} \tag{2.154}$$

Applying Fourier transformation to Eq. (2.153), we obtain

$$\left\{\frac{\omega^2}{c^2}\varepsilon^{\rm tr}(\omega,k) - k^2\right\}E(k) = E_x'(+0) + \imath k E_x(+0) + F(k), \tag{2.155}$$

where

$$F(k) = \int\limits_{-\infty}^{+\infty} dz \exp(-\imath kz)F_x(z).$$

The field inside plasma diminishes by moving away from the surface and such a decrease can be considered exponential for sufficiently large z. Therefore, $E(k)$, as a function of complex variable k, has no singularity in the lower half-plane k. Assuming that $K(|z|)$ falls down exponentially fast for large $|z|$ ($\sim \exp(-\alpha|z|)$), we find out that $\varepsilon^{\rm tr}(\omega, k)$, as a function of k, has no singularity in the complex variable plane inside a finite band $-\alpha <$ Im $k < +\alpha$ centered around the real axis. The same is true for $F(k)$ in the upper half-plane Im$k > -\alpha$.[29]

The roots of the dispersion equation

[29] Transverse dielectric permittivity, as a function of k, for the relativistic distribution of electrons given by Eq. (2.65) actually has no singularity in a finite band around the real axis. With taking collisions into account, we find

$$\frac{\omega^2}{c^2}\varepsilon^{\mathrm{tr}}(\omega,k) - k^2 = 0 \tag{2.156}$$

have vital meaning. Since, $\varepsilon^{\mathrm{tr}}(\omega, k)$ is an even function of k, then the roots of Eq. (2.156) are situated in pairs $\pm k$. In the band of $-\alpha <$ Im $k < \alpha$, the transverse dielectric permittivity has no singularity.[30] Now, we define the roots of Eq. (2.156) in such a band by $\pm k_r (r = 1, 2, \ldots n)$ and introduce

$$P_{2n}(k) = \left(k^2 - k_1^2\right)\ldots\left(k^2 - k_n^2\right). \tag{2.157}$$

Here, $P_{2n}(k) = 1$, if there is no root of Eq. (2.156) in the region $|$ Im k $| < \alpha$. Furthermore, we define

$$\tau(k) = \frac{\left(k^2 + a^2\right)^{n-1}}{P_{2n}(k)}\left\{k^2 - \frac{\omega^2}{c^2}\varepsilon^{\mathrm{tr}}(\omega,k)\right\}, \tag{2.158}$$

where $n = 0$ if there is no root in the region $|$ Im k $| < \alpha$. Function $\ln \tau(k)$ can be represented as

$$\ln\tau(k) = \frac{1}{2\pi\imath}\int\limits_{-\imath\beta-\infty}^{-\imath\beta+\infty} dk'\,\frac{\ln\tau(k')}{k'-k} - \frac{1}{2\pi\imath}\int\limits_{+\imath\beta-\infty}^{+\imath\beta+\infty} dk'\,\frac{\ln\tau(k')}{k'-k}, \tag{2.159}$$

where $\beta < \alpha$, because in the band of $|$ Im k $| \le \beta$, function $\ln\tau(k)$ has no singularity and vanishes at infinity. Thus, function (2.159) can be written as

$$\tau(k) = \frac{\tau_1(k)}{\tau_2(k)}, \tag{2.160}$$

where

$$\varepsilon^{\mathrm{tr}}(\omega,k) = 1 - \frac{2\pi e^2 N_e}{m\omega c}K_2^{-1}\left(\frac{mc^2}{\kappa T_e}\right)\left(\int\limits_{-\infty}^{-[(\omega+\imath\nu)/c]} + \int\limits_{[(\omega+\imath\nu)/c]}^{\infty}\right)\frac{dk'}{k'(k'-k)}$$

$$\times\exp\left(\frac{mc^2}{\kappa T_e}\frac{1}{\sqrt{1-(\omega+\imath\nu)^2/c^2k'^2}}\right)\left\{\left[1-\frac{(\omega+\imath\nu)^2}{c^2k'^2}\right]\frac{\kappa T_e}{mc^2} + \sqrt{1-\frac{(\omega+\imath\nu)^2}{c^2k'^2}}\right\}.$$

This function is analytical everywhere on the complex plane k with the exception of branch cuts beginning correspondingly from $-(\omega + \imath\nu)/c$ and $(\omega + \imath\nu)/c$ and going off to infinity. Therefore, in a band of width $k < 2\nu/c$ around the real axis, $\varepsilon^{\mathrm{tr}}(\omega, k)$ as a function of k has no singularity.

[30] In the equation of the previous footnote, $\alpha = \nu/c$.

$$\begin{cases} \tau_1(k) = \exp\left(\dfrac{1}{2\pi\imath}\displaystyle\int\limits_{-\imath\beta-\infty}^{-\imath\beta+\infty} dk' \dfrac{\ln\tau(k')}{k'-k}\right); \\ \tau_2(k) = \exp\left(\dfrac{1}{2\pi\imath}\displaystyle\int\limits_{+\imath\beta-\infty}^{+\imath\beta+\infty} dk' \dfrac{\ln\tau(k')}{k'-k}\right). \end{cases} \tag{2.161}$$

Function $\tau_1(k)$ is regular and has no zero in the half-plane of the complex variable $-\alpha < -\beta < \operatorname{Im} k$. Furthermore, function $\tau_2(k)$ is regular and has no zero in the half-plane $\alpha > \beta > \operatorname{Im} k$,

$$k^2 - \frac{\omega^2}{c^2}\varepsilon^{\mathrm{tr}}(\omega, k) = P_{2n}(k)\frac{\tau_1(k)}{\tau_2(k)}\frac{(k-\imath\alpha)^{-n+1}}{(k+\imath\alpha)^{n-1}}. \tag{2.162}$$

The latter expression allows us to write Eq. (2.155) in the following form:

$$\begin{aligned} &\{F(k) + E_x'(0) + \imath k E_x(0)\}\tau_1^{-1}(k)(k+\imath\alpha)^{n-1} \\ &= -E(k)P_{2n}(k)\tau_2^{-1}(k)(k-\imath\alpha)^{1-n}. \end{aligned} \tag{2.163}$$

The left side of relation (2.163) is regular in the half-plane $\operatorname{Im}k > -\alpha$ and $\operatorname{Im}k < 0$. Because of existence of overlapping of regularity regions of the left and right sides of relation (2.163), they determine a function, which is regular in the whole finite region of the complex variable plane k. Furthermore, when $k \to \infty$, from Eq. (2.155), we find

$$E(k) = -\frac{\imath}{k}E_x(+0) - \frac{1}{k^2}E_x'(+0) + O\left(\frac{1}{|k|^3}\right). \tag{2.164}$$

Therefore, the right side of relation (2.163) behaves as k^2 at infinity. In this connection, the function determined by the right and left sides of relation (2.163) is a polynomial of degree n, which is denoted by $Q_n(k)$. As a result,

$$E(k) = \frac{Q_n(k)\tau_2(k)}{P_{2n}(k)(k-\imath\alpha)^{1-n}}. \tag{2.165}$$

Since function $E(k)$ is regular for $\operatorname{Im}k < 0$, then

$$Q_n(k) \sim (k+k_1)\ldots(k+k_n), \tag{2.166}$$

where $\operatorname{Im}k_r > 0$. Finally, considering condition (2.164) for $k \to \infty$, we find

$$E(k) = -\imath \frac{E_x(+0)\tau_2(k)}{(k-k_1)\ldots(k-k_n)(k-\imath\alpha)^{1-n}}, \tag{2.167}$$

from which we obtain the field in plasma in the case of diffusive reflection of electrons from the boundary

$$E_x(z) = -\frac{\imath E_x(+0)}{2\pi} \int\limits_{-\imath\delta-\infty}^{-\imath\delta+\infty} dk \frac{\exp(\imath kz)\tau_2(k)}{(k-k_1)\ldots(k-k_n)(k-\imath\alpha)^{1-n}}, \tag{2.168}$$

where $\delta > 0$.

As mentioned before, it is necessary to know the effective penetration depth in the study of reflection and absorption of electromagnetic waves incident on the plasma surface. In the case of mirror reflection of electrons, such depth is determined by formula (2.151). Now, from formula (2.164), we find the corresponding formula for the diffusive reflection case:

$$E'_x(+0) = -\lim_{|k|\to\infty} \left\{k^2 E(k) + \imath k E_x(+0)\right\}. \tag{2.169}$$

Considering that for $|k| \to \infty$,

$$\tau_2(k) = 1 - \frac{1}{2\pi\imath k} \int\limits_{-\infty}^{+\infty} dk' \ln\tau(k') + O\left(\frac{1}{|k|^2}\right),$$

based on relations (2.167), and (2.169), we find

$$E'_x(+0) = E_x(+0)\left\{\imath(k_1 + \ldots + k_n) - \alpha(n-1) - \frac{1}{2\pi}\int\limits_{-\infty}^{+\infty} dk \ln\tau(k)\right\}. \tag{2.170}$$

Thus, from here, we obtain

$$\lambda_D = -\frac{E_x(+0)}{E'_x(+0)} = -\frac{1}{\frac{1}{2\pi}\int\limits_{-\infty}^{+\infty} dk \ln\tau(k) + \alpha(1-n) - \imath\Sigma_{k=1}^{n} k_r}. \tag{2.171}$$

Thus, λ_D is determined by the zeros of the dispersion equation (2.156), which are located inside the band of $0 \leq \operatorname{Im} k_r < \alpha$, and also by

$$\frac{1}{2\pi}\int\limits_{-\infty}^{+\infty} dk \ln \tau(k) = \frac{1}{\pi}\int\limits_{0}^{\infty} dk \ln \left\{ \frac{\left(k^2+\alpha^2\right)^{n-1}\left[k^2 - \frac{\omega^2}{c^2}\varepsilon^{\mathrm{tr}}(\omega,k)\right]}{\left(k^2-k_1^2\right)\ldots\left(k^2-k_n^2\right)} \right\}. \qquad (2.172)$$

Formulas (2.168) and (2.170) correspond to the formulas obtained in [55]. Their further simplifications were done in [61]. In this case, for Im $k < 0$ in formula (2.161), which determines $\tau_2(k)$, one can displace the integration contour to the real axis. Then, using

$$\int_{-\infty}^{+\infty} \frac{dk'}{k'-k} \ln\left(1 - \frac{k_r^2}{k'^2}\right) = 2\pi\imath \ln\left(\frac{k-k_r}{k}\right),$$

we find

$$\tau_2(k) = (k-k_1)\ldots(k-k_n)(k-\imath\alpha)^{1-n}\frac{1}{k}$$
$$\times \exp\left(\frac{1}{2\pi\imath}\int\limits_{-\infty}^{+\infty}\frac{dk'}{k-k'}\ln\left[1 - \frac{\omega^2}{c^2k'^2}\varepsilon^{\mathrm{tr}}(\omega,k')\right]\right).$$

Therefore, formula (2.167) can be written as

$$E(k) = -\frac{\imath}{k}E_x(0)\exp\left(\frac{1}{2\pi\imath}\int\limits_{-\infty}^{+\infty}\frac{dk'}{k-k'}\ln\left[1 - \frac{\omega^2}{c^2k'^2}\varepsilon^{\mathrm{tr}}(\omega,k')\right]\right), \qquad (2.173)$$

or, by making use of inverse Fourier transformations, as

$$E_x(z) = -\frac{\imath E_x(0)}{2\pi}\int\limits_{-\imath\delta-\infty}^{-\imath\delta+\infty}$$
$$\times \frac{dk}{k}\exp(\imath kz)\exp\left(\frac{1}{2\pi\imath}\int\limits_{-\infty}^{+\infty}\frac{dk'}{k-k'}\ln\left[1 - \frac{\omega^2}{c^2k'^2}\varepsilon^{\mathrm{tr}}(\omega,k')\right]\right). \qquad (2.174)$$

Finally, considering expression (2.169), from formula (2.173), we find the following expression for the effective penetration depth:

$$\lambda_D = -\frac{E_x(0)}{E'_x(0)} = \left\{ \frac{1}{\pi} \int_0^\infty dk \ln \left[1 - \frac{\omega^2}{c^2 k^2} \varepsilon^{\text{tr}}(\omega, k) \right] \right\}^{-1} . \tag{2.175}$$

2.9 Reflection and Absorption of Electromagnetic Waves in Semi-bounded Plasma

Reflection of electromagnetic waves from the plasma surface as well as the absorption of incident waves can be described by complex reflection coefficient r, which represents the ratio of the complex amplitudes of reflected and incident monochromatic waves $\exp(-\imath\omega t)$. Below, we will restrict our consideration to the normal incidence of a plane transverse monochromatic wave with non-zero field components (E_x, B_y) on the boundary surface of semi-bounded plasma $(z \geq 0)$. Then, using boundary conditions

$$\{E_x\}_{z=0} = \{B_y\}_{z=0} = 0,$$

we obtain

$$r = -\frac{1 - \frac{c}{4\pi} Z(\omega)}{1 + \frac{c}{4\pi} Z(\omega)}, \tag{2.176}$$

for the complex reflection coefficient, where $Z(\omega)$ is the surface impedance of semi-bounded plasma defined by the ratio of the electric field to the magnetic induction on the plasma surface at $z = 0$ [54, 62, 63]

$$Z(\omega) = \frac{4\pi}{c} \frac{E_x(0)}{B_y(0)} = \frac{4\pi\imath\omega}{c^2} \frac{E_x(0)}{E'_x(+0)} . \tag{2.177}$$

The x axis is along the electric field $\vec{E}$; $E'_x(0)$ is the derivative of the electric field with respect to z coordinate on the plasma surface. Here, we make use of the results of the previous section. In addition, note that the surface impedance is coupled to the so-called complex effective penetration depth of fields into plasma given by Eq. (2.151) as follows:

$$\lambda = -\frac{E_x(0)}{E'_x(+0)} = \frac{\int_0^\infty dz\, B_y(z)}{B_y(0)} = \imath \frac{c}{\omega} \frac{E_x(0)}{B_y(0)} = -\frac{c^2}{4\pi\imath\omega} Z(\omega). \tag{2.178}$$

The quantity $|r|^2$ is the ratio of the electromagnetic energy flux reflected from the plasma surface to the incident flux. Therefore, the quantity

$$A = 1 - |r|^2 \tag{2.179}$$

represents the fraction of the electromagnetic energy flux, which is absorbed in plasma. When the penetration depth or the characteristic length of field variation in plasma ~ $|\lambda|$ is much less than the radiation wavelength in vacuum ~c/ω, relation (2.179) takes the form

$$A = \frac{c}{\pi} R, \tag{2.180}$$

where $Z(\omega) = R + \imath\chi$, i.e., $R = \text{Re}\, Z(\omega)$.

Thus, the characteristics of the reflection and absorption processes in semi-bounded plasma is determined by the expression of the complex penetration depth λ. We will calculate this quantity for the problem of transverse waves reflection from a surface in the case of normal incidence. According to Eqs. (2.151) and (2.175), we obtain

$$\lambda = +\frac{1}{\pi} \int_{-\infty}^{+\infty} dk_z \frac{1}{k^2 - \frac{\omega^2}{c^2} \varepsilon^{\text{tr}}(\omega, k)}. \tag{2.181}$$

Let us begin the analysis of this expression from the case of long-wavelength high-frequency waves, $\omega \gg kv_{Te}$, when spatial dispersion can be neglected and $\varepsilon^{\text{tr}}(\omega, k) = \varepsilon(\omega)$. This case happens when the effective penetration depth of fields is much longer than the mean free path l and the average distance traversed by particles during the one period of field oscillation ($\sim v_T/\omega$, where v_T is the particle thermal velocity). Moreover, we suppose that the wave frequency ω is higher than the electron collision frequency ν and, therefore, $\varepsilon(\omega)$ can be written as

$$\varepsilon(\omega) = 1 - \frac{\omega_0^2}{\omega^2}\left(1 - \imath\frac{\nu}{\omega}\right). \tag{2.182}$$

Here, ω_0^2 is given by expressions (2.44) and (2.67): $\omega_0^2 = \omega_{\text{pe}}^2$ for the non-relativistic temperature of electrons $mc^2 \gg \kappa T_e$ and $\omega_0^2 = \omega_{0R}^2 = 4\pi e^2 N_e c^2/\kappa T_e$ for ultra-relativistic temperature $\kappa T_e \gg mc^2$. Substituting Eq. (2.182) into Eq. (2.181), in the absence of spatial dispersion, we obtain

$$\lambda = \imath \frac{c}{\omega} \frac{1}{\sqrt{\varepsilon(\omega)}}. \tag{2.183}$$

It must be noted that in this expression $Im\sqrt{\varepsilon(\omega)} > 0$. In particular, if $\omega < \omega_0$, from Eq. (2.183), it follows that λ~c/ω_0. It should be remarked that formula (2.183) can be simply obtained from field equations because in the absence of spatial dispersion, the field takes the form of $\exp\left[\imath z(\omega/c)\sqrt{\varepsilon(\omega)}\right]$. Strictly speaking, expression (2.183) is correct only for non-relativistic temperature. As it was shown in Sect. 2.4, for plasma with relativistic temperature, when T_e ~ mc^2 ($v_T \sim c$), the correction terms of the order of ~k^2, stipulated by the presence of the small parameter $(v_T/c)^2$, become important in expression (2.182) and in turn in the dispersion equation of transverse electromagnetic waves

$$k^2 - \frac{\omega^2}{c^2}\varepsilon^{\mathrm{tr}}(\omega, k) = 0, \tag{2.184}$$

which coincides with the poles of the integrand in Eq. (2.181). Therefore, in relativistic plasma, even in the limit $k \to 0$, the role of spatial dispersion is noticeable. To take into account such corrections, it is necessary to use the following form of $\varepsilon^{\mathrm{tr}}(\omega, k)$:

$$\varepsilon^{\mathrm{tr}}(\omega, k) = \varepsilon(\omega) - \alpha\frac{c^2k^2}{\omega^2}, \tag{2.185}$$

where $\varepsilon(\omega)$ is given by Eq. (2.182) and (see Sect. 2.4)

$$\alpha = \frac{\omega_{\mathrm{pe}}^2}{\omega^2}K_2^{-1}\left(\frac{mc^2}{\kappa T_e}\right)\int_0^1 dx x^2\sqrt{1-x^2} \times\left\{1+\frac{\kappa T_e}{mc^2}\sqrt{1-x^2}\right\}\exp\left(-\frac{mc^2}{\kappa T_e}\frac{1}{\sqrt{1-x^2}}\right). \tag{2.186}$$

In the limiting cases of non-relativistic and ultra-relativistic plasmas, the latter expression reduces to

$$\alpha = \begin{cases} \alpha_{\mathrm{NR}} = \dfrac{\omega_{pe}^2}{\omega^2}\dfrac{\kappa T_e}{mc^2}, & for \quad \kappa T_e \ll mc^2, \\ \alpha_{\mathrm{UR}} = \dfrac{1}{5}\dfrac{\omega_{0R}^2}{\omega^2}, & for \quad \kappa T_e \gg mc^2. \end{cases} \tag{2.187}$$

In the ultra-relativistic case, Eq. (2.184) takes the form of

$$\omega^2 = \omega_{0R}^2\left(1 - \imath\frac{\nu}{\omega}\right) + \frac{6}{5}c^2k^2.$$

Thus, the last term, which contains k^2, has been changed to 1/5 with respect to the corresponding expression obtained under the condition of neglecting spatial dispersion [compared to Eq. (2.75)]. By substituting Eq. (2.185) into Eq. (2.181), we obtain

$$\lambda_M = \frac{\imath c}{\omega\sqrt{\varepsilon(\omega)}}\frac{1}{\sqrt{1+\alpha}} = \frac{\imath c}{\sqrt{\omega^2 - \omega_0^2 + \imath\omega_0^2\frac{\nu}{\omega}}}\quad\frac{1}{\sqrt{1+\alpha}},$$
$$\lambda_D = \frac{\imath c}{\omega\sqrt{\varepsilon(\omega)}}\sqrt{1+\alpha} = \frac{\imath c\sqrt{1+\alpha}}{\sqrt{\omega^2 - \omega_0^2 + \imath\omega_0^2\frac{\nu}{\omega}}}. \tag{2.188}$$

Now, we can determine the conditions under which the corrections arising from weak spatial dispersion are small. In the frequency range $\omega \le \omega_0$, from Eqs. (2.185) and (2.188), it follows that the influence of spatial dispersion is small if

$$\mid \varepsilon(\omega) \mid \ll \begin{cases} \dfrac{mc^2}{\kappa T_e}, & \left(\omega^2 \gg \omega_{\text{pe}}^2\dfrac{\kappa T_e}{mc^2}, \quad mc^2 \gg \kappa T_e\right), \\ 1, & \left(|\omega^2 - \omega_0^2| \ll \omega_0^2, \quad mc^2 \le \kappa T_e\right). \end{cases} \tag{2.189}$$

Only under these conditions, the field penetration depth λ is larger than the distance gone through by particles in the period of field oscillation. On the other hand, we neglected the contribution stipulated by the branch points of $\varepsilon^{\text{tr}}(\omega, k)$ in calculation of λ. Therefore, when the mean free path of particles is small compared to the effective penetration depth, the above results are valid only if

$$\frac{\nu^2}{\omega^2} \gg \begin{cases} \dfrac{\kappa T_e}{mc^2} \mid \varepsilon(\omega) \mid, & (\kappa T_e \ll mc^2), \\ \mid \varepsilon(\omega) \mid, & (\kappa T_e \ge mc^2), \end{cases} \tag{2.190}$$

which is stronger than the condition given by (2.189). In the opposite limit, when the particle collisions are negligible but relation (2.189) holds, it becomes necessary to take into account the additional contribution in the field penetration depth given by relations (2.151), (2.171), and (2.181), which arises from the branch points of $\varepsilon^{\text{tr}}(\omega, k)$, to the contributions stipulated by the zeros of Eq. (2.184) [55, 64]. This additional contribution is equal to

$$\delta\lambda_M = -\frac{2\imath c}{\pi\omega}\left(1+\imath\frac{\nu}{\omega}\right)\int_1^\infty dx \frac{Im\varepsilon_+^{\mathrm{tr}}\left(\omega,\frac{\omega+\imath\nu}{c}x\right)}{\left[\mathrm{Re}\varepsilon_+^{\mathrm{tr}}\left(\omega,\frac{\omega+\imath\nu}{c}x\right)-\left(1+\imath\frac{\nu}{\omega}\right)^2x^2\right]^2+\left[Im\varepsilon_+^{\mathrm{tr}}\left(\omega,\frac{\omega+\imath\nu}{c}x\right)\right]^2},$$

$$\delta\lambda_D^{-1} = \frac{\imath\omega}{\pi c}\left(1+\imath\frac{\nu}{\omega}\right)^3\int_0^1 da\int_1^\infty dx \frac{x^2 Im\varepsilon_+^{\mathrm{tr}}\left(\omega,\frac{\omega+\imath\nu}{c}x\right)}{\left[a\mathrm{Re}\varepsilon_+^{\mathrm{tr}}\left(\omega,\frac{\omega+\imath\nu}{c}x\right)-\left(1+\imath\frac{\nu}{\omega}\right)^2x^2\right]^2+\left[aIm\varepsilon_+^{\mathrm{tr}}\left(\omega,\frac{\omega+\imath\nu}{c}x\right)\right]^2}.$$

(2.191)

Here, $\varepsilon_+^{\mathrm{tr}}(\omega, k)$ represents the generalization of expression (2.70) by taking into account the particle collisions. Therefore,

$$\mathrm{Re}\ \varepsilon_+^{\mathrm{tr}}\left(\omega, \frac{\omega+\imath\nu}{c}x\right) = 1 + \frac{2\pi e^2 N_e}{m\omega(\omega+\imath\nu)} K_2^{-1}\left(\frac{mc^2}{\kappa T_e}\right)\left(\int_{-\infty}^{-1} + \int_1^\infty\right)\frac{dx'}{x'}\mathcal{P}\frac{\sqrt{1-\frac{1}{x'^2}}}{x'-x}\times$$

$$\times\left[1+\frac{\kappa T_e}{mc^2}\sqrt{1-\frac{1}{x'^2}}\right]\exp\left(-\frac{mc^2}{\kappa T_e}\frac{1}{1-\frac{1}{x'^2}}\right),$$

$$Im\ \varepsilon_+^{\mathrm{tr}}\left(\omega, \frac{\omega+\imath\nu}{c}x\right) = \frac{2\pi^2 e^2 N_e}{m\omega(\omega+\imath\nu)} K_2^{-1}\left(\frac{mc^2}{\kappa T_e}\right)\frac{1}{x}\sqrt{1-\frac{1}{x^2}}\times$$

$$\times\left[1+\frac{\kappa T_e}{mc^2}\sqrt{1-\frac{1}{x^2}}\right]\exp\left(-\frac{mc^2}{\kappa T_e}\frac{x}{\sqrt{x^2-1}}\right).$$

(2.192)

Here, $\mathcal{P}$ denotes that at singular point $x = x'$ the principal value has to be taken into consideration.

For the case of non-relativistic temperature of plasma, from Eq. (2.191) it follows that

$$\delta\lambda_M = \imath\sqrt{\frac{8}{\pi}}\frac{c}{\omega}\frac{\omega_{\mathrm{pe}}^2}{\omega^2}\left(\frac{\kappa T_e}{mc^2}\right)^{\frac{3}{2}},$$

$$\delta\lambda_D^{-1} = -\frac{\imath}{\sqrt{2\pi}}\frac{\omega}{c}\frac{\omega_{\mathrm{pe}}^2}{\omega^2}\sqrt{\frac{\kappa T_e}{mc^2}}.$$

(2.193)

At the same time, in the limit of ultra-relativistic temperature $\kappa T_e \gg mc^2$, and in the frequency range $\omega \approx \omega_0 \gg \nu$, from Eq. (2.191), we obtain

$$\delta\lambda_M = \imath\frac{3c}{2\omega_0}\int\limits_1^\infty \frac{dx\cdot x^{-1}(1-x^{-2})}{\left\{1+\frac{3}{4\pi}\left[\left(1-\frac{1}{x^2}\right)\ln\frac{x-1}{x+1}-\frac{2}{x}\right]-x^2\right\}^2+\left(\frac{3\pi}{4x}\right)^2\left(1-\frac{1}{x^2}\right)^2} \approx \imath 0,09\frac{c}{\omega_0},$$

$$\delta\lambda_D^{-1} = -\imath\frac{\omega_0}{c}\frac{3}{4}\int\limits_0^1\frac{da}{a^2}$$

$$\times\int\limits_1^\infty \frac{dx\cdot x(1-x^{-2})}{\left\{1+\frac{3}{4\pi}\left[\frac{-2}{x}+\left(1-\frac{1}{x^2}\right)\ln\frac{x-1}{x+1}\right]-\frac{x^2}{a}\right\}^2+\left(\frac{3\pi}{4x}\right)^2\left(1-\frac{1}{x^2}\right)^2} \approx -\imath 0,18\frac{\omega_0}{c}.$$

(2.194)

Expressions (2.193) and (2.194) in addition to Eq. (2.188) determine the field penetration depth into semi-bounded plasma under the condition given by (2.189).

Let us now consider the field boundary value problem under the conditions of the so-called anomalous skin-effect, when the effective penetration depth is much less than the mean free path of particles and the average distance gone through in the field oscillation period $\tilde{}v_{\mathrm{Te}}/\omega$. Then, the approximate expression for the transverse dielectric permittivity,

$$\varepsilon^{\mathrm{tr}}(\omega,k) = \frac{4\pi\imath C}{|\,k\,|\,\omega}, \qquad (2.195)$$

obtained in the limit of $\omega+\imath\nu \ll kv_{\mathrm{Te}}$ is valid. Here, according to Eq. (2.192),

$$C = \frac{\pi}{2}\frac{e^2N_e}{mc}\left(1+\frac{\kappa T_e}{mc^2}\right)K_2^{-1}\left(\frac{mc^2}{\kappa T_e}\right)\exp\left(-\frac{mc^2}{\kappa T_e}\right). \qquad (2.196)$$

From this expression, we obtain

$$C = \begin{cases} C_{\mathrm{NR}} = \sqrt{\frac{\pi}{2}}\frac{e^2N_e}{m}\sqrt{\frac{m}{\kappa T_e}}, & mc^2 \gg \kappa T_e, \\ C_{\mathrm{UR}} = \frac{\pi e^2 N_e c}{4\kappa T_e}, & mc^2 \ll \kappa T_e, \end{cases} \qquad (2.197)$$

for the non-relativistic and ultra-relativistic temperatures [40]. Substituting Eq. (2.195) into Eq. (2.181), we obtain [55]

$$\begin{aligned}\lambda_M &= \frac{2}{3}\left(1+\frac{\imath}{\sqrt{3}}\right)\left(\frac{c^2}{4\pi\omega C}\right)^{\frac{1}{3}};\\ \lambda_D &= \frac{3}{4}\left(1+\frac{\imath}{\sqrt{3}}\right)\left(\frac{c^2}{4\pi\omega C}\right)^{\frac{1}{3}}.\end{aligned} \tag{2.198}$$

When $\omega \ll kv_{Te}$, or when the effective penetration depth of the field is small compared to the average distance gone through in the field oscillation period, expression (2.198) will be valid in the frequency range

$$\omega^2 \ll \begin{cases} \omega_{\mathrm{pe}}^2\left(\dfrac{\kappa T_e}{mc^2}\right), & mc^2 \gg \kappa T_e, \\ \omega_0^2, & mc^2 \le \kappa T_e, \end{cases} \tag{2.199}$$

which is in contrast with Eq. (2.189). It must be noticed that the fulfilment of the inequality $\omega \gg \nu$ is not necessary for the validity of Eq. (2.198) if $\nu \ll kv_{\mathrm{Te}}$.

Relations (2.198) have a simple physical meaning. It is known that, in the case of the normal skin-effect, the effective penetration depth is of the order of

$$\lambda \approx \frac{c}{\sqrt{2\pi\sigma_0\omega}},$$

where σ_0 is the static conductivity of the medium $\left(\sigma_0 = \omega_0^2/4\pi\nu\right)$. When field penetration depth is much smaller than the mean free path, only those particles moving at an angle less than $\sim \lambda/l \sim \nu/kv_{\mathrm{Te}} \ll 1$ will experience the collision in the field penetration layer. Therefore, conductivity σ_0 must be replaced by $(\lambda/l)\sigma_0$. Thus, it leads to the following estimation:

$$\lambda \approx \left(\frac{c^2 v_{\mathrm{Te}}}{\omega\omega_0^2}\right)^{1/3},$$

which is of the order of magnitude of Eq. (2.198). Such qualitative analysis of the anomalous skin-effect was first carried out in [65].

Let us return to expressions (2.193) and give their physical interpretation. For this purpose, we rewrite the relative absorbed energy given by relation (2.179) in this limit by taking into account Eqs. (2.182) and (2.183) $\left(\omega^2 < \omega_0^2 \simeq \omega_{\mathrm{pe}}^2\right)$:

$$A_M = \frac{2\nu}{\sqrt{\omega_{\text{pe}}^2 - \omega^2}} + 8\sqrt{\frac{2}{\pi}}\left(\frac{\omega_{\text{pe}}^2}{\omega^2} - 1\right)\left(\frac{\kappa T_e}{mc^2}\right)^{3/2},$$
$$A_D = \frac{2\nu}{\sqrt{\omega_{\text{pe}}^2 - \omega^2}} + \sqrt{\frac{8}{\pi}}\sqrt{\frac{\kappa T_e}{mc^2}}. \tag{2.200}$$

The first term in these expressions describes the collisional absorption of electromagnetic energy and it is not related to spatial dispersion. But, the second term is essentially related to spatial dispersion and characterizes the absorption when the collisions between plasma particles do not happen. It is determined by the effective collisions of particles with the plasma surface [66].

Actually, the particles moving and reflected from the plasma surface are affected by the electromagnetic field wave penetrating into plasma and the average work done by this field is non-zero due to the particle collisions with the surface. Therefore, this case essentially depends on the characteristics of the electron reflection from the plasma boundary.

The electrons moving toward the plasma surface experience a velocity change under the action of the electromagnetic field. In this case, the velocity component parallel to the electric field is

$$v_x = v_x^0 + \frac{e}{m}\int_{-\infty}^{t} E_x(t', z)dt', \qquad z = \mid t' - t_0 \| v_z^0 \mid \quad (t < t_0).$$

Here, v_x^0 and v_z^0 are the electron velocity components, unperturbed by the electromagnetic field, far from the surface and t_0 is the moment at which an electron collides with the surface. The velocity of the particles moving outward from the plasma surface depends on the boundary and reflection conditions. Thus, we have

$$v_{xM} = v_x^0 + \frac{e}{m}\int_{-\infty}^{t} E_x(t', z)dt'; \qquad z = \mid t' - t_0 \| v_z^0 \mid \quad (t > t_0),$$

for the case of mirror reflection, and

$$v_{xD} = \frac{e}{m}\int_{t_0}^{t} dt' E_x(t', z); \quad z = \mid t' - t_0 \| v_z^0 \mid \quad (t > t_0),$$

for the case of diffusion.

At $t \to +\infty$, when these particles go far away from the plasma surface, their energy will be different from the incident energy, because $v_x(t \to +\infty) \neq v_x^0$, i.e.,

$$\Delta W_M = \frac{mv_x^2(+\infty)}{2} - \frac{mv_x^{0^2}}{2}, \qquad \Delta W_D = \frac{m}{2}\left\{v_x^2(+\infty) + v_x^2(t_0 - 0) - v_x^{0^2}\right\}.$$

This expression must be averaged over all possible values of t_0 at which different electrons collide with the plasma surface. However, for such averaging of $\langle \Delta W \rangle_{t_0}$, it is necessary to have the exact expression for the electric field $E_x(t, z)$ in plasma. Under the condition given by (2.189), when spatial dispersion is weak, one can use the following expression:

$$E_x(t, z) = E_x(0) \exp(-\imath\omega t) \exp\left(-z\frac{\sqrt{\omega_0^2 - \omega^2}}{c}\right).$$

Then, it can be easily shown that, for both mirror and diffusion reflections, we have

$$\begin{aligned} W = \langle \Delta W_M \rangle_{t_0} &= \frac{e^2}{m}\frac{\omega_0^2 - \omega^2}{\omega^4}\frac{v_z^2}{c^2} E_x(0)E_x^*(0), \\ W = \langle \Delta W_D \rangle_{t_0} &= \frac{e^2}{2m\omega^2} E_x(0)E_x^*(0). \end{aligned} \tag{2.201}$$

Multiplying these expressions by the flux of electrons per unit area of the plasma surface,

$$2\pi v \cos\theta \sin\theta d\theta f_0(p)p^2 dp,$$

and integrating it over all incident angles θ $(0 \le \theta \le \pi/2)$, where θ is the angle between the normal to plasma and the direction of particles velocity, and over the momentum p for the non-relativistic Maxwellian distribution function $f_0(p)$, we obtain

$$W_M = \frac{2e^2 N_e}{mc^2}\frac{\omega_{\mathrm{pe}}^2 - \omega^2}{\omega^4}\left(\frac{\kappa T_e}{m}\right)^{3/2}\frac{1}{\sqrt{2\pi}} E_x(0)E_x^*(0);$$

$$W_D = \frac{N_e e^2}{2m}\frac{1}{\omega^2}\left(\frac{\kappa T_e}{m}\right)^{1/2}\frac{1}{\sqrt{2\pi}} E_x(0)E_x^*(0).$$

Finally, these expressions must be divided by the incident flux of electromagnetic energy,

$$\frac{c}{8\pi} E_x(-\infty)E_x^*(-\infty),$$

where $E_x(-\infty)$ is the amplitude of the incident wave. According to the boundary condition,

$$E_x(0) = (1 + r)E_x(-\infty),$$

where r is the reflective coefficient, which in our case is equal to (for $\omega_{pe} > \omega$)

$$r \simeq \frac{\imath\omega + \sqrt{\omega_{pe}^2 - \omega^2}}{\imath\omega - \sqrt{\omega_{pe}^2 - \omega^2}},$$

we obtain

$$A = \rho 8 \sqrt{\frac{2}{\pi}} \frac{\omega_{pe}^2 - \omega^2}{\omega^2} \left(\frac{\kappa T_e}{mc^2}\right)^{3/2} + (1 - \rho) \sqrt{\frac{8}{\pi}} \left(\frac{\kappa T_e}{mc^2}\right)^{1/2}. \tag{2.202}$$

This formula corresponds to those parts of expressions (2.200), which are not related to the particles collisions with each other.

Let us compare this quantity to the collisional absorption described by the first term in Eq. (2.200) when $| \omega_{pe}^2 - \omega^2 | \gg \nu^2$. We see that if

$$\frac{\nu}{\omega} < 4 \sqrt{\frac{2}{\pi}} \left(\frac{\kappa T_e}{mc^2}\right)^{3/2} \left(1 - \frac{\omega^2}{\omega_{pe}^2}\right)^{3/2}, \tag{2.203}$$

the collisionless dissipation energy dominates. It must be noticed that under this condition the second term in Eq. (2.200) becomes dominant. Since $\omega \gg \nu$, then one should use ν_{eff} given by Eq. (2.125) as the collision frequency. Thus, the dissipation determined by Eq. (2.202) turns out to be larger than the dissipation arising from the collisions given by

$$A_\nu = \frac{2\nu}{\sqrt{\omega_{pe}^2 - \omega^2}},$$

for mirror reflection if

$$\left(\frac{e^2 N_e^{1/3}}{\kappa T_e}\right) \left(\frac{mc^2}{\kappa T_e}\right) L^{2/3} < \left[1 - \left(\frac{\omega}{\omega_{pe}}\right)^2\right] \left(\frac{\omega_{pe}}{\omega}\right)^{4/3},$$

and for diffusion reflection if

$$\left(\frac{e^2 N_e^{1/3}}{\kappa T_e}\right)^3 \frac{mc^2}{\kappa T_e} L^2 < 1 - \left(\frac{\omega}{\omega_{pe}}\right)^2.$$

Here, L is the Coulomb logarithm given by Eq. (2.118). Furthermore, in the above relations, it was assumed that $\omega_{\text{pe}}^2 - \omega^2 \gg \nu^2$. These inequalities have a relatively wide range of validity. For frequencies not very close to ω_{pe}, the corresponding inequality is

$$25N_eL^2 < T_e^4,$$

where T_e is in degree $\overset{\circ}{c}$ and N_e is particle numbers in 1 cm^2.

The surface absorption can be described phenomenologically in the frame of macroscopic electrodynamics if in addition to the volume material equation

$$\vec{D} = \varepsilon(\omega)\vec{E}.$$

One can introduce the surface material equation [67, 68]:

$$\vec{i} = \gamma(\omega)\left\{\vec{E} - \vec{n}\left(\vec{n}\cdot\vec{E}\right)\right\}, \tag{2.204}$$

which relates the surface current density $\vec{i}$ to the longitudinal components of the electric field on the plasma surface. Here, $\vec{n}$ is the surface normal, and $\gamma(\omega)$ is the surface conductivity. For the Maxwellian distribution with non-relativistic temperature, we have

$$\gamma(\omega) = \sqrt{\frac{1}{2\pi}}\frac{\omega_{\text{pe}}^2\left(\omega_{\text{pe}}^2 - \omega^2\right)}{\omega^4}\left(\frac{\kappa T_e}{mc^2}\right)^{3/2}. \tag{2.205}$$

Such a phenomenological description is convenient for investigation of the problem of incident wave reflection from the plasma boundary. There exist two possibilities for the polarization of the incident wave: S-polarization with the electric vector parallel to the plasma surface and P-polarization with the electric vector oriented in the incidence plane. For the S-polarized wave, the complex reflection coefficient r_s is

$$r_s = \frac{\cos\theta - \sqrt{\varepsilon(\omega) - \sin^2\theta - \frac{4\pi\gamma(\omega)}{c}}}{\cos\theta + \sqrt{\varepsilon(\omega) - \sin^2\theta - \frac{4\pi\gamma(\omega)}{c}}} \equiv \frac{\cos\theta - \sqrt{n_s^2 - \sin^2\theta}}{\cos\theta + \sqrt{n_s^2 - \sin^2\theta}}. \tag{2.206}$$

Here, θ is the incidence angle (the angle between the wave vector of the incident wave and the surface normal $\vec{n}$), and n_s is the effective complex refractive index given by

$$n_s^2(\theta) = \varepsilon(\omega) + \frac{8\pi\gamma(\omega)}{c}\sqrt{\varepsilon(\omega) - \sin^2\theta + \left(\frac{4\pi\gamma(\omega)}{c}\right)^2}. \tag{2.207}$$

The square root in these expressions has a negative imaginary part. Note also that the last presentation of r_s in Eq. (2.206) coincides with the usual Fresnel's formula except here n_s is a function of the incidence angle θ. For the P-polarized wave,

$$r_p = \frac{\cos\theta\left[\sqrt{\varepsilon(\omega) - \sin^2\theta + \frac{4\pi\gamma(\omega)}{c}} + \frac{\sin^2\theta}{\sqrt{\varepsilon(\omega) - \sin^2\theta}}\right] - 1}{\cos\theta\left[\sqrt{\varepsilon(\omega) - \sin^2\theta + \frac{4\pi\gamma(\omega)}{c}} + \frac{\sin^2\theta}{\sqrt{\varepsilon(\omega) - \sin^2\theta}}\right] + 1}$$

$$\equiv \frac{n_p^2\cos\theta - \sqrt{n_s^2 - \sin^2\theta}}{n_p^2\cos\theta + \sqrt{n_s^2 - \sin^2\theta}}, \tag{2.208}$$

where

$$n_p^2(\theta) = \varepsilon(\omega) + \frac{8\pi\gamma(\omega)}{c}\sqrt{\varepsilon(\omega) - \sin^2\theta + \left(\frac{4\pi\gamma(\omega)}{c}\right)^2} + \frac{4\pi\gamma(\omega)}{c} \times \frac{\sin^2\theta}{\sqrt{\varepsilon(\omega) - \sin^2\theta}}. \tag{2.209}$$

In conclusion, it must be noticed that the material Eq. (2.204) does not take into account the possibility of longitudinal waves' existence in semi-bounded plasma [69].

2.10 Linear Electromagnetic Phenomena in Bounded Plasmas

The main goal of the present section is the study of the electromagnetic properties of bounded plasma-like media when the scale length of plasma parameters in the direction perpendicular to the plasma boundary is comparable or smaller than the inhomogeneity scale length of the electromagnetic fields. Moreover, we treat this problem in both non-degenerate or Maxwellian plasmas, when all charge carriers obey Maxwell–Boltzmann statistics, and degenerate plasmas, when charge carriers obey Fermi–Dirac statistics.

Furthermore, we study the low-frequency electromagnetic waves and relaxation processes of the small perturbations in the plasma-like media bounded by vacuum. In the frequency range at which spatial dispersion is important, the properties of these oscillations (their frequency spectrum and damping decrement or their growing increment) essentially depend on the boundary conditions. In this frequency range, the waves move with the phase velocity, which is smaller than the thermal velocity of charged particles and, hence, light speed. To a good degree of accuracy, the electromagnetic waves excited by such perturbations are potential. Therefore, we can restrict our analysis to the Poisson's equation describing the oscillations of charge carriers' density.

Boundaries of plasma-like media first act as a potential barrier which can prevent the outflow of charge carriers (electrons, ions, holes) from the media. However, charge carriers with sufficient thermal energy could overcome this barrier and form non-zero charge densities and currents on the surface of the medium. Furthermore, the real surface of the medium ceases to be smooth at the micro-level; therefore, the specular reflection of the particle will be changed to a chaotic diffusive one. Finally, on the boundary surface, there always exists an admixture of impurity particles absorbed from outside with which charge carriers can interact inelastically. Therefore, quite different phenomena such as adhesion, recombination, and ionization could arise. These phenomena impact not only the behavior of charge carriers but also electromagnetic fields as well.

Here, we do not intend to present a general solution for this problem, but we rather treat a model in which the surface of the plasma-like medium corresponds to an infinitely high and thin barrier near the boundary of media from which elastic (mirror) reflection of charge carriers takes place. Such a model suits for solid-state materials (semiconductors and metals) where charge carriers energy on the average is their thermal energies, which are much smaller than the work function. In this model, charged particles are assumed to be in an infinitely deep well with thickness of the order of the crystal lattice spacing (i.e., several Angstroms), which is much smaller than all plasma scale lengths such as Debye length, Larmor radii of charged particles and wavelength of oscillations, or field inhomogeneity scale lengths.

It should be noted that a very interesting wave absorption phenomenon in the inhomogeneous layer near the plasma surface is ignored here. This phenomenon known as plasma surface resonance is connected to the resonance absorption of the bulk (volume) waves when their frequency coincides with the local plasma frequency near the surface. Such absorption becomes important when this inhomogeneity scale length is comparable to the wavelength of the resonant plasma oscillations. Therefore, this resonance can be ignored if the inhomogeneity scale length of the plasma surface is much smaller than the wavelength of the absorbing waves.

In this way, our analysis in this section pertains to the problem of electromagnetic waves in semi-bounded plasma and plasma layers with mirror reflection of charged particles from the plasma boundary ($\rho = 1$).[31] On the basis of surface impedance and Poisson's equation, the dispersion equation and their spectra for surface waves are obtained.

2.10.1 Surface Electromagnetic Waves in Semi-Bounded Plasmas

In the previous sections we have dealt with short-wavelength electromagnetic oscillations in spatially inhomogeneous plasma, their wavelengths being much smaller than the characteristic dimension of the plasma inhomogeneity. To describe these oscillations, plasma can be treated as effectively infinite and the approximation of geometrical optics can be applied. Let us consider the opposite limit, when the scales of the plasma inhomogeneity are sufficiently sharp compared to the wavelength. We shall study only the simplest problems of the electrodynamics of plasmas with sharp inhomogeneities, i.e., surface electromagnetic waves in semi-bounded plasma conterminous with vacuum. Surface waves are qualitatively a new type of electromagnetic oscillations of a bounded medium. They are waves traveling along the medium surface and damping in the perpendicular direction.

It is obvious that the character of surface waves essentially depends on the properties of the plasma surface, or, more exactly, on the boundary conditions, considered in Sect. 1.1, which must supplement the field equations. From a variety of models of plasma surfaces only two cases of different physical nature will be studied. This section deals with the first model in which the existence of a sufficiently sharp plasma surface is presupposed. Here, all plasma quantities with the dimension of length (wavelengths, Debye and Larmor radii of particles, mean free paths, etc.) greatly exceed the size of the density variation near the plasma surface. A similar situation occurs, for example, in gaseous plasma confined in a glass vessel with electromagnetic properties insignificantly differing from the vacuum properties, or in solid-state plasma whose surface structure is defined by a crystal lattice. The second model assumes plasma to be confined by a strong external magnetic field, its surface being the boundary layer with thickness of the order of the Larmor radius of particles. This surface can be treated as sharp for surface waves much longer than the Larmor radius of particles. This situation is characteristic of high-temperature plasma in controlled thermonuclear fusion devices where plasma is isolated from the metal walls by the magnetic field and appears to be spatially confined and contiguous to vacuum. This model of a plasma surface will be studied in Sect. 3.10.

[31]The boundary problem in the case of diffusive reflection ($\rho = 0$) is mathematically very complicated. To solve field equations, one has to use the Wiener–Hopf method, which is too awkward [34, 55, 60, 61].

2.10.1.1 Solution of the Vlasov Equation for Semi-Bounded Isotropic Plasma

We shall limit our analysis of surface waves in plasma with a sharp boundary to the case of isotropic collisionless semi-bounded plasma in the absence of external electric and magnetic fields. As the unperturbed distribution function of particles of type α ($\alpha = e, i$), the non-relativistic Maxwellian distribution function

$$f_{0\alpha} = \frac{N_{0\alpha}}{(2\pi m_\alpha T_\alpha)^{3/2}} \exp\left(-\frac{m_\alpha v^2}{2T_\alpha}\right) \tag{2.210}$$

will be taken for non-degenerate plasma, or the Fermi distribution function

$$f_{0\alpha} = \begin{cases} \dfrac{2}{(2\pi\hbar)^3} & for \quad p < p_{F\alpha} = \left(3\pi^2\right)^{1/3} \hbar N_{0\alpha}^{1/3}, \\ 0 & for \quad p > p_{F\alpha}, \end{cases} \tag{2.211}$$

for a degenerate one. Here, $N_{0\alpha} =$ const for $x > 0$ (in plasma, see Fig. 2.5) and $N_{0\alpha} = 0$ for $x < 0$ (in vacuum). Then the solution of the kinetic equation for the perturbation of the distribution function

$$\frac{\partial \delta f_\alpha}{\partial t} + (\vec{v} \cdot \nabla)\delta f_\alpha + \frac{e_\alpha}{m_\alpha}\vec{E} \cdot \frac{\partial f_{0\alpha}}{\partial \vec{v}} = 0 \tag{2.212}$$

without any restrictions on generality can be sought in the form

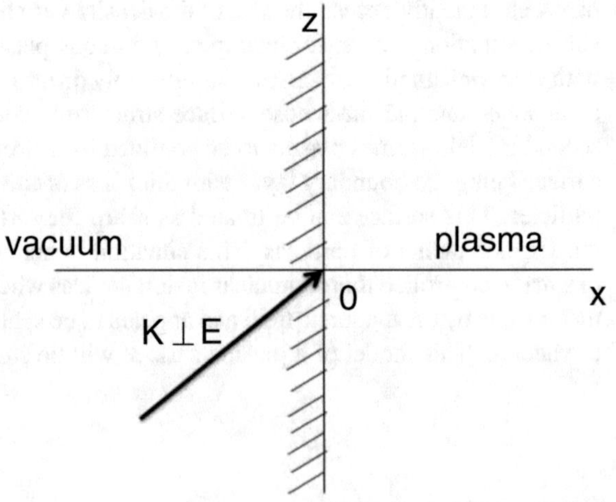

Fig. 2.5 Semi-bounded plasma with a sharp boundary

$$\delta f_\alpha = \delta f_\alpha(x) \exp\left(-\imath\omega t + \imath k_z z\right), \tag{2.213}$$

assuming the wave vector to be parallel to the plasma surface. As a result, from Eq. (2.212) we obtain

$$-\imath\left(\omega - \vec{k}\cdot\vec{v}\right)\delta f_\alpha + v_x \frac{\partial \delta f_\alpha}{\partial x} + \frac{e_\alpha}{m_\alpha}\vec{E}\cdot\frac{\partial f_{0\alpha}}{\partial \vec{v}} = 0. \tag{2.214}$$

To solve Eq. (2.214) it is necessary to set the boundary condition for $\delta f_\alpha(x)$ at $x = 0$, i.e., on the plasma surface. This very condition must contain all information on the character of interaction of charged particles with the surface, confining plasma. Henceforth, we shall assume charged particles to undergo a mirror reflection from this surface. This implies that at $x = 0$ the condition

$$\delta f_\alpha(0,\ v_x > 0) = \delta f_\alpha(0,\ v_x < 0) \tag{2.215}$$

is satisfied. Besides, when studying plasma surface waves (or the field penetration into plasma), we assume the electromagnetic fields $\vec{E}$ and $\vec{B}$ to vanish far from the plasma surface.

To solve Eq. (2.214), it is convenient to write

$$\delta f_\alpha(x, v_x) = \delta f_\alpha^+(x, v_x) + \delta f_\alpha^-(x, v_x),$$
$$\delta f_\alpha^\pm(x, v_x) = \begin{cases} \delta f_\alpha(x, v_x > 0), \\ \delta f_\alpha(x, v_x < 0). \end{cases} \tag{2.216}$$

Each of the quantities $\delta f_\alpha^\pm$ also satisfies Eq. (2.214). Therefore, using $\delta f_\alpha^-(x, v_x) \to 0$ for $x \to \infty$, we obtain

$$\delta f_\alpha^-(x, v_x) = \frac{e_\alpha}{m_\alpha}\frac{\partial f_{0\alpha}}{\partial \vec{v}}\frac{1}{v_x}\cdot\int_x^\infty dx' \vec{E}(x') \exp\left[\imath\frac{(x - x')}{v_x}(\omega - k_z v_z)\right]. \tag{2.217}$$

To find $\delta f_\alpha^+(x, v_x)$ let us rewrite the condition (2.215) in the form of

$$\delta f_\alpha^+(0, v_x) = \delta f_\alpha^-(0, -v_x). \tag{2.215a}$$

Then from Eq. (2.214) we have

$$\delta f_{\alpha}^{+}(x, v_x) = -\frac{e_\alpha}{m_\alpha}\frac{\partial f_0}{\partial \vec{v}}\frac{1}{v_x}\cdot\left\{\int_0^x dx'\vec{E}(x')\exp\left[\imath\frac{(x-x')}{v_x}(\omega - k_z v_z)\right] + \right. \tag{2.218}$$

$$\left. + \int_0^\infty dx'\vec{E}(x')\exp\left[\imath\frac{(x-x')}{v_x}(\omega + k_z v_z)\right]\right\}.$$

Substituting expressions (2.217) and (2.218) into the formula for the current density

$$\vec{j}(x) = \sum_\alpha e_\alpha \int d\vec{p}\,\vec{v}\,\delta f_\alpha(x, \vec{v})$$
$$= \sum_\alpha e_\alpha \int_{v_x>0} d\vec{p}\,\vec{v}\,\delta f_\alpha^+ + \int_{v_x<0} d\vec{p}\,\vec{v}\,\delta f_\alpha^-, \tag{2.219}$$

we have after simple calculations

$$j_i^{(x)} = \int_0^\infty dx' \left[K_{ij}(|x-x'|) + K_{ij}(|x+x'|)E_j(x')\right]. \tag{2.220}$$

Here,

$$K_{ij}(|x|) = -\sum_\alpha \frac{e_\alpha^2}{m_\alpha}\int_{v_x\geq 0} d\vec{p}\,\frac{v_i}{v_x}\frac{\partial f_{0\alpha}}{\partial v_j}\exp\left[\imath\frac{|x|}{v_x}(\omega - k_z v_z)\right]. \tag{2.221}$$

2.10.1.2 Solution of Field Equations

We can proceed to solve the field equations which in the present geometry can be written as

$$\frac{\imath\omega}{c}\vec{B} = \left\{\begin{array}{c} -\imath k_z E_y \\ \imath k_z E_x - \dfrac{\partial E_z}{\partial x} \\ \dfrac{\partial E_y}{\partial x}\end{array}\right\}, \qquad -\frac{\imath\omega}{c}\vec{E} + \frac{4\pi}{c}\vec{j} = \left\{\begin{array}{c} -\imath k_z B_y \\ \imath k_z B_x - \dfrac{\partial B_z}{\partial x} \\ \dfrac{\partial B_y}{\partial x}\end{array}\right\}. \tag{2.222}$$

This system of equations is valid for $x \geq 0$ (plasma) as well as for $x < 0$ (vacuum) since $\vec{j}(x < 0) = 0$. They follow from relations (2.220) and (2.221) accounting for the plasma dependence on x.

Noting the symmetry of the tensor $K_{ij}(|x|)$, it is easy to show that the system of Eq. (2.222) splits into two independent subsystems for the field components E_x, E_z, B_y and B_x, B_z, E_y, respectively. The latter admits no solutions in the form of surface waves. Therefore, we confine our analysis to the equations for the field components E_x, E_z, and B_y:

$$\begin{aligned}
&\frac{\partial E_z}{\partial x} - \imath k_z E_x + \imath\frac{\omega}{c}B_y = 0,\\
&\imath k_z B_y - \imath\frac{\omega}{c}E_x + \frac{4\pi}{c}j_x = 0,\\
&\frac{\partial B_y}{\partial x} + \imath\frac{\omega}{c}E_z - \frac{4\pi}{c}j_z = 0.
\end{aligned} \tag{2.223}$$

By integrating these equations over an infinitely narrow intermediate layer near the plasma-vacuum interface, we obtain the boundary conditions which link the fields $\vec{E}$ and $\vec{B}$ in vacuum and plasma. As a corollary to the finiteness of the fields $\vec{E}$, $\vec{B}$ and the current density $\vec{j}$, we obtain the general electrodynamic boundary condition, i.e., the continuity of the tangential components of the fields $\vec{E}$ and $\vec{B}$ at $x = 0$:

$$\{E_z\}_{x=0} = \{B_y\}_{x=0} = 0, \tag{2.224}$$

where the notation $\{A\}_{x=0} = A(x \to +0) - A(x \to -0)$ is introduced.

In order to solve this boundary problem, we shall apply the following method. We extend Eqs. (2.223) into the range $x < 0$ assuming

$$\begin{aligned}
&E_x(x) = -E_x(-x), \qquad && N_{0\alpha}(x) = N_{0\alpha}(-x),\\
&E_z(x) = E_z(-x), && B_y(x) = -B_y(-x).
\end{aligned} \tag{2.225}$$

Then in the whole range $-\infty \leq x \leq +\infty$, we have

$$j_i(x) = \int_{-\infty}^{+\infty} dx' \sigma_{ij}(|x - x'|)E_j(x'), \tag{2.226}$$

where

$$\sigma_{ij}(|x|) = -\sum_{\alpha} \frac{e_{\alpha}^2}{m_{\alpha}} \int d\vec{p}\, \frac{v_i}{v_x} \frac{\partial f_{0\alpha}}{\partial v_j} \exp\left[\imath \frac{|x|}{v_x} (\omega - k_z v_z)\right] \tag{2.227}$$

is the tensor connecting $\vec{j}(x)$ and $\vec{E}(x)$ in spatially infinite isotropic plasma. Its Fourier transform

$$\sigma_{ij}\left(\omega, \vec{k}\right) = \int_{-\infty}^{+\infty} dx\, \sigma_{ij}(|x|) e^{-\imath k_x x} = -\imath \sum_{\alpha} \frac{e_{\alpha}^2}{m_{\alpha}} \int \frac{d\vec{p}\, v_i}{\omega - \vec{k}\cdot\vec{v}} \frac{\partial f_{0\alpha}}{\partial v_j} \tag{2.228}$$

coincides with the conductivity tensor of isotropic plasma of Eq. (2.20).

Thus, the material equation

$$j_i\left(\vec{k}\right) = \sigma_{ij}\left(\omega, \vec{k}\right) E_j\left(\vec{k}\right) \tag{2.229}$$

connecting the Fourier components of the current and electric field densities, in the case of the mirror reflection of particles from the surface for semi-bounded plasma, is of the same form as for the spatially infinite isotropic case. This is clear, since for the mirror reflection of particles from the plasma surface, their motion is the same as in infinite plasma and thus the perturbation, developed in the system due to the electromagnetic field, is independent of the surface presence.

It should be noted that when applying the continuation relations (2.225), the components E_x and B_y (oddly continued) at $x = 0$ undergo an abrupt change. One can easily see this when substituting expression (2.226) into the system of Eq. (2.223) and integrating it over a thin layer near $x = 0$. At the same time if one proceeds from relation (2.220) and assumes $j(x) = 0$ for $x < 0$, then the integration of expression (2.226) over an intermediate layer near the plasma surface results in the continuity condition for the tangential components of the fields $\vec{E}$ and $\vec{B}$ at $x = 0$, shown by the boundary conditions (2.224). This is natural since the system of Eqs. (2.223) and (2.226) are valid only in the region of plasma with $x \geq 0$, the applied continuation being only a mathematical method for solving this problem. On the other hand, this fact should be accounted for when solving Eq. (2.223) which is a system of integro-differential equations with differential kernels after the substitution of the expressions for the current density (2.226). Such a system should be solved with the aid of the Fourier transformation

$$A\,(x) = \int_{-\infty}^{+\infty} dk_x\, e^{\imath k_x x} A(k_x), \qquad A(k_x) = \frac{1}{2\pi} \int_{-\infty}^{+\infty} dx\, e^{-\imath k_x x} A(x), \tag{2.230}$$

accounting for possible jumps of the quantities E_x and B_y at $x = 0$.

Fourier transforming Eqs. (2.223) and taking into account the continuity of the function $E_z(x)$ and the discontinuity of the function $B_y(x)$ at $x = 0$, we obtain the algebraic equations

$$\begin{cases} \imath k_x E_z\,(k_x) - \imath\, k_z E_x(k_x) + \imath \dfrac{\omega}{c} B_y(k_x) = 0, \\ \imath k_z B_y(k_x) - \imath \dfrac{\omega}{c} E_x(k_x) + \dfrac{4\pi}{c} j_x(k_x) = 0, \\ \imath\, k_x B_y\,(k_x) - \dfrac{1}{\pi} B_y(x=0) + \imath \dfrac{\omega}{c} E_z(k_x) - \dfrac{4\pi}{c} j_z(k_x) = 0. \end{cases} \tag{2.231}$$

Hence,

$$E_z\,(k_x) = \frac{-\imath c}{\pi \omega k^2} B_y(x=0) \left(\frac{k_z^2}{\varepsilon^l(\omega, k)} - \frac{k_x^2 \omega^2}{c^2 k^2 - \omega^2 \varepsilon^{\mathrm{tr}}(\omega, k)} \right). \tag{2.232}$$

Here $k^2 = k_x^2 + k_z^2$ and $\varepsilon^l\,(\omega, k)$ and $\varepsilon^{\mathrm{tr}}(\omega, k)$ are, respectively, the longitudinal and transverse dielectric permittivities of isotropic plasma, determined by expressions (2.59) and (2.60).

2.10.1.3 Surface Impedance

After substituting expression (2.232) into the expression for the Fourier transformation (2.230), we shall determine the so-called surface impedance of semi-bounded isotropic plasma:

$$Z_s = \frac{4\pi}{c} \frac{E_z(x=0)}{B_y(x=0)} = -\frac{8\imath\omega}{c^2} \int_0^\infty \frac{dk_x}{k^2} \left(\frac{k_z^2 c^2}{\omega^2 \varepsilon^l\,(\omega, k)} - \frac{k_x^2 c^2}{c^2 k^2 - \omega^2 \varepsilon^{\mathrm{tr}}(\omega, k)} \right). \tag{2.233}$$

The system of equations (2.223) may analogously be solved for vacuum, i.e., for $\vec{j}\,(x) = 0$. One must only continue the fields $\vec{E}$ and $\vec{B}$ into the region $x > 0$ according to Eq. (2.225). Thus, the surface impedance of the vacuum half-space is

$$Z_V = -\frac{4\pi}{c} \frac{E_z(x=0)}{B_y(x=0)} = \frac{4\pi\imath}{c} \sqrt{\frac{k_z^2 c^2}{\omega^2} - 1}. \tag{2.234}$$

For $\varepsilon^l = \varepsilon^{\mathrm{tr}} = 1$ this formula can also be obtained from Eq. (2.233).

2.10.1.4 Dispersion Equation for Surface Waves

Now we can apply the boundary conditions (2.224) and equate Z_s and Z_V. This equality yields the dispersion equation for surface waves in semi-bounded isotropic plasma:

$$\sqrt{\frac{k_z^2c^2}{\omega^2}-1}+\frac{2\omega}{\pi c}\int\limits_0^\infty \frac{dk_x}{k^2}\left(\frac{k_z^2c^2}{\omega^2\varepsilon^l(\omega,k)}-\frac{k_x^2c^2}{k^2c^2-\omega^2\varepsilon^{\text{tr}}(\omega,k)}\right)=0. \tag{2.235}$$

In the limit $c\to\infty$, from this equation we obtain a dispersion equation for longitudinal (potential) surface waves in semi-bounded plasma:

$$1+\frac{2}{\pi}\int\limits_0^\infty \frac{dk_x|k_z|}{k^2\varepsilon^l(\omega,k)}=0. \tag{2.236}$$

2.10.1.5 Surface Waves in Cold Semi-Bounded Plasma

Let us analyze the frequency spectra of the surface electromagnetic waves in semi-bounded plasma. We start with the case of cold plasma, neglecting the thermal motion of particles. Accounting for

$$\varepsilon^{\text{tr}}(\omega)=\varepsilon^l(\omega)=\varepsilon(\omega)=1-\frac{\omega_{\text{pe}}^2}{\omega^2}, \tag{2.237}$$

from the general dispersion equation (2.235), we obtain

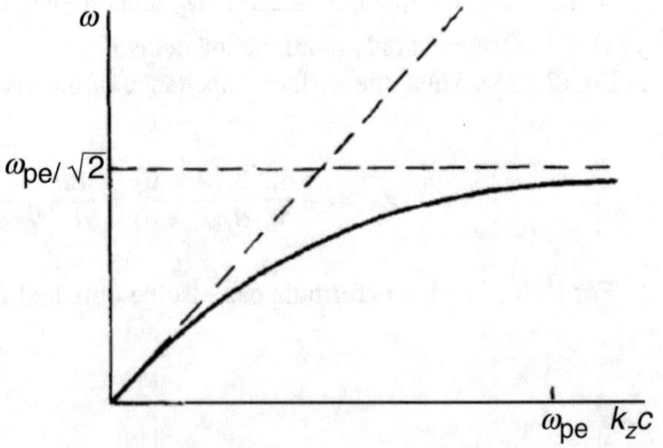

Fig. 2.6 Spectrum of surface electromagnetic waves in cold semi-bounded plasma

$$\sqrt{k_z^2c^2 + \omega_{\mathrm{pe}}^2 - \omega^2} + \left(1 - \frac{\omega_{\mathrm{pe}}^2}{\omega^2}\right)\sqrt{k_z^2c^2 - \omega^2} = 0. \tag{2.238}$$

It shows that in semi-bounded cold isotropic plasma, surface waves exist only in the frequency range $\omega < \omega_{\mathrm{pe}}$ and their phase velocity is always smaller than that of light, $\omega/k_z < c$. The general solution of Eq. (2.238) is graphically shown in Fig. 2.6. Here, the dashed continuation of the dispersion curve corresponds to the frequency range where the electron thermal motion $(\omega \sim k_z v_{\mathrm{Te}})$ becomes significant and oscillations are strongly damped. The analytical solution can easily be written in limiting cases of long and short wavelengths:

$$\omega^2 = \begin{cases} k_z^2c^2 & \text{for} \quad k_z^2c^2 \ll \omega_{\mathrm{pe}}^2, \\ \omega_{\mathrm{pe}}^2/2 & \text{for} \quad k_z^2c^2 \gg \omega_{\mathrm{pe}}^2. \end{cases} \tag{2.239}$$

In the long-wavelength limit, the phase velocity of the surface waves is near the velocity of light and oscillations are closely transverse, while in the short-wavelength limit their phase velocity is small compared to that of light, oscillations being almost longitudinal (potential). Potentiality of short (slow) surface waves in cold plasma is also verified by the solution of Eq. (2.236) which takes the form

$$\varepsilon(\omega) = 1 - \frac{\omega_{\mathrm{pe}}^2}{\omega^2} = -1, \tag{2.240}$$

and has a solution identical to the second expression of the solution (2.239).

2.10.1.6 Cherenkov Damping of Surface Waves

In the above, the thermal motion of particles and the collisionless Cherenkov damping of surface waves have been ignored. Slow waves, being closely longitudinal, are obviously most strongly damped. Thus, to analyze the collisionless damping of surface waves, we shall confine our analysis to Eq. (2.236). It can be analyzed only by means of numerical integration. However, in the range $\omega \gg k_z v_{\mathrm{Te}}$ we can approximate Eq. (2.236) as

$$2 - \frac{\omega_{\mathrm{pe}}^2}{\omega^2} + \imath\sqrt{\frac{2}{\pi}}|k_z| \int\limits_0^\infty \frac{dk_x \omega_{\mathrm{pe}}^2 \omega}{k^5 v_{\mathrm{Te}}^3} \exp\left(-\frac{\omega^2}{2k^2 v_{\mathrm{Te}}^2}\right) = 0. \tag{2.241}$$

Small values of the imaginary parts specified by the Cherenkov absorption of surface waves by plasma electrons have been accounted for when obtaining Eq. (2.241). Since for $\omega \gg k_z v_{\mathrm{Te}}$ the main contribution in the integral term of

Eq. (2.241) gives the range of large values of k_x, then k can be replaced by k_x in the integrand with a high degree of accuracy. As a result, we have from Eq. (2.241)

$$2 - \frac{\omega_{\mathrm{pe}}^2}{\omega^2}\left(1 - 2\imath\sqrt{\frac{2}{\pi}}\frac{|k_z|v_{\mathrm{Te}}}{\omega}\right) = 0. \tag{2.242}$$

Hence, Eq. (2.239) again gives the frequency spectrum of longitudinal surface waves. For the damping decrement ($\omega \to \omega + \imath\delta$) we obtain[32]

$$\delta = -\sqrt{\frac{2}{\pi}}|k_z|v_{\mathrm{Te}}. \tag{2.243}$$

Damping of high-frequency surface waves, in contrast to that of volume Langmuir waves (Sect. 2.3), is not exponentially small, though their phase velocity greatly exceeds the thermal velocity of electrons. This is a corollary to the fact that the resonance interaction of particles with a wave occurs for $\omega \gg k_z v_{\mathrm{Te}}$ in the frequency range $\omega = k_x v_{\mathrm{Te}}$. Since, according to Eq. (2.236), k_x has arbitrarily large values (in particular, $k_x > \omega/v_{\mathrm{Te}}$ for surface waves), the main bulk of electrons and not just an exponentially small portion of the particles (as in the case of volume Langmuir waves in unbounded plasma) takes part in their absorption.

2.10.1.7 Surface Ion-Acoustic Waves

Let us now proceed to the analysis of surface ion-acoustic waves in semi-bounded isotropic plasma. Volume ion-acoustic waves exist in unbounded plasma only when $T_e \gg T_i$ (Sect. 2.5). This is also necessary for surface waves, since similar to volumetric waves, surface ion-acoustic waves are longitudinal, described by Eq. (2.236). For non-degenerate plasma we have

$$\varepsilon^l(\omega, k) = 1 - \frac{\omega_{\mathrm{pi}}^2}{\omega^2} + \frac{\omega_{\mathrm{pe}}^2}{k^2 v_{\mathrm{Te}}^2}\left(1 + \imath\sqrt{\frac{\pi}{2}}\frac{\omega}{k v_{\mathrm{Te}}}\right). \tag{2.244}$$

Since$v_{\mathrm{Te}} \gg \omega/k \gg v_{\mathrm{Ti}}$, we can account only for the Cherenkov dissipation on electrons. The ion thermal motion is completely ignored. After substituting expression (2.244) and integrating over Eq. (2.236), we obtain

[32]In the damping decrement of surface waves an exact numerical solution of Eq. (2.236) gives the factor 0.125 instead of $\sqrt{2/\pi} \approx 0.8$.

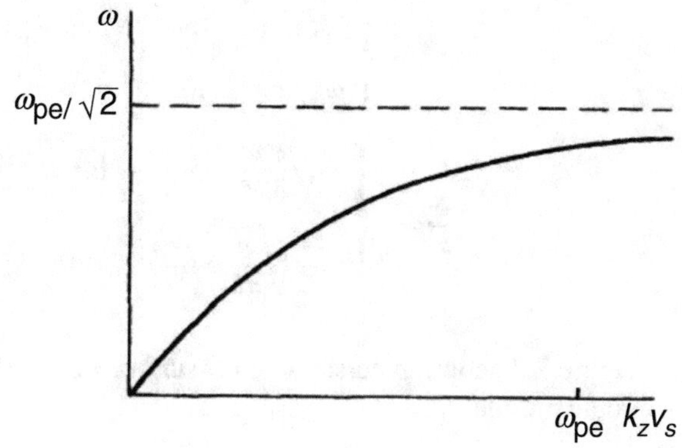

Fig. 2.7 Spectrum of surface ion-acoustic wave in semi-bounded isotropic plasma

$$\varepsilon_i + \left(1 + \frac{\omega_{\mathrm{pi}}^2}{k_z^2 v_s^2 \varepsilon_i}\right)^{-1/2} - \imath\sqrt{\frac{2m}{\pi M}}\,\frac{\omega_{\mathrm{pi}}^2\omega}{|k_z|^3 v_s^3}$$
$$\times \frac{1}{\varepsilon_i}\int\limits_0^\infty \frac{dx}{\sqrt{x^2+1}\left(x^2+1+\omega_{\mathrm{pi}}^2/k_z^2 v_s^2 \varepsilon_i\right)^2} = 0, \qquad (2.245)$$

where $v_s = \sqrt{T_e/M}$ and $\varepsilon_i = 1 - \omega_{\mathrm{pi}}^2/\omega^2$. Though the integral here may be calculated in terms of elementary functions to analyze the spectra of ion-acoustic waves, it is more convenient to leave it in this form. In Eq. (2.245), the imaginary part specified by the Cherenkov wave dissipation on plasma electrons is small in comparison to the real part. Thus, in the first-order approximation it can be neglected. Surface ion-acoustic waves are possible only in the frequency range where

$$\varepsilon_i(\omega) = 1 - \frac{\omega_{\mathrm{pi}}^2}{\omega^2} < -\frac{\omega_{\mathrm{pi}}^2}{k_z^2 v_s^2},$$

i.e., $\omega^2 < \omega_{\mathrm{pi}}^2$. The frequency spectrum of oscillations and their damping decrement (accounting for the imaginary part) can be obtained from the real part of Eq. (2.245). In general, the formulas for ω and $\delta(\omega \to \omega + \imath\delta)$ are rather unwieldy, but they take a simple form in the limiting cases of long $\left(k_z^2 v_s^2 \ll \omega_{\mathrm{pi}}^2\right)$ and short $\left(k_z^2 v_s^2 \gg \omega_{\mathrm{pi}}^2\right)$ surface waves:

$$\omega^2 = \begin{cases} k_z^2 v_s^2 & \text{for } k_z^2 v_s^2 \ll \omega_{\text{pi}}^2, \\ \omega_{\text{pi}}^2/2 & \text{for } k_z^2 v_s^2 \gg \omega_{\text{pi}}^2, \end{cases}$$

$$\frac{\delta}{\omega} = \begin{cases} -\sqrt{\dfrac{\pi\, m}{8\, M}} & \text{for } k_z^2 v_s^2 \ll \omega_{\text{pi}}^2, \\ -\dfrac{1}{6}\sqrt{\dfrac{m}{\pi M}}\dfrac{\omega_{\text{pi}}^3}{k_z^3 v_s^3} & \text{for } k_z^2 v_s^2 \gg \omega_{\text{pi}}^2. \end{cases} \tag{2.246}$$

Figure 2.7 shows ω versus $k_z v_s$ for surface ion-acoustic waves in semi-bounded isotropic plasma.

2.10.2 *Surface Waves on Plasma Layers*

Study of the mirror reflection model is important because it provides us with a better understanding of the role of the boundary plasma interface. This model describes the confinement of charged particles in metals and semiconductor plasmas. In this model, all oscillations of plasma in equilibrium are damping with time. Moreover, the finite size of the layer appears when the thickness of the layer is much smaller than the wavelength of the oscillation along the surface of the layer. The spectra of the ion-acoustic oscillations and diffusion modes as well as their dependency on the thickness of the layer are determined. For the ion-acoustic modes, this dependency essentially exists in the short-wavelength region of spectrum but when layer thickness is comparable to the Debye radius of charge carriers, diffusion has the monopolar character.

The problem of the boundary conditions in electrodynamics for the first time arose with the formulation of the Maxwell's equations for electromagnetic fields in material media. For the monochromatic field ~ exp $(-\imath\omega t)$, and in the absence of external sources and spatial dispersion, these equations can be written as [54, 70, 71]:

$$\begin{cases} \nabla \times \vec{E} = \imath \dfrac{\omega}{c} \vec{B}\,, \qquad \left(\nabla \times \vec{B}\right)_i = -\imath \dfrac{\omega}{c}\, \varepsilon_{ij}(\omega)\, E_j, \\ \nabla \cdot \left(\varepsilon_{ij}(\omega)\, \vec{E_j}\right) = 0, \qquad \nabla \cdot \vec{B} = 0. \end{cases} \tag{2.247}$$

The differential system (2.247) is of the same order in the medium and in vacuum $(\varepsilon_{ij}(\omega) \rightarrow \delta_{ij})$. This is the consequence of neglecting spatial dispersion. Therefore, boundary conditions can be obtained by integrating Eq. (2.247) over an infinitely thin layer near the interface of the medium. They will have the following form [54, 70, 71]:

$$\left\{\overrightarrow{E_t}\right\}_s = 0, \quad \left\{\overrightarrow{B_t}\right\}_s = 0, \quad \left\{\overrightarrow{D_n}\right\}_s = 0, \tag{2.248}$$

where $\vec{E}_t$ and $\vec{B}_t$ are the tangential components of fields $\vec{E}, \vec{B}$; $\vec{D}_n$ is the normal component of the induction vector $\vec{D}$; and s is the interface of the medium. Here, the notation $\{A\}_s = 0 = (A_2 - A_1)_s = 0$ means that the quantity A on the interface s between two media is identical.

Here, we restrict our study only to slow waves for which $\omega \ll kc$, where k is the wave vector or the inverse of the scale length of the field inhomogeneity. As a result, instead of the system of Maxwell's equations, we can restrict ourselves to Poisson's equation:

$$\frac{\partial}{\partial r_i}\left(\varepsilon_{ij}(\omega)\,\frac{\partial \phi}{\partial r_j}\right) = 0. \tag{2.249}$$

Here, it is assumed that the medium is bounded by vacuum because in vacuum there is no longitudinal (potential) field. Considering only the plane plasma layer and comparing these results with results obtained for volume (bulk) oscillations in unbounded plasma media, we can explain which new effects arise in a bounded medium with respect to the unbounded ones.

2.10.2.1 Potential Surface Waves on Thin Layers of Non-Dispersive Media

Let us start with the isotropic plasma medium where $\varepsilon_{ij}(\omega) = \varepsilon(\omega)\delta_{ij}$. Therefore, without loss of generality, the solution of Eq. (2.249) can be written in the form $\phi(x)\exp(\imath k_z z)$. As a result, we have

$$\frac{\partial}{\partial x}\,\varepsilon(\omega, x)\,\frac{\partial \phi}{\partial x} - \varepsilon(\omega, x)\,k_z^2 \phi = 0. \tag{2.250}$$

This equation is valid both inside and outside plasma. The plasma layer is assumed to occupy the region $0 \le x \le a$, i.e.,

$$n(x) = \begin{cases} n_0 = \text{const.} & 0 \le x \le a, \\ 0, & x \le 0, a \le x. \end{cases}$$

The solution of Eq. (2.250) should be written in the form of

$$\phi(x) = \begin{cases} C_1 \exp(|k_z|x), & x \leq 0, \\ C_2 \exp(|k_z|x) + C_3 \exp(-|k_z|x), & 0 \leq x \leq a, \\ C_4 \exp(-|k_z|x), & x \geq a. \end{cases} \tag{2.251}$$

The boundary conditions obtained by integrating Eq. (2.250) near the boundaries $x = 0$ and $x = a$ are

$$\{\phi\}_{x=0,\,a} = 0, \qquad \left\{\varepsilon(\omega, x)\,\frac{\partial \phi}{\partial x}\right\}_{x=0,\,a} = 0. \tag{2.252}$$

Substituting relation (2.251) into Eq. (2.252), we are led to the dispersion equation

$$\left[\varepsilon(\omega)^2 + 1\right]\left[\exp(|k_z|a) - \exp(-|k_z|a)\right] + 2\varepsilon(\omega)\left[\exp(|k_z|a) + \exp(-|k_z|a)\right] = 0. \tag{2.253}$$

When $a \to \infty$ (or $|k_z \mid a \gg 1$), this equation reduces to expression (2.240) when $\varepsilon(\omega)$ is given by Eq. (2.237). Now, we consider the opposite limit, i.e., the thin layer case when $|k_z \mid a \ll 1$. In this limit, Eq. (2.253) takes the form

$$\varepsilon(\omega) + \frac{2}{a \mid k_z \mid} = 1 - \frac{\omega_{\mathrm{pe}}^2}{\omega(\omega + \imath\nu_e)} + \frac{2}{a \mid k_z \mid} = 0. \tag{2.254}$$

From Eq. (2.254), when $\omega \gg \nu_e$, we find the frequency spectrum:

$$\omega^2 = \frac{\omega_{\mathrm{pe}}^2 \mid k_z \mid a}{2}, \quad \gamma = -\frac{\nu_e}{2}. \tag{2.255}$$

In the opposite limit, $\omega \ll \nu_e$, we have

$$\omega = -\imath\frac{\omega_{\mathrm{pe}}^2}{2\nu_e} \mid k_z \mid a = -2\pi\imath\sigma \mid k_z \mid a. \tag{2.256}$$

The first spectrum is well-known as the deep water oscillations spectrum and the second one could be called charge Maxwell relaxations in deep conducting fluids [71]. They are valid when spatial dispersion is neglected, i.e., when the following inequalities hold, respectively,

$$\begin{aligned} &\omega_{\mathrm{pe}}\sqrt{\mid k_z \mid a} \gg \nu_e \gg k_z v_{\mathrm{Te}}, k_z v_{\mathrm{Fe}}, \\ &\nu_e \gg \omega_{\mathrm{pe}}\sqrt{\mid k_z \mid a} \gg k_z v_{\mathrm{Te}}, k_z v_{\mathrm{Fe}}. \end{aligned} \tag{2.257}$$

It is worthwhile to mention that the spectra given by expressions (2.255) and (2.256) correspond to the integral equation of the electron density perturbations

$$\frac{\partial^2 n_e}{\partial t^2} - \nu_e \frac{\partial n_e}{\partial t} + \frac{a\omega_{\mathrm{pe}}^2}{2} \int dz' \frac{n_e(z')}{(z-z')^2} = 0, \tag{2.258}$$

indicating the non-local character of oscillations along the surface.

2.10.2.2 Surface Waves on Thin Layers of Dispersive Media

In this part, the results obtained in the previous part are generalized for low-frequency waves on a plasma layer when spatial dispersion is important. We use the model of mirror reflection of charge carriers from the surface layer as well as the calculation method developed in [71–73]. In isotropic media, without loss of generality, the components of electromagnetic fields may be written in the form of

$$A(x) = \sum_{n=-\infty}^{\infty} \exp\left(-\imath\omega t + \imath k_{\parallel} z\right) a(n) \exp\left(\imath n\pi x/a\right). \tag{2.259}$$

As a result, the Maxwell's equations can be written as

$$\begin{cases} \imath \dfrac{n\pi}{a} E_z - \imath k_{\parallel} E_x + \imath \dfrac{\omega}{c} B_y = 0, \\ \imath k_{\parallel} B_y - \imath \dfrac{\omega}{c} E_x + \dfrac{4\pi}{c} j_x = 0, \\ \imath \dfrac{n\pi}{a} B_y + \dfrac{2}{a}\left[B_y(a)\exp\left(-\imath n\pi\right) - B_y(0)\right] + \imath \dfrac{\omega}{c} E_z - \dfrac{4\pi}{c} j_z = 0, \end{cases} \tag{2.260}$$

where $B_y(0)$ and $B_y(a)$ are the boundary values of $B_y(x)$, and j_x and j_z are the Fourier transforms of the components of the current density and are non-zero only in the plasma medium. We have to mention that here $\vec{k} = \left(n\pi/a, 0, k_{\parallel}\right)$.

Equations (2.260) are valid both inside the plasma medium, in which $\vec{j} \neq 0$, and outside it where $\vec{j} = 0$. Therefore, we solve this system of equations with respect to $B_y(x)$ in three regions $x \leq 0$, $0 \leq x \leq a$, and $x \geq a$. Matching the solutions and using the discontinuity of $B_y(x)$ on the surfaces $x = 0$ and a, we obtain the general dispersion equation of oscillations

$$\left[\sqrt{k_{\parallel}^2-\frac{\omega^2}{c^2}}-\frac{2}{a}\sum\nolimits_{n\geq 0}^{\infty\prime}G_n\right]^2-\frac{4}{a^2}\left[\sum\nolimits_{n\geq 0}^{\infty\prime}\exp\left(\imath n\pi\right)G_n\right]^2=0, \tag{2.261}$$

where

$$G_n=\frac{1}{k^2}\left(\frac{k_{\parallel}^2}{\varepsilon^l(\omega,k)}-\frac{\omega^2\pi^2n^2/a^2}{k^2c^2-\omega^2\varepsilon^{\mathrm{tr}}(\omega,k)}\right), \tag{2.262}$$

and also $k^2=\pi^2n^2/a^2+k_{\parallel}^2$. The prime means that for $n=0$, G_n must be divided by 2. The analysis of this equation was done in [71–73].

In what follows, we restrict ourselves to the potential oscillations. Consequently, Eq. (2.261) takes the form

$$F^{\pm}=1+\frac{2}{a\mid k_{\parallel}\mid}\sum_{n=0}^{\infty\prime}[1\pm(-1)^n]\frac{k_{\parallel}^2}{k^2\varepsilon^l(k,\omega)}=0, \tag{2.263}$$

describing purely longitudinal (potential) surface oscillations.

For the short-wavelength oscillations, when $a|k_{\parallel}|\gg 1$ (or $a\to\infty$), the summation in Eq. (2.263) becomes an integral over $dk_x=\Delta n\pi/a$, where $\Delta n=1$. As a result, we obtain the dispersion equation of longitudinal (potential) surface oscillations on a semi-bounded medium given by Eq. (2.236)

$$1+\frac{2}{\pi}\int_0^{\infty}\frac{\mid k_{\parallel}\mid dk_x}{k^2\varepsilon^l(\omega,k)}=0,$$

where $k^2=k_{\parallel}^2+k_x^2$.

Thus, Eq. (2.263) describes surface-type oscillations in thin layers of plasma-like media, and, therefore, our problem is to find the solution of this equation for $\omega(k_{\parallel})$. We see that two branches of oscillations exist: symmetric (with even n) and asymmetric (with odd n). For the long-wavelength oscillations, when $a\mid k_{\parallel}\mid\leq 1$, these branches are well-separated, whereas in the short-wavelength limit, when $a\mid k_{\parallel}\mid\gg 1$, they coincide.

Now, we study the oscillations in a plasma thin layer of thickness a in the frequency range in which strong spatial dispersion exists. This is done by substituting

$$\varepsilon^l(k,\omega) = 1 - \frac{\omega_{\text{pi}}^2}{\omega^2}\left(1 - \imath\frac{\nu_i}{\omega}\right) + \frac{\omega_{\text{pe}}^2}{k^2 v_{\text{Te}}^2}\left(1 + \imath\left\{\begin{array}{l} \sqrt{\frac{\pi}{2}}\frac{\omega}{k v_{\text{Te}}} \\ \frac{\omega\nu_e}{k^2 v_{\text{Te}}^2} \end{array}\right.\right), \tag{2.264}$$

and

$$\varepsilon^l(k,\omega) = 1 + \sum_\alpha \frac{\omega_{p\alpha}^2}{k^2 v_{T\alpha}^2 - \imath\omega\nu_\alpha}, \tag{2.265}$$

into Eq. (2.263), respectively, for ion-acoustic

$$k v_{\text{Ti}}, \nu_i \ll \left\{\begin{array}{l} \nu_e \ll \omega \ll k v_{\text{Te}}, \\ \omega \ll \frac{k^2 v_{\text{Te}}^2}{\nu_e} \ll \nu_e, \end{array}\right. \tag{2.266}$$

and diffusion

$$\omega, k v_{T\alpha} \ll \nu_\alpha, \tag{2.267}$$

frequency ranges. We assume that layer thickness is larger than the Debye radius, i.e., $a^2 \gg v_s^2/\omega_{\text{pi}}^2$. Practically, this assumption is always fulfilled, although in this case it is also assumed that $a \mid k_z \mid \ll 1$. Under these conditions, from Eq. (2.263), it follows that the oscillation spectrum of asymmetric modes (with odd n) is not different from the bulk oscillations spectrum studied earlier. Considering symmetric modes for ion-acoustic oscillations, from Eq. (2.263), we obtain

$$\frac{2}{a \mid k_z \mid} - \frac{\omega_{\text{pi}}^2}{\omega(\omega + \imath\nu_i)} + \frac{\omega_{\text{pi}}^2}{k^2 v_s^2}\left(1 + \imath\alpha(k_\parallel)\right) = 0, \tag{2.268}$$

where

$$\alpha(k_\parallel) = \left\{\begin{array}{ll} \sqrt{\frac{\pi}{2}}\frac{\omega}{\mid k_\parallel \mid v_{\text{Te}}}, & k_\parallel v_{\text{Te}} \gg \nu_e, \\ \frac{\omega\nu_e}{k_\parallel^2 v_{\text{Te}}^2}, & \nu_e \gg k_\parallel v_{\text{Te}} \gg \sqrt{\omega\nu_e}. \end{array}\right. \tag{2.269}$$

From Eqs. (2.268) and (2.269), we find the frequency spectrum of the symmetric ion-acoustic oscillations in a plasma layer

$$\omega^2 = \frac{\omega_{\rm pi}^2}{\frac{2}{a|k_\parallel|} + \frac{\omega_{\rm pi}^2}{k_\parallel^2 v_s^2}}, \qquad \frac{\gamma}{\omega} = -\frac{\nu_i}{2\omega} - \alpha(k_\parallel)\frac{\omega^3}{k_\parallel^2 v_s^2}. \tag{2.270}$$

As seen from these formulas, the ion-acoustic oscillation spectrum strongly depends on layer thickness a, which becomes important in regions of relatively short-wavelength modes, when

$$1 \ll \frac{\omega_{\rm pi}^2 a^2}{v_s^2} \ll \frac{2}{a \mid k_\parallel \mid}. \tag{2.271}$$

For this reason, Poisson's equation becomes non-local and may be written as

$$\frac{\omega_{\rm pi}^2}{\omega(\omega + \imath\nu_e)} \nabla^2\phi + \frac{\omega_{\rm pi}^2}{v_s^2}\phi + \int dz' Q(z - z')\nabla^2\phi(z') = 0, \tag{2.272}$$

where

$$Q(z - z') = -\int dk_z \exp\left[\imath k_z(z - z')\right]\left\{\frac{2}{a \mid k_z \mid} + \frac{\imath\alpha(k_z)}{k_z^2 + n^2\pi^2/a^2}\right\}. \tag{2.273}$$

Now, we turn to the diffusion modes existing in the frequency range (2.267) where the dielectric permittivity tensor is expressed by Eq. (2.265). In the analysis of diffusion modes, it is more convenient to write Eq. (2.265) in the form

$$\varepsilon^l(k, \omega) = 1 + \sum_\alpha \frac{1}{k^2 r_{D\alpha}^2} \frac{\imath k^2 D_\alpha}{\omega + \imath k^2 D_\alpha}. \tag{2.274}$$

Here, D_α, the monopolar diffusion coefficient of α species, is related to conductivity $\sigma_\alpha = e_\alpha^2 N_\alpha / m_\alpha \nu_\alpha$ by

$$D_\alpha = \frac{T_\alpha}{e_\alpha^2 N_\alpha}\sigma_\alpha. \tag{2.275}$$

Unlike the ion-acoustic waves, it is possible to consider the diffusion modes of oscillations when layer thickness a is smaller than Debye radius of charge carriers. Namely, under these circumstances, monopolar diffusion takes place. Therefore, Eq. (2.263) for symmetric modes takes the form

$$1+\frac{2}{a\mid k_{\parallel}\mid}\frac{1-\frac{\imath\omega}{k_{\parallel}^2 D_\alpha}}{1-\frac{\imath\omega}{k_{\parallel}^2 D_\alpha}+\frac{1}{k_{\parallel}^2 {r_{D\alpha}}^2}}=0. \tag{2.276}$$

The solution of this equation,

$$\omega=-\imath D_\alpha\left(k_{\parallel}^2+\frac{a\mid k_{\parallel}\mid}{2r_{D_\alpha}^2}\right), \tag{2.277}$$

differs from the monopolar regime of volume diffusion, through the existence of the second term. This term becomes significant when $k_{\parallel}^2 r_{D_\alpha}^2 \ll a\mid k_{\parallel}\mid \ll 1$ and, as a result, Eq. (2.277) reduces to

$$\omega=-\imath\frac{D_\alpha}{2r_{D_\alpha}^2}\,a\mid k_{\parallel}\mid=-2\imath\pi\sigma_\alpha\,a\mid k_{\parallel}\mid, \tag{2.278}$$

which describes the Maxwell relaxation of charge carriers in a thin layer of a medium. As a result, the density diffusion equation in the spatial form acquires a non-local form

$$\frac{\partial n}{\partial t}+D_{\parallel a}\frac{\partial^2 n}{\partial z^2}+\frac{a\omega_{\rm pe}^2}{2}\int dz'\frac{n(z')}{(z-z')^2}=0. \tag{2.279}$$

From this equation, one can find relations (2.277) and (2.278).

Here, it should be noted that the similar analysis of low-frequency longitudinal oscillations in plasma-like media (colloidal crystals) was carried out in [71, 74, 75] where the hydrodynamic description was used. It is remarkable that the experimental results indicate the linear dependence of oscillations frequency on the wave number.

If layer thickness a greatly exceeds the Debye radius, diffusion becomes ambipolar. In this case, the solution of Eq. (2.263) for both symmetric and asymmetric modes gives the following frequency spectrum:

$$\omega=-\imath k_{\parallel}^2 D_i\left(1+\frac{T_e}{T_i}\right)\equiv-\imath k_{\parallel}^2 D_a. \tag{2.280}$$

In conclusion, it should be noted that all results obtained in this part can simply be generalized to the degenerate case by the exchanges $r_{D_\alpha}^2 \to 3v_{F\alpha}^2/\omega_{p\alpha}^2$ and $D_\alpha \to v_{F\alpha}^2/\nu_\alpha$.

2.11 Problems

2.11.1 Study the magnetic permeability of a strongly collisional non-degenerate electron gas.

Solution Using formulas (2.63), (2.135), and (2.136), in the general case, we find

$$1 - \frac{1}{\mu} = \frac{\omega^2}{k^2 c^2}\left[-\frac{\omega_{\text{pe}}^2}{k^2 v_{\text{Te}}^2 - \imath\omega\nu_e} + \imath\frac{\omega_{\text{pe}}^2}{\omega\nu_e}\right] = -\frac{\omega_{\text{pe}}^2 \omega v_{\text{Te}}^2 \left(\omega\nu_e - \imath k^2 v_{\text{Te}}^2\right)}{c^2 \nu_e \left(k^4 v_{\text{Te}}^4 + \omega^2 \nu_e^2\right)}. \tag{2.281}$$

Hence, we obtain

$$\mu(\omega, k) = -\frac{\omega_{\text{pe}}^2 \omega v_{\text{Te}}^2 \left(\omega\nu_e - \imath k^2 v_{\text{Te}}^2\right)}{c^2 \nu_e \left(k^4 v_{\text{Te}}^4 + \omega^2 \nu_e^2\right)}. \tag{2.282}$$

This means that this gas is weakly diamagnetic, and $\text{Im}\mu > 0$.

2.11.2 Find the potential of a uniformly moving point charge in non-isothermal plasma with $T_e \gg T_i$. For simplicity, restrict the motion of the charge along the radius vector $\vec{u} \parallel \vec{r}$.

Solution The sought potential can be written in the form

$$U\left(\vec{r}\right) = \int d^3\vec{k} \exp\left(\imath\vec{k}\cdot\vec{r}\right) U\left(\vec{k}\right)$$
$$= \frac{e}{(2\pi)^2} \int_0^\infty dk \int_{-1}^{+1} dx \frac{\exp\left(\imath k\left(r - u_0 t\right)\right)}{\varepsilon^l(k u_0 x, k)}. \tag{2.283}$$

For non-isothermal plasma, in the low-frequency range, we have

$$\varepsilon^l(k u_0 x, k) = 1 + \frac{1}{k^2 r_D^2}(1 + \imath\beta) - \frac{\omega_{\text{pi}}^2}{k^2 u_0^2 x^2}. \tag{2.284}$$

Here, $1/r_D^2 = \omega_{\text{pe}}^2/v_{\text{Te}}^2$, $\beta = \sqrt{\pi/2}\omega/k v_{\text{Te}} \ll 1$, in the non-degenerate plasma case; indeed, $\omega = k u_0 x$, $\omega_{\text{pe}} = \sqrt{4\pi e^2 n_e/m}$, where n_e is the density, m is the electron mass, $v_{\text{Te}} = \sqrt{T_e/m}$ is the thermal velocity, $\omega_{\text{pi}} = \sqrt{4\pi e_i^2 n_i/M}$, where M is the mass of ions, and n_i is their density. If the electrons of plasma are degenerate, then $1/r_D^2 = 3\omega_{\text{pe}}^2/v_{\text{Fe}}^2$, $\beta = \pi/2\omega/k v_{\text{Fe}} \ll 1$, where $v_{\text{Fe}} = \sqrt{2\epsilon_F/m}$ is the Fermi velocity and ϵ_F is the Fermi energy of electrons.

Substituting expression (2.284) in Eq. (2.283), after simple calculations, we obtain

$$U(r) = \frac{e}{(2\pi)^2} \int_0^\infty dk \int_{-1}^{1} dx \exp\left(\imath k(r - u_0 t)\right)$$
$$\times \left[\frac{k^2 r_D^2}{1 + k^2 r_D^2} + \frac{1 - \imath\beta}{(1 + 1/k^2 r_D^2)} \frac{1}{x^2 k^2 u_0^2/\omega_{pi}^2 (1 + 1/k^2 r_D^2) - 1 + \imath\beta}\right] \tag{2.285}$$

Now we consider the limit $\beta \to +0$. In this case, the second term in Eq. (2.285) turns into

$$\frac{1}{x^2 - \omega_{\rm pi}^2 (1 + \imath\beta)/k^2 u_0^2} = -\frac{\imath\pi}{2} \frac{k u_0 \sqrt{1 + 1/k^2 r_D^2}}{\omega_{\rm pi}}$$
$$\times \left[\delta\left(x - \frac{\omega_{\rm pi}}{k u_0 \sqrt{1 + 1/k^2 r_D^2}}\right) - \delta\left(x + \frac{\omega_{\rm pi}}{k u_0 \sqrt{1 + 1/k^2 r_D^2}}\right)\right].$$

Making use of this relation, we can write the potential (2.285) in the form of the sum of two terms:

$$U(r) = U_1(r) + U_2(r). \tag{2.286}$$

The first term is

$$U_1(r) = \frac{e}{|r - u_0 t|} \exp\left(-|r - u_0 t|/r_D\right), \tag{2.287}$$

while the second term after integration over x reduces to

$$U_2(r) = -\frac{e}{4\pi |r - u_0 t|} \int \frac{dk}{k} \frac{|r - u_0 t| \omega_{\rm pi}}{u_0 (1 + 1/k^2 r_D^2)} \sin \frac{|r - u_0 t| \omega_{\rm pi}}{u_0 \sqrt{1 + 1/k^2 r_D^2}}. \tag{2.288}$$

This expression is simplified in two opposite limits:

(a) For $k r_D \gg 1$ (or $r_D \gg |r - u_0 t|$):

$$U(r) \simeq \frac{e \omega_{\rm pi}}{4\pi u_0} L \sin \frac{|r - u_0 t| \omega_{\rm pi}}{u_0}, \tag{2.289}$$

where $L = \int_0^\infty \frac{dk}{k}$.

(b) For $k r_D \ll 1$ (or $r_D \ll |r - u_0 t|$):

$$U(r) = -\frac{e}{4\pi}\frac{1}{|r - u_0 t|}. \tag{2.290}$$

2.11.3 Calculate the electron distribution function for completely ionized stationary and homogenous plasma in a constant external electric field. Use the Lorentz approximation. Evaluate plasma conductivity for the Lorentz gas with infinitely heavy ions [54].

Solution Since the Lorentz gas approximation is valid for $Z = |e_i/e| \gg 1$ when electrons do not interact with each other, but interact with ions only, then the Landau kinetic equation for this gas with heavy ions in an external electric field can be written as

$$e\vec{E}\cdot\frac{\partial f_e}{\partial \vec{p}} = \frac{\partial}{\partial p_i} 2\pi e^2 e_i^2 N_i L \frac{v^2\delta_{ij} - v_i v_j}{v^3}\frac{\partial f_e}{\partial p_j}. \tag{2.291}$$

The solution of this equation can be expanded as

$$f_e = f_M + \frac{v\cdot\vec{f}_1(v)}{v}, \tag{2.292}$$

where $|f_M|$ is the Maxwellian distribution function. For rather weak fields, therefore

$$e\vec{E}\cdot\frac{\partial f_M}{\partial \vec{p}} = \frac{4\pi e^2 e_i^2 N_i L}{m^2}\frac{\vec{f}_1\cdot\vec{v}}{v^4}, \tag{2.293}$$

or

$$f_e = f_M + \frac{e\vec{E}}{m\nu(v)}\cdot\frac{\partial f_M}{\partial \vec{v}}, \tag{2.294}$$

where

$$\nu(v) = \frac{4\pi e^2 e_i^2 N_i L}{m^2 v^3}. \tag{2.295}$$

Hence, we obtain the density of the electron current in plasma

$$\vec{j} = e\int d\vec{p}\,\vec{v} f_e = \frac{32}{3\pi}\frac{e^2 N_e}{m\nu_{\text{eff}}}\vec{E} = \sigma\vec{E}, \tag{2.296}$$

and plasma conductivity in the Lorentz gas model

$$\sigma = \frac{32}{3\pi}\frac{e^2 N_e}{m\nu_{\rm eff}}, \tag{2.297}$$

where

$$\nu_{\rm eff} = \frac{4}{3}\sqrt{\frac{2\pi}{m}}\frac{e^2 e_i^2 N_i L}{T_e^{3/2}}.$$

Equation (2.297) is valid under the condition $u = eE/m\nu_{\rm eff} \ll v_{\rm Te}$, only. When $u > v_{\rm Te}$, the so-called *runaway electrons* appear. Therefore, the concept of plasma conductivity can be used only up to some critical value of the field determined by the condition

$$\frac{eE_{\rm cr}}{m\nu_{\rm eff}} = v_{\rm Te}, \tag{2.298}$$

which is called the Dreicer field. For $E > E_{\rm cr}$, there exists no stationary state of plasma.

2.11.4 Use the kinetic equation with the model BGK integral to calculate the electron distribution function and plasma heating in an external electric field [54].

Solution Taking into account only electron-neutral collisions, the kinetic equation with the BGK collision integral can be written as

$$e\vec{E}\cdot\frac{\partial f_e}{\partial \vec{p}} = -\nu_{\rm en}(f_e - N_e\Phi_{\rm en}), \tag{2.299}$$

where

$$\Phi_{\rm en} = \frac{1}{(2m\pi T_{\rm en})}\exp\left(-\frac{mv^2}{2T_{\rm en}}\right), \tag{2.300}$$

and

$$T_{\rm en} = \frac{mT_n + MT_e}{m+M}. \tag{2.301}$$

T_n is the temperature of the neutrals and M is their mass. The kinetic equation can be solved by the ansatz

$$f_e = f_0(v) + \frac{\vec{v}\cdot\vec{f}_1(v)}{v}, \tag{2.302}$$

assuming $\left|\vec{f}_1\right| \ll f_0$. Averaging over the angles of the velocity with respect to $\vec{E}$ we obtain two equations

$$\begin{aligned}&\frac{e}{3v^2 m}\frac{\partial}{\partial v}\left(v^2\vec{E}\cdot\vec{f}_1\right) = -\nu_{\rm en}(f_0 - N_e\Phi_{\rm en}),\\&\frac{e\vec{E}}{m}\frac{\partial f_0}{\partial v} = -\nu_{\rm en}\vec{f}_1.\end{aligned} \tag{2.303}$$

Substituting $\vec{f}_1$ from the second equation into the first one, we obtain

$$\frac{e^2E^2}{3m^2v^2\nu_{\rm eff}}\frac{\partial}{\partial v}\left(v^2\frac{\partial f_0}{\partial v}\right) + \nu_{\rm en}(f_0 - N_e\Phi_{\rm en}) = 0. \tag{2.304}$$

The solution of this equation is the Maxwellian distribution function with the temperature T_e given by

$$\frac{2e^2E^2}{m^2\nu_{\rm en}} - \frac{3\nu_{\rm en}}{m+M}(T_e - T_n) = 0, \tag{2.305}$$

or

$$T_e \approx T_n + \frac{2M}{3m}\frac{e^2E^2}{m\nu_{\rm en}^2}. \tag{2.306}$$

The stationary value of the temperature is the result of the balance between ohmic electron heating and the energy transfer from the electrons to the neutrals. Finally, plasma conductivity is determined from

$$\vec{j} = e\int d\vec{p}\,\vec{v}\,\frac{\vec{v}\cdot\vec{f}_1}{v} = \frac{e^2N_e}{m\nu_{\rm en}}\vec{E} = \sigma\vec{E}, \tag{2.307}$$

which gives $\sigma = e^2N_e/m\nu_{en}$.

2.11.5 Using the model of non-interacting particles calculate the average force (*Miller's force*) acting on the electrons in an external electric high-frequency field with and without an inhomogeneous magnetic field superimposed. Neglect relativistic effects (see problem 1.11.7) [54].

Solution Let us write the equation of motion of the electrons in the form

$$\frac{\partial \vec{V}}{\partial t} + \left(\vec{V} \cdot \nabla\right)\vec{V} = \frac{e\vec{E}\left(t, \vec{r}\right)}{m} + \frac{e}{mc}\left\{\vec{V} \times \left[\vec{B}_0 + \vec{B}\left(t, \vec{r}\right)\right]\right\}, \tag{2.308}$$

where $\vec{B}_0$ is the external homogeneous magnetic field and

$$\begin{aligned} \vec{B}\left(t, \vec{r}\right) &= \vec{B}\left(\vec{r}\right)\cos\omega_0 t, \\ \vec{B}\left(\vec{r}\right) &= \frac{c}{\omega_0}\nabla \times \vec{E}\left(\vec{r}\right). \end{aligned} \tag{2.309}$$

Assuming the field $\vec{E}\left(t, \vec{r}\right)$ and $\vec{B}\left(t, \vec{r}\right)$ (and consequently the velocity $\vec{V}$) to be small, we obtain in the linear approximation

$$\frac{d\vec{V}_0}{dt} = \frac{e\vec{E}\left(t, \vec{r}\right)}{m} + \frac{e}{mc}\left[\vec{V}_0 \times \vec{B}_0\right]. \tag{2.310}$$

Substituting $\vec{V}_0$ into the small non-linear terms of Eq. (2.308) and averaging them over the time, we obtain the average force

$$\vec{F}_{\text{av}} = -m\overline{\left(\vec{V}_0 \cdot \nabla\right)\vec{V}_0} + \frac{e}{c}\overline{\left(\vec{V}_0 \times \vec{B}_0\right)}. \tag{2.311}$$

In the absence of an external homogeneous magnetic field

$$\begin{aligned} \vec{V}_0 &= -\frac{e\vec{E}\left(\vec{r}\right)}{m\omega_0}\cos\omega_0 t, \\ \vec{F}_{\text{av}} &= -\frac{e^2}{2m\omega_0^2}\left\{\left(\vec{E} \cdot \nabla\right)\vec{E} + \left[\vec{E} \times \nabla \times \vec{E}\right]\right\} = -\frac{e^2}{4m\omega_0^2}\nabla E^2\left(\vec{r}\right). \end{aligned} \tag{2.312}$$

The average force ejects electrons (and consequently plasma) from the region of a strong high-frequency field. In the presence of an external homogeneous magnetic field the opposite situation where plasma is absorbed into the region of the strong high-frequency field is possible, too.

2.11.6 Using the model of non-interacting particles find the average force acting on the electron in the high-frequency electric field $\vec{E}\left(\vec{r}, t\right) = \vec{E}\left(\vec{r}\right)\sin\omega_0 t$ and constant magnetic field $\vec{B}_0$ (see problem 1.11.7) [54].

Solution According to the previous problem, this force is equal to

$$\vec{F}_{\text{av}} = -\frac{e^2}{4m\omega_0^2}\nabla\left\{\frac{\omega_0}{\omega_0+\Omega}E_\perp^{(+)2} + \frac{\omega_0}{\omega_0-\Omega}E_\perp^{(-)2} + E_\parallel^2\right\},$$

where $\Omega = e\vec{B}_0/(mc)$;$E_\parallel$ and $E_\perp^{(\pm)}$ are the longitudinal (i.e., along $\vec{B}_0$) and the right-hand and left-hand circular polarized transverse components of the electric field, respectively.

2.11.7 Use the Landau kinetic equation to derive the longitudinal and transverse permittivities of isotropic fully ionized non-degenerate plasma under the condition of frequent collisions $\nu_\alpha \gg \omega, kv_{T_\alpha}, \alpha = e, i$ [54].

Solution Under these conditions the electrons contribute dominantly to the induced current, whereas the ions remain unperturbed. The linearized electron kinetic equation has the form

$$e\vec{E}\cdot\frac{\partial f_{0e}}{\partial\vec{p}} = N_{0i}\frac{\partial}{\partial p_i}I_{ij}^{ei}\left(\vec{p}\right)\frac{\partial\delta f_e}{\partial p_j} + \frac{\partial}{\partial p_i}\int d\vec{p'}I_{ij}^{ee}\left(\vec{p},\vec{p'}\right)$$
$$\times\left[\frac{\partial f_{0e}}{\partial p_j}\delta f_e\left(\vec{p'}\right) + \frac{\partial\delta f_e}{\partial p_j}f_{0e}\left(\vec{p'}\right) - f_{0e}\left(\vec{p}\right)\frac{\partial f_{0e}}{\partial p'_j} - \delta f_e\left(\vec{p}\right)\frac{\partial f_{0e}}{\partial p'_j}\right]. \tag{2.313}$$

It is convenient to solve this equation by the Chapman–Enskog method, i.e., by expanding $\delta f_e\left(\vec{p}\right)$ in Sonin polynomials. We confine ourselves to two terms of the expansion

$$\delta f_e\left(\vec{p}\right) = \frac{\vec{v}\cdot\vec{E}}{E}\left[a_0 + a_1\left(\frac{5}{2} - \frac{v^2}{2v_{\text{Te}}^2}\right)\right]f_{0e}, \tag{2.314}$$

which we substitute into Eq. (2.313). Multiplying the resulting equation by the polynomials 1 and $\left[\frac{5}{2} - \frac{v^2}{2v_{\text{Te}}^2}\right]$ and integrating over the momentum, we obtain two equations for the expansion coefficients a_0 and a_1. For plasma with singly charged ions $e_i = |e|$ these equations are

$$\frac{eE}{T_e} = -\nu_{\text{eff}}\left(a_0 + \frac{3}{2}a_1\right), \quad \frac{3}{2}a_0 + \frac{13+4\sqrt{2}}{4}a_1 = 0, \tag{2.315}$$

which yields

$$a_0 = -\frac{13+4\sqrt{2}}{6}a_1, \quad a_1 = -\frac{3}{2+2\sqrt{2}}\frac{eE}{\nu_{\text{eff}}T_e}. \tag{2.316}$$

Then, the induced electron current density can be obtained from expression (2.314):

$$\vec{j} \approx \vec{j}_e \approx 1.96 \frac{e^2 N_{0e}}{m\nu_{\text{eff}}} \vec{E}. \tag{2.317}$$

Finally, dielectric permittivity follows

$$\varepsilon^l = \varepsilon^{\text{tr}} = 1 + \imath 1.96 \frac{\omega_{\text{pe}}^2}{\omega \nu_{\text{eff}}}. \tag{2.318}$$

Strictly speaking, expression (2.318) is valid only under the condition $\omega\nu_{\text{eff}} \gg k^2 v_{T_e}^2$. In the opposite case, we must take account of small space- and time-dependent terms on the left-hand side of Eq. (2.313). The analysis shows that these terms are significant only for longitudinal dielectric permittivity and do not contribute to transverse dielectric permittivity.

2.11.8 Verify that the particular solution of the linearized Vlasov equation (2.12) which corresponds to undamped longitudinal waves (the Van Kampen modes) is possible in isotropic plasma.

Solution Supposing Eq. (2.12) to be inhomogeneous, expression (2.14) should be supplemented by the solution of

$$\left(\omega - \vec{k} \cdot \vec{v}\right) \delta f_\alpha^{(1)} = 0,$$

where

$$\delta f_\alpha^{(1)} = n_{\alpha 1} k \delta\left(\omega - \vec{k} \cdot \vec{v}\right). \tag{2.319}$$

Here, $n_{\alpha 1}$ is an arbitrary constant. The solution of Eq. (2.319) corresponds to the one-velocity (i.e., $v = \omega/k$) modulated beam of particles of type α with the density $n_{\alpha 1}$.

As a result, the solution of Eq. (2.12) takes the form

$$\delta f_\alpha^{(0)} = -\frac{\imath e_\alpha \vec{E} \cdot \frac{\partial f_{0\alpha}}{\partial \vec{p}}}{\omega - \vec{k} \cdot \vec{v}} + n_{\alpha 1} k \delta\left(\omega - \vec{k} \cdot \vec{v}\right). \tag{2.320}$$

Introducing it into the Poisson's equation, we arrive at the dispersion equation for longitudinal oscillations of isotropic electron plasma [compared to Eq. (2.24)]:

$$\varepsilon^l(\omega, k) = 1 + \frac{4\pi e^2}{\omega k^2} \int d\vec{p} \frac{\left(\vec{k} \cdot \vec{v}\right)^2}{\omega - \vec{k} \cdot \vec{v}} \frac{\partial f_0}{\partial \epsilon} + \frac{4\pi e}{k^3} d_{e1} = 0. \tag{2.321}$$

Here, d_{e1} is a new constant uniquely related to n_{e1}.

The imaginary term in Eq. (2.321) can approach zero due to the matching of the constant n_{e1} or d_{e1}. This means that damping of such waves will be absent in plasma. These waves are called the Van Kampen modes after the Dutch physicist who was the first demonstrated the completeness of the solutions in the form of relation (2.320). This means that any perturbation in plasma can be expanded into the set of functions of the form of relation (2.320).

2.11.9 Analyze the frequency dependency of the penetration depth of a transverse field (normal and anomalous skin-effect) [54].

Solution The penetration depth of a transverse field with real frequency ω is determined by the roots $k(\omega)$ of the dispersion equation

$$k^2 = \frac{\omega^2}{c^2}\varepsilon^{\mathrm{tr}}(\omega, k), \tag{2.322}$$

and is given by

$$\lambda_{\mathrm{sk}} = \frac{1}{Im\{k(\omega)\}}. \tag{2.323}$$

For the discussion we distinguish different frequency ranges:

(a) In the range $\omega \gg \nu_e, kv_0$ (ν_e is the electron collision frequency, v_0 is the average velocity of the electron random motion, $v_0 = v_{\mathrm{Te}}$ for non-degenerate plasma, and $v_0 = v_{\mathrm{Fe}}$ for the degenerate one), we have

$$\varepsilon^{\mathrm{tr}} = 1 - \frac{\omega_{\mathrm{pe}}^2}{\omega^2}\left(1 - \imath\frac{\nu_e}{\omega}\right). \tag{2.324}$$

Here, $\nu_e = \nu_{\mathrm{en}}$ for weakly ionized non-degenerate and degenerate plasmas, $\nu_e = \nu_{\mathrm{eff}}$ for completely ionized non-degenerate plasma, and $\nu_e = \nu_{\mathrm{Fe}}$ for the degenerate one.

Substituting expression (2.324) into Eq. (2.322) yields

$$\lambda_{\mathrm{sk}} = \begin{cases} \dfrac{2c\omega^2}{\omega_{\mathrm{pe}}^2\nu_e}, & for \quad \omega \gg \omega_{\mathrm{pe}}, \\ \dfrac{c}{\omega_{\mathrm{pe}}}, & for \quad \omega_{\mathrm{pe}}\dfrac{v_0}{c} \ll \omega \ll \omega_{\mathrm{pe}}. \end{cases} \tag{2.325}$$

(b) In the range $kv_0 \gg \omega, \nu_e$, we have

$$\varepsilon^{\mathrm{tr}} = 1 + \imath\alpha\frac{\omega_{\mathrm{pe}}^2}{\omega k v_0}. \tag{2.326}$$

Here, $\alpha = \sqrt{\pi/2}$ for non-degenerate plasma and $\alpha = 3\pi/4$ for the degenerate one. In this frequency range, we obtain from Eq. (2.322)

$$\lambda_{\rm sk} = 2\left(\frac{c^2 v_0}{\alpha\omega_{\rm pe}^2\omega}\right)^{1/3}, \tag{2.327}$$

for

$\omega^* \ll \omega \ll \omega_{\rm pe} v_0/c$, with $\omega^* = \frac{\nu_e^3 c^2}{\omega_{\rm pe}^2 v_0^2}$.

Thus, the anomalous skin-effect given by expression (2.327) exists in the range $\nu_e \ll v_0\omega_{\rm pe}/c$, only.

The existence condition $\omega > \omega^*$ is necessary for any ratio between ω and ν_e.

(c) In the range $\nu_e \gg kv_0, \omega$, we have

$$\varepsilon^{\rm tr} = 1 + \imath\alpha_1\frac{\omega_{\rm pe}^2}{\omega\nu_e}, \tag{2.328}$$

with $\alpha_1 = 1$ for weakly ionized and $\alpha_1 = 1.96$ for fully ionized plasmas. The skin-depth is

$$\lambda_{\rm sk} = \left(\frac{2\nu_e c^2}{\alpha_1\omega_{\rm pe}^2\omega}\right)^{1/2}, \tag{2.329}$$

for $\omega \ll \omega^*, \nu_e$. Thus, the normal skin-effect (2.329) exists in the frequency range $\omega \ll \omega^*$. The ratio between ω^* and ν_e can be arbitrary. The results are shown schematically in Fig. 2.8.

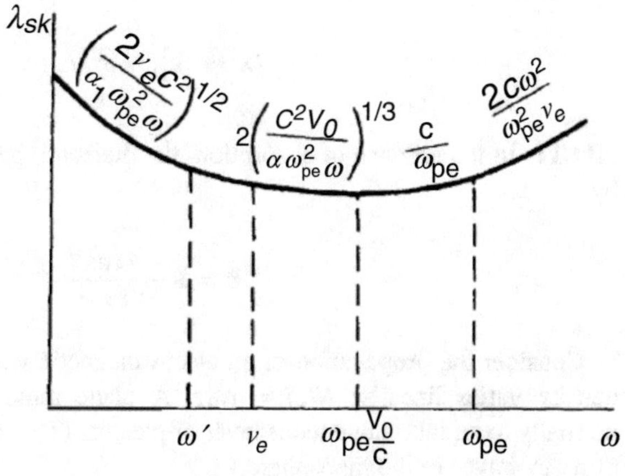

Fig. 2.8 Frequency dependence of the penetration depth of the transverse field

2.11.10 Using relation (2.135) for longitudinal dielectric permittivity analyze the diffusion spread of a small inhomogeneity of the density of charged particles in weakly ionized isotropic plasma [54].

Solution The diffusion spread is a slow process with characteristic time $\tau \sim 1/|\omega| \gg 1/\nu_e, 1/\nu_i$. Therefore, for its description it is necessary to analyze the low-frequency limit of Eq. (2.135)

$$\varepsilon^l(\omega, k) = 1 + \frac{\imath\omega_{\text{pe}}^2}{\omega\nu_{\text{en}} + \imath k^2 v_{\text{en}}^2} + \frac{\imath\omega_{\text{pi}}^2}{\omega\nu_{\text{in}} + \imath k^2 v_{\text{Ti}}^2}. \tag{2.330}$$

Here, $\omega \sim 1/\tau$ characterizes the time period of the spread of inhomogeneity with a space dimension $L \sim 1/k$. For low densities of charged particles, the diffusion of electrons and ions occurs independently (free diffusion) and is described by the poles in the electron and ion parts in expression (2.330):

$$\omega\nu_{\alpha n} + \imath k^2 v_{T\alpha}^2 = 0, \text{for } \alpha = e, i.$$

Hence we obtain the free diffusion coefficients for each of the components: $D_\alpha \approx v_{T\alpha}^2/\nu_{\alpha n}$. The zeros of $\varepsilon^l(\omega, k)$ describe the spread of the inhomogeneity in plasmas taking account of self-consistent interaction of electrons and ions, i.e., the process of ambipolar diffusion. When the characteristic dimension of inhomogeneity greatly exceeds the Debye lengths of electrons and ions, taking account of $\nu_{\text{in}} \gg \nu_{\text{en}} m/M$, we find

$$\omega = -\imath \frac{k^2\left(v_{\text{Ti}}^2 + v_s^2\right)}{\nu_{\text{in}}}. \tag{2.331}$$

Thus the ambipolar diffusion coefficient is

$$D_a \approx \left(v_{\text{Ti}}^2 + v_s^2\right)/\nu_{\text{in}}.$$

2.11.11 In the absence of absorption, the dielectric permittivity of plasma is given by

$$\varepsilon = 1 - \frac{4\pi e^2 N}{m\omega^2}.$$

Consider the propagation of an electromagnetic wave in plasma whose number density varies linearly: $N(z) = N_0 z$. A plane monochromatic wave is incident normally to an inhomogeneous layer of plasma. (This may occur in the propagation of radio waves in the ionosphere.)

Hint The equation for $E(z)$ can be solved by presenting the required function as a Fourier integral [76].

Solution For normal incidence of the wave to the inhomogeneous layer, the electric field is only a function of z and satisfies the equation

$$\frac{d^2E}{dz^2} = \frac{\omega^2}{c^2}\varepsilon(\omega, z)E = 0. \tag{2.332}$$

Let $m\omega^2/4\pi e^2 N_0 = z_1$, then $\varepsilon = 1 - z/z_1$. Introducing the new variable $\xi = (\omega^2/c^2 z_1)^{1/3}(z_1 - z)$, we transform Eq. (2.332) to the form

$$\frac{d^2E}{d\xi^2} + \xi E = 0. \tag{2.333}$$

The simplest way to solve Eq. (2.333) is to use the Fourier transformation. Let us expand $E(\xi)$ in a Fourier integral:

$$E(\xi) = \int_{-\infty}^{\infty} E(u)\, e^{\imath\xi u}\, du, \qquad E(u) = \frac{1}{2\pi}\int_{-\infty}^{\infty} E(\xi)\, e^{-\imath\xi u} d\xi.$$

Substituting this expansion into Eq. (2.333), we obtain the first-order differential equation for the amplitude $E(u)$,

$$\frac{dE(u)}{du} + \imath u^2 E(u) = 0. \tag{2.334}$$

As a result, we have a simpler first-order equation instead of the second-order equation. The solution of Eq. (2.334) is easily found by integrating

$$E(u) = A' e^{-\imath u^3/3}.$$

Then we have

$$E(\xi) = A' \int_{-\infty}^{\infty} \exp\left[-\imath\left(u^3/3 - \xi u\right] du.$$

Rewriting $\exp[-\imath(u^3/3 - \xi u)]$ in the form of sine and cosine sum, and noting that the integral of $\sin(u^3/3 - \xi u)$ is zero because the integrand is an odd function of u, we have

$$E(\xi) = \frac{A}{\sqrt{\pi}} \int_0^\infty \cos\left(\frac{u^3}{3} - \xi u\right) du. \tag{2.335}$$

The function

$$\Phi(\xi) = \frac{1}{\sqrt{\pi}} \int_0^\infty \cos\left(\frac{u^3}{3} + \xi u\right) du$$

is known as Airy's function (it can be expressed in terms of Bessel functions of index 1/3). Finally,

$$E(\xi) = A\Phi(-\xi).$$

The constant A should be determined from the boundary conditions.

Now consider the behavior of $E(\xi)$ at large $|\xi|$. Using the asymptotic formulas for $\Phi(\xi)$, we have for large positive ξ:

$$E(\xi) = \frac{A}{\xi^{1/4}} \sin\left(\frac{2}{3}\xi^{3/2} + \frac{\pi}{4}\right).$$

It is clear that the field has an oscillating character. For large (by modulus) negative ξ:

$$E(\xi) = \frac{A}{2|\xi|^{1/4}} e^{-\frac{2}{3}|\xi|^{3/4}},$$

that is, the field decays exponentially. The reason for this is that negative ξ corresponds to the negative values of the dielectric permittivity ε. However, at $\varepsilon < 0$, the wave vector $k = \frac{\omega}{c}\sqrt{\varepsilon}$ is purely imaginary, and the damping takes place. The damping in this case is not due to the conversion of electromagnetic energy into heat (since the dielectric permittivity is real and dissipation is absent) but due to the destructive interference of the incident and secondary fields.

2.11.12 An electromagnetic wave (E-wave) is incident on the surface of a semi-bounded isotropic medium ($z \geq 0$) without spatial dispersion. Study the reflection and absorption of the wave by the boundary interface (surface xz; Fersnal's problem).

Solution Maxwell's equation for the electromagnetic waves with frequency ω reduces to one equation for the component B_y:

$$\frac{d^2 B_y}{dz^2} + \left(\frac{\omega^2}{c^2}\,\varepsilon(\omega) - k_x^2\right) B_y = 0. \tag{2.336}$$

This equation is valid both in vacuum ($z \le 0$, $\varepsilon = 1$) and the medium ($z \ge 0$, $\varepsilon = \varepsilon(\omega)$). The other non-zero components are

$$E_x = -\frac{\imath c}{\omega\varepsilon(\omega)}\frac{dB_z}{dz}, \qquad E_z = \frac{k_x c}{\omega\varepsilon(\omega)} B_y, \tag{2.337}$$

where k_x is the wave vector along the medium surface.

In the region $z \le 0$, the solution of Eq. (2.336) consists of the reflected and incident waves, and

$$E_{x\,\mathrm{ref}}(0) = r E_{x\,\mathrm{inc}}(0), \tag{2.338}$$

where r is the complex coefficient of reflection, $|r|^2$ is the fraction of the reflected field energy, while $A = 1 - |r|^2$ is the fraction of the energy absorbed by the medium.

In the region $z \ge 0$, there is the refracted wave traveling (or damping) when moving away from the surface:

$$B_y(z) = B_y(0) e^{-k_0 z}, \qquad k_0^2 = k_x^2 - \frac{\omega^2}{c^2}\varepsilon. \tag{2.339}$$

We have to use the following boundary conditions:

$$\{E_x\}_{x=0} = \left\{\frac{1}{\varepsilon(\omega)}\frac{\partial B_z}{\partial z}\right\}_{x=0} = 0, \qquad \{B_z\}_{x=0} = 0. \tag{2.340}$$

For reflection and refraction of waves, after straightforward calculations, we finally find

$$r = \frac{1 - \frac{\omega}{4\pi k_0} z}{1 + \frac{\omega}{4\pi k_0} z}, \tag{2.341}$$

where z is the complex impedance

$$z = \imath \frac{4\pi k_0}{\omega\varepsilon(\omega)}. \tag{2.342}$$

From relation (2.341), it follows that for real $\varepsilon(\omega)$ and k_0^2 (or exactly when $\mathrm{Re}\, k_0^2 > 0$), $|r|^2 = 1$, i.e., the incident wave is completely reflected from the medium. Particularly, for $\varepsilon(\omega) < 0$ such a reflection occurs at any incidence angle (i.e., for every k_x).

It should be remarked that a good reflection occurs also for a purely imaginary, but large value of $\varepsilon(\omega)$, as is the case in metals, up to the optical frequency,

$$\varepsilon(\omega) \simeq \imath \frac{4\pi\sigma}{\omega}, \tag{2.343}$$

where σ is of the order of $\sigma \simeq 10^{15} - 10^{17} s^{-1}$. Under the conditions $|\varepsilon| \gg 1$, according to Eq. (2.342), the reflection coefficient is different from unity by a small amount of order of $\frac{1}{|\varepsilon|} \sim \frac{\omega}{4\pi\sigma} \ll 1$. Such an energy fraction of the incident wave is absorbed in a thin layer during field penetration into the conducting medium which is called skin layer:

$$\delta = \frac{c}{\sqrt{2\pi\sigma\omega}}. \tag{2.344}$$

2.11.13 Study potential surface waves on the boundary of two isotropic media [54].

Solution According to the general derivation of the dispersion equation for surface waves (Sect. 2.10) and on equalizing the surface impedance between two isotropic plasmas, for potential surface waves we obtain

$$\int_0^\infty \frac{dk_x}{k^2 \varepsilon_1^l(\omega, k)} + \int_0^\infty \frac{dk_x}{k^2 \varepsilon_2^l(\omega, k)} = 0, \tag{2.345}$$

where $\varepsilon_1^l(\omega, k)$ and $\varepsilon_2^l(\omega, k)$ are the longitudinal dielectric permittivities of the first and the second plasma medium, respectively. When spatial dispersion is neglected, i.e., $\varepsilon_{1,2}^l(\omega, k) = \varepsilon_{1,2}(\omega)$, Eq. (2.345) yields the known equation for high-frequency surface waves on the boundary between two media:

$$\varepsilon_1(\omega) + \varepsilon_2(\omega) = 0. \tag{2.346}$$

Let one plasma medium be non-degenerate cold plasma, i.e., $\omega \gg kv_{\mathrm{Te}}, \nu_{e1}$, and the other be dense degenerate plasma, where $\nu_{e2}, \omega \ll kv_{Fe2}$. Then

$$\begin{aligned} \varepsilon_1^l(\omega, k) &= 1 - \frac{\omega_{pe1}^2}{\omega^2}\left(1 - \imath\frac{\nu_{e1}}{\omega}\right), \\ \varepsilon_2^l(\omega, k) &= 1 + 3\frac{\omega_{pe2}^2}{k^2 v_{Fe2}^2}\left(1 + \imath\frac{\pi}{2}\frac{\omega}{kv_{Fe2}}\right). \end{aligned} \tag{2.347}$$

Substituting expression (2.347) into Eq. (2.345), we obtain

$$1-\frac{\omega_{pe1}^2}{\omega^2}+\sqrt{1+3\frac{\omega_{pe2}^2}{k_z^2 v_{Fe2}^2}+\imath\frac{\omega_{pe1}^2\nu_{e1}}{\omega^3}+3\frac{\omega_{pe2}^2\omega}{v_{Fe2}^3}\left(1+3\frac{\omega_{pe2}^2}{k_z^2 v_{Fe2}^2}\right)}$$
$$\times\int_0^{\infty}\frac{|k_z|dk}{k\left(k^2+3\frac{\omega_{pe2}^2}{v_{Fe2}^2}\right)^2}=0. \tag{2.348}$$

Finally, accounting for the small imaginary terms, we arrive at ($\omega \rightarrow \omega + \imath\delta$):

$$\omega^2=\frac{\omega_{\rm pe1}^2}{\sqrt{1+3\omega_{\rm pe2}^2/\left(k_z^2 v_{\rm Fe2}^2\right)+1}}. \tag{2.349}$$

In Eqs. (2.348) and (2.349), the integration may be carried out in a general form. The result is rather unwieldy; therefore, we analyze only the long- and short-wavelength limits:

$$\omega=\begin{cases}\omega_{pe1}\sqrt{\dfrac{|k_z|v_{Fe2}}{\sqrt{3}\omega_{pe2}}} & \text{for} \quad |k_z|v_{Fe2}\ll\omega_{pe2},\\[2ex] \dfrac{\omega_{pe1}}{\sqrt{2}} & \text{for} \quad |k_z|v_{Fe2}\gg\omega_{pe2},\end{cases} \tag{2.350}$$

$$\delta=-\frac{\nu_{e1}}{2}-\begin{cases}\dfrac{1}{6}\dfrac{\omega_{pe1}^2|k_z|v_{Fe2}}{\omega_{pe2}^2}\ln\dfrac{12\omega_{pe2}^2}{k_z^2v_{Fe2}^2} & \text{for} \quad |k_z|v_{Fe2}\ll\omega_{pe2},\\[2ex] \omega_{pe2}\dfrac{\omega_{pe1}^3}{4|k_z^3|v_{Fe2}^3} & \text{for} \quad |k_z|v_{Fe2}\gg\omega_{pe2}.\end{cases} \tag{2.351}$$

In the long-wavelength limit, the frequency spectrum of surface waves is $\omega \sim \sqrt{k_z}$. Consequently, phase and group velocities of these waves sharply increase when the wavelength grows. Therefore, the damping decrement δ, specified by the Cherenkov wave absorption by electrons of degenerate plasma, sharply decreases.

2.11.14 Find and analyze the dispersion equation of surface waves on the surface of a planar dielectric surrounded on both sides by different dielectric media.

Solution Dielectric permittivity of such a sandwich is written as

$$\varepsilon(\omega,x)=\begin{cases}\varepsilon_1(\omega), & x\le 0,\\ \varepsilon_3(\omega), & 0\le x\le a,\\ \varepsilon_2(\omega), & x\ge a.\end{cases} \tag{2.352}$$

Solution of the field equations is sought in the form of

$$f(x)\exp(-\imath\omega t + \imath k_z z). \tag{2.353}$$

We solve the equation for the E-wave,

$$\frac{\partial^2 E_z}{\partial x^2} - k_z^2 E_z + \frac{\omega^2}{c^2}\varepsilon(\omega, x)E_z = 0, \tag{2.354}$$

with non-zero field components E_z, and

$$E_x = -\frac{\imath k_z}{\kappa^2}\frac{\partial E_z}{\partial x}, \quad B_y = -\frac{\imath\omega}{c\kappa^2}\varepsilon\frac{\partial E_z}{\partial x}, \tag{2.355}$$

where $\kappa^2 = k_z^2 - \varepsilon\omega^2/c^2$.

The boundary conditions on the surface interface at $x = 0, a$ are

$$\{E_z\}_{x=0,a} = 0, \quad \left\{\frac{\varepsilon}{\kappa^2}\frac{\partial E_z}{\partial x}\right\}_{x=0,a} = 0. \tag{2.356}$$

Solving Eq. (2.354) in all regions and substituting solutions in the boundary conditions (2.356), we find four homogeneous equations for four integration coefficients. The solutions are written as

$$E_z = \begin{cases} C_1\exp(\kappa_1 x), & x \le 0, \\ C_2\exp(\kappa_3 x) + C_3\exp(-\kappa_3 x), & 0 \le x \le a, \\ C_4\exp(-\kappa_2 x), & x \ge a. \end{cases} \tag{2.357}$$

Here, $\kappa_i = \sqrt{k_z^2 - \omega^2\varepsilon_i/c^2}, \varepsilon_i = (\varepsilon_1, \varepsilon_2, \varepsilon_3)$. The system of homogeneous equations for coefficients C_i is written in the form

$$\begin{aligned} & C_1 = C_2 + C_3, \quad C_4\exp(-\kappa_2 a) = C_2\exp(\kappa_3 a) + C_3\exp(-\kappa_3 a) \\ & \frac{\varepsilon_1}{\kappa_1}C_1 = \frac{\varepsilon_3}{\kappa_3}(C_2 - C_3) = \frac{\varepsilon_1}{\kappa_1}(C_2 - C_3), \\ & \frac{\varepsilon_2}{\kappa_2}C_4\exp(-\kappa_2 a) = -\frac{\varepsilon_3}{\kappa_3}(C_2\exp(\kappa_3 a) - C_3\exp(-\kappa_3 a)) = \\ & = \frac{\varepsilon_2}{\kappa_2}(C_2 xp(\kappa_3 a) + C_3\exp(-\kappa_3 a)) \end{aligned} \tag{2.358}$$

From the solvability condition of this system, we find dispersion equation

$$\left(1-\frac{\kappa_1\varepsilon_3}{\kappa_3\varepsilon_1}\right)\left(1-\frac{\kappa_3\varepsilon_2}{\kappa_2\varepsilon_3}\right)\exp\left(-\kappa_3 a\right)$$
$$+\left(1+\frac{\kappa_1\varepsilon_3}{\kappa_3\varepsilon_1}\right)\left(1+\frac{\kappa_3\varepsilon_2}{\kappa_2\varepsilon_3}\right)\exp\left(\kappa_3 a\right)=0. \qquad (2.359)$$

Since in region I and II, the field should be damped by going away from the interface ($x = 0, a$), then Re$\kappa_{1,2} > 0$. As about κ_3, here it is possible both damping (when Re$\kappa_3 > 0$) and oscillating (when κ_3 is purely imaginary) solutions. However, in the limit $a \to 0$, region III is absent and from Eq. (2.355) we find the dispersion equation for surface waves on the surface interface between media I and II:

$$\kappa_2\varepsilon_1+\kappa_1\varepsilon_2=\varepsilon_1\sqrt{k_z^2-\varepsilon_2\omega^2/c^2}+\varepsilon_2\sqrt{k_z^2-\varepsilon_1\omega^2/c^2}=0, \qquad (2.360)$$

which coincides with Eq. (2.238). In the opposite limit, $R \to \infty$, the first term in Eq. (2.359) can be neglected and we find two equations for surface waves on $x = 0$, a, respectively,

$$\kappa_3\varepsilon_1+\kappa_1\varepsilon_3=\varepsilon_1\sqrt{k_z^2-\varepsilon_3\omega^2/c^2}+\varepsilon_3\sqrt{k_z^2-\varepsilon_1\omega^2/c^2}=0 \qquad (2.361a)$$

$$\kappa_3\varepsilon_2+\kappa_2\varepsilon_3=\varepsilon_2\sqrt{k_z^2-\varepsilon_3\omega^2/c^2}+\varepsilon_3\sqrt{k_z^2-\varepsilon_2\omega^2/c^2}=0. \qquad (2.361b)$$

In this case, condition Re$\kappa_3 > 0$ should hold.

Now we consider the case when oscillations may exist in medium III and for simplicity we take $\kappa_3 = \imath|\kappa_3|$. Then, from Eq. (2.359) we obtain

$$1+\frac{\kappa_1\varepsilon_2}{\kappa_2\varepsilon_1}=\left(\frac{\kappa_1\varepsilon_3}{|\kappa_3|\varepsilon_1}-\frac{|\kappa_3|\varepsilon_2}{\kappa_2\varepsilon_3}\right)\tan|\kappa_3|a=0. \qquad (2.362)$$

In the limit $|\kappa_3|a \to 0$, we find Eq. (2.360) if

$$|\kappa_3|a=(n+1)\pi/2, \qquad \tan|\kappa_3|a\to\infty.$$

Hence, from Eq. (2.362) we find dispersion equation

$$\kappa_1\kappa_2\varepsilon_3^2=\varepsilon_1\varepsilon_2\left(\frac{\pi(2n+1)}{2a}\right)^2, \qquad (2.363)$$

which describes the volumetric-surface wave with volumetric-surface character in the medium III and damps with distance from the surface interface$x = 0$, a in media I and II.

2.11.15 Find the dispersion equation of surface waves on an isotropic dielectric cylinder with dielectric permittivity $\varepsilon(\omega)$. Consider only symmetric modes with $l = 0$.

Solution Only E-type surface waves exist. Therefore, we write the solution of

$$\nabla_{\perp}^2 E_z + \kappa^2 \varepsilon E_z = 0, \qquad \nabla^2 B_z + \frac{\omega^2}{c^2} B_z = 0, \tag{2.364}$$

where

$$\nabla_{\perp}^2 \equiv \frac{1}{r}\frac{\partial}{\partial r} r \frac{\partial}{\partial r} - \frac{l^2}{r^2},$$

for components E_z inside the cylinder ($r < R$) which is finite on $r = 0$, and outside the cylinder ($r > R$) which damps in $r \to \infty$:

$$E_z = \begin{cases} C_1 I_0(\kappa_1 r), & r < R, \\ C_2 K_0(\kappa_0 r), & r > R, \end{cases} \tag{2.365}$$

where $\kappa_1^2 = k^2 - \omega^2\varepsilon/c^2$, $\kappa_0^2 = k^2 - \omega^2/c^2$. Solutions (2.365) satisfy the boundary condition

$$\{E_z\}_{r=R} = \{B_\varphi\}_{r=r} = 0. \tag{2.366}$$

Indeed, B_φ is determined by (for $l = 0$)

$$B_\varphi = -\kappa^{-2}\left(\imath \frac{\omega}{c}\frac{\partial E_z}{\partial r} - k_z \frac{l}{r} B_z\right).$$

Thus the second condition (2.366) reduces to

$$\left\{\frac{\varepsilon}{\kappa^2}\frac{\partial E_z}{\partial r}\right\}_{r=R} = 0. \tag{2.367}$$

Substituting solutions (2.365) into the boundary conditions (2.366) leads to the following equation for surface waves with $l = 0$ on the surface of the dielectric cylinder:

$$\kappa_1 I_0(\kappa_1 R) K_0'(\kappa_0 R) - \varepsilon \kappa_0 I_0'(\kappa_1 R) K_0(\kappa_0 R) = 0. \tag{2.368}$$

Here, we assumed that solutions (2.365) are valid for $\mathrm{Re}\kappa_1 > 0$, $\mathrm{Re}\ \kappa_0 > 0$. In the short-wave limit, for $\kappa_{0,\ 1} R \gg 1$, from Eq. (2.368) we find

$$\kappa_1 + \varepsilon(\omega)\kappa_0 = \sqrt{k_z^2 - \omega^2\varepsilon/c^2} + \varepsilon\sqrt{k_z^2 - \omega^2/c^2} = 0, \tag{2.369}$$

which coincides with Eq. (2.238) for the planar case with $\varepsilon_1 = \varepsilon$ and $\varepsilon_2 = 1$, when the dielectric is surrounded by vacuum.

In the opposite case, i.e., long-wave limit, $\kappa_{0,\,1}R \ll 1$, from Eq. (2.368) we find

$$\varepsilon(\omega)\ln\kappa_0 R - \frac{2}{\kappa_0^2 R^2} = \varepsilon(\omega)\ln R\sqrt{k_z^2 - \omega^2/c^2} - \frac{2}{R^2\left(k_z^2 - \omega^2/c^2\right)} = 0. \tag{2.370}$$

Hence, in contrast to the short-wavelength oscillations existing in $\varepsilon(\omega) \gg 1$, and $\varepsilon(\omega) \sim 1$, long-wavelength surface waves are possible only when $\varepsilon(\omega) \gg 1$. Phase velocity of such waves is close to the light speed in vacuum, $\omega \approx k_z c$.

2.11.16 Investigate the non-linear Langmuir waves in cold collisionless plasma.

Solution The frequency of plasma waves depends on the mass of electrons. Now, we would like to show how the nature of these oscillations changes with increasing wave amplitude while electron mass relativistically depends on its energy. The problem of relativistic oscillatory motion of electrons in the absence of the external magnetic field has already been investigated in the non-linear theory of the longitudinal oscillations of cold plasma [70, 77]. Now, we restrict our study on the one-dimensional case and write the equations of electron motion in the self-consistent model of the non-interacting particles and fields:

$$\begin{cases} \dfrac{d\vec{p}}{dt} = -e\vec{E}, \\ \dfrac{\partial E}{\partial x} = 4\pi e(n_0 - n_e), \\ \dfrac{\partial E}{\partial t} = 4\pi enu = 4\pi eu\left(n_0 - \dfrac{1}{4\pi e}\dfrac{\partial E}{\partial x}\right). \end{cases} \tag{2.371}$$

Here, u and p are the velocity and momentum of the electron, e is the charge, and $n_0 - n_e$ is the difference of equilibrium (ion) and electron concentrations. Relativistic momentum and velocity of the electron are related to its rest mass m_0 as

$$p = \frac{m_0 u}{\sqrt{1 - u^2/c^2}}, \qquad u = \frac{p/m_0}{\sqrt{1 + p^2/m_0^2c^2}}. \tag{2.372}$$

From relations (2.372) alongside with equations system (2.371), we find a second-order differential equation:

$$\frac{d^2p}{dt^2} = -\frac{4\pi e^2 n_0}{m_0} \frac{p}{\sqrt{1+\frac{p^2}{m_0^2c^2}}}. \tag{2.373}$$

Introducing dimensionless momentum and time

$$\tilde{p} \equiv \frac{p}{m_0 c}, \qquad \tau \equiv t \cdot \omega_p = t \cdot \sqrt{\frac{4\pi e^2 n_0}{m_0}},$$

we find the non-linear equation of motion in the form

$$\frac{d^2\tilde{p}}{d\tau^2} + \frac{\tilde{p}}{\sqrt{1+\tilde{p}^2}} = 0. \tag{2.374}$$

In order to solve this non-linear equation, we decrease the order of Eq. (2.374) by making use of new variable $s = \frac{d\tilde{p}}{d\tau}$ as follows:

$$\begin{gathered} s\,ds = -\frac{\tilde{p}\,d\tilde{p}}{\sqrt{1+\tilde{p}^2}}, \qquad \frac{s^2}{2} = -\sqrt{1+\tilde{p}^2} + C, \\ \frac{d\tilde{p}}{d\tau} = \pm\sqrt{2\left(C - \sqrt{1+\tilde{p}^2}\right)}. \end{gathered} \tag{2.375}$$

Making use of conditions $\tilde{p}|_{\tau=\tau_0} = \tilde{p}_0$, $\frac{d\tilde{p}}{d\tau}\Big|_{\tau=\tau_0} = 0$, we find $C = \sqrt{1+\tilde{p}_0^2}$ and

$$\int_{\tilde{p}_0}^{\tilde{p}} \frac{d\tilde{p}}{\pm\sqrt{2\left(C - \sqrt{1+\tilde{p}^2}\right)}} = \int_{\tau_0}^{\tau} d\tau = \tau - \tau_0.$$

Next, using new variable $\sqrt{1+\tilde{p}^2} = a$, $\tilde{p} = \pm\sqrt{a^2-1}$, $d\tilde{p} = \pm\frac{a\,da}{\sqrt{a^2-1}}$, we obtain

$$\pm\sqrt{2}(\tau - \tau_0) = \int_{\tilde{p}_0}^{\tilde{p}} \frac{d\tilde{p}}{\sqrt{C - \sqrt{1+\tilde{p}^2}}} = \frac{1}{\sqrt{-C}} \int_{C}^{a} \frac{a\,da}{\sqrt{(1-a^2)\left(1-\frac{a}{C}\right)}}.$$

The last integral is taken only when $C = \pm 1$. Therefore, we make use of another replacement $\frac{1+a}{2} = z^2$, $da = 4z\,dz$, and find

$$\pm(\tau - \tau_0) = -\imath\sqrt{\frac{2}{1+C}} \int\limits_{\sqrt{\frac{1+C}{2}}}^{z} \frac{(2z^2 - 1)dz}{\sqrt{(1 - z^2)\left(1 - \frac{2z^2}{1+C}\right)}}.$$

To further solve the problem, we will use special functions. Normal elliptic integral of the first kind is equal to

$$F(\phi, k) = \int\limits_{0}^{\sin\phi} \frac{dz}{\sqrt{(1 - z^2)(1 - k^2 z^2)}}.$$

When $\phi = \pi/2$, the first argument is omitted and the resulting quantity is called the complete normal elliptic integral of the first kind. The same agreement also applies to the other normal elliptic integrals. In addition to the elliptic integral of the first kind, we use the little-known elliptic integral

$$D(\phi, k) = \int\limits_{0}^{\sin\phi} \frac{z^2 dz}{\sqrt{(1 - z^2)(1 - k^2 z^2)}},$$

which is equal to the combination of normal elliptic integrals of the first and second kinds:

$$D(\phi, k) = \frac{1}{k^2}[F(\phi, k) - E(\phi, k)].$$

Then the analytic expression for momentum in the segment $0 \le \tilde{p} \le \tilde{p}_0$ is

$$\pm(\tau - \tau_0) = -\imath\sqrt{\frac{2}{1+C}}\left[2D\left(arcsin\sqrt{\frac{1+\sqrt{1+\tilde{p}^2}}{2}}, \sqrt{\frac{2}{1+C}}\right)\right.$$
$$-F\left(arcsin\sqrt{\frac{1+\sqrt{1+\tilde{p}^2}}{2}}, \sqrt{\frac{2}{1+C}}\right)$$
$$\left.+F\left(arcsin\sqrt{\frac{1+C}{2}}, \sqrt{\frac{2}{1+C}}\right) - 2D\left(arcsin\sqrt{\frac{1+C}{2}}, \sqrt{\frac{2}{1+C}}\right)\right],$$

where τ_0 is the time at which momentum takes the peak value $\tilde{p}_0$. The part of the momentum function for the segment $-\tilde{p}_0 \le \tilde{p} \le 0$ is reconstructed from the obtained graph by reflecting it with respect to the time axis τ and shifting the resulting reflection by the half-period of the function $\tilde{p}(\tau)$. To reduce heuristic

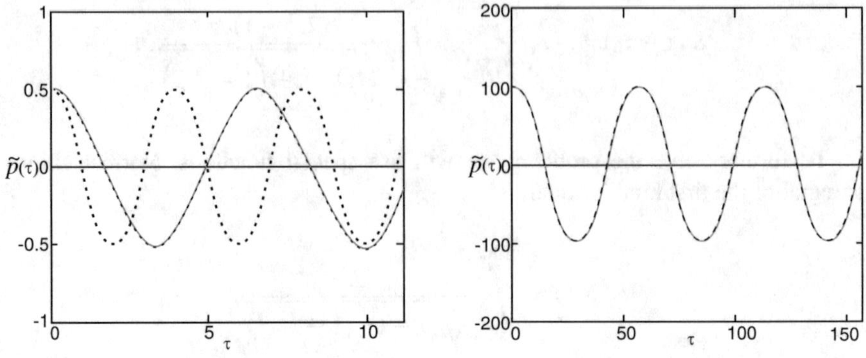

Fig. 2.9 Graphics of exact solution $\tilde{p}(\tau)$ (solid light curve), first Fourier harmonics (dotted dark), and limiting function (4-dimentional graphic), respectively, for $\tilde{p}_0 = 0.5$ (left) and $\tilde{p}_0 = 100$ (right)

operations to a minimum, the graph (Fig. 2.9) of the desired function $\tilde{p}(\tau)$ is still more convenient to be found numerically from the differential equation than from the last expression.

Evidently, the period T of function $\tilde{p}(\tau)$ is equal to

$$T = 2\int_{-\tilde{p}_0}^{\tilde{p}_0} \frac{d\tilde{p}}{\sqrt{2\left(C - \sqrt{1+\tilde{p}^2}\right)}} = 4\int_{0}^{\tilde{p}_0} \frac{d\tilde{p}}{\sqrt{2\left(C - \sqrt{1+\tilde{p}^2}\right)}},$$

and it is easily expressed in terms of elliptic integrals:

$$T = 4\imath\sqrt{\frac{2}{1+C}}\left[2D\left(arcsin\sqrt{\frac{1+C}{2}}, \sqrt{\frac{2}{1+C}}\right) - 2D\left(\sqrt{\frac{2}{1+C}}\right)\right.$$
$$\left. - F\left(arcsin\sqrt{\frac{1+C}{2}}, \sqrt{\frac{2}{1+C}}\right) + F\left(\sqrt{\frac{2}{1+C}}\right)\right].$$

When oscillation amplitude is small, $\tilde{p}_0 << 1$ (Fig. 2.10), the following relation holds:

$$T \approx 4\imath\left(1 - \frac{\tilde{p}_0^2}{8}\right)\left[2D\left(\frac{\pi}{2} - \imath\frac{\tilde{p}_0}{2}, 1 - \frac{\tilde{p}_0^2}{8}\right) - 2D\left(1 - \frac{\tilde{p}_0^2}{8}\right)\right.$$
$$\left. - F\left(\frac{\pi}{2} - \imath\frac{\tilde{p}_0}{2}, 1 - \frac{\tilde{p}_0^2}{8}\right) + F\left(1 - \frac{\tilde{p}_0^2}{8}\right)\right] \approx 2\pi,$$

and when the amplitude, on the contrary, is very large, $\tilde{p}_0 \gg 1$ (Fig. 2.10), the period is

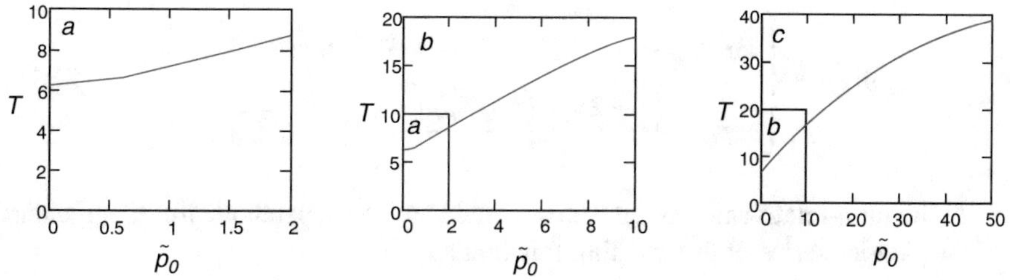

Fig. 2.10 Dependency of period T on peak amplitude $\tilde{p}_0$

$$T \approx 4\imath\sqrt{\frac{2}{\tilde{p}_0}}\left[2D\left(\frac{\pi}{2} - \imath \ln\sqrt{2\tilde{p}_0}, \sqrt{\frac{2}{\tilde{p}_0}}\right) - 2D\left(\sqrt{\frac{2}{\tilde{p}_0}}\right)\right.$$
$$\left. -F\left(\frac{\pi}{2} - \imath \ln\sqrt{2\tilde{p}_0}, \sqrt{\frac{2}{\tilde{p}_0}}\right) + F\left(\sqrt{\frac{2}{\tilde{p}_0}}\right)\right] \approx 4\sqrt{2\tilde{p}_0}.$$

Now let us consider some limiting forms of Eq. (2.374) in the lowest approximation of the equation with respect to the dimensionless parameter $\tilde{p}$. In the classical approximation $\tilde{p}_0 \ll 1$, it turns into the usual equation of harmonic oscillations with solution $\tilde{p}(\tau) = \tilde{p}_0 = \text{const}$ which in the graphic illustration is concentrated in a narrow band near τ axis. In the ultra-relativistic limit $\tilde{p}_0 \gg 1$, Eq. (2.374) degenerates and acquires parabolic solution $\tilde{p}(\tau) = \tilde{p}_0 - \operatorname{sgn}(\tilde{p}_0) \cdot \tau^2/2$ which is nothing more than a tangent parabola which can be carried on at each extremum of the periodic solution. An attempt to clarify these limiting cases for a larger number of terms in the expansion in terms of dimensionless parameter $\tilde{p}$ again encounters difficulty in the form of operating with elliptic and even hyperelliptic integrals.

Now we consider the expansion of the function $\tilde{p}(\tau)$in the Fourier series:

$$\tilde{p}(\tau) = \frac{A_0}{2} + \sum_{n=1}^{\infty}\left[A(n)\cos\left(\frac{2\pi n\tau}{T}\right) + B(n)\sin\left(\frac{2\pi n\tau}{T}\right)\right], A_0 = \frac{2}{T}\int_{-T/2}^{T/2}\tilde{p}(\tau)d\tau,$$

$$A(n) = \frac{2}{T}\int_{-T/2}^{T/2}\tilde{p}(\tau)\cos\left(\frac{2\pi n\tau}{T}\right)d\tau, B(n) = \frac{2}{T}\int_{-T/2}^{T/2}\tilde{p}(\tau)\sin\left(\frac{2\pi n\tau}{T}\right)d\tau.$$

In view of symmetry, $\tilde{p}(\tau) = \tilde{p}(-\tau)$, we have $A_0 = 0$ and $B(n) = 0$ for each natural n. As expected, the largest contribution to the solution $\tilde{p}(\tau)$ is made by the first harmonic with the fundamental period T. In the limit of large amplitude $\tilde{p}_0 \gg 1$, the graph of the solution tends to its limiting function (Fig. 2.9), consisting of branches of parabola $\left(-\sqrt{2\tilde{p}_0} \le \tau/2 \le \sqrt{2\tilde{p}_0}\right)$:

$$\tilde{p}(\tau) = \begin{cases} \tilde{p}_0 - \dfrac{\tau^2}{2}, & |\tau| \le \sqrt{2\tilde{p}_0}, \\ -\tilde{p}_0 + \dfrac{\left(\tau - 2\,\mathrm{sgn}\,(\tau)\sqrt{2\tilde{p}_0}\right)^2}{2}, & |\tau| > \sqrt{2\tilde{p}_0}. \end{cases} \tag{2.376}$$

In the ultra-relativistic case, the electric field has also a parabolic file. Coefficients of the Fourier series of this limiting function are

$$A_0 = 0, \qquad A(n) = \frac{32\tilde{p}_0}{\pi^3 n^3} \sin\left(\frac{\pi n}{2}\right), \qquad B(n) = 0.$$

In view of $\tilde{p}(0) = \tilde{p}_0$, we obtain the following numerical series:

$$\sum_{n=1}^{\infty} \frac{32}{\pi^3 n^3} \sin\left(\frac{\pi n}{2}\right) = 1.$$

It is possible to simulate the account of thermal losses by introducing the coefficient ν into the first equation of system (2.371):

$$\frac{d\vec{p}}{dt} = -e\vec{E} - \nu\vec{p}.$$

In this case, instead of Eq. (2.374), we find

$$\frac{d^2\tilde{p}}{d\tau^2} + \tilde{\nu}\frac{d\tilde{p}}{d\tau} + \frac{\tilde{p}}{\sqrt{1+\tilde{p}^2}} = 0,$$

where $\tilde{\nu} \equiv \nu m_0 c \omega_p \ll 1$. Classical approximation $\tilde{p}_0 \ll 1$ gives the well-known exponentially damping solution $\tilde{p} = \tilde{p}_0 \exp\left(-\frac{\tilde{\nu}}{2}\right) \cos\left(\tau\sqrt{1 - \frac{\tilde{\nu}^2}{4}}\right)$. At the distance from the time axis τ, i.e., in the region where the ultra-relativistic approximation $\tilde{p}_0 \gg 1$ works, an inhomogeneous second-order equation arises whose solution obviously is

$$\tilde{p}(\tau) = C_1 \mathrm{e}^{-\tilde{\nu}\tau} + C_2 - \frac{\tau}{\tilde{\nu}}, \qquad C_1 = -\frac{1}{\tilde{\nu}}\left[\tilde{p}'(\tau_0) + \frac{1}{\tilde{\nu}}\right] e^{\tilde{\nu}\tau_0},$$
$$C_2 = \tilde{p}(\tau_0) + \frac{\tau_0}{\tilde{\nu}} + \frac{1}{\tilde{\nu}}\left[\tilde{p}'(\tau_0) + \frac{1}{\tilde{\nu}}\right].$$

The value of function $\tilde{p}$ and its derivative $\tilde{p}'$ at the point τ_0 should be given. Non-algebraic dependence in the latter function makes the dissipative ultra-relativistic case (presence of thermal losses) radically different from the non-dissipative case. In fact, the field oscillation period increases with increasing field amplitude (Fig. 2.10) although the harmonicity is preserved. Besides, from Fig. 2.9, we can see that for large amplitude the field temporal behavior practically coincides with the limiting function behavior, while for small amplitudes they are completely different.

References

1. R. Balescu, Phys. Fluids **3**, 52 (1960)
2. A. Lenard, Ann. Phys. **3**, 390 (1960)
3. O.V. Konstantinov, V.I. Perel, J. Exp. Theor. Phys. **39**, 861 (1960)
4. V.P. Silin, J. Exp. Theor. Phys. **40**, 6 (1961)
5. N. Rostoker, Phys. Fluids **3**, 922 (1960)
6. V.P. Silin, Fiz. Met. Metalloved. **11**(5), 805 (1961)
7. V.M. Eleonskii, P.S. Zyrianov, V.P. Silin, Fiz. Met. Metalloved. **11**(6), 955 (1961)
8. Y.A. Alpert, V.L. Ginzburg, E.L. Feinberg, *Radio Wave Propagation* (Gostechizdat, Moscow, 1953)
9. V.P. Silin, J. Exp. Theor. Phys. **38**, 1771 (1960)
10. J. Gibbs, *Elementary Principles in Statistical Mechanics* (Dover Publication Inc, New York, 1960)
11. A.A. Vlasov, J. Exp. Theor. Phys. **8**, 291 (1938)
12. L.D. Landau, J. Exp. Theor. Phys. **7**, 203 (1936)
13. S.T. Beliaev, *Plasma Physics and Problem of Controlled Thermonuclear Reactions*, vol 3 (Izd-vo Academy Science of USSR, Moscow, 1958)
14. F.D. Gakhov, *Boundary Problems* (Fizmatgiz, Moscow, 1958)
15. M.E. Gertsenshtein, J. Exp. Theor. Phys. **22**(10), 303 (1952)
16. V.D. Shafranov, J. Exp. Theor. Phys. **34**, 1475 (1958)
17. V.L. Ginzburg, V.V. Zheleznìakov, Astron. Zh. **35**, 694 (1958)
18. B.A. Trubnikov, *Plasma Physics and Problem of Controlled Thermonuclear Reactions*, vol 4 (Izd-vo Academy Science of USSR, Moscow, 1958)
19. V.V. Zhelezniakov, Radiofizika **2**, 1 (1959)
20. V.L. Feinberg, Usp. Fiz. Nauk **69**, 537 (1959)
21. V.L. Ginzburg, *Electromagnetic Waves in Plasma* (Fizmatgiz, Moscow, 1960)
22. R. Becker, *Theory of Electricity [Russian translation]*, vol 2 (Gostechizdat, Moscow, 1941)
23. L. Rayleigh, Philos. Mag. **11**, 117 (1906)
24. I. Langmuir, Proc. Natl. Acad. Sci. U. S. A. **14**, 627 (1928)
25. L. Tonks, I. Langmuir, Phys. Rev. **33**, 195 (1926)
26. A.A. Vlasov, Lect. Note Moscow State Univ. **75**(2) (1945)
27. L.D. Landau, J. Exp. Theor. Phys. **16**, 574 (1946)
28. N.G. Van Kampen, Physica **21**, 949 (1955)
29. K.M. Case, Ann. Phys. **7**, 349 (1959)
30. V.A. Fock, *Radio Wave Diffraction Around Earth Surface* (Izd-vo Academy Science of USSR, Moscow, 1946)
31. G.V. Gordeev, J. Exp. Theor. Phys. **22**, 230 (1952)
32. V.N. Fadeeva, N.M. Terentev, *Table of Probability Integrals* (Gostekhizdat, Moscow, 1954)
33. G.V. Gordeev, J. Exp. Theor. Phys. **24**, 445 (1953)
34. J. Doetsch, *Theorie und Wending der Laplace-Transformation* (Dover Publication, New York, 1948)
35. D. Bohm, E.P. Gross, Phys. Rev **75**, 1851 (1949)
36. Y.L. Klimontovich, Dokl. Acad. Sci. USSR **87**, 927 (1952)

37. P.C. Clemmow, A.J. Wilson, Proc. Camb. Philol. Soc. **53**, 222 (1958)
38. P.C. Clemmow, A.J. Wilson, Proc. R. Soc. Lond. A Math. Phys. Sci. **237**, 117 (1956)
39. V.M. Galitskii, A.B. Migdal, *Plasma Physics and Problem of Thermonuclear Fusion*, vol 1 (Izd-vo Academy Science of USSR, Moscow, 1958)
40. V.P. Silin, J. Exp. Theor. Phys. **38**, 1577 (1960)
41. L.D. Landau, E.M. Lifshitz, *Course of Theoretical Physics, Vol. 5: Statistical Physics*, 3rd edn. (Nauka, Moscow, 1976).; Pergamon Press, Oxford, 1980), Part 1
42. G.V. Gordeev, J. Exp. Theor. Phys. **27**, 18 (1954)
43. B.D. Fried, R.W. Jould, Phys. Fluids **4**, 139 (1961)
44. V.P. Silin, J. Exp. Theor. Phys. **23**, 649 (1959)
45. S.I. Braginskii, A.P. Kazantsev, *Plasma Physics and Problem of Controlled Thermonuclear Reactions* (Izd-vo Academy Science of USSR, Moscow, 1958)
46. G. Hilbert, *Grundzüge einer allgemeinen Theorie der linearen Integralgleichungen* (Chelsea Publishing Company, New York, 1953)
47. S. Chapman, T. Cowling, *The Mathematical Theory of Nou-uniform Gases* (Cambridge University Press, Cambridge, 1951)
48. N.N. Bogolubov, *Problem of Dynamic Theory of Statistical Physics* (Gostekhizdat, Moscow, 1946)
49. B.N. Gershman, V.L. Ginzburg, N.G. Denisov, Usp. Fiz. Nauk **61**, 561 (1957)
50. L. Spitzer, *Physics of Completely Ionized Gas* (Translated from English Edition) (Mir, M, 1965)
51. V.P. Silin, Bull. Lebedev Inst. Acad. Sci. USSR **6**, 200 (1955)
52. V.L. Ginzburg, A.V. Gurevich, Usp. Fiz. Nauk **71**, 201 (1960)
53. A.V. Gurevich, J. Exp. Theor. Phys. **35**, 392 (1958)
54. A.F. Alexandrov, L.S. Bogdankevich, A.A. Rukhadze, *Principles of Plasma Electrodynamics* (Springer, Heidelberg, 1984)
55. G.E.H. Reuter, E.H. Sondheimer, Proc. R. Soc. Lond. A Math. Phys. Sci. **195**, 336 (1948)
56. K.N. Stepanov, J. Exp. Theor. Phys. **36**, 1457 (1959)
57. V.P. Silin, J. Exp. Theor. Phys. **40**, 616 (1961)
58. Y.N. Dnestrovskii, D.P. Koctomarov, J. Exp. Theor. Phys. **39**, 845 (1960)
59. V.I. Kurilko, J. Tech. Phys. **31**, 71 (1961)
60. V.I. Smirnov, *Course of Higher Mathematics*, vol 4 (Gostechizdat, Moscow, 1941)
61. R.B. Dingle, Physica **19**, 311 (1953)
62. L.D. Landau, E.M. Lifshitz, *Electrodynamics of Continuous Media*, 2nd edn. (Pergamon, New York, 1984)
63. V.L. Ginzburg, G.P. Motulevich, Usp. Fiz. Nauk **55**, 469 (1955)
64. R.B. Dingle, Appl. Sci. Res. **B2**, 69 (1953)
65. A.B. Pippard, Physica **15**, 45 (1949)
66. T. Holstein, Phys. Rev. **88**, 1427 (1952)
67. V.P. Silin, J. Exp. Theor. Phys. **35**, 1001 (1958)
68. V.P. Silin, J. Exp. Theor. Phys. **36**, 1443 (1959)
69. V.P. Silin, E.P. Fetisov, J. Exp. Theor. Phys. **41**(1), 159 (1961)
70. A.I. Akhiezer, I.A. Akhiezer, R.V. Polovin, A.G. Sitenko, K.N. Stepanov, *Plasma Electrodynamics* (Pergamon, New York, 1975)
71. B. Shokri, Phys. Plasmas **7**, 3867 (2000)
72. A.A. Rukhadze, B. Shokri, J. Tech. Phys. **142**, 12 (1997)
73. A.A. Rukhadze, B. Shokri, Phys. Scr. **57**, 127 (1998)
74. B.M. Bolotovskii, Usp. Fiz. Nauk **62**, 201 (1957)
75. J. Jelly, *Cherenkov Radiation* (Pergamon Press, New York, 1958)
76. I.N. Toptygin, *Electromagnetic Phenomena in Matter: Statistical and Quantum Approaches*, 1st edn. (Wiley-VCH, Weinheim, 2015)
77. A.I. Akhiezer, G.Y. Lubarskii, Dokl. Akad. Nauk USSR **80**, 193 (1951)

Chapter 3
Anisotropic Plasma

3.1 Dielectric Permittivity of Collisionless Plasma in a Constant Magnetic Field

In this chapter, we will consider electromagnetic properties of anisotropic plasmas. An example of such plasmas is plasma in the external homogeneous electric and magnetic fields. Under certain conditions the distribution functions of particles in such plasma are supposed to be isotropic and Maxwellian. Another example of anisotropic plasmas is plasma with anisotropic distribution functions of particles. Furthermore, in the equilibrium state, there are no current and charge distributions in plasma and, as a result, plasma is homogeneous in space.

Let us begin from the derivation of the dielectric permittivity of plasma embedded in an external magnetic field. For this aim, we consider a small deviation from the equilibrium state and represent the distribution function of the particles of the type α in the following form:

$$f_\alpha\left(\vec{p},\vec{r},t\right)=f_{0\alpha}\left(\vec{p}\right)+\delta f_\alpha\left(\vec{p},\vec{r},t\right), \tag{3.1}$$

where $f_{0\alpha}\left(\vec{p}\right)$ is the equilibrium distribution function which is homogeneous in space and time and $\delta f_\alpha\left(\vec{p},\vec{r},t\right)$ is a small deviation from the equilibrium state. Suppose that, in the equilibrium state, an external constant magnetic field with the magnetic induction $\vec{B}_0$ exists. In the deviation from the equilibrium state, non-stationary perturbations of the electric and magnetic fields $\vec{E}\left(\vec{r},t\right)$ and $\vec{B}\left(\vec{r},t\right)$, which are induced by the currents and charges, appear. Moreover, the electromagnetic force acting on a particle of the type α can be written as

B. Shokri, A. A. Rukhadze, *Electrodynamics of Conducting Dispersive Media*, Springer Series on Atomic, Optical, and Plasma Physics 111,
https://doi.org/10.1007/978-3-030-28968-3_3

$$\vec{F}_{\alpha} = e_{\alpha}\left\{\vec{E} + \frac{1}{c}\left[\vec{v} \times \left(\vec{B} + \vec{B}_0\right)\right]\right\}. \tag{3.2}$$

Using expressions (3.1) and (3.2), from the kinetic equation (2.3), we obtain the following linear equation for the perturbation $\delta f_{\alpha}\left(\vec{p}, \vec{r}, t\right)$ in magneto-active collisionless plasma:

$$\frac{\partial}{\partial t}\delta f_{\alpha} + \vec{v} \cdot \frac{\partial}{\partial \vec{r}}\delta f_{\alpha} + \frac{e_{\alpha}}{c}\left[\vec{v} \times \vec{B}_0\right] \cdot \frac{\partial}{\partial \vec{p}}\delta f_{\alpha}$$
$$= -e_{\alpha}\left\{\vec{E} + \frac{1}{c}\left[\vec{v} \times \vec{B}\right]\right\} \cdot \frac{\partial}{\partial \vec{p}} f_{0\alpha}. \tag{3.3}$$

It should be noted that, in the presence of an external constant electric field $\vec{E}_0$ in magneto-active plasma mentioned above, a drift of particles takes place, which in the collisionless limit results in

$$\vec{E}_0 = -\frac{1}{c}\vec{V}_0 \times \vec{B}_0,$$

where $\vec{V}_0$ is the drift velocity. In the coordinates system in which $\vec{V}_0$ is zero or parallel to $\vec{B}_0$, the electric field $\vec{E}_0$ is zero. Thus, it is possible to exclude $\vec{E}_0$ from the kinetic equations such as Eq. (3.3) by the transformation of the coordinate system. The characteristic system of differential equation (3.3) can be written as

$$\frac{d\vec{p}}{dt} = \frac{e_{\alpha}}{c}\left[\vec{v} \times \vec{B}_0\right], \qquad \frac{d\vec{r}}{dt} = \vec{v}. \tag{3.4}$$

Thus, we can find the solution of kinetic equation (3.3) in terms of the characteristics of Eq. (3.4). For the initial condition $\delta f_{\alpha} = 0$ at $t = -\infty$, corresponding to adiabatic switching of the interaction at infinity, we obtain

$$\delta f_{\alpha}\left(\vec{P}, \vec{r}, t\right) =$$
$$-e_{\alpha}\int_{-\infty}^{t} dt' \frac{\partial f_{0\alpha}\left(\vec{P}(t-t')\right)}{\partial \vec{P}(t-t')}\left\{\vec{E}\left(\vec{r}+\vec{R}(t-t'),t'\right)+\frac{1}{c}\left[\vec{U}(t-t')\times\vec{B}\left(\vec{r}+\vec{R}(t-t'),t'\right)\right]\right\}, \tag{3.5}$$

where

$$\vec{P}(\tau)=\frac{\epsilon_\alpha}{c^2}\vec{U}(\tau),\ \vec{U}(\tau)=\frac{\vec{B}_0\left(\vec{v}.\vec{B}_0\right)}{B_0^2}+\frac{\vec{B}_0\times\left(\vec{v}\times\vec{B}_0\right)}{B_0^2}\cos\Omega_\alpha\tau-\frac{(\vec{v}\times B_0)}{B_0}\sin\Omega_\alpha\tau,$$

$$\vec{R}(\tau)=-\int_0^\tau\vec{U}(\tau')d\tau',\quad \Omega_\alpha=\frac{e_\alpha cB_0}{\epsilon_\alpha}. \tag{3.6}$$

In deriving Eqs. (3.5) and (3.6), it was considered that, in accordance to Eq. (3.4), the energy of a charged particle in the homogeneous magnetic field is constant:

$$\epsilon_\alpha=\frac{m_\alpha c^2}{\sqrt{1-v^2/c^2}}=\sqrt{c^2p^2+m^2c^4}.$$

From Eq. (3.5), an expression for the density of current induced in plasma can be easily found. As in the equilibrium state, there is no current, then

$$j_i\left(\vec{r},t\right)=\sum_\alpha e_\alpha\int v_i\delta f_\alpha d\vec{p}=-\sum_\alpha e_\alpha^2\int_{-\infty}^{t}dt'\int d\vec{r'}\int d\vec{p}v_i\frac{\partial f_{0\alpha}\left(\vec{P}(t-t')\right)}{\partial P_j(t-t')}\times$$
$$\times\left\{E_j\left(\vec{r'},t'\right)+\frac{1}{c}\left[\vec{U}(t-t')\times\vec{B}(r',t')\right]_j\right\}\delta\left[\vec{R}(t-t')+\vec{r}-\vec{r'}\right]. \tag{3.7}$$

The kernel of this integral relation is a function of differences $\vec{r}-\vec{r'}$ and $t-t'$. Therefore, representing the electromagnetic field with the help of Fourier expansion as a sum of plane waves $\exp\left(-\imath\omega t+\imath\vec{k}\cdot\vec{r}\right)$ and using the Maxwell's equation

$$\left[\vec{k}\times\vec{E}\right]=\frac{\omega}{c}\vec{B},$$

from Eq. (3.7), one can obtain

$$j_i\left(\omega,\vec{k}\right)=\sigma_{ij}\left(\omega,\vec{k}\right)E_j\left(\omega,\vec{k}\right), \tag{3.8}$$

where $\sigma_{ij}\left(\omega,\vec{k}\right)$ is the complex tensor of conductivity

$$\sigma_{ij}\left(\omega,\vec{k}\right) = -\sum_{\alpha} e_{\alpha}^{2} \int_{0}^{\infty} d\tau \int d\vec{p}\, v_{i} \exp\left[\imath \vec{k}\cdot\vec{R}(\tau) + \imath\omega\tau\right]$$
$$\left\{\left(1 - \frac{\vec{k}\cdot\vec{U}(\tau)}{\omega}\right)\delta_{jl} + \frac{U_{j}(\tau)k_{l}}{\omega}\right\}\frac{\partial f_{0\alpha}\left(\vec{P}(\tau)\right)}{\partial P_{l}(\tau)}. \tag{3.9}$$

Then, according to Eq. (1.46), for the dielectric permittivity, we obtain[1]

$$\varepsilon_{ij}\left(\omega,\vec{k}\right) = \delta_{ij} + \frac{4\pi\imath}{\omega}\sigma_{ij}\left(\omega,\vec{k}\right).$$

In the absence of any external magnetic field, we have

$$\vec{U}(\tau) = \vec{v}, \qquad \vec{R}(\tau) = -\vec{v}\tau, \qquad \vec{P}(\tau) = \vec{p}.$$

In this case, we find

$$\varepsilon_{ij}\left(\omega,\vec{k}\right) = \delta_{ij} + \sum_{\alpha}\frac{4\pi e_{\alpha}^{2}}{\omega}\int d\vec{p}\,\frac{v_{i}}{\omega - \vec{k}\cdot\vec{v}}\,\frac{\partial f_{0\alpha}\left(\vec{p}\right)}{\partial p_{l}}$$
$$\left\{\left(1 - \frac{\vec{k}\cdot\vec{v}}{\omega}\right)\delta_{jl} + \frac{v_{j}k_{l}}{\omega}\right\}, \tag{3.10}$$

which coincides with Eq. (2.21) if the distribution function of particles in the equilibrium state $f_{0\alpha}\left(\vec{p}\right)$ depends only on $|\vec{p}| = p$. It must be noted that, in this case, the expression (3.9) for the conductivity of magneto-active plasma is also simplified significantly. Actually, in this case, as the energy of charged particles in the magnetic field is constant, we have

$$v_{i}\frac{\partial f_{0\alpha}(P(\tau))}{\partial P_{j}(\tau)} = U_{j}(\tau)\frac{\partial f_{0\alpha}(p)}{\partial p_{i}}.$$

Thus, from Eq. (3.9), for the isotropic distribution function $f_{0\alpha}(p)$, we have

$$\varepsilon_{ij}\left(\omega,\vec{k}\right) = \delta_{ij} - \sum_{\alpha}\frac{4\pi\imath e_{\alpha}^{2}}{\omega}\int_{0}^{\infty} d\tau \int d\vec{p}\,\frac{\partial f_{0\alpha}(p)}{\partial p_{i}}U_{j}(\tau)\exp\left[\imath\vec{k}\cdot\vec{R}(\tau) + \imath\omega\tau\right]. \tag{3.11}$$

[1]Here, as in Sect. 2.2, we suppose that ω has an infinitely small positive imaginary part.

In the long-wave limit $\left(\vec{k} \to 0\right)$, when the spatial dispersion of the dielectric permittivity can be neglected, from Eq. (3.11), supposing that the external field $\vec{B}_0$ is parallel to the z-axis, we find

$$\varepsilon_{ij}(\omega, 0) = \begin{pmatrix} \varepsilon_1 & \imath g & 0 \\ -\imath g & \varepsilon_1 & 0 \\ 0 & 0 & \varepsilon_2 \end{pmatrix}, \tag{3.12}$$

where

$$\begin{aligned}
\varepsilon_1 &= 1 + \frac{2\pi}{3\omega}\sum_\alpha e_\alpha^2 \int d\vec{p}\, v^2 \frac{\partial f_{0\alpha}}{\partial \epsilon_\alpha}\left\{\mathcal{P}\left(\frac{1}{\omega - \Omega_\alpha} + \frac{1}{\omega + \Omega_\alpha}\right) - \imath\pi[\delta(\omega - \Omega_\alpha) + \delta(\omega + \Omega_\alpha)]\right\}, \\
g &= \frac{2\pi}{3\omega}\sum_\alpha e_\alpha^2 \int d\vec{p}\, v^2 \frac{\partial f_{0\alpha}}{\partial \epsilon_\alpha}\left\{\mathcal{P}\left(\frac{1}{\omega - \Omega_\alpha} - \frac{1}{\omega + \Omega_\alpha}\right) - \imath\pi[\delta(\omega - \Omega_\alpha) - \delta(\omega + \Omega_\alpha)]\right\}, \\
\varepsilon_2 &= 1 + \frac{4\pi}{3\omega^2}\sum_\alpha e_\alpha^2 \int d\vec{p}\, v^2 \frac{\partial f_{0\alpha}}{\partial \epsilon_\alpha}.
\end{aligned} \tag{3.13}$$

The symbol $\mathcal{P}$denotes that the principal value of the integrals has to be taken at $\omega \pm \Omega_\alpha = 0$.

Formulas (3.12) and (3.13) determine the dielectric permittivity of magneto-active plasma, taking only frequency dispersion into account. In the absence of the magnetic field, when $B_0 \to 0$, tensors (3.12) become diagonal, corresponding to expression (2.26).

In conclusion, let us represent one more form of the tensor of the dielectric permittivity for magneto-active plasma, which is convenient for further integration, when the explicit function $f_{0\alpha}\left(\vec{p}\right)$ is known. Without loss of generality, we may assume that the external magnetic field $\vec{B}_0$ is oriented along the z-axis and the wave vector $\vec{k}$ lies along the xzplane, i.e., $\vec{k} = (k_\perp, 0, k_z)$. Besides, it is convenient to use the cylindrical system of coordinates in the velocity space $(v_\perp, \varphi, v_z)$ defined by $v_x = v_\perp \cos\varphi,\ \ v_y = v_\perp \sin\varphi$. Using these coordinates from Eq. (3.11), we obtain[2]

[2]The dielectric permittivity for magneto-active plasma in the forms of Eqs. (3.14) and (3.32) was first obtained by Trubnikov [1, 2]. In the absence of spatial dispersion ($\vec{k} \to 0$), when thermal motion of particles is neglected, it was given in Ginzburg's book [3, 4]. The thermal motion of particles was firstly considered in [5, 6] (see also [4, 7–14]).

$$\varepsilon_{ij}\left(\omega,\vec{k}\right)=\delta_{ij}-\sum_{\alpha}\frac{4\pi\imath e_{\alpha}^{2}}{\omega}\int_{0}^{\infty}d\tau\int d\vec{p}\frac{\partial f_{0\alpha}}{\partial\epsilon_{\alpha}}T_{ij}\exp\left\{\imath(\omega-k_{z}v_{z})\tau-\frac{\imath k_{\perp}v_{\perp}}{\Omega_{\alpha}}[\sin(\Omega_{\alpha}\tau-\varphi)+\sin\varphi]\right\},\tag{3.14}$$

where

$$T_{ij}=\begin{pmatrix} v_{\perp}^{2}\cos\varphi\cos\left(\Omega_{\alpha}\tau-\varphi\right), & v_{\perp}^{2}\cos\varphi\sin\left(\Omega_{\alpha}\tau-\varphi\right), & v_{\perp}v_{z}\cos\left(\Omega_{\alpha}\tau-\varphi\right)\\ -v_{\perp}^{2}\sin\varphi\cos\left(\Omega_{\alpha}\tau-\varphi\right), & -v_{\perp}^{2}\sin\varphi\sin\left(\Omega_{\alpha}\tau-\varphi\right), & -v_{\perp}v_{z}\sin\left(\Omega_{\alpha}\tau-\varphi\right)\\ v_{z}v_{\perp}\cos\left(\Omega_{\alpha}\tau-\varphi\right), & v_{\perp}v_{z}\sin\left(\Omega_{\alpha}\tau-\varphi\right), & v_{z}^{2}\end{pmatrix}.$$

Using

$$\exp\left(\pm\imath z\sin\varphi\right)=\sum_{s=-\infty}^{+\infty}J_{s}(z)\exp\left(\imath s\varphi\right),$$

where $J_s(z)$ and $J_s'(z)$ are the Bessel function and its derivatives of the order of s, respectively; integration over τ and φ reduces relation (3.14) to

$$\varepsilon_{ij}\left(\omega,\vec{k}\right)=\delta_{ij}+\sum_{\alpha}\frac{4\pi e_{\alpha}^{2}}{\omega}\int d\vec{p}\,\frac{\partial f_{0\alpha}(p)}{\partial\epsilon_{\alpha}}\sum_{s=-\infty}^{+\infty}\Pi_{ij}^{(s)}\left\{\mathcal{P}\frac{1}{\omega-k_{z}v_{z}-s\Omega_{\alpha}}-\imath\pi\delta(\omega-k_{z}v_{z}-s\Omega_{\alpha})\right\},\tag{3.15}$$

where

$$\Pi_{ij}^{(s)}=\begin{pmatrix} v_{\perp}^{2}\left(s\dfrac{J_{s}(b_{\alpha})}{b_{\alpha}}\right)^{2}, & \imath v_{\perp}^{2}\left(s\dfrac{J_{s}(b_{\alpha})}{b_{\alpha}}\right)J_{s}'(b_{\alpha}), & v_{\perp}v_{z}\left(s\dfrac{J_{s}(b_{\alpha})}{b_{\alpha}}\right)J_{s}(b_{\alpha})\\ -\imath v_{\perp}^{2}\left(s\dfrac{J_{n}(b_{\alpha})}{b_{\alpha}}\right)J_{s}'(b_{\alpha}), & v_{\perp}^{2}\left(J_{s}'(b_{\alpha})\right)^{2}, & -\imath v_{\perp}v_{z}J_{s}(b_{\alpha})J_{s}'(b_{\alpha})\\ v_{\perp}v_{z}\left(s\dfrac{J_{s}(b_{\alpha})}{b_{\alpha}}\right)J_{s}(b_{\alpha}), & \imath v_{\perp}v_{z}J_{s}(b_{\alpha})J_{s}'(b_{\alpha}), & v_{z}^{2}J_{s}^{2}(b_{\alpha})\end{pmatrix},$$

and $b_\alpha = k_\perp v_\perp/\Omega_\alpha$. This representation of the dielectric permittivity of magneto-active plasma shows that cyclotron resonances are associated with the zeros of $\omega - k_z v_z - s\Omega_\alpha$ in the denominator. The principal value of the integrands determines the Hermitian part of tensor $\varepsilon_{ij}\left(\omega,\vec{k}\right)$ and the terms containing the δ-function contribute to the anti-Hermitian part responsible for the wave absorption. Hence, in magneto-active plasma, only the particles satisfying the condition $\omega = k_z v_z + s\Omega_\alpha$ can absorb waves. This condition replaces the Cherenkov resonance condition

$\omega = \vec{k} \cdot \vec{v}$ of non-magnetized plasma (see session 11). The case of $s = 0$ and $s \neq 0$ corresponds to the purely Cherenkov and cyclotron absorption, respectively.

Finally, let us derive an expression for the so-called longitudinal dielectric permittivity of magneto-active plasma, which describes the electrostatic field in such plasma (for the electrostatic field of a point charge, for example, see Sect. 1.8). This permittivity is determined by

$$\varepsilon\left(\omega, \vec{k}\right) = \frac{k_i k_j}{k^2} \varepsilon_{ij}\left(\omega, \vec{k}\right)$$

$$= 1 - \sum_\alpha \frac{4\pi e_\alpha^2}{k^2} \int d\vec{p}\, \frac{\partial f_{0\alpha}}{\partial \epsilon_\alpha} \left[1 - \sum_{s=-\infty}^{+\infty} \frac{\omega J_s^2(b_\alpha)}{\omega - k_z v_z - s\Omega_\alpha}\right]. \tag{3.16}$$

In the static limit ($\omega \to 0$), from this expression, it follows that

$$\varepsilon(0, k) = 1 - \sum_\alpha \frac{4\pi e_\alpha^2}{k^2} \int d\vec{p}\, \frac{\partial f_{0\alpha}}{\partial \epsilon_\alpha}.$$

This expression coincides with the expression (2.27), which means that, in both magneto-active and non-magnetized plasmas, the screening of the electrostatic field takes place and the screening radius is the same as formula (2.28).

For Maxwellian distributions (2.47) or (3.18), dielectric permittivity of magneto-active plasma given by Eq. (3.16) takes the form

$$\varepsilon\left(\omega, \vec{k}\right) = 1 + \sum_\alpha \frac{\omega_{p\alpha}^2}{k^2 v_{T\alpha}^2} \left[1 - \sum_n \frac{\omega}{\omega - n\Omega_\alpha} A_n(z_\alpha) I_+(\beta_{n\alpha})\right]. \tag{3.17}$$

3.2 Electromagnetic Oscillations of Non-relativistic Plasma in a Constant Magnetic Field

In this section, we will investigate the electromagnetic waves in magneto-active pure electron plasma.[3] But before this, let us obtain the dielectric permittivity of Maxwellian plasma with the distribution function

[3] About the electromagnetic properties of magneto-active electron plasma, one can see [3–7, 9–48].

$$f_{0\alpha}(p) = \frac{N_\alpha}{(2\pi m_\alpha \kappa T_\alpha)^{\frac{3}{2}}} \exp\left(-\frac{p^2}{2m_\alpha \kappa T_\alpha}\right). \tag{3.18}$$

Substituting this distribution into Eq. (3.14) and after integration over the momentum $\vec{p}$, one can obtain

$$\varepsilon_{ij}\left(\omega, \vec{k}\right) = \delta_{ij} + \imath\frac{\omega_{pe}^2}{\omega}\int_0^\infty d\tau T_{ij} \exp\left\{\imath\omega\tau - \frac{\kappa T_e}{2m}\left[k_z^2\tau^2 + \frac{2k_\perp^2}{\Omega_e^2}(1-\cos\Omega_e\tau)\right]\right\},$$

where

$$T_{ij} = \begin{pmatrix} \cos\Omega_e\tau & \sin\Omega_e\tau & 0 \\ -\sin\Omega_e\tau & \cos\Omega_e\tau & 0 \\ 0 & 0 & 1 \end{pmatrix} -$$

$$-\frac{\kappa T_e}{m\Omega_e^2}\times\begin{pmatrix} k_\perp^2\sin^2\Omega_e\tau & k_\perp^2\sin\Omega_e\tau(1-\cos\Omega_e\tau) & k_\perp k_z\Omega_e\tau\sin\Omega_e\tau \\ -k_\perp^2\sin\Omega_e\tau(1-\cos\Omega_e\tau) & -k_\perp^2(1-\cos\Omega_e\tau)^2 & -k_\perp k_z\Omega_e\tau(1-\cos\Omega_e\tau) \\ k_\perp k_z\Omega_e\tau\sin\Omega_e\tau & k_\perp k_z\Omega_e\tau(1-\cos\Omega_e\tau) & k_z^2\Omega_e^2\tau^2 \end{pmatrix}.$$

After performing the integration, we can express the dielectric tensor in terms of the tabulated functions. As a result, we obtain

$$\begin{aligned}
\varepsilon_{xx} &= 1 - \sum_\alpha\sum_s \frac{s^2\omega_{p\alpha}^2}{\omega(\omega - s\Omega_\alpha)}\frac{A_s(z_\alpha)}{z_\alpha}I_+(\beta_{s\alpha}),\\
\varepsilon_{xy} &= -\varepsilon_{yx} = -\imath\sum_\alpha\sum_s \frac{s\omega_{p\alpha}^2}{\omega(\omega - s\Omega_\alpha)}A_s'(z_\alpha)I_+(\beta_{s\alpha}),\\
\varepsilon_{yy} &= \varepsilon_{xx} + 2\sum_\alpha\sum_s \frac{\omega_{p\alpha}^2 z_\alpha}{\omega(\omega - s\Omega_\alpha)}A_s'(z_\alpha)I_+(\beta_{s\alpha}),\\
\varepsilon_{xz} &= \varepsilon_{zx} = -\sum_\alpha\sum_s \frac{\omega_{p\alpha}^2 s k_\perp}{\omega\Omega_\alpha k_z}\frac{A_s(z_\alpha)}{z_\alpha}I_+(\beta_{s\alpha}),\\
\varepsilon_{yz} &= -\varepsilon_{zy} = \imath\sum_\alpha\sum_s \frac{\omega_{p\alpha}^2 k_\perp}{\omega\Omega_\alpha k_z}A_s'(z_\alpha)I_+(\beta_{s\alpha}),\\
\varepsilon_{zz} &= 1 + \sum_\alpha\sum_s \frac{\omega_{p\alpha}^2}{k_\alpha^2 v_{T\alpha}^2}\left[1 - \frac{s\Omega_\alpha}{\omega}A_s(z_\alpha)I_+(\beta_{s\alpha})\right],
\end{aligned} \tag{3.19}$$

where $I_s(z_\alpha)$ is the Bessel function of an imaginary argument and

$$A_s(z_\alpha) = \exp(-z_\alpha) I_s(z_\alpha), \quad z_\alpha = \frac{k_\perp^2 v_{T\alpha}^2}{\Omega_\alpha^2}, \quad \beta_{s\alpha} = \frac{\omega - s\Omega_\alpha}{| k_z | v_{T\alpha}}.$$

Here, the prime sign means derivative with respect to the argument. In the absence of the external magnetic field, $\vec{B}_0 = \vec{0}$, expressions (3.19) reduce to the longitudinal and transverse permittivities of isotropic Maxwellian plasma given by Eqs. (2.81) and (2.82). Now, let us study the electromagnetic waves in magneto-active pure electron plasma using the dispersion equation (1.108),

$$\left| k^2 \delta_{ij} - k_i k_j - \frac{\omega^2}{c^2} \varepsilon_{ij}\left(\omega, \vec{k}\right) \right| = 0. \tag{3.20}$$

We begin the analysis from the long-wave limit $\left(\vec{k} \to 0\right)$ when spatial dispersion can be neglected. The dielectric tensor for pure electron plasma in this limit looks as

$$\varepsilon_{ij}(\omega, 0) = \begin{pmatrix} \varepsilon_1 & \imath g & 0 \\ -\imath g & \varepsilon_1 & 0 \\ 0 & 0 & \varepsilon_2 \end{pmatrix}, \tag{3.21}$$

where

$$\varepsilon_1 = 1 - \frac{\omega_{\text{pe}}^2}{\omega^2 - \Omega_e^2}, \quad \varepsilon_2 = 1 - \frac{\omega_{\text{pe}}^2}{\omega^2}, \quad g = -\frac{\omega_{\text{pe}}^2 \Omega_e}{\omega\left(\omega^2 - \Omega_e^2\right)}.$$

From Eq. (3.21), it follows that magneto-active electron plasma, even in the absence of spatial dispersion, is a gyrotropic medium. Therefore, sometimes plasma in the external magnetic field is called gyrotropic plasma.

Substituting Eq. (3.21) into Eq. (3.20), we obtain a quadratic equation with respect to n^2:

$$L(n^2) = n^4\left(\varepsilon_1 \sin^2\vartheta + \varepsilon_2 \cos^2\vartheta\right) - n^2\left[\left(\varepsilon_1^2 - g^2 - \varepsilon_1\varepsilon_2\right)\sin^2\vartheta + 2\varepsilon_1\varepsilon_2\right] + \varepsilon_2\left(\varepsilon_1^2 - g^2\right) = 0, \tag{3.22}$$

where $n = c/\omega k$ is the refractive index and ϑ is the angle between the wave vector $\vec{k}$ and the magnetic field $\vec{B}_0$. In the absence of the external magnetic field, $\vec{B}_0 \to 0$, we have

$$\varepsilon(\omega) = \varepsilon_1 = \varepsilon_2 = 1 - \frac{\omega_{\rm pe}^2}{\omega^2}, \qquad g = 0.$$

Equation (3.22), in this limit, splits into three equations (two of them are identical)

$$\varepsilon(\omega) = 0, \qquad n^2 = \varepsilon(\omega), \tag{3.23}$$

corresponding to the dispersion equation of longitudinal and transverse waves in isotropic electron plasma, respectively. In contrast to isotropic plasma, from Eq. (3.22), it follows that the longitudinal and transverse waves are not independent in magneto-active plasma. Generally, Eq. (3.22) does not split into independent equations. Only for purely longitudinal propagation when $\vartheta = 0$, this separation takes place. Then, from Eq. (3.22) we obtain three solutions

$$\varepsilon_2(\omega) = 0, \tag{3.24}$$

corresponding to the longitudinal oscillations and

$$n_{1,2}^2(\omega) = \varepsilon_1(\omega) \pm g(\omega), \tag{3.25}$$

corresponding to the two branches of transverse waves. For an arbitrary angle ϑ, from Eq. (3.22) it follows [8]

$$n_{1,2}^2 = \frac{1}{2(\varepsilon_1 \sin^2\vartheta + \varepsilon_2 \cos^2\vartheta)}$$
$$\times \left\{ \left(\varepsilon_1^2 - g^2 - \varepsilon_1\varepsilon_2\right) \sin^2\vartheta_+ 2\varepsilon_1\varepsilon_2 \pm \sqrt{\left(\varepsilon_1^2 - g^2 - \varepsilon_1\varepsilon_2\right)^2 \sin^4\vartheta + 4\varepsilon_2^2 g^2 \cos^2\vartheta} \right\}. \tag{3.26}$$

Thus, in magneto-active plasma in the case of arbitrary ϑ, there exist two branches of electromagnetic waves. The wave with the refractive index n_2 is called the ordinary wave, whereas the wave with the refractive index n_1 is called the extraordinary wave.

It can be easily shown that for the ordinary and extraordinary waves described by relation (3.26) the following relations hold between the field components E_x and E_y:

$$\frac{E_{y_{1,2}}}{E_{x_{1,2}}} = -\frac{ıg}{n_{1,2}^2 - \varepsilon_1}. \tag{3.27}$$

If $\vartheta = 0$, then Eq. (3.27) reduces to

$$\frac{E_{y_{1,2}}}{E_{x_{1,2}}} = \mp \imath.$$

Then, it follows that these waves possess circular polarization. So, the ordinary wave is left-handed, whereas the extraordinary wave is right-handed. The rotation angle of the polarization plane of the electromagnetic wave per unit length is equal to (see Sect. 1.7)

$$\psi = \frac{\omega}{2c}(n_1 - n_2) = \frac{\omega}{2c}\left(\sqrt{\varepsilon_1 + g} - \sqrt{\varepsilon_1 - g}\right).$$

When $\vartheta \neq 0$, the wave's polarization, according to Eq. (3.27), is elliptical.

In the case of strictly transverse propagation ($\vartheta = \pi/2$), the analysis of Eq. (3.22) is simple. It splits into two equations:

$$n_1^2 = \varepsilon_1 - \frac{g^2}{\varepsilon_1}, \qquad n_2^2 = \varepsilon_2. \tag{3.28}$$

Finally, let us consider the question when the waves are longitudinal in cold magneto-active plasma in the absence of spatial dispersion. We have already mentioned above that only waves propagating along the external magnetic field are strictly longitudinal. However, in the frequency range, when $n^2 \to \infty$, the phase velocity of the wave tends to zero ($k \to \infty$). In this case, spatial dispersion becomes essential. Such waves are potential, i.e., $\vec{E} = -\imath \vec{k}\phi$ and, hence, they are longitudinal. From Eq. (3.20), then, we obtain a condition for the existence of longitudinal waves in the case of cold plasma,

$$\frac{k_i k_j}{k^2}\varepsilon_{ij}\left(\omega, \vec{k}\right) = \varepsilon_1 \sin^2\vartheta + \varepsilon_2 \cos^2\vartheta = 0. \tag{3.29}$$

From Eq. (3.26), it follows that just under the condition given by Eq. (3.29) we have $n_{1,2}^2 \to \infty$. Using the expressions for ε_1, ε_2 [see Eq. (3.21)], we obtain the spectrum of the longitudinal wave

$$\omega^2 = \frac{\omega_{\text{pe}}^2 + \Omega_e^2}{2} \pm \frac{1}{2}\sqrt{\left(\omega_{\text{pe}}^2 + \Omega_e^2\right) - 4\omega_{\text{pe}}^2\Omega_e^2 \cos^2\vartheta}. \tag{3.30}$$

When $\vartheta = 0$, we have

$$\omega_1^2 = \omega_{pe}^2, \qquad \omega_2^2 = \Omega_e^2,$$

whereas when $\vartheta \to \pi/2$, we find

$$\omega_1^2 = \omega_{\text{pe}}^2 + \Omega_e^2, \qquad \omega_2^2 \to 0.$$

Let us now investigate the effects arising in magneto-active plasma when spatial dispersion is taken into account. We will restrict our study only to weak spatial dispersion, using the expansion of the dielectric permittivity in powers of k^2. Then, we obtain

$$\varepsilon_{ij}(\omega, k) = \varepsilon_{ij}(\omega, 0) + \frac{c^2k^2}{\omega^2}\eta_{ij}(\omega, \vartheta). \tag{3.31}$$

Tensor $\eta_{ij}(\omega, \vartheta)$, as $\varepsilon_{ij}(\omega, 0)$, in this approximation, is Hermitian and its components are

$$\eta_{11} = -\frac{\kappa T_e}{mc^2}\left\{\frac{3\omega_{\text{pe}}^2\omega^2\sin^2\vartheta}{\left(\omega^2 - \Omega_e^2\right)\left(\omega^2 - 4\Omega_e^2\right)} + \frac{\omega_{\text{pe}}^2\omega^2\left(\omega^2 + 3\Omega_e^2\right)\cos^2\vartheta}{\left(\omega^2 - \Omega_e^2\right)^3}\right\},$$

$$\eta_{12} = -\eta_{21} = \imath\frac{\kappa T_e}{mc^2}\left\{\frac{6\omega_{\text{pe}}^2\omega\Omega_e\sin^2\vartheta}{\left(\omega^2 - \Omega_e^2\right)\left(\omega^2 - 4\Omega_e^2\right)} + \frac{\omega_{\text{pe}}^2\omega\Omega_e\left(3\omega^2 + \Omega_e^2\right)\cos^2\vartheta}{\left(\omega^2 - \Omega_e^2\right)^3}\right\},$$

$$\eta_{13} = \eta_{31} = -2\frac{\kappa T_e}{mc^2}\frac{\omega_{\text{pe}}^2\omega^2\sin\vartheta\cos\vartheta}{\left(\omega^2 - \Omega_e^2\right)^2},$$

$$\eta_{22} = \eta_{11} + 2\frac{x T_e}{mc^2}\frac{\omega_{\text{pe}}^2\sin^2\vartheta}{\omega^2 - \Omega_e^2},$$

$$\eta_{23} = -\eta_{32} = \imath\frac{\kappa T_e}{mc^2}\frac{\omega_{\text{pe}}^2\Omega_e\left(3\omega^2 - \Omega_e^2\right)\sin\vartheta\cos\vartheta}{\omega\left(\omega^2 - \Omega_e^2\right)^2},$$

$$\eta_{33} = -\frac{\kappa Te}{mc^2}\left(3\frac{\omega_{\text{pe}}^2}{\omega^2}\cos^2\vartheta + \frac{\omega_{\text{pe}}^2\sin^2\vartheta}{\omega^2 - \Omega_e^2}\right). \tag{3.32}$$

Substituting Eq. (3.31) into Eq. (3.20), we obtain a cubic equation with respect to n^2.

$$\begin{aligned} &n^6\left[\eta_{11}\sin^2\vartheta + \eta_{33}\cos^2\vartheta + 2\eta_{13}\sin\vartheta\cos\vartheta\right] + n^4\left[\left(\varepsilon_1\sin^2\vartheta + \varepsilon_2\cos^2\vartheta\right)(1 - \eta_{22}) + \right.\\ &+ 2\imath g\left(\eta_{23}\sin\vartheta\cos\vartheta - \eta_{12}\sin^2\vartheta\right) - \eta_{33}\varepsilon_1\left(1 + \cos^2\vartheta\right) - \eta_{11}\left(\varepsilon_2 + \varepsilon_1\sin^2\vartheta\right) - \\ &\left.- 2\varepsilon_1\eta_{13}\sin\vartheta\cos\vartheta\right] - n^2\left[2\varepsilon_1\varepsilon_2 + \left(\varepsilon_1^2 - g^2 - \varepsilon_1\varepsilon_2\right)\sin^2\vartheta - \eta_{33}\left(\varepsilon_1^2 - g^2\right) - 2\imath\varepsilon_2 g\eta_{12} + \right.\\ &\left. + \varepsilon_1\varepsilon_2(\eta_{11} + \eta_{22})\right] + \varepsilon_2\left(\varepsilon_1^2 - g^2\right) = 0. \end{aligned} \tag{3.33}$$

The coefficients of the terms n^4 and n^2 in this equation are approximately equal to the coefficients of Eq. (3.22). They differ from the small terms of the order of

$\kappa T_e/mc^2$. Therefore, one can find the approximate solutions of Eq. (3.33). In this approximation, the first two solutions of Eq. (3.33) are different from solutions (3.26) in correction terms of the order of $\kappa T_e/mc^2$. However, a new solution also appears, which can be written as

$$n_3^2 = -\frac{\varepsilon_1 \sin^2\vartheta + \varepsilon_2 \cos^2\vartheta}{\eta_{11}\sin^2\vartheta + \eta_{33}\cos^2\vartheta + 2\eta_{13}\sin\vartheta\cos\vartheta}. \tag{3.34}$$

In the absence of the magnetic field, $B_0 \to 0$, this solution coincides with Eq. (2.54), which describes the longitudinal wave in isotropic non-magnetized plasma. By this reason, the solution (3.34) is called the plasma wave in magneto-active plasma. Moreover, in the limit of $T_e \to 0$, which corresponds to the case of cold plasma, when spatial dispersion can be neglected, Eq. (3.29) follows from Eq. (3.34). Thus, we see that consideration of spatial dispersion in both isotropic and magneto-active plasma results in the appearance of the longitudinal plasma waves instead of plasma oscillations taking place in the case of cold plasma. In fact, spatial dispersion can dramatically affect the properties of plasma waves. At the same time, the influence of spatial dispersion on the ordinary and extraordinary waves is very small, i.e., it is of the order of v_{Te}^2/c^2. But, it must be noted that this conclusion is correct only sufficiently far from resonance frequencies $\omega = \Omega_e$ and $\omega = 2\Omega_e$. If next terms in the expansion of the dielectric permittivity in powers of the wave vector are taken into account, the same situation will happen at the multiples of gyro-frequencies Ω_e, i.e., $\omega = s\Omega_e$. These terms correspondingly lead to the corrections of the order of $(v_{\mathrm{Te}}/c)^{2s}$. In the range of such resonance frequencies, the quantities $\eta_{ij}(\omega, \vartheta)$ in expressions (3.32) diverge, meaning that the expansion (3.31) becomes incorrect. Below, this question will be considered in particular.

Expressions (3.26) and (3.34) for the square of the refractive index, in specific conditions, can be negative or complex. However, since the dissipation processes were not considered in the derivation of Eqs. (3.21) and (3.31), the damping of corresponding waves is not related to energy absorption in plasma.

Now, let us discuss the problem of wave absorption in magneto-active collisionless plasma. As mentioned above, the dissipative processes were completely neglected in plasma and, as a result, the dielectric tensor (3.31) became Hermitian. Now, we will calculate the anti-Hermitian part of the dielectric permittivity. In the case of isotropic plasma, the damping decrements of waves given by Eqs. (2.35) and (1.39) are determined in terms of the imaginary parts of longitudinal and transverse permittivities, directly. However, for magneto-active plasma, it is more convenient to express the damping decrement in terms of the complex wave vector or complex refractive coefficient, which must be determined from the dispersion equation. Introducing $k = k'(\omega) + \imath k''(\omega)$ or $n = n'(\omega) + \imath n''(\omega)$, then for a weakly absorbing medium we can write [36]

$$\gamma = \frac{k''(\omega)}{\frac{dk'(\omega)}{d\omega}} = \frac{\omega n''(\omega)}{\frac{d}{d\omega}(\omega n'(\omega))}. \tag{3.35}$$

Thus, to calculate the damping decrement of waves apart from the refractive index $n'(\omega)$, we have to determine the imaginary part of the wave vector $k''(\omega)$ or absorption coefficient $n''(\omega)$.

First, we analyze the general dispersion equation (3.20) in the long-wave limit $\left(\vec{k} \to 0\right)$, when the thermal effects are weak. The dielectric tensor (3.31) turns out to be Hermitian and, consequently, plasma is non-absorbing. Consideration of the Cherenkov and cyclotron dissipative mechanisms due to the weak thermal motion of particles results in a small and anti-Hermitian correction of tensor (3.21), which, according to Eq. (3.19), for Maxwellian purely electron plasma is

$$\varepsilon_{ij}^{0a}(\omega, k) = \begin{pmatrix} \varepsilon_1^a & \imath g^a & 0 \\ -\imath g^a & \varepsilon_1^a & 0 \\ 0 & 0 & \varepsilon_2^a \end{pmatrix}, \tag{3.36}$$

where

$$\begin{aligned} \varepsilon_1^a &= \imath \sqrt{\frac{\pi}{8}} \frac{\omega_{\mathrm{pe}}^2}{\omega \mid k_z \mid v_{\mathrm{Te}}} \left(\exp\left[-\frac{(\omega - \Omega_e)^2}{2k_z^2 v_{\mathrm{Te}}^2} \right] + \exp\left[-\frac{(\omega + \Omega_e)^2}{2k_z^2 v_{\mathrm{Te}}^2} \right] \right), \\ g^a &= \imath \sqrt{\frac{\pi}{8}} \frac{\omega_{\mathrm{pe}}^2}{\omega \mid k_z \mid v_{\mathrm{Te}}} \left(\exp\left[-\frac{(\omega - \Omega_e)^2}{2k_z^2 v_{\mathrm{Te}}^2} \right] - \exp\left[-\frac{(\omega + \Omega_e)^2}{2k_z^2 v_{\mathrm{Te}}^2} \right] \right), \\ \varepsilon_2^a &= \imath \sqrt{\frac{\pi}{2}} \frac{\omega \omega_{\mathrm{pe}}^2}{\mid k_z^3 \mid v_{\mathrm{Te}}^3} \exp\left(-\frac{\omega^2}{2k_z^2 v_{\mathrm{Te}}^2} \right). \end{aligned} \tag{3.37}$$

Having derived these expressions, from Eqs. (3.19) and (3.31) we considered only the terms of $s = 0$ and $s = \pm 1$. Therefore, they are not correct for the description of wave absorption in the range of the higher cyclotron harmonics, $s = 2$, 3, 4, ... In order to describe the absorption when $\omega = 2\Omega_e$, one must take into account the terms of $s = \pm 2$ in the anti-Hermitian part of the dielectric tensor (3.37). All other corrections of the order of v_T^2/c^2 may be neglected. In this approximation, we obtain

$$\delta\varepsilon_{ij}^a = \begin{pmatrix} \delta\varepsilon_1^a & \imath \delta g^a & 0 \\ -\imath \delta g^a & \delta\varepsilon_1^a & 0 \\ 0 & 0 & 0 \end{pmatrix}, \tag{3.38}$$

where

$$\delta\varepsilon_1^a = \imath\sqrt{\frac{\pi}{8}}\frac{\omega_{\mathrm{pe}}^2 k_\perp^2 v_{\mathrm{Te}}}{\omega\Omega_e^2 \mid k_z \mid}\left(\exp\left[-\frac{(\omega-2\Omega_e)^2}{2k_z^2 v_{\mathrm{Te}}^2}\right]+\exp\left[-\frac{(\omega+2\Omega_e)^2}{2k_z^2 v_{\mathrm{Te}}^2}\right]\right),$$
$$\delta g^a = \imath\sqrt{\frac{\pi}{8}}\frac{\omega_{\mathrm{pe}}^2 k_\perp^2 v_{\mathrm{Te}}}{\omega\Omega_e^2 \mid k_z \mid}\left(\exp\left[-\frac{(\omega-\Omega_e)^2}{2k_z^2 v_{\mathrm{Te}}^2}\right]-\exp\left[-\frac{(\omega+\Omega_e)^2}{2k_z^2 v_{\mathrm{Te}}^2}\right]\right). \tag{3.39}$$

It should be noted that the elements of tensor (3.38) are proportional to $2\Omega_e - \omega$. On the other hand, the corrections caused by this tensor are significant only when $\omega = 2\Omega_e$. Therefore, these elements can be neglected.

Now, we can write the total dielectric permittivity of magneto-active plasma as a sum of expressions (3.31) (Hermitian part) and (3.36), (3.38) (anti-Hermitian part):

$$\varepsilon_{ij}\left(\omega,\vec{k}\right) = \varepsilon_{ij}(\omega,0) + \frac{c^2k^2}{\omega^2}\eta_{ij}(\omega,\vartheta) + \varepsilon_{ij}^{0a}\left(\omega,\vec{k}\right) + \delta\varepsilon_{ij}^a, \tag{3.40}$$

where $\varepsilon_{ij}(\omega,0)$ is given by Eq. (3.21), η_{ij} by Eq. (3.32), $\varepsilon_{ij}^{0a}\left(\omega,\vec{k}\right)$ by Eq. (3.37), and $\delta\varepsilon_{ij}^a$ by Eq. (3.39). It must be noted that Eq. (3.40) is valid only in the frequency range of weak spatial dispersion when

$$\frac{v_T^2\omega^2 s^2 \sin^2\vartheta}{c^2\Omega^2} \ll 1, \qquad \frac{c}{\omega}\frac{(\omega - s\Omega_e)}{v_T\cos\vartheta} \gg 1. \tag{3.41}$$

Moreover, this expression allows us to investigate only the zero, first, and second harmonics of cyclotron resonance ($s = 0, s = \pm 1, s = \pm 2$) and only harmonics sufficiently far from the resonances, when the second inequality of relation (3.41) is satisfied. The opposite limit will be considered below.[4]

Substituting Eq. (3.40) into dispersion equation (3.20) and considering that the anti-Hermitian part of the dielectric permittivity is small, we find

$$n_\alpha^2 \approx n_\alpha'^2 + 2\imath n_\alpha' n_\alpha''. \tag{3.42}$$

Here, $n_\alpha = n_\alpha' + \imath n_\alpha''$ ($\alpha = 1, 2, 3$) are the complex refractive coefficients for ordinary, extraordinary and longitudinal waves, respectively. The real parts of these quantities, in this approximation, are determined by Eqs. (3.26) and (3.34) without considering the anti-Hermitian components of the dielectric permittivity. The imaginary parts $n_\alpha''(\omega)$ representing the absorption coefficients of waves are defined by

[4]About electron cyclotron absorption of waves in magneto-active plasma, see [14, 30–32, 36, 38, 42, 45, 47, 49].

$$2n'_{1,2}n''_{1,2} = -n'^2_{1,2}\frac{\varepsilon''_1 \sin^2\vartheta + \varepsilon''_2 \cos^2\vartheta}{\varepsilon_1 \sin^2\vartheta + \varepsilon_2 \cos^2\vartheta} - \frac{1}{2(\varepsilon_1 \sin^2\vartheta + \varepsilon_2 \cos^2\vartheta)} \times$$

$$\times\Bigg\{ 2(\varepsilon_1\varepsilon''_2 + \varepsilon_2\varepsilon''_1) + (2\varepsilon_1\varepsilon''_1 - 2gg'' - \varepsilon_1\varepsilon''_2 - \varepsilon_2\varepsilon''_1)\sin^2\vartheta \pm$$

$$\left.\frac{(\varepsilon_1^2 - g^2 - \varepsilon_1\varepsilon_2)(2\varepsilon_1\varepsilon''_1 - 2gg'' - \varepsilon_1\varepsilon''_2 - \varepsilon_2\varepsilon'')\sin^2\vartheta + 4\varepsilon_2 g(\varepsilon_2 g'' + g\varepsilon''_2)\cos^2\vartheta}{\sqrt{(\varepsilon_1^2 - g^2 - \varepsilon_1\varepsilon_2)^2 \sin^4\vartheta + 4g^2\varepsilon_2^2\cos^2\vartheta}}\right\}, \tag{3.43}$$

$$2n'_3 n''_3 = \frac{\varepsilon''_1 \sin^2\vartheta + \varepsilon''_2 \cos^2\vartheta_2}{\eta_{11}\sin^2\vartheta + \eta_{33}\cos^2\vartheta + 2\eta_{13}\sin\vartheta\cos\vartheta}, \tag{3.44}$$

where

$$\imath\varepsilon''_1 = \varepsilon_1^a + \delta\varepsilon_1^a, \quad \imath\varepsilon''_2 = \delta\varepsilon_2^a, \qquad \imath g'' = g^a + \delta g^a.$$

In addition, it must also be noticed that $\varepsilon''_1, \varepsilon''_2$ and g'' in Eqs. (3.43) and (3.44) are functions of $n'(\omega)$; moreover, the anti-Hermitian part of the dielectric tensor $\varepsilon^a_{ij}\left(\omega, \vec{k}\right) = \varepsilon^{0a}_{ij}\left(\omega, \vec{k}\right) + \delta\varepsilon^a_{ij}$ is a function only of the real part of $\vec{k}$.

Using absorption coefficients given by Eqs. (3.43)–(3.45), one can calculate the damping decrements of waves. For longitudinal waves, the expression of damping decrement is relatively simple

$$\gamma_3(\vartheta) = \frac{\varepsilon''_1 \sin^2\vartheta + \varepsilon''_2 \cos^2\vartheta}{\frac{d}{d\omega}(\varepsilon_1 \sin^2\vartheta + \varepsilon_2 \cos^2\vartheta)}. \tag{3.45}$$

In this expression, all corrections stipulated by spatial dispersion, i.e., the terms of $\sim\eta_{ij}$ are neglected. Analogous expressions in the case of arbitrary propagation angle ϑ become unwieldy for the ordinary and extraordinary waves. However, they become relatively simple in the limits $\vartheta = 0$ and $\vartheta = \pi/2$. For example, when $\vartheta = 0$, and spatial dispersion is neglected, the refractive indexes of these waves are given by Eq. (3.25).

$$n'^2_{1,2} = \varepsilon_1(\omega) \pm g(\omega).$$

Then, for damping decrements, we obtain

$$2n'_{1,2}n''_{1,2} = \varepsilon''_1 \pm g'', \tag{3.46}$$

where, in accordance to Eq. (3.37), we have

$$\varepsilon_1'' = \frac{\omega_{\text{pe}}^2}{\omega k v_{\text{Te}}}\sqrt{\frac{\pi}{8}}\left[\exp\left\{-\frac{(\Omega_e-\omega)^2}{2k^2v_{\text{Te}}^2}\right\}+\exp\left\{-\frac{(\Omega_e+\omega)^2}{2k^2v_{\text{Te}}^2}\right\}\right],$$
$$g'' = \frac{\omega_{\text{pe}}^2}{\omega k v_{\text{Te}}}\sqrt{\frac{\pi}{8}}\left[\exp\left\{-\frac{(\Omega_e-\omega)^2}{2k^2v_{\text{Te}}^2}\right\}-\exp\left\{-\frac{(\Omega_e+\omega)^2}{2k^2v_{\text{Te}}^2}\right\}\right]. \quad (3.47)$$

Regarding the case of $\vartheta = \pi/2$, when the wave's refractive indices are given by Eq. (3.28), as follows from Eqs. (3.37) and (3.39), one can find that the damping decrements of electromagnetic waves in non-relativistic Maxwellian collisionless plasma are exactly equal to zero.

Finally, this fact should be noted that absorption becomes significant at frequencies $\omega = \Omega_e, 2\Omega_e$, as follows from Eqs. (3.37) and (3.47). Besides, when $\omega = \Omega_e$, absorption is basically determined by ε_1^a and g^a, whereas when $\omega = 2\Omega_e$, it is determined by $\delta\varepsilon_1^a$ and δg^a.

Let us now consider the problem of cyclotron wave's propagation and absorption in magneto-active collisionless plasma. As it was noted, the above consideration of wave's propagation is correct only in the frequency range sufficiently far from the cyclotron frequencies $n\,\Omega_e$. Now, we consider the cyclotron waves more precisely, only for the case of $\vartheta = 0$. In this special case, the cyclotron absorption is easily investigated because the resonance takes place only at the first cyclotron harmonic $\omega \simeq \Omega_e$. Dispersion equation (3.20) for the ordinary (left-handed polarized) and extraordinary (right-handed polarized) waves propagating along the external magnetic field looks as:

$$k^2c^2 - \omega^2(\varepsilon_{xx} \pm \imath\varepsilon_{xy}) = k^2c^2 - \omega^2\left[1 - \frac{\omega_{\text{pe}}^2}{\omega(\omega \mp \Omega_e)} I_+\left(\frac{\omega \mp \Omega_e}{k v_{T_e}}\right)\right] = 0. \quad (3.48)$$

It is sufficient to discuss one of these equations since the other one is obtained by the replacement $\Omega_e \rightleftharpoons -\Omega_e$.

We consider the cyclotron waves only near the electron cyclotron frequency $\omega \simeq \Omega_e$ (more exactly $|\omega - \Omega_e| \ll \Omega_e$). If $|\omega - \Omega_e| \gg k v_{\text{Te}}$, as it was assumed above, it means that the frequency ω lies outside the resonance absorption line, then Eq. (3.48) coincides with Eq. (3.25) and, therefore, for the complex refractive index $n = n' + \imath n''$, we obtain

$$n'^2 = 1 - \frac{\omega_{\text{pe}}^2}{\omega(\omega - \Omega_e)}, \qquad n'' = \sqrt{\frac{\pi}{8}}\frac{\omega_{\text{pe}}^2 c}{n'^2\omega^2 v_{\text{Te}}}\exp\left[-\frac{(\omega-\Omega_e)^2c^2}{2n'^2\omega^2 v_{T_e}^2}\right]. \quad (3.49)$$

Consequently, the absorption of electron waves is exponentially small in collisionless plasma outside the resonance line, when $|\omega - \Omega_e|^3 \gg v_{\text{Te}}^2\omega_{\text{pe}}^2\Omega_e/c^2$. When the frequency ω approaches the cyclotron frequency Ω_e, the absorption grows. In this case, inside the absorption line, when $|\omega - \Omega_e|^3 \ll v_{\text{Te}}^2\omega_{\text{pe}}^2\Omega_e/c^2$, the wave becomes strongly damped

$$n = n' + \imath n'' = \frac{\imath + \sqrt{3}}{2} \left(\sqrt{\frac{\pi}{2}} \frac{\omega_{\mathrm{pe}}^2 c}{\omega^2 v_{\mathrm{Te}}} \right)^{1/3}. \tag{3.50}$$

From Eq. (3.50), it follows that the absorption length of cyclotron waves inside the resonance line is of the order of $\simeq c/\omega n'' \approx \left(c^2 v_{\mathrm{Te}}/\omega\omega_{\mathrm{pe}}^2\right)^{1/3}$. Comparing this quantity with the depth of the anomalous skin-effect in isotropic collisionless plasma (see Sect. 2.9), we see that they are equal. This result is physically demonstrative. Since the electrons rotate with the Larmor frequency, they can be regarded as oscillators with the natural frequency Ω_e and produce a field of the same frequency. Consequently, all the peculiarities of isotropic plasma at the frequency ω must appear in magneto-active plasma at combination frequencies $\omega \mp \Omega_e$. In particular, the anomalous skin-effect of the transverse electromagnetic field which occurs in isotropic plasma for $|\,\omega\,| \ll kv_{\mathrm{Te}}$ is shifted to the range $|\,\omega - \Omega_e\,| \ll kv_{\mathrm{Te}}$.

In conclusion, we will consider the low-frequency waves in pure electron magneto-active plasma. The frequency of these waves is much less than the cyclotron frequency, $\omega \ll \Omega_e$, but much higher than the ion frequencies, which allows us to neglect the ion contributions to the dielectric permittivity of plasma. The simplest case is longitudinal propagation, when $\vartheta = 0$ and, in general, Eq. (3.25) is valid. We rewrite this equation in the low-frequency range in the following form:

$$k^2 = \frac{\omega^2}{c^2} \left(1 \pm \frac{\omega_{\mathrm{pe}}^2}{\omega\,\Omega_e} \right). \tag{3.51}$$

We see that if $\omega \ll \omega_{\mathrm{pe}}^2/\Omega_e$, then only one branch of these oscillations can exist in plasma, i.e., only one branch with $k^2 > 0$ (or $n^2 > 0$). The frequency spectrum of these circularly polarized waves is quadratic with respect to the wave vector

$$\omega = \frac{k^2 c^2 \Omega_e}{\omega_{\mathrm{pe}}^2}, \tag{3.52}$$

and the rotation of electric field's polarization coincides with the electron Larmor rotation. If $\vartheta \neq 0$, from Eq. (3.26) we obtain

$$\omega = \frac{k^2 c^2 \Omega_e}{\omega_{\mathrm{pe}}^2} \,|\, \cos\vartheta \,|\,, \tag{3.53}$$

which is the generalization of Eq. (3.52). Moreover, it is easy to consider the dissipation connected to the imaginary part of g. This leads to the following expression for the damping decrement:

$$\gamma = -g'' \frac{\omega^2 \Omega_e}{\omega_{\text{pe}}^2} = -\sqrt{\frac{\pi}{2}} \frac{kc^2 \Omega_e}{\omega_{\text{pe}}^2 v_{\text{Te}}} \exp\left[-\frac{(\omega - \Omega_e)^2}{2k_z^2 v_{\text{Te}}^2} \right]. \tag{3.54}$$

The oscillations with spectrum (3.53) and damping decrement (3.54) are known as Helicons (whistlers) or spiral waves.

3.3 Relativistic Electron Plasma in the Magnetic Field

In Sect. 2.4, it was noted that the relativistic consideration of wave's propagation in isotropic plasma is necessary when the temperature is sufficiently high and, as a result, the thermal velocity of particles is of the order of the light velocity, or when the phase velocity of waves is comparable to the light velocity. In magneto-active plasma, the relativistic effects are essential near the cyclotron frequencies when $\omega \simeq n\Omega_e$. As it was shown in the previous section, in this range, the cyclotron absorption is significant in the non-relativistic case, but for the correct description, the relativistic effect must be considered.

To calculate the dielectric tensor of magneto-active relativistic electron plasma, we will use general expression (3.14) with the equilibrium distribution function (2.65)

$$f_{0e}(p) = \frac{N_e}{4\pi (mc)^3} \frac{mc^2}{\kappa T_e} K_2^{-1}\left(\frac{mc^2}{\kappa T_e}\right) \exp\left(-\frac{\sqrt{c^2p^2 + m^2c^4}}{\kappa T_e}\right). \tag{3.55}$$

Substituting this expression into Eq. (3.14), after simple calculations, we obtain the following convenient form of the dielectric tensor [1, 2, 6]:

$$\varepsilon_{ij}\left(\omega, \vec{k}\right) = \delta_{ij} + \imath \frac{\omega_{\text{pe}}^2}{\omega} \left(\frac{mc^2}{\kappa T_e}\right)^2 K_2^{-1}\left(\frac{mc^2}{\kappa T_e}\right) \int_0^\infty d\tau \left\{ \frac{K_2\left(\sqrt{R}\right)}{R} T_{ij}^{(1)} - \frac{K_3\left(\sqrt{R}\right)}{R^{3/2}} T_{ij}^{(2)} \right\}, \tag{3.56}$$

where

$$R = \left(\frac{mc^2}{\kappa T_e} - \imath\omega\tau\right)^2 + 2\frac{c^2 k_\perp^2}{\Omega_e^2}(1 - \cos\Omega_e\tau) + c^2 k_z^2 \tau^2,$$

$$T_{ij}^{(1)} = \begin{pmatrix} \cos\Omega_e\tau & \sin\Omega_e\tau & 0 \\ -\sin\Omega_e\tau & \cos\Omega_e\tau & 0 \\ 0 & 0 & 1 \end{pmatrix},$$

$$T_{ij}^{(2)} = \frac{c^2}{\Omega_e^2}\begin{pmatrix} k_\perp^2 \sin^2\Omega_e\tau & k_\perp^2 \sin\Omega_e\tau(1-\cos\Omega_e\tau) & k_\perp k_z\Omega_e\tau\sin\Omega_e\tau \\ -k_\perp^2 \sin\Omega_e\tau(1-\cos\Omega_e\tau) & -k_\perp^2(1-\cos\Omega_e\tau)^2 & -k_\perp k_z\Omega_e\tau(1-\cos\Omega_e\tau) \\ k_\perp k_z\Omega_e\tau\sin\Omega_e\tau & k_\perp k_z\Omega_e\tau(1-\cos\Omega_e\tau) & k_z^2\Omega_e^2\tau^2 \end{pmatrix}.$$

Here, $K_n(x)$ is the MacDonald function of the order of n. In the non-relativistic limit, when $mc^2 \gg \kappa T_e$, it must be considered that for $|x| \gg 1$,

$$K_n(x) \simeq \sqrt{\frac{\pi}{2x}}\exp(-x).$$

Then, from Eq. (3.56), we obtain Eq. (3.19). In the absence of the external magnetic field, $\vec{B}_0 = 0$, from Eq. (3.56), the dielectric tensor of isotropic electron plasma is obtained

$$\varepsilon^{\rm tr}(\omega,k) = 1 + \imath\frac{\omega_{\rm pe}^2}{\omega}\left(\frac{mc^2}{\kappa T_e}\right)^2 K_2^{-1}\left(\frac{mc^2}{\kappa T_e}\right)\int_0^\infty d\tau\frac{K_2(\sqrt{R_0})}{R_0},$$

$$\varepsilon^l(\omega,k) = 1 + \imath\frac{\omega_{\rm pe}^2}{\omega}\left(\frac{mc^2}{\kappa T_e}\right)^2 K_2^{-1}\left(\frac{mc^2}{\kappa T_e}\right)\int_0^\infty d\tau\left[\frac{K_2(\sqrt{R_0})}{R_0} - c^2k^2\tau^2\frac{K_3(\sqrt{R_0})}{R_0^{3/2}}\right], \tag{3.57}$$

where

$$R_0 = \left(\frac{mc^2}{\kappa T_e} - \imath\omega\tau\right)^2 + c^2k^2\tau^2.$$

Using the integral representation

$$\frac{K_1(\sqrt{s^2+r^2})}{\sqrt{s^2+r^2}} = \frac{1}{4\pi}\int\frac{d\vec{p}}{\sqrt{1+p^2}}\exp\left(-s\sqrt{1+p^2} - \imath\vec{r}\cdot\vec{p}\right),$$

and its derivatives, one can reduce Eq. (3.57) to Eqs. (2.69) and (2.70).

For the analysis of electromagnetic wave's propagation in relativistic magnetoactive plasma, we begin from the long-wave limit $\left(\vec{k} \to 0\right)$ when spatial dispersion can be neglected. Then, from Eq. (3.56) it follows that

$$\varepsilon_{ij}(\omega,0) = \begin{pmatrix} \varepsilon_1 & \imath g & 0 \\ -\imath g & \varepsilon_1 & 0 \\ 0 & 0 & \varepsilon_2 \end{pmatrix}, \tag{3.58}$$

where

$$\varepsilon_1 = 1 - \frac{\omega_{\text{pe}}^2}{\omega^2}\frac{mc^2}{\kappa T_e}K_2^{-1}\left(\frac{mc^2}{\kappa T_e}\right)\int_1^\infty \frac{dz}{z^2}K_2\left(\frac{mc^2}{\kappa T_e}z\right)\cosh\left[\frac{mc^2}{\kappa T_e}\frac{\Omega_e}{\omega}(z-1)\right],$$

$$\varepsilon_2 = 1 - \frac{\omega_{\text{pe}}^2}{\omega^2}\frac{mc^2}{\kappa T_e}K_2^{-1}\left(\frac{mc^2}{\kappa T_e}\right)\int_1^\infty \frac{dz}{z^2}K_2\left(\frac{mc^2}{\kappa T_e}z\right),$$

$$g = \frac{\imath\omega_{\text{pe}}^2}{\omega^2}\frac{mc^2}{\kappa T_e}K_2^{-1}\left(\frac{mc^2}{\kappa T_e}\right)\int_1^\infty \frac{dz}{z^2}K_2\left(\frac{mc^2}{\kappa T_e}z\right)\sinh\left[\frac{mc^2}{\kappa T_e}\frac{\Omega_e}{\omega}(z-1)\right].$$

In the non-relativistic limit ($mc^2 \gg \kappa T_e$), these expressions coincide with Eq. (3.21).

The structure of the dielectric tensor (3.58) is the same as that of Eq. (3.21). Therefore, the dispersion equation obtained from its substitution in Eq. (3.20) is similar to Eq. (3.26).[5] All of the above results formally hold in this case as well. But, the essential difference is in the frequency dependency of ε_1, ε_2, and g and consequently the refractive indices of the waves. First, it must be noted that ε_1 and g remain finite when $\omega = \Omega_e$, whereas in the non-relativistic plasma they grow unlimitedly. According to Eq. (3.58), ε_1 and g have some peaks at $\omega = \Omega_e$, but when the electron's temperature, these peaks become flatter and they shift to the lower frequencies. In the limit $\omega \gg \Omega_e$, expression (3.58) coincides with the expression of the dielectric permittivity of isotropic relativistic plasma in the absence of spatial dispersion. The other difference is more principal. It regards to the anti-Hermitian part of the dielectric tensor, which is related to the absorption of electromagnetic waves caused by non-collisional dissipative processes. In non-relativistic collisionless plasma, as we were convinced by the results of the previous section, the anti-Hermitian part of the dielectric tensor is exactly zero in the absence of spatial dispersion when $\vec{k} \to 0$. When the relativistic effects are taken into account, the anti-Hermitian part of $\varepsilon_{ij}(\omega,0)$ becomes non-zero. Actually, from Eq. (3.15) for the anti-Hermitian part of the dielectric tensor, in the limit of $\vec{k} \to 0$, we obtain

[5] About the oscillations of magneto-active relativistic plasma, one can see [1, 35–39, 50–57].

$$\varepsilon^a_{ij}(\omega,0)=\begin{pmatrix}\varepsilon^a_1 & \imath g^a & 0\\ -\imath g^a & \varepsilon^a_1 & 0\\ 0 & 0 & 0\end{pmatrix}, \tag{3.59}$$

where

$$\varepsilon^a_1=g^a=\begin{cases}\imath\pi\dfrac{\omega^2_{\text{pe}}\Omega^3_e}{6\omega^5}\left(\dfrac{mc^2}{\kappa T_e}\right)^2 K_2^{-1}\left(\dfrac{mc^2}{\kappa T_e}\right)\left(1-\dfrac{\omega^2}{\Omega^2_e}\right)^{3/2}\exp\left(-\dfrac{mc^2}{\kappa T_e}\dfrac{\Omega_e}{\omega}\right), & \omega<\Omega_e,\\ 0, & \omega>\Omega_e.\end{cases} \tag{3.60}$$

In the non-relativistic limit, when $mc^2 \gg \kappa T_e$, from Eq. (3.60) and the asymptote of the McDonald function at large argument it follows that

$$\varepsilon^a_1=g^a=\begin{cases}\imath\sqrt{\dfrac{\pi}{2}}\dfrac{\omega^2_{\text{pe}}\Omega^3_e}{3\omega^5}\left(\dfrac{mc^2}{\kappa T_e}\right)^{5/2}\left(1-\dfrac{\omega^2}{\Omega^2_e}\right)^{3/2}\exp\left[\left(1-\dfrac{\Omega_e}{\omega}\right)\dfrac{mc^2}{\kappa T_e}\right], & \omega<\Omega_e,\\ 0, & \omega>\Omega_e.\end{cases} \tag{3.61}$$

Formally, when $c\to\infty$, these quantities tend to zero, which is in agreement with the results of Sect. 3.2, showing the absence of absorption in non-relativistic plasma when $k=0$. But really, even in non-relativistic plasma, quantities (3.61) result in the significant absorption of cyclotron waves.

In the ultra-relativistic limit, when $\kappa T_e \gg mc^2$, from Eq. (3.60), we obtain

$$\varepsilon^a_1=g^a=\begin{cases}\imath\pi\dfrac{\omega^2_{\text{pe}}\Omega^3_e}{12\omega^5}\left(\dfrac{mc^2}{\kappa T_e}\right)^4\left(1-\dfrac{\omega^2}{\Omega^2_e}\right)^{3/2}\exp\left(-\dfrac{mc^2}{\kappa T_e}\dfrac{\Omega_e}{\omega}\right), & \omega<\Omega_e,\\ 0, & \omega>\Omega_e.\end{cases} \tag{3.62}$$

Finally, based on Eq. (3.60), when $\vec{B}_0\to 0$, absorption of the wave with $\vec{k}=0$ in relativistic plasma tends to zero, which corresponds to the results of Sect. 2.4.

Let us now write the concrete formulas for the refractive indices and absorption coefficients for relativistic magneto-active plasma when $\vec{k}\to 0$ and spatial dispersion is neglected. Such relations become simple for longitudinal ($\vartheta=0$) and transverse ($\vartheta=\pi/2$) propagation of waves. For longitudinal propagation, according to Eqs. (3.24), (3.25), and (3.43), the extraordinary electromagnetic waves are undamped. Furthermore, the ordinary waves are damped and for them

$$n'_1 n''_1=\varepsilon''_1=-\imath\varepsilon^a_1, \qquad n''_2=0. \tag{3.63}$$

If $\vartheta=\pi/2$, then, in view of Eq. (3.43) [and also Eq. (3.28)], we have

$$2n_1'n_1'' = \varepsilon_1''\left(1+\frac{g}{\varepsilon_1}\right)^2, \qquad n_2''=0, \tag{3.64}$$

where

$$\imath\varepsilon_1'' = \varepsilon_1^a = g^a.$$

In addition, it must be noted that, in the case of $\vartheta = \pi/2$, the longitudinal waves are also damped, which is easily seen from Eq. (3.45) being valid for arbitrary ϑ. From this equation, it follows the damping decrement

$$\gamma_3(\vartheta) = \frac{\varepsilon_1''\sin^2\vartheta}{\frac{\partial}{\partial\omega}\left[\varepsilon_1'(\omega)\sin^2\vartheta + \varepsilon_2'(\omega)\cos^2\vartheta\right]}, \tag{3.65}$$

where $\varepsilon_1'(\omega)$ and $\varepsilon_2'(\omega)$ are given by the real parts of Eq. (3.21).

In conclusion, let us consider wave's propagation in magneto-active relativistic plasma by considering the effect of spatial dispersion. This problem needs unwieldy mathematical calculations. Relatively simple case is the limit of weak spatial dispersion when dielectric tensor expansion in powers of k^2 is possible. In this case, in the high-frequency range, when $\omega \gg \Omega_e, \omega_{pe}$, the wave's refractive index in plasma is close to unity and the dielectric tensor slightly differs from the unity tensor δ_{ij}:

In the limit of weak spatial dispersion, we can expand the dielectric tensor (3.56) in power of k^2. As a result, we obtain [compare to Eq. (3.31)]:

$$\varepsilon_{ij}\left(\omega,\vec{k}\right) = \varepsilon_{ij}(\omega,0) + \frac{c^2k^2}{\omega^2}\eta_{ij}(\omega,\vartheta), \tag{3.66}$$

where tensor $\eta_{ij}(\omega,\vartheta)$ is Hermitian and its components are

$$\eta_{11}=\frac{\omega_{pe}^2}{\Omega_e^2}K_2^{-1}\left(\frac{mc^2}{\kappa T_e}\right)\int_1^\infty\frac{dz}{z^3}K_3\left(\frac{mc^2}{\kappa T_e}z\right)\left\{\left[\sin^2\vartheta\left(1-\cosh\left[\frac{mc^2}{\kappa T_e}\frac{\Omega_e}{\omega}(z-1)\right]\right)-\right.\right.$$

$$\left.\left.-\frac{1}{2}\cos^2\vartheta\frac{\Omega_e^2}{\omega^2}\left(\frac{mc^2}{\kappa T_e}\right)^2(z-1)^2\right]\cosh\left[\frac{mc^2}{\kappa T_e}\frac{\Omega_e}{\omega}(z-1)\right]-\sin^2\vartheta\sinh^2\left[\frac{mc^2}{\kappa T_e}\frac{\Omega_e}{\omega}(z-1)\right]\right\};$$

$$\eta_{12}=-\eta_{21}=\imath\frac{\omega_{pe}^2}{\Omega_e^2}K_2^{-1}\left(\frac{mc^2}{\kappa T_e}\right)\int_1^\infty\frac{dz}{z^3}K_3\left(\frac{mc^2}{\kappa T_e}z\right)\left\{\left[\sin^2\vartheta\left(1-\cosh\left[\frac{mc^2}{\kappa T_e}\frac{\Omega_e}{\omega}(z-1)\right]\right)-\right.\right.$$

$$\left.-\frac{1}{2}\cos^2\vartheta\frac{\Omega_e^2}{\omega^2}\left(\frac{mc^2}{\kappa T_e}\right)^2(z-1)^2\right]\sinh\left[\frac{mc^2}{\kappa T_e}\frac{\Omega_e}{\omega}(z-1)\right]+$$

$$\left.+\sin^2\vartheta\sinh\left[\frac{mc^2}{\kappa T_e}\frac{\Omega_e}{\omega}(z-1)\right]\cosh\left[\frac{mc^2}{\kappa T_e}\frac{\Omega_e}{\omega}(z-1)\right]\right\};$$

$$\eta_{13}=\eta_{31}=-\frac{\omega_{pe}^2}{2\Omega_e\omega}K_2^{-1}\left(\frac{mc^2}{\kappa T_e}\right)\int\limits_1^\infty\frac{dz}{z^3}K_3\left(\frac{mc^2}{\kappa T_e}z\right)\sin2\vartheta\times\frac{mc^2}{\kappa T_e}(z-1)\sinh\left[\frac{mc^2}{\kappa T_e}\frac{\Omega_e}{\omega}(z-1)\right];$$

$$\eta_{22}=\eta_{11}-2\frac{\omega_{pe}^2}{\Omega_e^2}K_2^{-1}\left(\frac{mc^2}{\kappa T_e}\right)\int\limits_1^\infty\frac{dz}{z^3}K_3\left(\frac{mc^2}{\kappa T_e}z\right)\sin^2\vartheta\left(1-\cosh\left[\frac{mc^2}{\kappa T_e}\frac{\Omega_e}{\omega}(z-1)\right]\right);$$

$$\eta_{23}=-\eta_{32}=-\imath\frac{\omega_{pe}^2}{2\Omega_e\omega}K_2^{-1}\left(\frac{mc^2}{\kappa T_e}\right)\int\limits_1^\infty\frac{dz}{z^3}K_3\left(\frac{mc^2}{\kappa T_e}z\right)\sin2\vartheta\times$$

$$\times\frac{mc^2}{\kappa T_e}(z-1)\left(1-\cosh\left[\frac{mc^2}{\kappa T_e}\frac{\Omega_e}{\omega}(z-1)\right]\right);$$

$$\eta_{33}=\frac{\omega_{pe}^2}{\Omega_e^2}K_2^{-1}\left(\frac{mc^2}{\kappa T_e}\right)\int\limits_1^\infty\frac{dz}{z^3}K_3\left(\frac{mc^2}{\kappa T_e}z\right)\times$$

$$\times\left\{\sin^2\vartheta\left(1-\cosh\left[\frac{mc^2}{\kappa T_e}\frac{\Omega_e}{\omega}(z-1)\right]\right)-\frac{3}{2}\cos^2\vartheta\frac{\Omega_e^2}{\omega^2}\left(\frac{mc^2}{\kappa T_e}\right)^2(z-1)^2\right\}. \tag{3.67}$$

These expressions are similar to Eq. (3.32) and, therefore, all results given by Eqs. (3.33) and (3.34) formally remain unchanged in the considered case as well. However, in relativistic plasma a new qualitative difference from non-relativistic plasma appears. This difference is based on the dependency of tensor $\eta_{ij}(\omega,\vartheta)$ on frequency. As it was mentioned above, the components of dielectric tensor $\varepsilon_{ij}(\omega,0)$ and tensor $\eta_{ij}(\omega,\vartheta)$ are finite at $\omega=\Omega_e$ and $\omega=2\Omega_e$. However, it should be noted that, in this range of frequencies, the components of tensor $\eta_{ij}(\omega,\vartheta)$ at $\omega\approx\Omega_e$ and $\omega\approx 2\Omega_e$have maximum values and when the electron's temperature grows, these maximums become flatter and they shift to the lower frequencies.

In the limit of ultra-relativistic plasma and the relatively weak magnetic field, when $\Omega_e\ll\omega\kappa T_e/mc^2$, by using the formula

$$K_n(x)\approx\frac{(n-1)!}{2}\left(\frac{2}{x}\right)^n,\qquad(|x|\ll 1),$$

from Eqs. (3.66) and (3.67), with accuracy up to terms of the order of $mc^2/\kappa T_e$, we obtain

$$\varepsilon_{ij}\left(\omega,\vec{k}\right)=\left(1-\frac{1}{3}\frac{\omega_{\rm pe}^2}{\omega^2}\frac{mc^2}{\kappa T_e}\right)\delta_{ij}-\frac{1}{15}\frac{\omega_{\rm pe}^2}{\omega^4}\frac{mc^2}{\kappa T_e}c^2k^2\delta_{ij}-\frac{2}{15}\frac{\omega_{\rm pe}^2}{\omega^4}\frac{mc^2}{\kappa T_e}c^2k_ik_j. \tag{3.68}$$

This expression does not depend on the magnetic field and, as it may be expected, coincides with the dielectric tensor of isotropic ultra-relativistic plasma in the frequency range of weak spatial dispersion (see Sect. 2.4).

The other case which is also relatively simple for the analysis of wave's propagation in relativistic magneto-active plasma is the case of very high-frequency range when $\omega \gg \Omega_e$.[6] In this case, $\varepsilon_{ij}\left(\omega, \vec{k}\right)$ can be represented as

$$\varepsilon_{ij}\left(\omega, \vec{k}\right) = \delta_{ij} + \varepsilon_{ij}^{(1)}\left(\omega, \vec{k}\right),$$

where $\varepsilon_{ij}^{(1)}\left(\omega, \vec{k}\right) \ll 1$. Substituting this expression into dispersion equation (3.20) and considering that the components of $\varepsilon_{ij}^{(1)}\left(\omega, \vec{k}\right)$ are small, we obtain the complex refractive indices of ordinary and extraordinary waves [1, 2, 58]

$$n_{1,2} = 1 + \frac{1}{4}\left\{(E_1 + E_2) \pm \sqrt{(E_1 - E_2)^2 - 4E_3^2}\right\}, \tag{3.69}$$

where for $\alpha = 1, 2, 3$,

$$E_\alpha = \imath \frac{\omega_{\text{pe}}^2}{\omega}\left(\frac{mc^2}{\kappa T_e}\right)^2 K_2^{-1}\left(\frac{mc^2}{\kappa T_e}\right)\int_0^\infty d\tau\left\{\frac{K_2\left(\sqrt{R}\right)}{R}\psi_\alpha^{(1)} - \frac{K_3\left(\sqrt{R}\right)}{R^{3/2}}\psi_\alpha^{(2)}\right\}. \tag{3.70}$$

Here, we introduce the following notation:

$$\psi_1^{(1)} = \cos^2\vartheta\cos\Omega_e\tau + \sin^2\vartheta, \quad \psi_2^{(1)} = \cos\Omega_e\tau, \quad \psi_3^{(1)} = \cos\vartheta\sin\Omega_e\tau;$$

$$\psi_1^{(2)} = -\left(\frac{\omega}{\Omega_e}\right)^2 \sin^2\vartheta\cos^2\vartheta(\Omega_e\tau - \sin\Omega_e\tau)^2, \quad \psi_2^{(2)} = -\left(\frac{\omega}{\Omega_e}\right)^2 \sin^2\vartheta(1 - \cos\Omega_e\tau),$$

$$\psi_3^{(2)} = -\left(\frac{\omega}{\Omega_e}\right)^2 \cos\vartheta\sin^2\vartheta(1 - \cos\Omega_e\tau)(\Omega_e\tau - \sin\Omega_e\tau),$$

$$R = \left(\frac{mc^2}{\kappa T_e} - \imath\omega\tau\right)^2 + 2\frac{\omega^2}{\Omega_e^2}(1 - \cos\Omega_e\tau) + \omega^2\tau^2.$$

In deriving the above expressions, we have taken into account that in the considered approximation $n_{1,2}^2 \simeq 1$ (or $k^2 \simeq \omega^2/c^2$).

From Eq. (3.69), one can easily calculate the absorption coefficients of waves. In the case of weakly absorbing plasma, when $\text{Im}E_\alpha \ll \text{Re}\, E_\alpha$, they are

[6] For more details see [1, 2, 13, 14, 54, 58].

$$Im\, n_{1,2} = n''_{1,2} = \frac{1}{4}\left\{ Im\,(E_1 + E_2) \pm \frac{\text{Re}\,(E_1 - E_2) Im\,(E_1 - E_2) - 4(\text{Re}\,E_3)(Im\,E_3)}{\sqrt{(\text{Re}\,E_1 - \text{Re}\,E_2)^2 - 4(\text{Re}\,E_3)^2}} \right\}. \tag{3.71}$$

For an arbitrary angleϑ, expression (3.71) will be very complicated. However, for $\vartheta = \pi/2$, expressions (3.69)–(3.71) are substantially simplified. Since, in this case $E_z = 0$, then

$$\begin{cases} n_1 = 1 + \dfrac{E_1}{2}, & n_2 = 1 + \dfrac{E_2}{2}, \\ n''_1 = Imn_1 = \dfrac{1}{2} ImE_1, & n''_2 = Imn_2 = \dfrac{1}{2} ImE_2. \end{cases} \tag{3.72}$$

Simple expression is obtained from Eq. (3.71) for the sum of the absorption coefficients of both electromagnetic waves:

$$n''_1 + n''_2 = Im(n_1 + n_2) = \frac{1}{2} Im(E_1 + E_2) =$$
$$= \frac{\omega_{\text{pe}}^2}{\omega}\left(\frac{mc^2}{\kappa T_e}\right)^2 K_2^{-1}\left(\frac{mc^2}{\kappa T_e}\right) \text{Re} \int_0^\infty d\tau \left\{ \frac{K_2(\sqrt{R})}{R}\left(\psi_1^{(1)} + \psi_2^{(1)}\right) - \frac{K_3(\sqrt{R})}{R^{3/2}}\left(\psi_1^{(2)} + \psi_2^{(2)}\right) \right\}.$$

For plasma with non-relativistic and weakly relativistic temperatures ($mc^2 \gg \kappa T_e$), an approximate calculation of the integral in the latter equation results in [59]

$$n''_1 + n''_2 = Im\,n_1 + Im\,n_2 \approx -\frac{1}{2}\frac{\omega_{\text{pe}}^2}{\omega \Omega_e}\sqrt{\pi}\left(\frac{mc^2}{\kappa T_e}\right)^{5/2} \frac{1}{R_0}\sqrt{\frac{R_0}{R''(\xi_0)}} \exp\left(\frac{mc^2}{\kappa T_e} - \sqrt{R_0}\right),$$

where

$$R_0 = \left[\frac{\omega}{\Omega_e} \sin\vartheta(\cos\xi_0 - 1)\right]^2 - \left(\frac{mc^2}{\kappa T_e}\right)^2 coth^2\vartheta \gg 1,$$

and ξ_0 is determined from

$$\xi_0 - \sin\xi_0 \approx \imath \frac{mc^2}{\kappa T_e}\frac{\Omega_e}{\omega \sin\vartheta}.$$

3.4 Electron-Ion Plasma in the External Magnetic Field

In previous sections, we considered electromagnetic wave's propagation in pure electron magneto-active plasma. In other words, the ion's motion was completely neglected by assuming that ions are infinitely heavy. Indeed, ions are much heavier than electrons. However, in many cases such an assumption is incorrect. For example, as it was shown in the previous chapter (see Sects. 2.5 and 2.6), if plasma is non-isothermal and the electron's temperature is much higher than ion's temperature $T_e \gg T_i$, the low-frequency slow longitudinal electromagnetic waves called the ion-acoustic waves appear. Below, we will clear up the conditions under which the consideration of ion's motion becomes necessary in magneto-active plasma and will consider the effects stipulated by the ion's motion in low-frequency electromagnetic wave's propagation. For simplicity, we will restrict our study only on non-relativistic Maxwellian ion-electron plasma consisting of electrons and only one type of ions. Then, the equilibrium distribution functions of particles are given by Eqs. (2.80) and (3.18) and, therefore, the dielectric tensor is defined by Eq. (3.19).[7] Finally, we suppose that equilibrium plasma is neutral, or

$$eN_e + e_i N_i = 0.$$

Here, we shall indicate that, in the limit of $B_0 \to 0$, this dielectric permittivity coincides with the dielectric permittivity of isotropic electron-ion plasma given by Eqs. (2.81) and (2.82). The analysis of the wave's propagation in electron-ion plasma as usual is begun from the long-wave limit, when $\vec{k} \to 0$, meaning that spatial dispersion is neglected. Then, from Eq. (3.19) we obtain [3, 4, 7]

$$\varepsilon_{ij} = \begin{pmatrix} \varepsilon_1 & \imath g & 0 \\ -\imath g & \varepsilon_1 & 0 \\ 0 & 0 & \varepsilon_2 \end{pmatrix}, \tag{3.73}$$

where

$$\varepsilon_1 = 1 - \frac{\omega_{\mathrm{pe}}^2}{\omega^2 - \Omega_e^2} - \frac{\omega_{\mathrm{pi}}^2}{\omega^2 - \Omega_i^2},$$

$$g = -\frac{\omega_{\mathrm{pe}}^2 \Omega_e}{\omega\left(\omega^2 - \Omega_e^2\right)} - \frac{\omega_{\mathrm{pi}}^2 \Omega_i}{\omega\left(\omega^2 - \Omega_i^2\right)},$$

---

[7] For more details about the derivation of the dielectric tensor, one can see [4, 6]; propagation of electromagnetic waves in electron-ion plasma was investigated in [3, 4, 7–9, 42, 45, 60–72].

$$\varepsilon_2 = 1 - \frac{\omega_{\rm pe}^2}{\omega^2} - \frac{\omega_{\rm pi}^2}{\omega^2}\,. \tag{3.74}$$

Expressions (3.73) and (3.74) can be obtained from expressions (3.12) and (3.13) by making use of the Maxwell distribution (2.80) for electrons and ions. Furthermore, this tensor looks as tensor (3.21). However, it considers not only the electron but also the ion contributions. This is the very reason of difference arising in the low-frequency range of wave's propagation. The difference becomes much more significant when $\omega \le \Omega_i$. For example, if $\omega \ll \Omega_i$, from Eq. (3.74), we obtain

$$\varepsilon_1 = 1 + \frac{\omega_{\rm pi}^2}{\Omega_i^2} = 1 + \frac{c^2}{v_A^2}, \qquad g = 0, \qquad \varepsilon_2 \simeq 1 - \frac{\omega_{\rm pe}^2}{\omega^2}\,. \tag{3.75}$$

Here, $v_A = \left(B_0^2/4\pi M N_i\right)^{1/2}$ is the Alfven velocity. Then, in view of Eq. (3.26), we obtain the refractive indices of ordinary and extraordinary waves

$$n_1^2 = \varepsilon_1 = 1 + \frac{c^2}{v_A^2},$$

$$n_2^2 = \frac{\varepsilon_1 \varepsilon_2}{\varepsilon_1 \sin^2\vartheta + \varepsilon_2 \cos^2\vartheta} = \frac{\left(1 - \dfrac{\omega_{\rm pe}^2}{\omega^2}\right)\left(1 + \dfrac{\omega_{\rm pi}^2}{\Omega_i^2}\right)}{1 + \dfrac{\omega_{\rm pi}^2}{\Omega_i^2}\sin^2\vartheta - \dfrac{\omega_{\rm pe}^2}{\omega^2}\cos^2\vartheta}\,. \tag{3.76}$$

It is convenient to represent these relations as $\omega = \omega(k)$, where $k = n\omega/c$. If, in addition, we assume that $\omega^2 \ll \omega_{\rm pi}^2$, then, from Eqs. (3.76), for arbitrary angle ϑ we obtain, respectively,

$$\omega^2 = \frac{k^2 v_A^2}{1 + \frac{v_A^2}{c^2}}, \qquad \omega^2 = \frac{k_z^2 v_A^2}{1 + \frac{v_A^2}{c^2}}\,. \tag{3.77}$$

The first relation is known as the spectrum of magnetohydrodynamic waves, whereas the second relation corresponds to the spectrum of Alfven waves.

It must be noted that the ion contribution may be essential not only when $\omega \le \Omega_i$, but also in higher frequency ranges when $\omega \gg \Omega_i$. In fact, ions play an important role up to the frequency $\Omega_i\sqrt{M/m} \sim \Omega_e\sqrt{m/M}$. To prove, let us consider the frequency band $\Omega_i \ll \omega \ll \Omega_e$. In this range, Eq. (3.74) takes the form

$$\varepsilon_1 = 1 + \frac{\omega_{\rm pe}^2}{\Omega_e^2} - \frac{\omega_{\rm pi}^2}{\omega^2}, \qquad g = \frac{\omega_{\rm pe}^2}{\omega \Omega_e}, \qquad \varepsilon_2 = 1 - \frac{\omega_{\rm pe}^2}{\omega^2}\,. \tag{3.78}$$

Substituting these expressions into Eq. (3.26) for wave's propagation across the magnetic field $\vartheta = \pi/2$, we obtain

$$n_1^2 = \frac{1 - \left(\frac{\omega_{\mathrm{pe}}^2}{\omega\Omega_e}\right)^2}{1 + \frac{\omega_{\mathrm{pe}}^2}{\Omega_e^2} - \frac{\omega_{\mathrm{pi}}^2}{\omega^2}}, \qquad n_2^2 = 1 - \frac{\omega_{\mathrm{pe}}^2}{\omega^2}. \tag{3.79}$$

We see that if $\omega \leq \Omega_e\sqrt{m/M}$, the ion contribution to the first relation of Eq. (3.79) is significant. Thus, the inequality

$$\omega \gg \Omega_e\sqrt{\frac{m}{M}}, \tag{3.80}$$

represents the condition of the validity of the pure electron approximation for description of wave's propagation in magneto-active plasma.

When the waves propagate along the magnetic field (or $\vartheta \neq \pi/2$), the restriction is softer. Really, by substituting Eq. (3.78) into Eq. (3.25), it becomes clear that the ion contribution can be neglected, meaning that the inequality $\omega \gg \Omega_i$ is sufficient for the pure electron description of magneto-active plasma in this case.

Special attention should be paid to the ion cyclotron waves. In fact, ion cyclotron absorption, in electron-ion plasma, takes place when $|\omega - \Omega_i| \ll \Omega_i \ll \omega_{\mathrm{pe}}$. In contrast to the electron cyclotron waves, which are pure electron waves, the ion cyclotron waves essentially depend on the electron's motion. Sufficiently far from the center of the resonance line, when $|\omega - \Omega_i| \gg k_z v_{\mathrm{Ti}}$, for a description of ion cyclotron waves, from Eq. (3.74), one can obtain the following expressions[8]:

$$\varepsilon_1 = -\frac{\omega_{\mathrm{pi}}^2}{\omega^2 - \Omega_i^2}, \qquad g = -\frac{\omega_{\mathrm{pi}}^2\Omega_i}{\omega\left(\omega^2 - \Omega_i^2\right)}, \qquad \varepsilon_2 = -\frac{\omega_{\mathrm{pe}}^2}{\omega^2} = -\frac{\omega_{\mathrm{pi}}^2}{\omega^2}\frac{M}{m}. \tag{3.81}$$

Substituting these expressions into Eq. (3.26) and considering that $|\varepsilon_2| \gg |\varepsilon_1| \sim |g|$, we obtain the refractive indices of electromagnetic waves near $\omega \sim |\Omega_i|$ when $\frac{\left(\omega^2 - \Omega_i^2\right)}{\omega^2} \ll 1$ [8]:

[8]Ion cyclotron absorption in the region $|\omega - \Omega_i| \leq k_z v_{Ti}$, was investigated in [73].

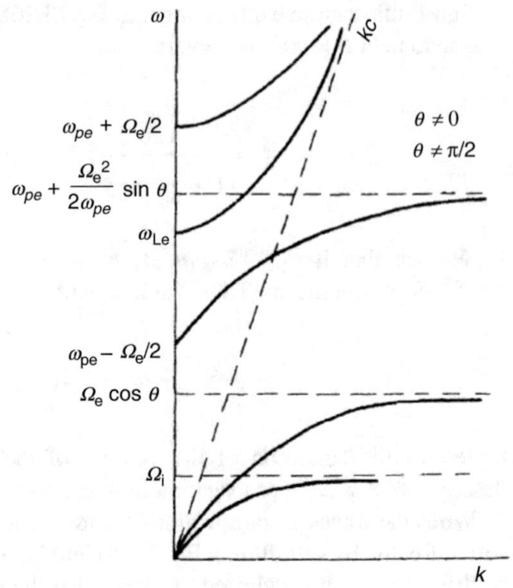

Fig. 3.1 Spectra of electromagnetic wave in cold magneto-active plasma

$$n_1^2 = -\frac{\varepsilon_1\varepsilon_2(1+\cos^2\vartheta)}{\varepsilon_1\sin^2\vartheta+\varepsilon_2\cos^2\vartheta} = \frac{\omega_{\rm pe}^2}{\omega^2}\frac{(1+\cos^2\vartheta)}{\sin^2\vartheta+\dfrac{\omega_{\rm pe}^2}{\omega_{\rm pi}^2}\dfrac{\omega^2-\Omega_i^2}{\omega^2}\cos^2\vartheta},$$

$$n_2^2 = \frac{\varepsilon_1^2-g^2}{\varepsilon_1(1+\cos^2\vartheta)} = -\frac{\omega_{\rm pi}^2}{\omega^2}\frac{1}{1+\cos^2\vartheta}. \tag{3.82}$$

We see that when $\omega\to\Omega_i$, the refractive index n_1 increases and becomes much greater than n_2. Moreover, for longitudinal propagation (if $\vartheta \ll m/M$) $n_1^2\to\infty$ when $\omega\to\Omega_i$. Just in this case, the most intensive absorption of the electromagnetic waves takes place.

It should be noted that the dispersion equation of magneto-active electron-ion plasma has the same form as that of electron plasma. In this case, tensor $\eta_{ij}(\omega,\vartheta)$, apart from electron terms, which are of the order of $\frac{kT_e}{mc^2}$, has ion terms of the order of $\frac{kT_i}{Mc^2}$. However, because of the resonance dependence of tensor $\eta_{ij}(\omega,\vartheta)$ on the frequency ω, the ion's contribution to the region $\omega\sim\Omega_i$ may become essential.

In Fig. 3.1, the frequency dependence of all types of electromagnetic waves in cold magneto-active plasma is shown for oblique ($\vartheta\neq 0,\pi/2$) propagation, assuming $\omega_{\rm pe}^2>\Omega_e^2$. It must be noticed that, in this case, when the thermal motion of particles can be neglected, the frequencies ω_l, at which $n^2(\omega_l)\to\infty$, correspond to the quasi-longitudinal oscillations of plasma. They are described by the dispersion equation

$$\varepsilon_1 \sin^2\vartheta + \varepsilon_2 \cos^2\vartheta = 1 - \left[\frac{\omega_{pe}^2}{\omega^2 - \Omega_{e^2}} + \frac{\omega_{pe}^2}{\omega^2 - \Omega_i^2}\right] \sin^2\vartheta - \frac{\omega_{pe}^2}{\omega^2} \cos^2\vartheta = 0. \quad (3.83)$$

It is easy to solve this equation and obtain the longitudinal wave spectra

$$\omega_{1,2}^2 = \frac{\omega_{pe}^2 + \Omega_e^2}{2} \pm \frac{1}{2}\sqrt{\left(\omega_{pe}^2 + \Omega_e^2\right)^2 - 4\omega_{pe}^2\Omega_e^2 \cos^2\vartheta},$$

$$\omega_3^2 = \left(1 - \frac{\omega_{pi}^2}{\omega_{pe}^2} \tan^2\vartheta\right)\Omega_i^2. \quad (3.84)$$

These expressions are invalid near $\vartheta = \pi/2$, for $\cos^2\vartheta < m/M$. In this case, we have

$$\omega_1^2 = \omega_{pe}^2 + \Omega_e^2, \quad \omega_2^2 = \frac{\Omega_e^2\omega_{pi}^2}{\omega_{pe}^2 + \Omega_e^2} \simeq \Omega_e\Omega_i, \quad \omega_3^2 \to 0. \quad (3.85)$$

One can see that these oscillations spectra essentially depend on the ion's motion in this case.

In the literature, the oscillations corresponding to the first solution are called the upper hybrid modes and the oscillations corresponding to the second solution are the lower hybrid modes. The ϑ dependence of these spectra is shown in Fig. 3.2. It must be noted that the frequency of longitudinal oscillations corresponds to the poles of $n^2(\omega, 0)$ (see Fig. 3.1).

Let us now take into account the spatial dispersion of the dielectric permittivity and investigate the wave absorption in magneto-active electron-ion plasma. As we know, the wave absorption is determined by the anti-Hermitian part of the dielectric tensor of the form of Eq. (3.36) for pure electron plasma in the range of weak spatial dispersion. From general expressions (3.19), it can be easily shown that relations

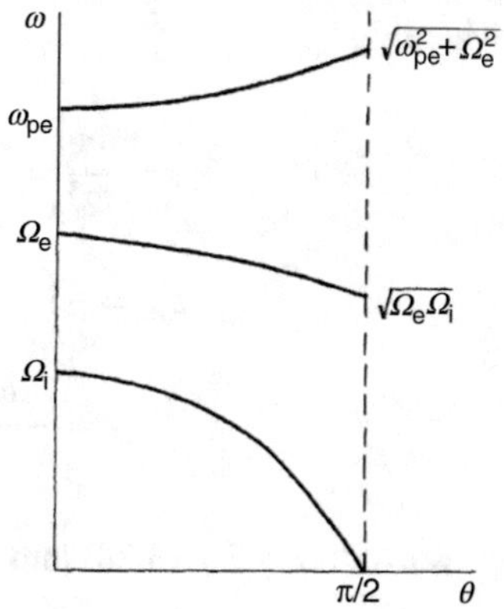

Fig. 3.2 Spectra of longitudinal oscillations in cold magneto-active plasma

(3.36), (3.37) stay also valid for electron-ion plasma except for a frequency band near the ion cyclotron frequency, $\omega \simeq \Omega_i$. In this region [8]

$$\varepsilon_1^a = g^a = \imath\sqrt{\frac{\pi}{8}\frac{\omega_{\rm pi}^2}{\omega}}\sqrt{\frac{M}{\kappa T_i}}\frac{1}{k|\cos\vartheta|}\exp\left[-\frac{M}{2\kappa T_i}\frac{(\Omega_i-\omega)^2}{k^2\cos^2\vartheta}\right],$$
$$\varepsilon_2^a = \imath\sqrt{\frac{\pi}{2}}\frac{\omega_{\rm pe}^2\omega}{k^3|\cos^3\vartheta|}\left(\frac{m}{\kappa T_e}\right)^3\exp\left(-\frac{m}{2\kappa T_e}\frac{\omega^2}{k^2\cos^2\vartheta}\right). \quad (3.86)$$

The quantity $k = n\omega/c$ in these expressions is determined by Eqs. (3.82) for ordinary and extraordinary waves, respectively. Considering expressions (3.86), we obtain the following imaginary corrections to the real parts of the refractive indices given by Eqs. (3.82) in the region of $\omega \sim \Omega_i$:

$$n_1'' = \frac{n_1'^3}{2(1+\cos^2\vartheta)}\left[\frac{\varepsilon_2''}{\varepsilon_2^2}\sin^2\vartheta + \frac{\varepsilon_1''}{\varepsilon_1^2}\sin^2\vartheta\right], \qquad n_2'' = \frac{\varepsilon_1''\left(1+\frac{g}{\varepsilon_1}\right)^2}{2n_2'(1+\cos^2\vartheta)}, \quad (3.87)$$

where $n'_{1,2}$ are given by Eq. (3.82). It must be noted that these expressions are valid only if $n'^2_{1,2} > 0$, or in other words, in the frequency range of plasma transparency.

In conclusion, let us consider the electromagnetic waves in magneto-active plasma under the conditions of strong spatial dispersion. Besides, we assume that the phase velocity of waves is less than electron's thermal velocity and higher than ion thermal velocity, $v_{\rm Ti} \ll \omega/k_z \ll v_{\rm Te}$. As we know, just in this range, the ion-acoustic waves will exist if plasma is non-isothermal $T_e \gg T_i$. For simplicity, we also assume $\omega \ll \Omega_i \ll \omega_{\rm pi}$. Confining our interest to the wavelengths longer than the ion Larmor radius, $k_\perp^2 v_{Ti}^2 \ll \Omega_i^2$, the dielectric tensor is obtained from Eq. (3.19)

$$\varepsilon_{ij}\left(\omega,\vec{k}\right) = \begin{pmatrix} \varepsilon_{xx} & 0 & 0 \\ 0 & \varepsilon_{yy} & \varepsilon_{yz} \\ 0 & \varepsilon_{zy} & \varepsilon_{zz} \end{pmatrix}, \quad (3.88)$$

where

$$\varepsilon_{xx} = \frac{\omega_{\rm pi}^2}{\Omega_i^2} = \frac{c^2}{v_A^2}, \qquad \varepsilon_{yy} = \frac{c^2}{v_A^2} + \imath\sqrt{2\pi}\frac{\omega_{\rm pe}^2 k_\perp^2 v_{\rm Te}}{\Omega_e^2\omega|k_z|},$$
$$\varepsilon_{yz} = -\varepsilon_{zy} = -\imath\frac{\omega_{\rm pe}^2 k_\perp}{\omega\Omega_e k_z}\left(1+\imath\sqrt{\frac{\pi}{2}}\frac{\omega}{|k_z|v_{\rm Te}}\right),$$
$$\varepsilon_{zz} = -\frac{\omega_{\rm pi}^2}{\omega^2} + \frac{\omega_{\rm pe}^2}{k_z^2 v_{\rm Te}^2}\left(1+\imath\sqrt{\frac{\pi}{2}}\frac{\omega}{|k_z|v_{\rm Te}}\right). \quad (3.89)$$

Substituting Eq. (3.88) into Eq. (3.20), we obtain two separate equations

$$k_z^2c^2 - \omega^2\varepsilon_{xx} = 0, \quad \left(k^2c^2 - \omega^2\varepsilon_{yy}\right)\varepsilon_{zz} + \omega^2\varepsilon_{yz}^2 = 0. \tag{3.90}$$

The first equation describes Alfven waves which remain undamped in the considered an approximation and their spectrum coincides with the second expression of Eq. (3.77) (in the limit of $v_A^2 \ll c^2$). Thus, the branch of fast Alfven waves $v_A \gg v_{\mathrm{Te}}$ with the spectrum (3.77) extends into the range $v_{\mathrm{Ti}} \ll v_A \ll v_{\mathrm{Te}}$.

From the second equation of Eqs. (3.90), we obtain the so-called fast and slow magnetosonic waves with frequency spectra and damping decrement as follows:

$$\begin{aligned} \omega_\pm^2 &= \frac{k^2}{2}\left\{v_A^2 + v_s^2 \pm \sqrt{\left(v_A^2 + v_s^2\right)^2 - 4v_A^2v_s^2\cos^2\vartheta}\right\}, \\ \gamma_\pm &= -\sqrt{\frac{\pi\, m}{8\, M}}\frac{kv_s}{2\mid \cos\vartheta\mid}\left(1 \pm \frac{\left(v_s^2\cos^2\vartheta - v_A^2\right)\cos 2\vartheta}{\sqrt{v_A^4 + v_s^4 - 2v_s^2v_A^2\cos 2\vartheta}}\right), \end{aligned} \tag{3.91}$$

where $v_s = \sqrt{Z\,\kappa T_e/M}$ is the ion sound velocity. When the plasma pressure is low $\beta = v_s^2/v_A^2 \ll 1$, the spectrum (3.91) takes an especially simple form

$$\begin{aligned} \omega_+^2 &= k^2v_A^2, \qquad \gamma_+ = -\sqrt{\frac{\pi\, m}{8\, M}}\frac{v_s}{v_A}\frac{\sin^2\vartheta}{\mid \cos\vartheta\mid}\omega_+, \\ \omega_-^2 &= k^2v_s^2\cos^2\vartheta \qquad \gamma_- = -\sqrt{\frac{\pi\, m}{8\, M}}\,\omega_-. \end{aligned} \tag{3.92}$$

We see that the fast magnetosonic wave extends the branch of the fast magnetohydrodynamic (MHD) wave with the spectrum (3.77) (the first expression in the limit of $v_A^2 \ll c^2$) into the range of small phase velocities $\omega/k_z \ll v_{\mathrm{Te}}$. At the same time, the slow magnetosonic wave becomes purely longitudinal in low-pressure plasma $\beta \ll 1$ and represents the long-wave limit of the ion-acoustic waves of non-isothermal plasma with $T_e \gg T_i$.

3.5 Particle Collisions in Magneto-Active Plasma

The account of particle collisions in completely ionized plasma, in general, presents a very difficult problem since it is coupled to the integro-differential equation with the Landau collision integral. Therefore, as in the case of isotropic non-magnetized plasma (see Sect. 2.7), we will analyze only some limiting cases, allowing to find a rather simple solution of the kinetic equation (2.3). In fact, such a situation takes place when inequalities

$$\omega \gg \nu_\alpha, \qquad | \omega \pm \Omega_\alpha | \gg \nu_\alpha, \tag{3.93}$$

are satisfied for both ions and electrons. These inequalities mean that the terms of collision integral of electrons and ions in the kinetic equation are small and, as a result, successive approximation can be applied.

Besides inequalities (3.93), we suppose that spatial dispersion is weak or, in other words, the thermal motion of particles is negligible. This means that the inequalities

$$\frac{k_z v_{T\alpha}}{\omega \pm \Omega_\alpha} \ll 1, \qquad \frac{k_\perp^2 v_{T\alpha}^2}{\Omega_\alpha^2} \ll 1, \tag{3.94}$$

hold as well. Under the conditions (3.93) and (3.94), the linearized kinetic equation (2.3) for the particles of the type α (electrons and ions) with the Landau collision integral can be represented as

$$-\imath\omega\delta f_\alpha + \frac{e_\alpha}{c}\left(\vec{v} \times \overrightarrow{B_0}\right) \cdot \frac{\partial \delta f_\alpha}{\partial \vec{p}} = -e_\alpha \vec{E} \cdot \frac{\partial f_{0\alpha}}{\partial \vec{p}} + \left(\frac{\partial f_\alpha}{\partial t}\right)_c, \tag{3.95}$$

where

$$\left(\frac{\partial f_\alpha}{\partial t}\right)_c = \frac{\partial}{\partial p_i} \int d\vec{p'} \sum_\beta I_{ij}^{\alpha\beta}\left(\frac{\vec{p}}{m_\alpha} - \frac{\vec{p'}}{m_\beta}\right)$$
$$\times \left\{ f_{0\beta}\left(\vec{p'}\right) \frac{\partial \delta f_\alpha\left(\vec{p}\right)}{\partial p_j} + \delta f_\beta\left(\vec{p'}\right) \frac{\partial f_{0\alpha}\left(\vec{p}\right)}{\partial p_j} - f_{0\alpha}\left(\vec{p}\right) \frac{\partial \delta f_\beta\left(\vec{p'}\right)}{\partial p'_j} - \delta f_\alpha\left(\vec{p}\right) \frac{\partial f_{0\beta}\left(\vec{p'}\right)}{\partial p'_j} \right\}. \tag{3.96}$$

The tensor quantity $I_{ij}^{\alpha\beta}(\vec{v})$ in this expression is defined by Eq. (2.118) and the equilibrium distribution function $f_{0\alpha}(p)$ is supposed to be Maxwellian.

In view of Eq. (3.93), the method of successive approximation already used in Sect. 2.7 can be applied here for Eq. (3.95). This is the expansion in terms of the collision integral, which allows us to represent δf_α as

$$\delta f_\alpha\left(\vec{p}\right) = \delta f_\alpha^{(1)}\left(\vec{p}\right) + \delta f_\alpha^{(2)}\left(\vec{p}\right), \tag{3.97}$$

where $\delta f_\alpha^{(1)}\left(\vec{p}\right)$ is the solution of Eq. (3.95) when the collision integral is neglected; $\delta f_\alpha^{(2)}\left(\vec{p}\right)$ is the correction which accounts the collisions.

Using Eq. (3.5) under the conditions (3.94), i.e., in the long-wave limit $\left(\vec{k} \to 0\right)$, we obtain

$$\delta f_{\alpha}^{(1)}\left(\vec{p}\right)=\frac{-\imath e_{\alpha}}{\omega}\frac{\partial f_{0\alpha}}{\partial p}\frac{\vec{E}}{p}\cdot\left[\frac{\vec{B}_0\left(\vec{p}.\vec{B}_0\right)}{B_0^2}+\frac{\vec{B}_0\times\left(\vec{p}\times\vec{B}_0\right)}{B_0^2}\frac{\omega^2}{\omega^2-\Omega_{\alpha}^2}-\frac{\vec{p}\times\vec{B}_0}{B_0}\frac{\imath\omega\Omega_{\alpha}}{\omega^2-\Omega_{\alpha}^2}\right]. \tag{3.98}$$

In the absence of the external magnetic field ($\vec{B}_0\rightarrow 0$), this expression reduces to Eq. (2.121)

Substituting the first approximated solution $\delta f_{\alpha}^{(1)}\left(\vec{p}\right)$ given by Eq. (3.98) into the collision integral in the right-hand side of Eq. (3.95), we obtain

$$-\imath\omega\delta f_{\alpha}^{(2)}+\frac{e_{\alpha}}{c}\left(\vec{v}\times\vec{B}_0\right)\cdot\frac{\partial\delta f_{\alpha}^{(2)}}{\partial\vec{p}}=\left(\frac{\partial f_{\alpha}\left(\vec{p}\right)}{\partial p}\right)_c^{(1)}, \tag{3.99}$$

for the second order correction $\delta f_{\alpha}^{(2)}\left(\vec{p}\right)$. Here,

$$\left(\frac{\partial f_{\alpha}}{\partial p}\right)_c^{(1)}=\frac{\partial}{\partial p_i}\int d\vec{p}'\sum_{\beta}I_{ij}^{\alpha\beta}\left(\frac{\vec{p}}{m_{\alpha}}-\frac{\vec{p}'}{m_{\beta}}\right)$$
$$\times\left[f_{0\beta}\left(\vec{p}'\right)\frac{\partial\delta f_{\alpha}^{(1)}\left(\vec{p}\right)}{\partial p_j}+\delta f_{\beta}^{(1)}\left(\vec{p}'\right)\frac{\partial f_{0\alpha}\left(\vec{p}\right)}{\partial p_j}-f_{0\alpha}\left(\vec{p}\right)\frac{\partial\delta f_{\beta}^{(1)}\left(\vec{p}'\right)}{\partial p_j'}-\delta f_{\alpha}^{(1)}\left(\vec{p}\right)\frac{\partial f_{0\beta}\left(\vec{p}'\right)}{\partial p_j'}\right]. \tag{3.100}$$

Equation (3.99) can be solved by the quite similar method used for the solution of Eq. (3.95) in the absence of particle collisions. Using Eq. (3.5), we obtain

$$\delta f_{\alpha}^{(2)}\left(\vec{p}\right)=\int_0^{\infty}d\tau\exp\left(\imath\omega\tau\right)\left(\frac{\partial f_{\alpha}\left(\vec{P}(\tau)\right)}{\partial t}\right)_c^{(1)}, \tag{3.101}$$

where $\vec{P}(\tau)$ is defined by expression (3.6).

Now, we can calculate the collision correction of the dielectric tensor. For this aim, we must evaluate the collisional contribution to the induced current. Having in mind that in the high-frequency region, ion's motion is negligible, we have to consider only the electron current. The ion current is inversely proportional to its mass and, as a result, it is small compared to the electron current. For this reason, in the collision of ions with electrons, the ion distribution function can be Maxwellian. Neglecting small terms of the order of *m*/*M*, we find the collision correction to the current density [4, 74, 75]:

$$\delta j_i^{(2)} = \sum_\alpha e_\alpha \int v_i \delta f^{(2)}\left(\vec{p}\right) d\vec{p} = \delta\sigma_{ij}(\omega) E_j, \tag{3.102}$$

where

$$\delta\sigma_{ij}(\omega) = \begin{pmatrix} \sigma_1 & \imath g_1 & 0 \\ -\imath g_1 & \sigma_1 & 0 \\ 0 & 0 & \sigma_2 \end{pmatrix}.$$

Here,

$$\sigma_1 = \frac{e^2 N_e\left(\omega^2 + \Omega_e^2\right)}{m\left(\omega^2 - \Omega_e^2\right)^2}\nu_{\text{eff}}, \quad \sigma_2 = \frac{e^2 N_e}{m\omega^2}\nu_{\text{eff}}, \quad g_1 = \frac{2e^2 N_e \omega \Omega_e}{m\left(\omega^2 - \Omega_e^2\right)^2}\nu_{\text{eff}}.$$

The effective collision frequency ν_{eff} in magneto-active plasma is [compare to Eq. (2.125)]

$$\nu_{\text{eff}} = \frac{4}{3}\sqrt{\frac{2\pi}{m}}\frac{e^2 L}{\left(\kappa T_e\right)^{3/2}}\sum_\beta e_\beta^2 N_\beta.$$

These formulas are valid in the high-frequency range when

$$|\omega - \Omega_e| \gg \nu_{\text{eff}}.$$

As noted in Sect. 2.7, it should be remarked that the above expression for ν_{eff} is correct for isotropic plasma only if $\omega \ll \omega_{\text{pe}}$. In magneto-active plasma, it is also necessary that $\omega_{\text{pe}} \gg \Omega_e$. Particle collisions in the region of high frequencies, when $\omega \gg \omega_{\text{pe}}$, and in the strong magnetic fields, when $\Omega_e \gg \omega_{\text{pe}}$, were investigated in [16, 76]. In this case, the collision integral is

$$\frac{\partial}{\partial P_{\alpha i}}\sum_\beta \int d\vec{P}_\beta d\vec{r}_\beta \frac{\partial U_{\alpha\beta}\left(\vec{r}_\alpha - \vec{r}_\beta\right)}{\partial r_{\alpha i}} \int\limits_{-\infty}^{0} d\tau \frac{\partial U_{\alpha\beta}\left(\vec{R}_\alpha - \vec{R}_\beta\right)}{\partial r_{\alpha j}}$$
$$\times \left[\frac{\partial}{\partial P_{\alpha j}} - \frac{\partial}{\partial P_{\beta j}}\right] f_\alpha\left(\vec{P}_\alpha, \vec{R}_\alpha, t+\tau\right) f_\beta\left(\vec{P}_\beta, \vec{R}_\beta, t+\tau\right),$$

where $U_{\alpha\beta} = \frac{e_\alpha e_\beta}{r}$; $\vec{P}_\alpha$, $\vec{R}_\alpha$ are the momentum and coordinate of the particle species α at the moment $t + \tau$. The collisional correction to the induced current given by Eq. (3.102) provides a purely anti-Hermitian term to tensor (3.73)

$$\delta\varepsilon_{ij}(\omega) = \frac{4\pi\imath}{\omega}\delta\sigma_{ij}(\omega) = \begin{pmatrix} \delta\varepsilon_1^a & \imath\delta g_1^a & 0 \\ -\imath\delta g_1^a & \delta\varepsilon_1^a & 0 \\ 0 & 0 & \delta\varepsilon_2^a \end{pmatrix}, \tag{3.103}$$

where[9]

$$\begin{gathered} \delta\varepsilon_1^a = \imath\frac{\omega_{pe}^2\omega\nu_{eff}}{\Omega_e^2}\left[\left(\frac{\Omega_e}{\omega^2-\Omega_e^2} - \frac{\Omega_i}{\omega^2-\Omega_i^2}\right)^2 + \left(\frac{\omega}{\omega^2-\Omega_e^2} - \frac{\omega}{\omega^2-\Omega_i^2}\right)^2\right], \\ \delta g_1^a = 2\imath\frac{\omega_{pe}^2\omega^2\nu_{eff}}{\Omega_e^2}\left(\frac{\Omega_e}{\omega^2-\Omega_e^2} - \frac{\Omega_i}{\omega^2-\Omega_i^2}\right)\left(\frac{1}{\omega^2-\Omega_e^2} - \frac{1}{\omega^2-\Omega_i^2}\right), \\ \delta\varepsilon_2^a = \imath\frac{\omega_{pe}^2\nu_{eff}}{\omega^3}. \end{gathered} \tag{3.104}$$

We see that the collisional contribution to the anti-Hermition part of the dielectric tensor has the same form as Eqs. (3.36), (3.37), and (3.86). Therefore, it is obvious that the corrections of absorption coefficients (3.43), (3.44), and (3.33) are stipulated by the particle collisions in magneto-active completely ionized plasma.

Let us consider now the effect of particle collisions on electromagnetic wave's propagation in weakly ionized magneto-active plasma. The kinetic equation with the BGK collision model allows a very clear treatment of this effect. Moreover, it is possible to obtain the general dielectric tensor without any restrictions on the frequencies and wavelengths from the BGK kinetic model. The linearized kinetic equation for the particles of the type α, in this model, is given by (see Sect. 2.7)

$$-\imath\left(\omega - \vec{k}\cdot\vec{v}\right)\delta f_\alpha + e_\alpha\vec{E}\cdot\frac{\partial f_{0\alpha}}{\partial\vec{p}} - \Omega_\alpha\frac{\partial\delta f_\alpha}{\partial\varphi} = -\nu_{\alpha n}(\delta f_\alpha - \eta_\alpha f_{0\alpha}), \tag{3.105}$$

where $N_\alpha\eta_\alpha = \int d\vec{p}\,\delta f_\alpha$, and $f_{0\alpha}(p)$ is the equilibrium non-relativistic Maxwellian distribution function. The method of solution of Eq. (3.105) is similar to that used for collisionless plasma (see Sect. 3.1). The general solution is

[9] The influence of particle collisions on plasma conductivity in the range of electron cyclotron resonance, $\omega \sim \Omega_e$, was investigated in [4, 74, 75, 77–79].

$$\delta f_\alpha(\vec{p}) = \frac{1}{\Omega_\alpha} \int\limits_\infty^\varphi d\varphi' \left(e_\alpha \vec{E} \cdot \frac{\partial f_{0\alpha}}{\partial \vec{p}} - \nu_{\alpha n} \eta_\alpha f_{0\alpha} \right) \exp\left[\frac{1}{\Omega_\alpha} \int\limits_\varphi^{\varphi'} d\varphi'' \left(\omega - \vec{k} \cdot \vec{v} + \imath \nu_{\alpha n} \right)_{\varphi''} \right]$$

$$= \frac{\imath e_\alpha}{T_\alpha} f_{0\alpha} \sum_s \frac{\exp\left(-\imath s\varphi + \imath b_\alpha \sin\varphi \right)}{\omega + \imath \nu_{\alpha n} - k_z v_z - s\Omega_\alpha}$$

$$\times \left[\frac{s\Omega_\alpha}{k_\perp} J_n(b_\alpha) E_x + \imath v_\perp J'_n(b_\alpha) E_y + \left(v_z E_z + \nu_{\alpha n}\eta_\alpha \frac{T_\alpha}{e_\alpha} \right) J_n(b_\alpha) \right],$$

$$\tag{3.106}$$

where $b_\alpha = k_\perp v_\perp / \Omega_\alpha$. Using this expression to calculate the induced current density

$$\vec{j}_\alpha = e_\alpha \int d\vec{p}\, \vec{v}\, \delta f_\alpha,$$

and eliminating η_α by means of the continuity equation

$$e_\alpha N_\alpha \omega \eta_\alpha = \vec{k} \cdot \vec{j},$$

we find the dielectric tensor of weakly ionized magneto-active plasma:

$$\varepsilon_{ij}\left(\omega, \vec{k}\right) = \delta_{ij} + \sum_\alpha \left(\delta_{i\mu} + \frac{\imath \nu_{\alpha n} G_{\alpha i} k_\mu}{\omega - \imath \nu_{\alpha n} \vec{k} \cdot \vec{G}_\alpha} \right) \frac{\omega + \imath \nu_{\alpha n}}{\omega} \left[\varepsilon^\alpha_{\mu j}\left(\omega, \vec{k}\right) - \delta_{\mu j} \right]. \tag{3.107}$$

Tensor $\varepsilon^\alpha_{ij}\left(\omega, \vec{k}\right)$ is the contribution of the particles of the type α to the dielectric tensor of collisionless magneto-active plasma at the shifted argument $\omega \to \omega + \imath \nu_{\alpha n}$. The vector $\vec{G}_\alpha$ is defined as

$$\vec{G}_\alpha = \begin{cases} G_x = \dfrac{\Omega_\alpha}{k_\perp} \displaystyle\sum_s \dfrac{s A_s(z_\alpha)}{\omega + \imath \nu_{\alpha n} - s\Omega_\alpha} I_+(\beta_{\alpha s}), \\[3mm] G_y = -\imath \dfrac{\Omega_\alpha}{k_\perp} \displaystyle\sum_s \dfrac{z_\alpha A'_s(z_\alpha)}{\omega + \imath \nu_{\alpha n} - s\Omega_\alpha} I_+(\beta_{\alpha s}), \\[3mm] G_z = -\dfrac{1}{k_z} \displaystyle\sum_s A_s(z_\alpha)[1 - I_+(\beta_{\alpha s})], \end{cases} \tag{3.108}$$

where

$$z_\alpha = \frac{k_\perp^2 v_{T\alpha}^2}{\Omega_\alpha^2}, \qquad \beta_{\alpha s} = \frac{\omega + \imath \nu_{\alpha n} - s\Omega_\alpha}{k_z v_{T\alpha}}.$$

Under the conditions (3.93) and (3.94), the dielectric tensor (3.107) takes the form of Eq. (3.103), where

$$\delta\varepsilon_{\perp}^{a}=\imath\sum_{\alpha}\frac{\omega_{p\alpha}^{2}\nu_{\alpha n}\left(\omega^{2}+\Omega_{\alpha}^{2}\right)}{\omega\left(\omega^{2}-\Omega_{\alpha}^{2}\right)^{2}},\quad \delta g_{1}^{a}=\imath\sum\frac{2\omega_{p\alpha}^{2}\Omega_{\alpha}\nu_{\alpha n}}{\left(\omega^{2}-\Omega_{\alpha}^{2}\right)^{2}},\quad \delta\varepsilon_{\parallel}^{a}=\imath\frac{\omega_{pe}^{2}\nu_{en}}{\omega^{3}}. \tag{3.109}$$

This means that all above conclusions made for completely ionized plasma remain also valid for weakly ionized plasma.

Finally, we consider the problem of inhomogeneity being in collisional plasma. This case is equivalent to the diffusion process and can be described by the dispersion equation of slow longitudinal waves in plasma

$$\varepsilon\left(\omega,\vec{k}\right)=\frac{k_{i}k_{j}}{k_{2}}\varepsilon_{ij}\left(\omega,\vec{k}\right)=0. \tag{3.110}$$

We will consider this equation for weakly ionized plasma because Eq. (3.107) is valid without any restrictions on the ratio of ω, Ω_{α}, ν_{α} and $kv_{T\alpha}$. The diffusion process is a low-frequency $\omega\ll\nu_{\alpha n}$, Ω_{α} and long-wave $k_{z}v_{T\alpha}\ll\nu_{\alpha n}$, $k_{\perp}v_{T\alpha}\ll\Omega_{\alpha}$ process and, as a result, Eq. (3.110) can be written as

$$\varepsilon\left(\omega,\vec{k}\right)=1+\sum_{\alpha}\frac{\omega_{p\alpha}^{2}}{k^{2}v_{T\alpha}^{2}}\frac{\frac{k_{\perp}^{2}v_{T\alpha}^{2}\nu_{\alpha n}}{\Omega_{\alpha}^{2}}+\frac{k_{z}^{2}v_{T\alpha}^{2}}{\nu_{\alpha n}}}{-\imath\omega+\frac{k_{\perp}^{2}v_{T\alpha}^{2}\nu_{\alpha n}}{\Omega_{\alpha}^{2}}+\frac{k_{z}^{2}v_{T\alpha}^{2}}{\nu_{\alpha n}}}=0. \tag{3.111}$$

For very weakly ionized plasma and small-size inhomogeneities, when $k^{2}v_{T\alpha}^{2}\gg\omega_{p\alpha}^{2}$, the solution of this equation is determined by the poles of the dielectric permittivity $\varepsilon\left(\omega,\vec{k}\right)$, i.e.,

$$\omega=-\imath\left(\frac{k_{\perp}^{2}v_{T\alpha}^{2}}{\Omega_{\alpha}^{2}}\nu_{\alpha n}+\frac{k_{z}^{2}v_{T\alpha}^{2}}{\nu_{\alpha n}}\right). \tag{3.112}$$

It is obvious that this expression is equivalent to the differential equation

$$\frac{\partial n_{\alpha}}{\partial t}-D_{\perp\alpha}\nabla_{\perp}^{2}n-D_{\parallel\alpha}\nabla_{\parallel}^{2}n=0, \tag{3.113}$$

where $\nabla_{\perp}^{2}$ and $\nabla_{\parallel}^{2}$ are the transverse and longitudinal Laplace operators. This equation represents the diffusion equation for very rarefied weakly ionized magneto-active plasma and

$$D_{\perp\alpha} \equiv \frac{v_{T\alpha}^2 \nu_{\alpha n}}{\Omega_\alpha^2}, \qquad D_{\|\alpha} = \frac{v_{T\alpha}^2}{\nu_{\alpha n}}, \tag{3.114}$$

is the transverse and longitudinal (relative to the direction of $\vec{B}_0$) diffusion coefficients of the particles of the type α in such plasma, respectively.

In the opposite limit, when plasma is very dense and the inhomogeneity size is large, then from Eq. (3.111), we obtain

$$\omega = -\imath \frac{\left(k_\perp^2 D_{\perp \mathrm{i}} + k_z^2 D_{\| i}\right)\left(1 + \frac{T_e}{T_i}\right)}{1 + \frac{T_e}{T_i}\frac{k_\perp^2 D_{\perp \mathrm{i}} + k_z^2 D_{\| i}}{k_\perp^2 D_{\perp \mathrm{e}} + k_z^2 D_{\| e}}}. \tag{3.115}$$

If $k_\perp = 0$, then from this expression, we obtain

$$\omega \simeq -\imath k_z^2 D_{\| i}\left(1 + \frac{T_e}{T_i}\right), \tag{3.116}$$

corresponding to the equation

$$\frac{\partial n}{\partial t} - D_{\| a}\nabla_\|^2 n = 0, \tag{3.117}$$

for longitudinal diffusion in magneto-active plasma. Here, $D_{\| a} = D_{\| i}(1 + T_e/T_i)$ is called the coefficient of longitudinal ambipolar diffusion and is the same as the coefficient of ambipolar diffusion in isotropic plasma. If $k_z = 0$, then from Eq. (3.115), we obtain

$$\omega \simeq -\imath k_\perp^2 \frac{T_i}{T_e} D_{\perp \mathrm{e}},$$

corresponding to the equation

$$\frac{\partial}{\partial t} n - D_{\perp \mathrm{a}}\nabla_\perp^2 n = 0,$$

for transverse diffusion in magneto-active plasma. Here, $D_{\perp a} = (T_i/T_e)\, D_{\perp e}$ is known as the coefficient of transverse ambipolar diffusion.

3.6 Magnetohydrodynamics of Collisionless Plasma

The magnetohydrodynamic description, which strictly speaking is valid only for conducting fluids, is often successfully applied to rarefied collisionless plasma [80, 81]. Therefore, many authors tried to find out the possibility of the derivation of the MHD equations to describe collisionless plasma [61, 63, 64, 81–86]. There exist two approaches to study this problem. The first approach was developed by Chu, Goldberger, and Lou, which will be considered in the first part of this section [81–85]. The other way will be discussed in the second part.

Let us begin from the kinetic equation of ions in collisionless plasma

$$\frac{\partial f_i}{\partial t} + \vec{v} \cdot \frac{\partial f_i}{\partial \vec{r}} + e_i \left\{ \vec{E} + \frac{1}{c}\left(\vec{v} \times \vec{B}\right) \right\} \cdot \frac{\partial f_i}{\partial \vec{p}} = 0. \tag{3.118}$$

We will assume that the characteristic time scales are much larger than the Larmor rotation period of ions $T_B = 2\pi/\Omega_i$, and the characteristic length scales are greater than the ion Larmor radius v/Ω_i. Then, the solution of Eq. (3.118) can be represented as an expansion in powers of T_B [87]

$$f = f_0 + f_1 + f_2 \cdots \tag{3.119}$$

In this case, from Eq. (3.118), it follows that

$$\left(\vec{E} + \frac{1}{c}\left(\vec{v} \times \vec{B}\right)\right) \cdot \frac{\partial f_0}{\partial \vec{p}} = 0, \tag{3.120}$$

$$\left(\frac{\partial}{\partial t} + \vec{v} \cdot \frac{\partial}{\partial \vec{r}}\right) f_0 + e_i \left(\vec{E} + \frac{1}{c}\left(\vec{v} \times \vec{B}\right)\right) \cdot \frac{\partial f_1}{\partial \vec{p}} = 0, \tag{3.121}$$

$$\left(\frac{\partial}{\partial t} + \vec{v} \cdot \frac{\partial}{\partial \vec{r}}\right) f_1 + e_i \left(\vec{E} + \frac{1}{c}\left(\vec{v} \times \vec{B}\right)\right) \cdot \frac{\partial f_2}{\partial \vec{p}} = 0. \tag{3.122}$$

Suppose that electric and magnetic fields are perpendicular to each other. Such an assumption is very essential.[10] It can be justified by the high mobility of electrons, because the electron's motion in plasma in the electric field parallel to the magnetic induction results in the elimination of such a field configuration. According to this assumption, we can introduce the quantity $\vec{V}_{\perp}$ perpendicular to $\vec{E}$ and $\vec{B}$ fields

---

[10] At the same time, it is the most vulnerable point of this method [see Eqs. (3.135) and (3.152)].

$$\vec{E} = -\frac{1}{c}\left(\vec{V}_\perp \times \vec{B}\right), \quad \vec{V}_\perp = c\frac{\left(\vec{E} \times \vec{B}\right)}{B^2}. \tag{3.123}$$

Then, Eq. (3.120) takes the form of

$$\vec{B} \cdot \left[\left(\vec{v} - \overrightarrow{V_\perp}\right) \times \frac{\partial}{\partial \vec{p}}\right] f_0 = 0. \tag{3.124}$$

Therefore, we can write the general solution of this equation as

$$f_0\left(\left(\vec{v} - \overrightarrow{V_\perp}\right)^2, \left(\vec{v} \times \vec{B}\right), \vec{r}, t\right). \tag{3.125}$$

To derive the hydrodynamic equations, let us integrate Eq. (3.121) over momentum and multiply it by the ion mass which results in the continuity equation

$$\frac{\partial W\left(\vec{r}, t\right)}{\partial t} + \nabla \cdot W\left(\overrightarrow{r, t}\right)\vec{V}\left(\vec{r}, t\right) = 0, \tag{3.126}$$

where

$$W\left(\vec{r}, t\right) = M\int d\vec{p} f_0, \quad W\vec{V} = M\int d\vec{p}\ \vec{v} f_0. \tag{3.127}$$

It must be noted that the quantity $\vec{V}_\perp$ given by Eq. (3.123) describes the drift motion of particles. The equation for $\vec{V}_\parallel$ can be obtained from Eq. (3.121) by multiplying by $\vec{p}$ and integrating over it. Considering Eq. (3.123), we obtain

$$\frac{\partial WV_i}{\partial t} = -\frac{\partial \Pi_{ij}}{\partial r_j} + \frac{e_i}{c}\left[\int d\vec{p}\left(\vec{v} - \vec{V}\right) f_1 \times \vec{B}\right]_i, \tag{3.128}$$

where

$$\Pi_{ij} = WV_iV_j + M\int d\vec{p}\left(\vec{v} - \vec{V}\right)_i\left(\vec{v} - \vec{V}\right)_j f_0. \tag{3.129}$$

According to Eq. (3.125), this tensor can be represented as

$$\Pi_{ij} = WV_iV_j + P_n b_i b_j + P_s\left(\delta_{ij} - b_i b_j\right), \tag{3.130}$$

where $\vec{b} = \vec{B}/B$. From Eq. (3.130), we see that the pressure in plasma may be anisotropic, namely, longitudinal (along the magnetic field $\vec{B}$) pressure may differ from transverse pressure.

Let us now consider the field equations. According to Eq. (3.123), the system of field equations can be written as

$$-\frac{1}{c}\nabla\cdot\left(\vec{V}\times\vec{B}\right) = 4\pi\int d\vec{p}\left[e_i(f_0+f_1)+ef_e\right], \tag{3.131}$$

$$\nabla\cdot\vec{B}=0, \tag{3.132}$$

$$-\frac{\partial}{\partial t}\frac{1}{c^2}\left(\vec{V}\times\vec{B}\right) = \nabla\times\vec{B} - \frac{4\pi}{c}\int d\vec{p}\;\vec{v}\left[e_i(f_0+f_1)+ef_e\right], \tag{3.133}$$

$$\nabla\times\left(\vec{V}\times\vec{B}\right) = \frac{\partial\vec{B}}{\partial t}, \tag{3.134}$$

where f_e is the electron distribution function.

Multiplying Eq. (3.131) by $\vec{V}$ and taking its sum with Eq. (3.133), by considering Eq. (3.125), we obtain

$$\begin{aligned}&\frac{\vec{V}}{c}\nabla\cdot\left(\vec{V}\times\vec{B}\right)+\frac{\partial}{\partial t}\frac{1}{c^2}\left(\vec{V}\times\vec{B}\right)+\nabla\times\vec{B}\\&=\frac{4\pi}{c}e_i\int d\vec{p}\left(\vec{v}-\vec{V}\right)f_1+\frac{4\pi}{c}e\int d\vec{p}\left(\vec{v}-\vec{V}\right)f_e.\end{aligned} \tag{3.135}$$

The average velocity of electrons perpendicular to $\vec{B}$ is equal to $\vec{V}_\perp$ because of mutual perpendicularity of electric and magnetic fields. Therefore, the last term of the right side of Eq. (3.135) is parallel to $\vec{B}$. In this case, Eq. (3.128) can be written as

$$\begin{aligned}\frac{\partial}{\partial t}WV_i = &-\frac{\partial\Pi_{ij}}{\partial r_j}+\frac{1}{4\pi}\left(\nabla\times\vec{B}\times\vec{B}\right)_i+\frac{1}{4\pi c}\left(\vec{V}\times\vec{B}\right)_i\nabla\cdot\left(\vec{V}\times\vec{B}\right)\\&-\frac{1}{c^2}\left[\vec{B}\times\frac{\partial}{\partial t}\left(\vec{V}\times\vec{B}\right)\right]_i.\end{aligned} \tag{3.136}$$

Equations (3.126), (3.132), (3.134), and (3.136) are the system of MHD equations for collisionless plasma. But unfortunately, the pressure quantity Π_{ij} in this system is not expressed in terms of the hydrodynamic quantities $W, \vec{V}, \vec{B}$ and, hence, this system is not closed. However, this system becomes closed when the magnetic pressure $(1/4\pi)\left[\left(\nabla\times\vec{B}\right)\times\vec{B}\right]$ is much larger than the kinetic pressure of particles so that the latter can be neglected. Such a situation was just investigated in [63].

One more possibility of obtaining the closed system of hydrodynamic equations consists of multiplying Eq. (3.121) by $\left(\vec{v}-\vec{V}\right)_i\left(\vec{v}-\vec{V}\right)_j$ and integrating over the momentum. As a result, we obtain

$$\frac{dP_n}{dt} = -P_n\nabla\cdot\vec{V} - 2P_n\vec{b}\left(\vec{b}\cdot\frac{\partial}{\partial\vec{r}}\right)\vec{V} - \nabla\cdot(q_n+q_s)\vec{b} - 2\vec{b}\cdot\frac{\partial}{\partial\vec{r}}q_s, \quad (3.137)$$

$$\frac{dP_s}{dt} = -2P_s\nabla\cdot\vec{V} + P_n\vec{b}\left(\vec{b}\cdot\frac{\partial}{\partial\vec{r}}\right)\vec{V} - \nabla\cdot q_s\vec{b} - q_s\nabla\cdot\vec{b}, \quad (3.138)$$

where

$$\frac{d}{dt} = \frac{\partial}{\partial t} + \left(\vec{V}\cdot\frac{\partial}{\partial\vec{r}}\right).$$

Quantities q_s and q_n are defined by

$$M\int d\vec{p}\left(\vec{v}-\vec{V}\right)_i\left(\vec{v}-\vec{V}\right)_j\left(\vec{v}-\vec{V}\right)_k f_0$$
$$= q_n b_i b_j b_k + q_s\left(\delta_{ij}b_k + \delta_{ik}b_j + \delta_{jk}b_i\right), \quad (3.139)$$

indicating that these quantities are determined by the distribution function and not by Hydrodynamic quantities. To avoid this, one can also derive another equation with quantities defined by the distribution function and so on. Generally, this chain of equations is not terminated. Therefore, it will be concluded that for collisionless plasma it is impossible to receive a closed system of hydrodynamic equations. However, it turns out that in the specific cases, quantity (3.139) is sufficiently small and can be neglected. Then, Eqs. (3.137) and (3.138) take the form of

$$\frac{dP_n}{dt} = -P_n\nabla\cdot\vec{V} - 2P_n\vec{b}\left(\vec{b}\cdot\frac{\partial}{\partial\vec{r}}\right)\vec{V}, \quad (3.140)$$

$$\frac{dP_s}{dt} = -2P_s\nabla\cdot\vec{V} + P_s\vec{b}\left(\vec{b}\cdot\frac{\partial}{\partial\vec{r}}\right)\vec{V}. \quad (3.141)$$

Considering Eqs. (3.126), (3.132), and (3.134), we find [82, 85, 88]

$$\begin{cases} \dfrac{d}{dt}\left(\dfrac{P_n B^2}{W^3}\right) = 0, \\ \dfrac{d}{dt}\left(\dfrac{P_s}{WB}\right) = 0. \end{cases} \quad (3.142)$$

These equations called the state equations are used for investigating the problem of plasma stability [85, 88].[11] From the above hydrodynamic equations, it was shown that plasma with anisotropic pressure is unstable and the growth increments of plasma oscillations were calculated.

At the same time, the stability problem of such plasma was analyzed by using the kinetic equation in [89, 90]. The results of this analysis are qualitatively the same, but quantitatively different. Therefore, below, we will try to derive the MHD equations of collisionless plasma used in the literature [89, 90].

For simplicity, we will restrict our study to the consideration of plasma with isotropic pressure. Moreover, to be close to the usual magnetohydrodynamics of fluids, we will consider non-isothermal plasma with the electron's temperature much higher than the ion's temperature [80, 86, 91].

As it was shown above, (see Sects. 2.3, 2.6, and 3.4), there exist slowly damped ion-acoustic and magnetosonic oscillations in such plasma and, therefore, one can hope to derive the closed system of MHD equations.

In non-isothermal plasma with $T_e \gg T_i$, one can neglect the thermal motion of ions. Then, from the kinetic equation of collisionless ions, we find the continuity equation (3.126) and the Newton's equation of motion

$$W\frac{d\vec{V}}{dt} \equiv W\frac{\partial \vec{V}}{\partial t} + W\left(\vec{V}\cdot\frac{\partial}{\partial \vec{r}}\right)\vec{V} = \rho_i\left[\vec{E} + \frac{1}{c}\left(\vec{V}\times\vec{B}\right)\right], \tag{3.143}$$

where $\rho_i = (e_i/M)\,W$ is the charge density of ions.

Since the thermal velocity of electrons is not small, for description of electrons we will use the kinetic equation

$$\frac{\partial f_e}{\partial t} + \vec{v}\cdot\frac{\partial f_e}{\partial \vec{r}} + e\left\{\vec{E} + \frac{1}{c}\left(\vec{v}\times\vec{B}\right)\right\}\cdot\frac{\partial f_e}{\partial \vec{p}} = 0. \tag{3.144}$$

As above, we assume that the characteristic frequencies are less than the ion Larmor frequency. Then, in Eq. (3.143), which can be written in the form of

$$\vec{E} = -\frac{1}{c}\left(\vec{V}\times\vec{B}\right) + \frac{M}{e_i}\frac{d\vec{V}}{dt},$$

the ratio of the last term in the right side of this equation to the first term is of the order of the ratio of the characteristic frequency of the hydrodynamic quantity variation to the ion Larmor frequency. Therefore, neglecting it, we find

---

[11] Note that the system of Eqs. (3.126), (3.132), (3.134), and (3.142) has the following integral: $\frac{1}{2}\int d\vec{r}\left\{WV^2 + (2P_s + P_n) + \frac{B^2}{2\pi} + \frac{\left(\vec{V}\times\vec{B}\right)^2}{4\pi c^2}\right\}$, which was treated as the full energy of the system in [88].

$$\vec{E} = -\frac{1}{c}\left(\vec{V} \times \vec{B}\right),$$

which at once results in Eq. (3.134).

Thus, for closing the system of MHD equations, when the thermal motion of electrons is taken into account, it is necessary to express the electric field $\vec{E}$ in Eq. (3.143) in terms of hydrodynamic quantities W, $\vec{V}$and B.[12] For this aim, we will use the field equations. Assuming that the plasma density is sufficiently high and the Langmuir frequency is much higher than the characteristic frequency, these equations can be written as

$$\vec{j} = \rho_i \vec{V} + e \int d\vec{p}\, \vec{v} f_e = \frac{c}{4\pi} \nabla \times \vec{B}, \tag{3.145}$$

$$\rho = \rho_i + e \int d\vec{p} f_e = 0. \tag{3.146}$$

The first equation corresponds to the field equation $\nabla \times \vec{B} = (1/c)\partial \vec{E}/\partial t + (4\pi/c)\,\vec{j}$, in which the displacement current is neglected, whereas the second equation is the consequence of the equation $\nabla \cdot \vec{E} = 4\pi\rho$.

It is evident that for calculating the current and charge densities of electrons the knowledge of their distribution functions is necessary. In the following, we consider only the case in which the state is near equilibrium and the distribution function f_e slightly differs from the Maxwellian. Then, for the small perturbation of the distribution, we can use Eq. (3.5), which leads to the following relation for the Fourier transform of the electron current density

$$j_i^e\left(\omega, \vec{k}\right) = \sigma_{ij}\left(\omega, \vec{k}\right) E_j,$$

where $\sigma_{ij}\left(\omega, \vec{k}\right)$ is given by Eq. (3.9). In the low-frequency range, when $\omega \ll \omega_{\mathrm{pe}}$ and $\omega/k \ll v_{\mathrm{Te}}$, this relation can be written as

[12] If the thermal motion of electrons can be neglected, then one can use the equation such as Eq. (3.143) for them. In this case, it is possible to obtain the equation of plasma motion in the form [63]

$W\frac{d\vec{V}}{dt} = -\frac{1}{4\pi}\left(\vec{B} \times \nabla \times \vec{B}\right).$

$$\vec{j}^{e}\left(\omega,\vec{k}\right)=-\imath\frac{\omega_{\mathrm{pi}}^{2}\omega\vec{B}_0\left(\vec{B}_0\cdot\vec{E}\right)}{4\pi v_s^{2}\left(\vec{k}\cdot\vec{B}_0\right)^{2}}-\frac{\omega_{\mathrm{pi}}^{2}B_0\left(\vec{E}\times\vec{k}\right)}{4\pi\Omega_i\left(\vec{k}\cdot\vec{B}_0\right)}$$
$$+\tau\left(\vec{k}\right)\frac{\omega_{\mathrm{pi}}^{2}}{4\pi}\left\{\frac{\omega^{2}}{v_s^{2}}\frac{\vec{B}_0\left(\vec{B}_0\cdot\vec{E}\right)}{\left(\vec{k}\cdot\vec{B}_0\right)^{2}}+\imath\omega\frac{\vec{B}_0\left[\left(\vec{B}_0\times\vec{k}\right)\cdot\vec{E}\right]-\left(\vec{B}_0\times\vec{k}\right)\left(\vec{B}_0\cdot\vec{E}\right)}{\Omega_iB_0\left(\vec{k}\cdot\vec{B}_0\right)}+2\frac{v_s^{2}}{\Omega_i^{2}}\frac{\left(\vec{k}\times\vec{B}_0\right)\left[\left(\vec{k}\times\vec{B}_0\right)\cdot\vec{E}\right]}{B_0^{2}}\right\}.$$
(3.147)

Here, $\vec{B}_0$ is the constant (equilibrium) magnetic induction, $v_s=\sqrt{|e_i/e|\kappa T_e/M}$ is the sound velocity (see Sect. 2.6), and

$$\tau\left(\vec{k}\right)=\int_0^{\infty}dt\exp\left[-\frac{\kappa T_e}{2m}\left(\frac{\vec{k}\cdot\vec{B}_0}{B_0}\right)^{2}t^{2}-\imath\omega t\right]\simeq\sqrt{\frac{\pi m}{2\kappa T_e}}\frac{B_0}{\left|\vec{k}\cdot\vec{B}_0\right|}. \tag{3.148}$$

It must be noted that the term proportional to $\tau\left(\vec{k}\right)$ in Eq. (3.147), describes the dissipation in plasma and is a small term.

From Eq. (3.147), in view of Eqs. (3.145) and (3.146), we find

$$\rho_i\vec{E}\left(\vec{r},t\right)=\frac{1}{c}\left[\vec{j}^{e}\times\vec{B}_0\right]-v_s^{2}\frac{\partial}{\partial\vec{r}}\delta W+\vec{F}^{(\mathrm{dis})}, \tag{3.149}$$

where δW is the mass density perturbation and

$$\vec{F}_i^{(\mathrm{dis})}=\frac{W_0v_s^{2}}{B_0^{2}}\times\left\{\left(B_{0i}\left(\vec{B}_0\cdot\frac{\partial}{\partial\vec{r}}\right)+\left[\vec{B}_0\times\left(\vec{B}_0\times\frac{\partial}{\partial\vec{r}}\right)\right]_i\right)\frac{1}{B_0^{2}}\left(\vec{B}_0\cdot\frac{\partial}{\partial\vec{r}}\right)B_{0j}\right.$$
$$\left.-\left[\vec{B}_0\times\left(\vec{B}_0\times\frac{\partial}{\partial\vec{r}}\right)\right]_i\frac{\partial}{\partial r_j}\right\}\cdot\int d\vec{r}'Q\left(\vec{r}-\vec{r}'\right)V_j\left(\vec{r}',t\right),$$
(3.150)

$$Q\left(\vec{r}\right)=\frac{1}{(2\pi)^{3}}\int d\vec{k}\exp\left(\imath\vec{k}\cdot\vec{r}\right)\tau\left(\vec{k}\right). \tag{3.151}$$

Substituting Eq. (3.149) into Eq. (3.143), we obtain

$$\frac{d\vec{V}}{dt}=-\frac{1}{4\pi W_0}\left[\vec{B}_0\times\nabla\times\vec{B}_0\right]-v_s^{2}\frac{\partial}{\partial\vec{r}}\frac{\delta W}{W_0}+\frac{1}{W_0}\vec{F}^{(\mathrm{dis})}. \tag{3.152}$$

It must be noted that this equation consists only of the hydrodynamic variables; without the dissipative (last) term, this equation is similar to the usual MHD equation of the ideal fluid and in this sense it generalizes Eq. (2.110). The essential difference of this equation from the usual MHD equation is stipulated by the non-local dissipative term meaning that the particles (electrons) in the entire space region of plasma take part in the absorption of perturbations propagating with the velocity smaller than the thermal velocity of electrons.

In the case, when spatial gradients are parallel to the external magnetic field $\vec{B}_0$, Eq. (3.152) leads to the equation of motion in the absence of the magnetic field (see Sect. 2.6),

$$\frac{d\vec{V}}{dt} = -\frac{v_s^2}{W_0}\frac{\partial \delta W}{\partial \vec{r}} + v_s^2 \frac{\partial}{\partial \vec{r}} \int d\vec{r}' Q_0\left(\vec{r} - \vec{r}'\right) \nabla \cdot \vec{V}\left(\vec{r}', t\right). \quad (3.153)$$

The function $Q_0\left(\vec{r}\right)$ differs from (3.151) by replacing $\left|\left(\vec{k} \cdot \vec{B}_0\right)\right|/B_0 \to \left|\vec{k}\right|$ in the expression $\tau\left(\vec{k}\right)$ given by Eq. (3.148).

In contrast to Eq. (3.136), the plasma pressure is isotropic $P_s = P_n$ in Eq. (3.152), which is the consequence of selecting the Maxwellian equilibrium distribution. For the perturbed pressure, we also have

$$\delta P_s = \delta P_n = \delta P = v_s^2 \delta W. \quad (3.154)$$

This equation differs from Eq. (3.142), where we had the explicit dependence of the state equations on the magnetic induction. Such a situation is the consequence of our assumption in deriving Eq. (3.136) when the instantaneous redistribution of electrons is practically possible and, as a result, the electric field stays perpendicular to the magnetic induction. In contrast, Eq. (3.152) was obtained without such a restriction. Moreover, the electron pressure in this equation appears as a result of the longitudinal electric field action (see Sect. 2.6), providing plasma neutrality in perturbations.

Using Eq. (3.152), let us examine the propagation and absorption of MHD and magnetosonic waves in plasma. As noticed above, when the dissipation is neglected, this equation reduces to the usual MHD equations [80, 91]. Therefore, the spectrum of oscillations must coincide with the well-known MHD spectrum. Actually, if the velocity $\vec{V}$ in the MHD waves is parallel to the vector $\left(\vec{k} \times \vec{B}_0\right)$, then we obtain [25, 29, 36, 60–65]

$$\omega^2 = \frac{\left(\vec{B}_0 \cdot \vec{k}\right)^2}{4\pi W_0}. \quad (3.155)$$

The vector of the dissipative force, as it is seen from Eq. (3.150), lies in the plane $(\vec{k}, \vec{B}_0)$ and, therefore, the MHD waves with the spectrum (3.155) will be undamped.

If the velocity $\vec{V}$ of the magnetosonic waves lies in the plane $(\vec{k}, \vec{B}_0)$, then from Eq. (3.152), with account of field equations (3.132), (3.134) and the continuity equation (3.126), we obtain the following frequency spectra [80, 91]:

$$\omega_{\pm}^{2}=\frac{1}{2}k^{2}\left\{v_{s}^{2}+v_{A}^{2}\pm\sqrt{\left(v_{A}^{2}+v_{s}^{2}\right)^{2}-4v_{A}^{2}v_{s}^{2}\cos^{2}\vartheta}\right\}, \tag{3.156}$$

and damping decrements [70]

$$\gamma_{\pm}=\frac{\gamma_{0}}{2}\left\{1\pm\frac{(\cos 2\vartheta-X)\cos 2\vartheta}{\sqrt{1+X^{2}-2X\cos 2\vartheta}}\right\}\frac{1}{|\cos\vartheta|}, \tag{3.157}$$

where ϑ is the angle between $\vec{k}$ and $\vec{B}_0$; $v_A=\sqrt{B_0^2/4\pi W_0}$ is the Alfven velocity, $X=1/\beta=v_A^2/v_s^2$and $\gamma_0=\sqrt{(\pi/8)|e_i/e|(m/M)}\,kv_s$ is the damping decrement of ion-acoustic waves in the absence of the external magnetic field, which is caused by the Cherenkov absorption by electrons (see Sect. 2.5). These expressions, as it is expected, coincide with Eq. (3.91).

For propagation along $\vec{B}_0(\vartheta=0)$, from Eqs. (3.156) and (3.157), we find

$$\omega_{+}^{2}=k^{2}v_{A}^{2},\quad \gamma_{+}=0;\quad \omega_{-}^{2}=k^{2}v_{s}^{2},\quad \gamma_{-}=\gamma_{0}.$$

Here, $v_A^2>v_s^2$.

For propagation nearly perpendicular to $\vec{B}_0(\vartheta\approx\pi/2)$, we have

$$\omega_{+}^{2}=k^{2}\left(v_{A}^{2}+v_{s}^{2}\right),\qquad \gamma_{+}=\frac{\gamma_{0}}{|\cos\alpha|};$$

$$\omega_{-}^{2}=k^{2}\frac{v_{A}^{2}v_{s}^{2}}{v_{A}^{2}+v_{s}^{2}}\cos^{2}\alpha,\quad \gamma_{-}=\gamma_{0}\,|\cos\alpha|\left\{1+\frac{v_{s}^{4}}{\left(v_{A}^{2}+v_{s}^{2}\right)^{2}}\right\}.$$

From this equation, it follows that for $\vartheta\approx\pi/2$ the damping decrement γ_+ grows fast. When the damping decrement is higher than the frequency, all above results are invalid. Thus, expression for γ_+ is valid when

$$\cos^{2}\alpha<\frac{\pi}{8}\frac{m}{M}\left|\frac{e_{i}}{e}\right|\frac{v_{s}^{2}}{v_{A}^{2}+v_{s}^{2}}.$$

It is evident that, with increasing v_A or magnetic induction, the prohibited region of angles is decreased.

For the weak magnetic field, when $\left(B_0^2/8\pi\right)\ll N_e\kappa T_e$, corresponding to $v_A^2\ll v_s^2$, from Eqs. (3.156) and (3.157), we find

$$\omega_+^2 = k^2 v_s^2, \qquad \gamma_+ = \gamma_0 \left(1 - 2 \sin^2\alpha \cos^2\alpha\right) \frac{1}{|\cos\alpha|};$$
$$\omega_-^2 = k^2 v_A^2 \cos^2\alpha, \qquad \gamma_- = 2\gamma_0 \sin^2\alpha \, |\cos\alpha|.$$

For strong magnetic fields, when $\left(B_0^2/8\pi\right) \gg N_e \kappa T_e$, we have

$$\omega_+^2 = k^2 v_A^2, \qquad \gamma_+ = \gamma_0 \frac{\sin^2\alpha}{|\cos\alpha|};$$
$$\omega_-^2 = k^2 v_s^2 \cos^2\alpha, \qquad \gamma_- = \gamma_0 \, |\cos\alpha|.$$

Here, we applied $\left(B_0^2/8\pi\right) \ll (M/m) N_e \kappa T_e$ because in deriving Eq. (3.147) we assumed $\omega/k \ll v_{\mathrm{Te}}$.

3.7 Interaction of Straight Neutralized Beams of Charged Particles with Plasma

As it was shown in Chap. 1, a charged particle traveling in the material medium can radiate the electromagnetic waves. If a dense beam of charged particles travels in the medium, then the coherent radiation will be possible which leads to the increase of the wave amplitude in time. In other words, this means that the system consisting of the medium and beam of charged particles becomes unstable. Therefore, the excitation of electromagnetic waves takes place in the system.

In this section, we will consider the interaction of electron beams with spatially unbounded collisionless plasma.[13] Moreover, we will restrict this study by consideration of straight neutralized beams traveling along the external magnetic field (or in the absence of such a field). Furthermore, we assume that these beams are spatially homogeneous and unbounded. Therefore, in the ground state of plasma, there is no electric current and inhomogeneous charge distribution. Such beams behave as moving plasma. Thus, the problem is reduced to the stability problem of non-equilibrium multi-stream plasma. According to the general theory, the growth rate of small perturbations is determined by the dispersion equation of the linear theory,

$$\left| k^2 \delta_{ij} - k_i k_j - \frac{\omega^2}{c^2} \varepsilon_{ij}\left(\omega, \vec{k}\right) \right| = 0, \tag{3.158}$$

[13]This problem was first considered by Akhiezer and Fainberg [92–94] and Gross and Bohm [95]. Today, an enormous number of papers are dedicated to this problem [89, 90, 92–129].

where $\varepsilon_{ij}\left(\omega, \vec{k}\right)$ is the dielectric permittivity of spatially homogeneous non-equilibrium plasma, which is of our interest. If only one of the roots $\omega_n\left(\vec{k}\right)$ of this equation has a positive imaginary part, $Im\,\omega_n\left(\vec{k}\right) > 0$, the corresponding perturbation will increase with time and the system will be unstable.

Let us suppose that the distribution function of the particles of the type α in the inertial system moving along with the particles is Maxwellian with non-relativistic temperature. In this case, the contribution of the particles of the type α to the dielectric tensor was already calculated in Chap. 2. Using these expressions, we can obtain the dielectric tensor of multi-component plasma in the laboratory frame without solving the kinetic equation, only by applying the Lorentz transformation formulas. Actually, the total induced current in plasma is the sum of the currents of its charged particle components

$$j_i\left(\omega, \vec{k}\right) = \sum_\alpha j_{\alpha i} = \sum_\alpha \sigma_{ij}^{(\alpha)}\left(\omega, \vec{k}\right)E_j, \tag{3.159}$$

where $\sigma_{ij}^{(\alpha)}\left(\omega, \vec{k}\right)$ is the contribution of the particles of the type α to the conductivity tensor in the laboratory frame. To determine the tensor $\sigma_{ij}^{\alpha}\left(\omega, \vec{k}\right)$, we pass over to the frame moving with the velocity $\vec{u}_\alpha$ (the directed velocity of the particle species α). In this frame, we have

$$j'_{i\alpha}\left(\omega'_\alpha, \vec{k'}_\alpha\right) = \sigma_{ij}^{(\alpha)}\left(\omega'_\alpha, \vec{k'}_\alpha\right)E'_{j\alpha}\left(\omega'_\alpha, \vec{k'}_\alpha\right), \tag{3.160}$$

where ω'_α and $\vec{k'}_\alpha$ are the Lorentz transformed frequency ω and wave vector $\vec{k}$, respectively,

$$\omega'_\alpha = \left(\omega - \vec{k}\cdot\vec{u}_\alpha\right)\gamma_\alpha,$$

$$\vec{k'_\alpha} = \vec{k} + \vec{u}_\alpha\gamma_\alpha\left[\frac{\vec{k}\cdot\vec{u}_\alpha}{u_\alpha^2}\left(1 - \frac{1}{\gamma_\alpha}\right) - \frac{\omega}{c^2}\right],$$

$$\gamma_\alpha = \left(1 - \frac{u_\alpha^2}{c^2}\right)^{-\frac{1}{2}}. \tag{3.161}$$

Here, $\sigma_{ij}^{(\alpha)}\left(\omega'_\alpha, \vec{k'}_\alpha\right)$ is the conductivity tensor of the α particle component in its respective moving frame; $\vec{E'}_\alpha$ and $\vec{j'}_\alpha$ are the electric field and current densities in this frame, which are related to $\vec{E}$ and $\vec{j}_\alpha$ by the Lorentz transformation

$$j_{i\alpha} = \alpha_{ij}\left(\vec{u}_\alpha\right) j'_{j\alpha}, \quad E'_{i\alpha} = \beta_{ij}(u_\alpha) E_j, \tag{3.162}$$

where

$$\begin{aligned} \alpha_{ij}\left(\vec{u}_\alpha\right) &= \delta_{ij} + \gamma_\alpha \left[\frac{u_{\alpha i} u_{\alpha j}}{u_\alpha^2}\left(1 - \frac{1}{\gamma_\alpha}\right) + \frac{k'_{\alpha j} u_{\alpha i}}{\omega'_\alpha}\right], \\ \beta_{ij}\left(\vec{u}_\alpha\right) &= \frac{\omega'_\alpha}{\omega} \alpha_{ji}\left(\vec{u}_\alpha\right) = \frac{\omega'_\alpha}{\omega} \delta_{ij} + \gamma_\alpha \left[\frac{u_{\alpha i} u_{\alpha j}}{u_\alpha^2}\left(\frac{1}{\gamma_\alpha} - 1\right) + \frac{k_i u_{\alpha j}}{\omega}\right]. \end{aligned} \tag{3.163}$$

It is easy to derive the dielectric tensor of multi-component plasma in the laboratory frame from Eqs. (3.159) and (3.163):

$$\begin{aligned} \varepsilon_{ij}\left(\omega, \vec{k}\right) &= \delta_{ij} + \frac{4\pi\imath}{\omega} \sum_\alpha \alpha_{i\mu}\left(\vec{u}_\alpha\right) \sigma^{(\alpha)}_{\mu\nu}\left(\omega'_\alpha, \vec{k'}_\alpha\right) \beta_{\nu j}\left(\vec{u}_\alpha\right) \\ &= \delta_{ij} + \sum_\alpha \frac{\omega'_\alpha}{\omega} \alpha_{i\mu}\left(\vec{u}_\alpha\right) \left[\varepsilon^{(\alpha)}_{\mu\nu}\left(\omega'_\alpha, \vec{k'}_\alpha\right) - \delta_{\mu\nu}\right] \beta_{\nu j}\left(\vec{u}_\alpha\right). \end{aligned} \tag{3.164}$$

Here, $\varepsilon^{(\alpha)}_{ij}\left(\omega'_\alpha, \vec{k'}_\alpha\right)$ denotes the dielectric tensor of the particles of the type α in their respective moving frame, which is known [see Eq. (3.19)]. Using this expression, one should keep in mind that not only the frequency and the wave vector but also the particle density (due to the volume contraction) and the mass must be transformed into $\varepsilon^{(\alpha)}_{ij}\left(\omega'_\alpha, \vec{k'}_\alpha\right)$. Hence, the Langmuir frequency is $\omega'_{p\alpha} = \omega_{p\alpha}\gamma_\alpha^{-1/2}$ where $\omega_{p\alpha} = \sqrt{4\pi e_\alpha^2 N_{0\alpha}/m_\alpha}$ remains invariant and the Larmor frequency is transformed as $\Omega'_\alpha = \Omega_\alpha \gamma_\alpha^{-1} = e_\alpha B_0/m_\alpha c\, 1/\gamma_\alpha$; $N_{0\alpha}$ is the density of the particles of the type α in the laboratory frame and m_α is their rest mass. Here, it should mention that the following relation holds for the dielectric permittivity which describes the longitudinal oscillations:

$$\varepsilon\left(\omega, \vec{k}\right) = \frac{k_i k_j}{k^2} \varepsilon_{ij}\left(\omega, \vec{k}\right) = 1 + \sum_\alpha \left[\varepsilon^{(\alpha)}\left(\omega'_\alpha, \vec{k'}_\alpha\right) - 1\right],$$

where $\varepsilon^{(\alpha)}\left(\omega'_\alpha, \vec{k'}_\alpha\right) = \frac{k_i k_j}{k^2} \varepsilon^{(\alpha)}_{ij}\left(\omega'_\alpha, \vec{k'}_\alpha\right)$ is the dielectric permittivity of the component of α species particles in its intrinsic frame.

It should be noted that Eqs. (3.162) and (3.163) generalize the well-known Minkowski material relations for moving isotropic media to the case of anisotropic media by taking into account spatial and frequency dispersions. These relations become much simpler in the non-relativistic limit when $u_\alpha \ll c$:

$$\omega'_\alpha=\omega-\vec{k}\cdot\vec{u}_\alpha,\ \vec{k'}_\alpha=\vec{k},\ \alpha_{ij}\left(\vec{u}_\alpha\right)=\delta_{ij}+\frac{k_j u_{\alpha i}}{\omega-\vec{k}.\vec{u}_\alpha},\ \beta_{ij}\left(\vec{u}_\alpha\right)=\frac{\left(\omega-\vec{k}\cdot\vec{u}_\alpha\right)\delta_{ij}+k_i u_{\alpha j}}{\omega},$$

$$\varepsilon_{ij}\left(\omega,\vec{k}\right)=\delta_{ij}+\sum_\alpha\frac{\omega-\vec{k}\cdot\vec{u}_\alpha}{\omega}\alpha_{i\mu}\left(\vec{u}_\alpha\right)\left[\varepsilon^{(\alpha)}_{\mu\nu}\left(\omega'_\alpha,\vec{k}\right)-\delta_{\mu\nu}\right]\beta_{\nu j}\left(\vec{u}_\alpha\right).\tag{3.165}$$

In summary, for a system consisting of a charged particle beam with velocity u and rest plasma, we have

$$\varepsilon_{ij}\left(\omega,\vec{k}\right)=\varepsilon^{(b)}_{ij}\left(\omega,\vec{k}\right)+\varepsilon^{(p)}_{ij}\left(\omega,\vec{k}\right)-\delta_{ij},\tag{3.166}$$

where $\varepsilon^{(b)}_{ij}\left(\omega,\vec{k}\right)$, $\varepsilon^{(p)}_{ij}\left(\omega,\vec{k}\right)$, and $\varepsilon_{ij}\left(\omega,\vec{k}\right)$ are the dielectric permittivity of the beam, plasma at rest, and the total beam-plasma system in the laboratory frame, respectively. Then,

$$\begin{aligned}\varepsilon_{ij}\left(\omega,\vec{k}\right)=&\,\varepsilon^{(p)}_{ij}\left(\omega,\vec{k}\right)+\frac{\omega'^2}{\omega^2}\left[\varepsilon^{(b)'}_{ij}\left(\omega',\vec{k'}\right)-\delta_{ij}\right]+\\&+\frac{\gamma^2}{\omega^2}\left\{\varepsilon^{(b)l}\left(\omega',\vec{k'}\right)-1-\frac{\omega'^2}{c^2k'^2}\left[\varepsilon^{(b)l}\left(\omega',\vec{k'}\right)-\varepsilon^{(b)\mathrm{tr}}\left(\omega',\vec{k'}\right)\right]\right\}\times\\&\times\left[\frac{\omega'}{\gamma}\left(u_ik_j+k_iu_j\right)+\left(k^2-\frac{\omega^2}{c^2}\right)u_iu_j\right],\end{aligned}\tag{3.167}$$

where $\varepsilon^{(b)'}_{ij}$ is the dielectric permittivity of the beam in the moving frame. If, in the latter expression, we pass to the limit $c\to\infty$, we find the dielectric permittivity of this beam-plasma system for the non-relativistic beam velocity u

$$\begin{aligned}\varepsilon_{ij}\left(\omega,\vec{k}\right)=&\,\varepsilon^{(p)}_{ij}\left(\omega,\vec{k}\right)+\frac{\omega'^2}{\omega^2}\left[\varepsilon'^{(b)}_{ij}\left(\omega',\vec{k}\right)-\delta_{ij}\right]\\&+\frac{1}{\omega^2}\left[\varepsilon^{(b)l}\left(\omega',\vec{k}\right)-1\right]\left[\omega'\left(u_ik_j+k_iu_j\right)+k^2u_iu_j\right],\end{aligned}\tag{3.168}$$

where $\omega'=\omega-\vec{k}.\vec{u}$ and $\vec{k}=\vec{k'}$. This relation is directly obtained from Eq. (2.19) by making use of the non-relativistic Maxwellian distribution function for plasma and beam particles.

Substituting the expression (3.167) into the dispersion equation (3.158), we find

$$k^2-\frac{\omega^2}{c^2}\left\{\varepsilon^{(p)\mathrm{tr}}\left(\omega,\vec{k}\right)+\frac{\omega'^2}{\omega^2}\left[\varepsilon^{(b)\mathrm{tr}}\left(\omega',\vec{k'}\right)-1\right]\right\}=0,\tag{3.169}$$

$$\left\{k^2-\frac{\omega^2}{c^2}\left(\varepsilon^{(p)\mathrm{tr}}\left(\omega,\vec{k}\right)+\frac{\omega'^2}{\omega^2}\left[\varepsilon^{(b)\mathrm{tr}}\left(\omega',\vec{k'}\right)-1\right]\right)\right\}\left[\varepsilon^{(b)l}\left(\omega',\vec{k'}\right)+\varepsilon^{(p)l}\left(\omega,\vec{k}\right)-1\right]-$$
$$-\frac{\gamma^2u^2k_\perp^2}{c^2}\left\{\varepsilon^{(b)l}\left(\omega',\vec{k'}\right)-1+\frac{\omega'^2}{c^2k'^2}\left[\varepsilon^{(b)\mathrm{tr}}\left(\omega',\vec{k'}\right)-\varepsilon^{(b)l}\left(\omega',\vec{k'}\right)\right]\right\}\times$$
$$\times\left\{\varepsilon^{(p)l}\left(\omega,\vec{k}\right)-1+\frac{\omega^2}{c^2k^2}\left[\varepsilon^{(p)\mathrm{tr}}\left(\omega,\vec{k}\right)-\varepsilon^{(p)l}\left(\omega,\vec{k}\right)\right]\right\}=0. \tag{3.170}$$

When $\vec{u}=0$, the equation system (3.169) and (3.170) reduces to the dispersion equation of longitudinal and transverse waves in plasma at rest. In the limiting case, when the density of plasma at rest is zero and we practically have only a beam, Eqs. (3.169) and (3.170) show the dispersion equation of electromagnetic waves in moving plasma [130].

Here, it should be noted that in what follows, we will consider only the beam-plasma system. However, the above relations can be applied to the case of the interaction of two plasma jets, when in general, they move relative to each other. This problem is of our interest in astrophysics and space plasma. Here, with the same method used for the beam-plasma system, we can examine some simple examples of the plasma-plasma system. In general, Eqs. (3.158), (3.164)–(3.170) can be used for this aim. Below, we will consider some examples of applications of the above relations.

3.7.1 Interaction of a Straight Monoenergetic Electron Beam with Cold Plasma: Cherenkov Instability

Let us begin the analysis of the interaction of a monoenergetic straight electron beam with cold plasma in the absence of the external magnetic field. In this case, for the particles of the type α, we have

$$\varepsilon^{(\alpha)}(\omega,k)=1-\frac{\omega_{p\alpha}^2}{\omega^2}. \tag{3.171}$$

For non-relativistic beam velocity, $u\ll c$, from Eq. (3.170), it follows that [92–94, 98, 99, 103, 104, 108, 109]

$$\varepsilon^{(b)l}\left(\omega',\vec{k'}\right)+\varepsilon^{(p)l}\left(\omega,\vec{k}\right)-1=0. \tag{3.172}$$

When $u\ll c$, the electric and magnetic fields are invariant under the transformation of coordinate system, and since here $\vec{k}=\vec{k'}$, Eq. (3.172) corresponds to the longitudinal plasma waves.

In the high-frequency range of ω and ω', when spatial dispersion is negligible, in the beam-plasma system, from Eq. (3.172), we find [92–94]

$$1-\frac{\omega_{\mathrm{pe}}^2}{\omega^2}-\frac{\omega_b^2}{\left(\omega-\vec{k}\cdot\vec{u}\right)^2}=0. \tag{3.173}$$

Since here the thermal motion of particles has been neglected, then Eq. (3.173) is valid only when $u \gg v_{\mathrm{Te}}, v_{\mathrm{Tb}}$. In the region $\omega \gg \vec{k}\cdot\vec{u}$, Eq. (3.173) has a solution corresponding to the undamped wave. In the opposite case, when $\omega \ll \vec{k}\cdot\vec{u}$, we find

$$\omega=\pm\frac{\omega_{\mathrm{pe}}^2\left(\vec{k}\cdot\vec{u}\right)}{\sqrt{\left(\vec{k}\cdot\vec{u}\right)^2-\omega_b^2}}. \tag{3.174}$$

From here, it is clear that when $\vec{k}\cdot\vec{u}<\omega_b$, the oscillation grows in time and the system is unstable. Equation (3.174) is valid only when $N_p \ll N_b$. In the opposite limit, when $N_b \ll N_p$, from Eq. (3.173), we find [92–94]

$$\omega=\vec{k}\cdot\vec{u}\pm\frac{\omega_b}{\sqrt{1-\frac{\omega_{\mathrm{pe}}^2}{\left(\vec{k}\cdot\vec{u}\right)^2}}}. \tag{3.175}$$

It is evident that the waves with the wavelengths $\vec{k}\cdot\vec{u}<\omega_{\mathrm{pe}}$ grow with time. Furthermore, there is another region of instability for Eq. (3.173), when $\omega \ll \omega_{\mathrm{pe}}$. In this case, we have

$$\omega=\frac{\left(\vec{k}\cdot\vec{u}\right)\left(1\pm\imath\sqrt{\frac{N_b}{N_p}}\right)}{1+\frac{N_b}{N_p}}. \tag{3.176}$$

This expression characterizes the waves growing in time for arbitrary beam density. The condition $\omega<\omega_{\mathrm{pe}}$ determines the applicability range of this expression:

$$\vec{u}\cdot\vec{k}\ll\sqrt{1+\frac{N_b}{N_p}}\omega_{\mathrm{pe}}.$$

In the limiting case $N_b \gg N_p$ and $N_b \ll N_p$, expression (3.176) reduces to expressions (3.175) and (3.174), respectively. Considering this expression and using Eqs. (3.163) and (3.164), from the dispersion equation (3.158), we obtain two equations for the plasma-beam system:

$$k^2c^2 - \omega^2\left(1 - \frac{\omega_{\rm pe}^2}{\omega^2} - \frac{\omega_b^2}{\gamma\omega^2}\right) = 0,$$

$$\left(k^2c^2 - \omega^2 + \omega_{\rm pe}^2 + \omega_b^2\gamma^{-1}\right)\left(1 - \frac{\omega_{\rm pe}^2}{\omega^2} - \frac{\omega_b^2}{\gamma^3\left(\omega - \vec{k}\cdot\vec{u}\right)^2}\right) - \frac{k_\perp^2 u^2}{\omega^2}\frac{\omega_{\rm pe}^2\omega_b^2}{\gamma\left(\omega - \vec{k}\cdot\vec{u}\right)^2} = 0. \quad (3.177)$$

Here, $\omega_{\rm pe}$ and ω_b are the Langmuir frequencies of plasma and of the beam electrons, respectively, in the laboratory frame; $\vec{u}$ is the beam directed velocity and $\gamma = (1 - u^2/c^2)^{-1/2}$. We have neglected the ion terms in Eq. (3.177), i.e., we consider only the interaction of the relativistic electron beam with the high-frequency plasma waves. Moreover, in the following, it will be assumed that the beam density N_b is much less than the plasma density N_b, i.e., $N_e \gg N_b$.

The first equation of Eq. (3.177) describes stable waves with real frequency, which is approximately equal to the frequency of transverse plasma waves. The contribution of the beam electrons is negligibly small due to their low density. One can easily understand why these waves are stable. The electric field of these longitudinal waves is oriented along the y axis. Thus, the exchange of energy with the beam electrons is impossible, $\vec{E}\cdot\vec{u} = 0$.

The second equation of Eq. (3.177) describes the longitudinal-transverse waves with field's polarization $\vec{u}\cdot\vec{E} \neq 0$. Therefore, the electric field of waves can affect the beam electrons. The beam can be decelerated by the field, transferring part of its energy to the wave vector. As a result, excitation of waves takes place, meaning that the beam-plasma system is unstable.

Two types of instability exist. The first type takes place when $\omega \simeq \vec{k}\cdot\vec{u}$, which is called Cherenkov instability. For simplicity, let us consider the purely longitudinal propagation of waves, assuming $k_\perp = 0$. Then, from the second equation of Eq. (3.177), we obtain

$$1 - \frac{\omega_{\rm pe}^2}{\omega^2} - \frac{\omega_b^2\gamma^{-3}}{(\omega - ku)^2} = 0. \quad (3.178)$$

Representing the solution of this equation as $\omega = ku + \delta$, it can be easily shown that the maximum growth rate (maximum Imδ) is obtained when $\omega^2 \simeq k^2u^2 \simeq \omega_p^2$. In this case, there are three roots

$$\begin{aligned}\delta_{1,2} &= \frac{-1 \pm \imath\sqrt{3}}{2}\omega_{\text{pe}}\left(\frac{N_b}{2N_p}\frac{1}{\gamma^3}\right)^{1/3},\\ \delta_3 &= \omega_{\text{pe}}\left(\frac{N_b}{2N_p}\frac{1}{\gamma^3}\right)^{1/3}.\end{aligned} \tag{3.179}$$

One of these solutions has the positive imaginary part, $\text{Im}\delta_1 > 0$. This solution corresponds to the unstable oscillations. This instability occurs when the beam velocity practically coincides (it slightly exceeds) with the phase velocity of plasma waves. Just by this reason, this instability is known as resonance Cherenkov instability.

It should be noted that the transverse component of the wave vector $k_\perp \neq 0$ practically does not affect the considered instability in the absence of the external magnetic field. Another quite similar situation takes place in the presence of such a field. To demonstrate this, let us consider the case when the external magnetic field is infinitely large and parallel to the axis z. In this case,

$$\varepsilon_{ij}^{(\alpha)}(\omega, k) = \begin{pmatrix} 1 & 0 & 0 \\ 0 & 1 & 0 \\ 0 & 0 & \varepsilon_{11}^{(\alpha)} \end{pmatrix}, \tag{3.180}$$

where

$$\varepsilon_{11}^{(\alpha)} = 1 - \frac{\omega_{p\alpha}^2}{\omega^2}. \tag{3.181}$$

Using these expressions, from the dispersion equation (3.158), we obtain

$$k^2c^2 - \omega^2 = 0, \qquad k_\perp^2 c^2 + \left(k_z^2 c^2 - \omega^2\right)\left[1 - \frac{\omega_{\text{pe}}^2}{\omega^2} - \frac{\omega_b^2 \gamma^{-3}}{\left(\omega - k_z u\right)^2}\right] = 0. \tag{3.182}$$

The first equation describes a purely transverse wave $\left(\vec{E} \parallel oy\right)$ with the phase velocity $\omega/k_z > c$; the beam does not interact with it. The second equation corresponds to longitudinal-transverse waves with the non-zero field components E_x and E_z so that $\left(\vec{E} \cdot \vec{u}\right) \neq 0$. These waves are interacting with the beam. In the first approximation, let us neglect the beam term in Eq. (3.182). Thus, without the beam term, this equation gives two branches of oscillations:

$$\omega_{1,2}^2 = \frac{1}{2}\left[\omega_{\mathrm{pe}}^2 + k^2c^2 \pm \sqrt{\left(\omega_{\mathrm{pe}}^2 + k^2c^2\right)^2 - 4\omega_{\mathrm{pe}}^2 k_z^2 c^2}\right], \tag{3.183}$$

corresponding to fast and slow waves, respectively. The electron beam can resonantly interact only with the slow wave. In other words, the instability will take place if $\omega = k_z u + \delta = \omega_2 + \delta$, and only under the condition

$$\omega_{\mathrm{pe}}^2 > k_\perp^2 u^2 \gamma^2, \tag{3.184}$$

which determines the threshold of instability in the plasma-beam system.

The solution of the second equation of Eq. (3.182), corresponding to time-increasing oscillations, has the form ($\delta \ll \omega/[2(\gamma^2 - 1)]$)

$$\begin{aligned} \omega &= \sqrt{\omega_{\mathrm{pe}}^2 - k_\perp^2 u^2 \gamma^2}, \\ \delta &= \frac{-1 + \imath\sqrt{3}}{2} \frac{\omega}{\gamma}\left(\frac{N_b}{2N_p}\right)^{\frac{1}{3}}\left[1 + \frac{k_\perp^2 u^2 \gamma^2(\gamma^2 - 1)}{\omega_{\mathrm{pe}}^2}\right]^{-\frac{1}{3}}. \end{aligned} \tag{3.185}$$

When $\omega \gg \vec{k} \cdot \vec{u}$, Eq. (3.177) in the high-frequency has no unstable solution. In the opposite limit, when $\omega \ll \vec{k} \cdot \vec{u}$, from Eq. (3.177), it follows that

$$\omega^2 = P_1 \pm \sqrt{P_1^2 - P_2}, \tag{3.186}$$

where

$$P_1 = \frac{1}{2}\frac{\left(k^2c^2 + \omega_{\mathrm{pe}}^2 + \gamma^{-1}\omega_b^2\right)\left[\left(\vec{k}\cdot\vec{u}\right)^2\left(1 + \omega_{\mathrm{pe}}^2\right) - \gamma^{-3}\omega_b^2\right]}{\left(\vec{k}\cdot\vec{u}\right)^2 - \gamma^{-3}\omega_b^2},$$

$$P_2 = \frac{\omega_{\mathrm{pe}}^2\left(\vec{k}\cdot\vec{u}\right)^2\left(k^2c^2 + \omega_{\mathrm{pe}}^2 + \gamma^{-1}\omega_b^2\right) + \gamma^{-1}\omega_b^2\omega_{\mathrm{pe}}^2 k_\perp^2 u^2}{\left(\vec{k}\cdot\vec{u}\right)^2 - \gamma^{-3}\omega_b^2}.$$

If $N_b \gg N_p$, the right side of the expression (3.186) may be complex, which corresponds to the wave growing in time. When $(u/c) \to 0$, this expression reduces to Eq. (3.174).

If $N_b \ll N_p$, from Eq. (3.186), we find the following approximation solution:

$$\omega = \vec{k} \cdot \vec{u} \pm \eta,$$

$$\eta = \frac{\gamma^{-3}\omega_b^2\left[k^2 + \frac{\omega_{\text{pe}}^2 + \gamma^{-1}\omega_b^2 - \left(\vec{k}\cdot\vec{u}\right)^2}{c^2}\right] + \frac{\gamma^{-1}\omega_b^2\omega_{\text{pe}}^2 k_\perp^2 u^2}{\left(\vec{k}\cdot\vec{u}\right)^2 c^2}}{\sqrt{\left[k^2 + \frac{\omega_{\text{pe}}^2 + \gamma^{-1}\omega_b^2 - \left(\vec{k}\cdot\vec{u}\right)^2}{c^2}\right]\left((1 - \frac{\omega_{\text{pe}}^2}{\left(\vec{k}\cdot\vec{u}\right)^2}\right)}}. \tag{3.187}$$

It is evident that when $\omega_{\text{pe}} > \left(\vec{k}\cdot\vec{u}\right)$, the beam-plasma system is unstable. It should be noticed that Eq. (3.187) is the generalization of Eq. (3.175) for the relativistic beam velocity $\vec{u}$.

When $\omega \ll \omega_b$, from Eq. (3.177), we find

$$\omega = \frac{\omega_{\text{pe}}^2\left(\vec{k}\cdot\vec{u}\right) \pm \imath\gamma^{-1}\omega_b\omega_{\text{pe}}\sqrt{\gamma^{-3}\left(\vec{k}\cdot\vec{u}\right)^2 + \frac{k_\perp^2 u^2\left(\gamma^{-3}\omega_b^2+\omega_{\text{pe}}^2\right)}{k^2c^2+\gamma^{-1}\omega_b^2+\omega_{\text{pe}}^2}}}{\gamma^{-3}\omega_b^2 + \omega_{\text{pe}}^2}. \tag{3.188}$$

This expression shows the waves growing in time and reduces to Eq. (3.176), in the limit $(u/c) \to 0$.

Beside the Cherenkov instabilities, considered above in the plasma-beam system, another type of instability exists, which takes place when $\vec{k}\cdot\vec{u} = 0$, i.e., when the purely transverse propagation of the waves occurs. Actually, in this case, from the second equation of Eq. (3.177), we find

$$\left(k^2c^2 - \omega^2 + \omega_{\text{pe}}^2 + \omega_b^2\gamma^{-1}\right)\left(1 - \frac{\omega_{\text{pe}}^2}{\omega^2} - \frac{\omega_b^2\gamma^{-3}}{\omega^2}\right) - \frac{k^2u^2\omega_{\text{pe}}^2\omega_b^2}{\omega^4\gamma} = 0. \tag{3.189}$$

It can be easily shown that the unstable solution of this equation exists in the low-frequency range, $\omega^2 \ll \omega_{\text{pe}}^2$:

$$\omega^2 \simeq -\frac{k^2u^2\omega_b^2\gamma^{-1}}{k^2c^2 + \omega_{\text{pe}}^2} \approx -\frac{u^2}{c^2}\frac{\omega_b^2}{\gamma}. \tag{3.190}$$

This instability is stipulated by the anisotropic distribution of the beam-plasma system as a whole. In this case, it is connected to the filamentation of the beam due to the perturbation of the magnetic field. This instability, known as interchange instability, corresponds to the lamination of the electron beam into separate current-carrying filaments with radii $r_0 < (c/\omega_{\text{pe}})$. It must be noted that the external magnetic field leads to the stabilization of this instability. In the infinite magnetic field this instability is absent.

3.7.2 *Effect of Thermal Motion on the Cherenkov Instability*

Let us now investigate the role of the thermal motion of the beam and the plasma electrons in the development of the Cherenkov instability, which was completely ignored above. For this reason, the above results are valid only when the phase velocities of the waves exceed the thermal velocity of the plasma electrons in the laboratory frame and the velocity spread of the beam electrons in their intrinsic frame, or

$$\frac{\omega}{k_z} \geq u \gg v_{\mathrm{Te}}, \qquad \frac{\omega'}{k'_z} \simeq \frac{\gamma^2\delta}{k_z} \simeq u\gamma\left(\frac{N_b}{2N_p}\right)^{\frac{1}{3}} \gg v_{\mathrm{Tb}}. \tag{3.191}$$

Here, v_{Te} is the thermal velocity of the plasma electrons and v_{Tb} is the non-relativistic velocity spread of the beam electrons in the intrinsic frame.

In order to find out how the Cherenkov instability in the plasma-beam system is modified when conditions (3.191) are violated, we will study the plasma-beam interaction in the simplest case of purely longitudinal waves propagating along the directed velocity of the beam. In this case, dispersion equation (3.158) does not depend on the strength of the external magnetic field and looks as (here $k' \approx k\gamma^{-1}$)

$$\varepsilon\left(\omega,\vec{k}\right) = \frac{k_i k_j}{k^2}\varepsilon_{ij}\left(\omega,\vec{k}\right) = 1 - \frac{\omega_{\mathrm{pe}}^2}{k^2 v_{\mathrm{Te}}^2}\left[1 - I_+\left(\frac{\omega}{k v_{\mathrm{Te}}}\right)\right] - \frac{\omega_b^2\gamma}{k^2 v_{\mathrm{Tb}}^2}\left[1 - I_+\left(\left(\frac{\omega - \vec{k}\cdot\vec{u}}{k v_{\mathrm{Tb}}}\right)\gamma^2\right)\right] = 0. \tag{3.192}$$

We see that, under the conditions (3.191), which are actually the conditions for the negligibility of the thermal motion in the system, this equation reduces to Eq. (3.177), which justifies the above consideration.

When the thermal motion of the particles is neglected, the Cherenkov beam-plasma instability is often called the hydrodynamic instability, which stresses that it is non-dissipative and can be described by the hydrodynamic equation of cold plasma. As it was shown above, this instability is developed for $\vec{k}\cdot\vec{u} \leq \omega_{\mathrm{pe}}$. We will show below that the Cherenkov instability also develops for $\vec{k}\cdot\vec{u} > \omega_{\mathrm{pe}}$ if the thermal motion of the particles is accounted for. In contrast to the hydrodynamic beam instability the instability for $\vec{k}\cdot\vec{u} > \omega_{\mathrm{pe}}$ is dissipative. It is caused by a change in the sign of the Cherenkov wave absorption and, therefore, it is called kinetic instability.

The kinetic beam instability develops when the opposite limit of the second inequality of inequalities (3.191) holds. In addition, we suppose that the first inequality remains. Then, the beam contribution to the real part of Eq. (3.192) can be neglected and may be written as

$$1-\frac{\omega_{\mathrm{pe}}^2}{\omega^2}\left(1+3\frac{k^2 v_{\mathrm{Te}}^2}{\omega^2}\right)+\imath\sqrt{\frac{\pi}{2}}\frac{\omega_{\mathrm{pe}}^2\omega}{k^3 v_{\mathrm{Te}}^3}\exp\left(-\frac{\omega^2}{2k^2 v_{\mathrm{Te}}^2}\right)$$
$$+\imath\sqrt{\frac{\pi}{2}}\frac{\omega_b^2\left(\omega-\vec{k}\cdot\vec{u}\right)}{k^3 v_{\mathrm{Tb}}^3}\gamma^3\exp\left[-\frac{\left(\omega-\vec{k}\cdot\vec{u}\right)^2\gamma^4}{2k^2 v_{\mathrm{Tb}}^2}\right]=0. \tag{3.193}$$

Assuming $\omega \to \omega + \delta$, the solution of this equation is

$$\omega^2=\omega_{\mathrm{pe}}^2+3k^2 v_{\mathrm{Te}}^2,$$
$$\delta=-\imath\sqrt{\frac{\pi}{8}}\omega_{\mathrm{pe}}^2\left\{\frac{\omega_b^2\gamma^3}{k^3 v_{\mathrm{Tb}}^3}\left(1-\frac{\vec{k}\cdot\vec{u}}{\omega_{\mathrm{pe}}}\right)\exp\left[-\frac{\left(\omega-\vec{k}\cdot\vec{u}\right)^2\gamma^4}{2k^2 v_{\mathrm{Tb}}^2}\right]+\frac{\omega_{\mathrm{pe}}^2}{k^3 v_{\mathrm{Te}}^3}\exp\left[-\frac{3}{2}-\frac{\omega_{\mathrm{pe}}^2}{2k^2 v_{\mathrm{Te}}^2}\right]\right\}. \tag{3.194}$$

If Im$\delta > 0$, the beam-plasma system is unstable, which is possible in the range $\vec{k}\cdot\vec{u} > \omega_{\mathrm{pe}}$. This instability is caused by a change in the sign of the Landau damping decrement of the beam term when the Cherenkov condition of particle radiation $u > \omega/k$ is satisfied. Note that the kinetic beam instability, regarding Eq. (3.194), has a maximum increment near the frequency of the Cherenkov resonance when $\vec{k}\cdot\vec{u} \simeq \omega \simeq \omega_{\mathrm{pe}}\left[1+(3/2)\,k^2 r_{\mathrm{De}}^2\right] > \omega_{\mathrm{pe}}$. The threshold of instability is determined by the condition $\delta = 0$ and the increment decreases with growing distance from the resonance frequency.

As was done above, we studied the effect of the thermal motion on the longitudinal wave excitation in the beam-plasma system. Now, we examine the same problem in the transverse waves in the non-relativistic limit.

When we take the thermal motion of particles into account, the expression of the dielectric permittivity will obtain large imaginary parts, indicating the strong damping of the waves. However, since the imaginary parts of the dielectric permittivity of the beam and plasma at rest are odd functions of ω and ω', then it is possible that the entire imaginary part of Eq. (3.169) becomes small, leading to the weakly damped waves. This possibility happens only when the thermal motion of particles is taken into account in both beam and plasma at rest. Thus, for non-relativistic temperature when $T_e \gg T_i$, and

$$\begin{cases} u \ll v_{\mathrm{Tb}} \left(1 + \dfrac{N_p}{N_b}\sqrt{\dfrac{T_e}{T_b}}\right), \\ u \ll v_{\mathrm{Te}} \left(1 + \dfrac{N_p}{N_b}\sqrt{\dfrac{T_e}{T_b}}\right), \end{cases} \tag{3.195}$$

where T_b is the beam temperature and v_{Tb} is the beam electron's thermal velocity, we have

$$\begin{cases} \varepsilon^{(b)\mathrm{tr}}\left(\omega', \vec{k}\right) = 1 + \imath\sqrt{\dfrac{\pi}{2}}\dfrac{\omega_b^2}{\omega' k v_{\mathrm{Tb}}}, \\ \varepsilon^{(p)\mathrm{tr}}\left(\omega, \vec{k}\right) = 1 + \imath\sqrt{\dfrac{\pi}{2}}\dfrac{\omega_{\mathrm{pe}}^2}{\omega k v_{\mathrm{Te}}}. \end{cases} \tag{3.196}$$

Substituting the latter expression in Eq. (3.169), we find

$$\omega = \frac{\vec{k}\cdot\vec{u} - \imath\sqrt{\frac{2}{\pi}} k v_{\mathrm{Tb}} \frac{k^2c^2}{\omega_b^2}}{1 + \frac{N_p}{N_b}\sqrt{\frac{T_e}{T_b}}}. \tag{3.197}$$

From the above relation, it is evident that this wave is weakly damped if

$$\frac{v_{\mathrm{Tb}}}{u}\frac{k^2c^2}{\omega_b^2} \ll 1.$$

3.7.3 *Current-Driven Instabilities in Plasma: Bunemann Instability*

Now, we will apply the method introduced above to the stability problem of current-driven plasmas.[14] In current-driven plasmas, the electrons move relative to the motionless ions with velocity $\vec{u}$, which is assumed to be parallel to the external magnetic field (if it exists). Such an assumption allows applying the method based on the Lorentz transform which was introduced above. Then, for the analysis of instabilities, we can use dispersion equation (3.158), in which $\varepsilon_{ij}\left(\omega, \vec{k}\right)$ may be calculated by Eqs. (3.163), (3.164).

[14]It is difficult to imagine plasma without any current. For example, plasma in an external constant electric field is current-driven. The drift velocity of electron u_e, very often may be regarded as constant and then all the results obtained in this section are valid for description of the stability of current-driven plasma.

Let us begin from non-magnetized current-driven plasma and consider the case when the electron's drift velocity is much greater than their thermal velocity. Then, the dispersion equation (3.158) takes the form

$$\left(k^2 - \frac{\omega^2}{c^2} + \frac{\omega_{\text{pe}}^2 \gamma^{-1} + \omega_{\text{pi}}^2}{c^2}\right)\left(1 - \frac{\omega_{\text{pe}}^2 \gamma^{-3}}{\left(\omega - \vec{k} \cdot \vec{u}\right)^2} - \frac{\omega_{\text{pi}}^2}{\omega^2}\right) - \frac{k_\perp^2 u^2 \omega_{\text{pe}}^2 \gamma^{-1} \omega_{\text{pi}}^2}{\omega^2 c^2 \left(\omega - \vec{k} \cdot \vec{u}\right)^2} = 0. \tag{3.198}$$

Only under the condition

$$\omega_{\text{pe}}^2 \geq \left(\vec{k} \cdot \vec{u}\right)^2 \gamma^3, \tag{3.199}$$

this equation has unstable solutions and the maximum increment is determined by

$$\delta = \frac{1 + \imath\sqrt{3}}{2} \left(\frac{m}{2M}\right)^{\frac{1}{3}} \left[1 + \frac{k_\perp^2 \gamma^2 u^2}{c^2 \left(k_\perp^2 + k_z^2 \gamma^2\right)}\right]^{\frac{1}{3}} \vec{k} \cdot \vec{u} \gamma, \tag{3.200}$$

which is obtained in the resonance case when expression (3.199) holds with the equality sign. This instability is known as Bunemann instability.[15]

A strong external magnetic field affects the stability of current-driven plasma, in particular, the Bunemann instability; thus, we have to generalize the above results for magneto-active plasma. For simplicity, we confine our analysis to sufficiently strong magnetic fields where the electrons are magnetized $\left(\Omega_e^2 \gg \omega_{\text{pe}}^2\right)$, but the ions, on the contrary, are not magnetized $\left(\omega_{\text{pi}}^2 \gg \Omega_i^2\right)$. Actually, such conditions correspond to the real situation.

Under these restrictions, Eq. (3.158) reduces to

$$k_\perp^2 \left(1 - \frac{\omega_{\text{pi}}^2}{\omega^2}\right) + \left[k_z^2 - \frac{\omega^2}{c^2}\left(1 - \frac{\omega_{\text{pi}}^2}{\omega^2}\right)\right]\left[1 - \frac{\omega_{\text{pe}}^2 \gamma^{-3}}{\left(\omega - k_z u\right)^2} - \frac{\omega_{\text{pi}}^2}{\omega^2}\right] = 0. \tag{3.201}$$

The unstable solutions of this equation exist only under the condition

---

[15] It is easily seen from Eq. (3.198) that in the non-resonance case, when relation (3.199) is satisfied with the inequality sign, the increment of Bunemann instability would be significantly smaller: $Im\omega \approx (m/2M)^{1/2} \gamma \vec{k} \cdot \vec{u}$.

$$\omega_{\mathrm{pe}}^2 \geq \left(k_\perp^2 + k_z^2\right) u^2 \gamma^3. \tag{3.202}$$

The instability, as considered above, is periodic since $\omega \ll k_z u$ and its increment becomes maximum for the resonance case when expression (3.202) holds with the equality sign:

$$\delta = \frac{1 + \iota\sqrt{3}}{2} \left[\frac{m}{2M} \left(1 + \frac{k_\perp^2}{k_z^2} \right) \right]^{\frac{1}{3}} k_z u \gamma. \tag{3.203}$$

Comparison of relations (3.202) and (3.203) with relations (3.199) and (3.200) shows that the external magnetic field hinders the development of the high-frequency Bunemann instability. The applicability range of expression (3.202) in which the instability exists is much narrower than the applicability range of expression (3.199). Furthermore, the increment of the instability in magneto-active plasma given by Eq. (3.203) is smaller than the increment given by Eq. (3.200) in the absence of the magnetic field.

3.7.4 *Current-Driven Instabilities in Plasma: Ion-Acoustic Instability*

As mentioned above, the considered high-frequency Bunemann instability occurs if $u \gg v_{\mathrm{Te}}$. Below, we will show that the current-driven instability is also possible for $u \ll v_{\mathrm{Te}}$. At these small drift velocities of electrons, the oscillations excited in current-driven plasma are obviously longitudinal with a high degree of accuracy and, therefore, Eq. (3.158) can be replaced by [131]

$$\begin{aligned}\varepsilon\left(\omega,\vec{k}\right) = \frac{k_i k_j}{k^2}\varepsilon_{ij}\left(\omega,\vec{k}\right) = 1 + \frac{\omega_{\mathrm{pe}}^2}{k^2 v_{\mathrm{Te}}^2}\left[1 - \sum_s \frac{\omega - \vec{k}\cdot\vec{u}}{\omega - \vec{k}\cdot\vec{u} - s\Omega_e} I_+\left(\frac{\omega - \vec{k}\cdot\vec{u} - s\Omega_e}{k_z v_{\mathrm{Te}}}\right) A_s\left(\frac{k_\perp^2 v_{\mathrm{Te}}^2}{\Omega_e^2}\right)\right] \\ + \frac{\omega_{\mathrm{pi}}^2}{k^2 v_{\mathrm{Ti}}^2}\left[1 - \sum_s \frac{\omega}{\omega - s\Omega_i} I_+\left(\frac{\omega - n\Omega_i}{k_z v_{\mathrm{Ti}}}\right) A_s\left(\frac{k_\perp^2 v_{\mathrm{Ti}}^2}{\Omega_i^2}\right)\right] = 0.\end{aligned} \tag{3.204}$$

Let us begin the analysis of Eq. (3.204) from the case when the external magnetic field is absent, i.e., $\Omega_{e,i} = 0$. Moreover, for $u \ll v_{\mathrm{Te}}$, unstable solutions of Eq. (3.204) can be expected in the frequency range $k v_{\mathrm{Ti}} \ll \omega \ll k v_{\mathrm{Te}}$. Then, under this condition, from Eq. (3.204), it follows

$$1+\frac{\omega_{\mathrm{pe}}^2}{k^2 v_{\mathrm{Te}}^2}\left(1+\imath\sqrt{\frac{\pi}{2}}\frac{\omega-\vec{k}\cdot\vec{u}}{k v_{\mathrm{Te}}}\right)-\frac{\omega_{\mathrm{pi}}^2}{\omega^2}+\imath\sqrt{\frac{\pi}{2}}\frac{\omega_{\mathrm{pi}}^2\omega}{k^3 v_{\mathrm{Ti}}^3}\exp\left(-\frac{\omega^2}{2k^2 v_{\mathrm{Ti}}^2}\right)=0. \quad (3.205)$$

In this equation, the imaginary terms arising from the Cherenkov dissipation of the wave by the particles (electrons and ions) are small compared to the real terms. Therefore, the solution can be obtained in the form of $\omega \to \omega + \delta$,

$$\omega^2=\frac{\omega_{pi}^2}{1+\dfrac{\omega_{pe}^2}{k^2 v_{Te}^2}},$$
$$\frac{\delta}{\omega}=-\imath\sqrt{\frac{\pi}{8}\frac{M}{m}}\frac{\omega^3}{k^3 v_{Te}^3}\left(1-\frac{u}{v_{ph}}\cos\vartheta\right)-\imath\sqrt{\frac{\pi}{8}}\frac{\omega^3}{k^3 v_{Ti}^3}\exp\left(-\frac{\omega^2}{2k^2 v_{Ti}^2}\right). \quad (3.206)$$

Here, $v_{\mathrm{ph}} = \omega/k$ is the phase velocity of waves and v is the angle between the vectors $\vec{u}$ and $\vec{k}$.

When $\vec{u} = 0$, spectrum (3.206) represents the ion-acoustic oscillations, occurring in non-isothermal plasma with $T_e \gg T_i$ (see Sect. 2.5), which in this limit are damped. The second relation of Eq. (3.206) shows that the damping decrement decreases for the non-zero electron's drift velocity $\vec{u}$. For $u > u_{\mathrm{cr}}$, when $\delta > 0$, these oscillations become unstable. Thus, this instability leading to the buildup of ion-acoustic oscillations is a purely Cherenkov instability and, therefore, it is called the ion-acoustic instability of current-driven plasma.

In conclusion, let us consider the influence of the external magnetic field on the development of the ion-acoustic instability. For simplicity, we consider the low-frequency oscillations, $\omega \ll \Omega_e$. In this case, instability can be expected in the phase velocity range $v_{\mathrm{Ti}} \ll \omega/k \ll v_{\mathrm{Te}}$. Then, from Eq. (3.204), for small velocities of electron's drift $u \ll v_{\mathrm{Te}}$, we obtain

$$1+\frac{\omega_{\mathrm{pe}}^2}{k^2 v_{\mathrm{Te}}^2}\left(1+\imath\sqrt{\frac{\pi}{2}}\frac{\omega-\vec{k}\cdot\vec{u}}{|k_z| v_{\mathrm{Te}}}\right)-\frac{\omega_{\mathrm{pi}}^2}{\omega^2}+\imath\sqrt{\frac{\pi}{2}}\frac{\omega_{\mathrm{pi}}^2\omega}{k^3 v_{\mathrm{Ti}}^3}\exp\left(-\frac{\omega^2}{2k^2 v_{\mathrm{Ti}}^2}\right)=0. \quad (3.207)$$

As before, the ions are considered non-magnetized and the electrons are strongly magnetized in the derivations.

Equation (3.207) differs from the analogous equation (3.205) only by a small imaginary term originating from Cherenkov absorption of waves by the electrons. Thus, the above analysis remains valid and the resulting expressions (3.206) must be modified only by the following replacement:

$$1 - \frac{u}{v_{\mathrm{ph}}} \cos \vartheta \to \frac{1}{\mid \cos \vartheta \mid} \left(1 - \frac{u}{v_{\mathrm{ph}}} \cos \vartheta \right). \tag{3.208}$$

As a result, the instability threshold determined by the condition $\delta \geq 0$ becomes smaller and the increment of the instability will be larger than that in the absence of the magnetic field. Consequently, the external magnetic field facilities the development of the ion-acoustic instability in current-driven plasma.

3.8 Dielectric Tensor of Weakly Inhomogeneous Magnetized Plasmas in the Approximation of Geometrical Optics

We now come into the explicit calculation of the dielectric tensor of weakly inhomogeneous plasmas $\varepsilon_{ij}\left(\omega, \vec{k}, x\right)$, and start the analysis with collisionless magneto-active plasma. As commonly done, we apply the kinetic equation with a self-consistent field (Vlasov's equation) for particles of the typeα:

$$\frac{\partial f_\alpha}{\partial t} + \vec{v} \cdot \frac{\partial f_\alpha}{\partial \vec{r}} + e_\alpha \left\{ \vec{E} + \frac{1}{c} \left(\vec{v} \times \vec{B} \right) \right\} \cdot \frac{\partial f_\alpha}{\partial \vec{p}_\alpha} = 0. \tag{3.209}$$

3.8.1 Distribution Function of Equilibrium Inhomogeneous Plasma

The external magnetic field $\vec{B}_0$ is assumed to be oriented along the z-axis and the gradient of the plasma inhomogeneity is taken along the x-axis, i.e., across the field $\vec{B}_0$. The distribution function $f_{0\alpha}(\vec{v}, x)$ of the stationary state, where $\vec{E}_0 = 0$, $B_0 \parallel oz$, should be determined first. In non-relativistic plasma where $\vec{p}_\alpha = m_\alpha \vec{v}$, we obtain from Eq. (3.209) for $f_{0\alpha}(\vec{v}, x)$

$$v_\perp \cos \phi \frac{\partial f_{0\alpha}}{\partial x} - \Omega_\alpha(x) \frac{\partial f_{0\alpha}}{\partial \phi} = 0. \tag{3.210}$$

We used cylindrical coordinates in the velocity space $\left(v_x = v_\perp \cos \phi, v_y = v_\perp \sin \phi, v_z\right)$ and introduced $\Omega_\alpha(x) = e_\alpha B_0(x)/(m_\alpha c)$, the inhomogeneous cyclotron frequency of particles of the type α.

Any function of the characteristics ϵ_α and C_α,

$$f_{0\alpha}(\vec{v},x) = f_{0\alpha}(\epsilon_\alpha, C_\alpha),$$

is a general solution of Eq. (3.210). Here, $\epsilon_\alpha = m_\alpha v^2/2$ is the energy and C_α follows from the characteristic equation

$$\frac{dx}{v_\perp \cos\phi} = -\frac{d\phi}{\Omega_a(x)}, \tag{3.211}$$

which has solutions of the form

$$C_\alpha = v_\perp \sin\phi + \int^x \Omega_a(x')\,dx'. \tag{3.212}$$

Hence,

$$f_{0\alpha} = f_{0\alpha}\left[\epsilon_\alpha, v_y + \int^x \Omega_a(x')\,dx'\right]. \tag{3.213}$$

In real plasmas, the characteristic length of the inhomogeneity significantly exceeds the Larmor radius of the particles. This allows us to introduce the small parameter

$$v_{T\alpha}/(\Omega_a L_0) \ll 1 \tag{3.214}$$

and to expand solution (3.213) in powers of it. We can write

$$f_{0\alpha}(\epsilon_a, C_a) = \left(1 + \frac{v_\perp \sin\phi}{\Omega_\alpha}\frac{\partial}{\partial x}\right) F_\alpha(\epsilon_\alpha, x), \tag{3.215}$$

where $F_\alpha(\epsilon_\alpha, x)$ is an arbitrary function of ϵ_α depending, in addition, parametrically on x. For non-degenerate plasma, it is natural to choose the local Maxwellian distribution function with inhomogeneous density and temperature and to write

$$F_\alpha(\epsilon_\alpha, x) = \frac{N_\alpha(x)}{[2\pi m_\alpha T_\alpha(x)]^{3/2}} \exp\left(-\frac{\epsilon_\alpha}{T_\alpha(x)}\right). \tag{3.216}$$

Due to the inhomogeneity, plasma in local equilibrium obtains principally new properties which we are studying in more detail now. We calculate the density of the electric charge ρ_0 and of the current $\vec{j}_0$ in local equilibrium. Assuming $\vec{E}_0 = 0$, we have

$$\rho_0 = \sum_\alpha e_\alpha \int f_{0\alpha} d\vec{p} = \sum_\alpha e_\alpha \int F_\alpha d\vec{p} = \sum_\alpha e_\alpha N_\alpha(x) = 0, \tag{3.217}$$

which is the condition for plasma quasi-neutrality.

The current density in local density in local equilibrium is

$$\vec{j}_0 = \sum_\alpha e_\alpha \int d\vec{p}\,\vec{v} f_{0\alpha}. \tag{3.218}$$

From Eq. (3.215) it follows for $f_{0\alpha}$ that only the second term, proportional to $v_\perp \sin\phi = v_y$, contributes to the current. The current flows parallel to the oy-axis explicitly we have

$$j_{0x} = j_{0z} = 0,$$

$$j_{0y} = \sum_\alpha e_\alpha \int v_y f_{0\alpha} d\vec{p} = \sum_\alpha e_\alpha \int d\vec{p}\, \frac{v_y^2}{\Omega_\alpha} \frac{\partial F_\alpha}{\partial x} = \sum_\alpha \frac{e_\alpha}{\Omega_\alpha m_\alpha} \frac{\partial N_\alpha T_\alpha}{\partial x}$$
$$= \frac{c}{B_0} \sum_\alpha \frac{\partial N_\alpha T_\alpha}{\partial x} = \frac{c}{B_0} \frac{\partial p_0}{\partial x}, \tag{3.219}$$

where $p_0 = \sum_\alpha N_\alpha T_\alpha$ is the total plasma pressure.

3.8.2 Magnetic Confinement of Inhomogeneous Plasma

Substituting relations (3.219) into the Maxwell's equation

$$\nabla \times B_0 = \frac{4\pi}{c} \vec{j}_0$$

gives the following condition for the plasma equilibrium:

$$\frac{\partial}{\partial x}\left(\frac{B_0^2}{8\pi} + p_0\right) = 0. \tag{3.220}$$

This is the well-known MHD condition for the ideal plasma equilibrium which has a clear physical meaning: inhomogeneous plasma is confined by the magnetic field pressure, and as a result the gradients of the magnetic and hydrodynamic pressure compensate each other. There is no need to analyze the corollaries, following from this condition, in detail. Note, however, that one of these corollaries is a conclusion about the possibility to confine hot plasmas in various thermonuclear devices. This conclusion is of great practical importance. In the following we need only the equilibrium condition (3.220), which implies that for low-pressure plasmas with

$$\beta = \frac{8\pi p_0}{B_0^2} \ll 1$$

the characteristic length of the inhomogeneity of the magnetic field L_B must greatly exceed the characteristic inhomogeneity scale of the kinetic pressure L_p(or L_N, L_T). Actually, Eq. (3.220) can be written as

$$\frac{d}{dx} \ln \frac{B_0^2}{8\pi} + \beta \frac{d}{dx} \ln p_0 = 0. \tag{3.221}$$

Hence, it follows that $(L_p/L_B) \sim \beta$ since $d\left[\ln\left(B_0^2/8\pi\right)\right]/dx \sim 1/L_B$and $d(\ln p_0)/dx \sim L_p^{-1}$. Low β plasmas $\beta \ll 1$ occur in a large number of practical cases, for example, in many devices for thermonuclear fusion, in many gas discharges and ionospheric plasmas, and in degenerate solid-state plasmas when the magnetic field is strong. Below we restrict ourselves to these low β plasmas, neglecting the magnetic field inhomogeneity as compared with the particle density and temperature scales.

3.8.3 Dielectric Tensor of Weakly Inhomogeneous Magnetized Plasma

Admitting a small perturbation δf_α of the local equilibrium in the form of an oscillation

$$\delta f_\alpha = \delta f_\alpha(x) \exp\left(-\imath\omega t + \imath k_y y + \imath k_z z\right), \tag{3.222}$$

we obtain from Eq. (3.209)

$$\begin{aligned}&\left(\omega - k_y v_y - k_z v_z\right)\delta f_\alpha + \imath v_x \frac{\partial \delta f_\alpha}{\partial x} - \imath \Omega_\alpha \frac{\partial \delta f_\alpha}{\partial \phi}\\ &= -\imath e_\alpha \left\{\vec{E} + \frac{1}{c}\left(\vec{v} \times \vec{B}\right)\right\} \cdot \frac{\partial f_{0\alpha}}{\partial \vec{p}_\alpha}.\end{aligned} \tag{3.223}$$

Note that the characteristic of this inhomogeneous partial differential equation is given by relation (3.212). Neglecting the magnetic field inhomogeneity ($\beta \ll 1$) thus yields

$$v_\perp \sin\phi + \Omega_\alpha x = v_\perp \sin\phi' + \Omega_\alpha x' = C_\alpha. \tag{3.224}$$

With account of this relation the general solution of Eq. (3.223) can be written as

$$\delta f_\alpha = \frac{e_\alpha}{m_\alpha \Omega_\alpha} \int\limits_{\infty}^{\phi} d\phi' \left\{ \vec{E}(x') + \frac{1}{c}\left[\vec{v} \times \vec{B}(x')\right] \right\} \cdot \frac{\partial f_{0\alpha}(x', \phi')}{\partial \vec{v}(\phi')} \times \exp\left[\frac{\imath}{\Omega_\alpha} \int\limits_{\phi}^{\phi'} d\phi'' \left(\omega - k_y v_\perp \sin\phi'' - k_z v_z\right) \right]. \tag{3.225}$$

Here, the coordinate x' depends on x and ϕ through the characteristic equation (3.224).

The function $\delta f_\alpha(x)$ and the fields $\vec{E}(x)$,$\vec{B}(x)$ can be presented in the form of $\exp\left[\imath \int^{x} k_x(x')\, dx'\right]$, too. Confining our consideration to the zero-order approximation of geometrical optics, i.e., when differentiating, taking account of the term proportional to $k_x(x)$ only, and ignoring the terms proportional to the spatial derivatives $k_x'(x)$, we can proceed further. Taking the derivative $\partial/\partial x$ of expression (3.225) and of the field equations and multiplying these expressions by k_x, we can eliminate due to the field equation $\partial\vec{B}/\partial t = -c\nabla \times \vec{E}$ the magnetic induction $\vec{B}(x')$ from expression (3.225) which gives forδf_α:

$$\delta f_\alpha\left(\vec{k}, x\right) = \frac{e_\alpha}{\Omega_\alpha} \int\limits_{\infty}^{\phi} d\phi' \left[\left(1 - \frac{\vec{k}\cdot\vec{v}}{\omega}\right)\delta_{ij} + \frac{v_i k_j}{\omega}\right]_{\phi'} \times \frac{\partial f_{0\alpha}(x', \phi')}{\partial p_{\alpha j}} E_i\left(\vec{k}, \omega\right) \exp\left[\frac{\imath}{\Omega_\alpha} \int\limits_{\phi}^{\phi'} d\phi'' \left(\omega - \vec{k}\cdot\vec{v}\right)_{\phi''}\right]. \tag{3.226}$$

Here, the vector $\vec{k}$ is three-dimensional $\vec{k} = \left(k_x, k_y, k_z\right)$.
Taking into account that

$$\frac{\partial f_{0\alpha}}{\partial p_{\alpha j}} = \frac{\partial f_{0\alpha}}{\partial \epsilon_{\alpha j}} v_j + \frac{\partial f_{0\alpha}}{\partial C_\alpha}\frac{\partial C_\alpha}{\partial p_{\alpha j}} = \frac{\partial f_{0\alpha}}{\partial \epsilon_\alpha} v_j + \frac{\delta_{yj}}{m_\alpha \Omega_\alpha}\frac{\partial f_{0\alpha}}{\partial x}, \tag{3.227}$$

and substituting relation (3.227) into expression (3.226), we obtain by simple transformations completely analogous to those made for homogeneous plasmas

$$\delta f\left(\vec{k},x\right) = \delta f_1\left(\vec{k},x\right) + \delta f_2\left(\vec{k},x\right). \tag{3.228}$$

For inhomogeneous Maxwellian plasma distributed according to expressions (3.215) and (3.216) we have

$$\delta f_1\left(\vec{k},x\right) = \frac{1}{T}\left(1 - \frac{k_y v_T^2}{\Omega\omega}\frac{\partial}{\partial x}\right)T\delta f^{(0)}\left(\vec{k},x\right),$$

$$\delta f_2\left(\vec{k},x\right) = -\frac{\imath e E_y}{m\Omega\omega}\frac{\partial F(x,\vec{v})}{\partial x}\sum_{s,n} J_s\left(\frac{k_\perp v_\perp}{\Omega}\right)J_n\left(\frac{k_\perp v_\perp}{\Omega}\right)e^{\imath(n-s)(\xi-\phi)}. \tag{3.229}$$

The species index α is omitted for simplicity here, and the following notations are introduced: ξ is the polar angle of the vector $\vec{k}$, i.e., $\vec{k} = (k_x = k_\perp\cos\xi, k_y = k_\perp\sin\xi, k_z)$; δf_0 is a function coinciding in form with the correction of the equilibrium distribution function of homogeneous Maxwellian plasma:

$$\delta f_\alpha = \frac{\gamma e_\alpha}{\Omega_\alpha}\int_\infty^{\varphi'}\left(\vec{E}\cdot\frac{\partial f_{0\alpha}}{\partial \vec{p}_\alpha}\right)_{\varphi'}\exp\left[\frac{\imath\gamma}{\Omega_\alpha}\int_\varphi^{\varphi'} d\varphi''\left(\omega - \vec{k}\cdot\vec{v}\right)_{\varphi'}\right],$$

the only difference being that N and T are now assumed to depend on the coordinate x.

It is easy to show that δf_2 does not contribute to the current density induced in plasma. Due to the plasma quasi-neutrality, after summation over the species index of the charged particles the contribution of δf_2 to the density of the induced charge density vanishes, too. Thus, the densities of the space charge and the current and therefore the dielectric tensor are determined by the correction $\delta f_1\left(\vec{k},x\right)$ alone. From the explicit form of expression (3.229) of $\delta f_1\left(\vec{k},x\right)$, the dielectric tensor of weakly inhomogeneous non-degenerate plasma follows:

$$\varepsilon_{ij}\left(\omega,\vec{k},x\right) = \delta_{ij} + \sum_\alpha \frac{1}{T_\alpha}\left(1 - \frac{k_y v_{T\alpha}^2}{\omega\Omega_\alpha}\frac{\partial}{\partial x}\right)T_\alpha\left[\varepsilon_{ij}^\alpha\left(\omega,\vec{k},x\right) - \delta_{ij}\right]. \tag{3.230}$$

Here, $\varepsilon_{ij}^\alpha\left(\omega,\vec{k},x\right)$ is the partial contribution of the particles of type α to the dielectric tensor. It coincides normally with the dielectric tensor of homogeneous plasma of Eq. (3.19).

However, N_α and T_α are functions of the coordinate x, here.

3.8.4 Larmor Drift Frequency

It should be noted that tensor (3.19) is used in a coordinate system where the wave vector has the components $\vec{k} = (k_\perp, 0, k_z)$. In Eq. (3.230) the orientation of $\vec{k}$ is arbitrary, however: $\vec{k} = \left(k_x, k_y, k_z\right)$. Therefore, tensor (3.19) must be transformed according to the general transformation rule applicable when the frame of coordinates is rotated. In general, however, this is not necessary as we will see later. Note that the longitudinal dielectric permittivity is invariant with respect to the transformation of the coordinate system. Thus, in order to calculate the longitudinal dielectric permittivity we can use Eq. (3.17), keeping in mind that we have to apply the operator

$$\sum_\alpha \frac{1}{T_\alpha}\left(1 - \frac{k_y v_{T\alpha}^2}{\omega\Omega_\alpha}\frac{\partial}{\partial x}\right)T_\alpha \tag{3.231}$$

to the components of this tensor in the case of inhomogeneous plasma. As a result, we obtain

$$\begin{aligned}\varepsilon\left(\omega, \vec{k}, x\right) = 1 + \sum_\alpha \frac{\omega_{p\alpha}^2}{k^2 v_{T\alpha}^2}\Bigg\{1 - \sum_n \frac{\omega}{\omega - n\Omega_\alpha}\Bigg[1 - \frac{k_y v_{T\alpha}^2}{\omega\Omega_\alpha} \\ \times\left(\frac{\partial \ln N_\alpha}{\partial x} + \frac{\partial T_\alpha}{\partial x}\frac{\partial}{\partial T_\alpha}\right)\Bigg] A_n\left(\frac{k_\perp^2 v_{T\alpha}^2}{\Omega_\alpha^2}\right) I_+\left(\frac{\omega - n\Omega_\alpha}{k_z v_{T\alpha}}\right)\Bigg\}.\end{aligned} \tag{3.232}$$

Here, $k_\perp$ is given by $k_\perp = \sqrt{k_x^2 + k_y^2}$.

As it is seen from expressions (3.230) and (3.232), there appears a new characteristic frequency

$$\omega_{\mathrm{dr}\alpha} = k_y v_{\mathrm{dr}\alpha} \sim \frac{k_y v_{T\alpha}^2}{\Omega_\alpha L_0} \tag{3.233}$$

in inhomogeneous Maxwellian plasma with the characteristic inhomogeneity scale L_0, which is called the Larmor drift frequency (we shall analyze the meaning and the physical nature of $\omega_{\mathrm{dr}\alpha}$ later). For high frequencies $\omega \gg \omega_{\mathrm{dr}\alpha}$ the terms of the tensor components $\varepsilon_{ij}\left(\omega, \vec{k}, x\right)$ which contain space derivatives can be neglected. Then the remaining components exactly coincide with the corresponding expressions for the components of the dielectric permittivity of homogeneous plasma with space dependent N_α and T_α, however. Moreover, from the derivation of expression (3.230) it follows that in the limit $\omega \gg \omega_{\mathrm{dr}\alpha}$ the relations for the dielectric tensor components, taking account of particle collisions, are also valid (Sect. 3.5). This holds true as well for the longitudinal and for the transverse dielectric permittivity of isotropic plasma without external fields (Sects. 2.3 and 2.7).

3.9 Spectra of HF and Larmor Oscillations in Weakly Inhomogeneous Plasmas

We apply the general results, obtained above, to the analysis of high-frequency ($\omega \gg \omega_{\mathrm{dr}\,\alpha}$) and low-frequency oscillations of weakly inhomogeneous plasma. We are especially interested in the limiting cases where analytical relations for the frequency spectra can be obtained.

3.9.1 *Transverse Oscillations of Weakly Inhomogeneous Isotropic Plasma*

To begin with, the transverse oscillations in isotropic plasma are analyzed. In the limit of sufficiently infrequent particle collisions $\omega \gg \nu_e$, using relations (2.127) and (2.182) for $\varepsilon^{\mathrm{tr}}\left(\omega, \vec{k}, x\right)$, we get the eikonal equation for these oscillations in the form of

$$k^2 - \frac{\omega^2}{c^2}\left[1 - \frac{\omega_{\mathrm{pe}}^2(x)}{\omega^2}\left(1 - \imath\frac{\nu_e(x)}{\omega}\right)\right] = 0. \tag{3.234}$$

The collision frequency is $\nu_e = \nu_{\mathrm{en}}$ or $\nu_e = \nu_{\mathrm{eff}}$ for the weakly or completely ionized plasma, respectively. Hence, we find

$$k_x^2(\omega, x) = -k_y^2 - k_z^2 + \frac{\omega^2}{c^2}\left(1 - \frac{\omega_{\mathrm{pe}}^2}{\omega^3} + \imath\frac{\omega_{\mathrm{pe}}^2\nu_e}{\omega^3}\right). \tag{3.235}$$

To calculate the frequency spectrum and the damping decrement of these oscillations according to the technique described above, one has to determine $\mathrm{Re}\{k_x\}$ and $\mathrm{Im}\{k_x\}$ from Eq. (3.235) and to substitute the result into Eq. (1.278). Then, taking account of $\mathrm{Re}\{k_x\} \gg \mathrm{Im}\,\{k_x\}$, we obtain the dispersion equation of the transverse electromagnetic waves in isotropic inhomogeneous plasma in the approximation of geometrical optics:

$$\begin{aligned}
&\int dx\,\mathrm{Re}\,\{k_x\} = \int dx\left(-k_y^2 - k_z^2 + \frac{\omega^2}{c^2} - \frac{\omega_{\mathrm{pe}}^2}{c^2}\right)^{1/2} = \pi n, \\
&\delta = -\frac{1}{2}\int \frac{dx}{\mathrm{Re}\,\{k_x\}}\frac{\omega_{\mathrm{pe}}^2\nu_e}{\omega^2}\left(\int \frac{dx}{\mathrm{Re}\,\{k_x\}}\right)^{-1}.
\end{aligned} \tag{3.236}$$

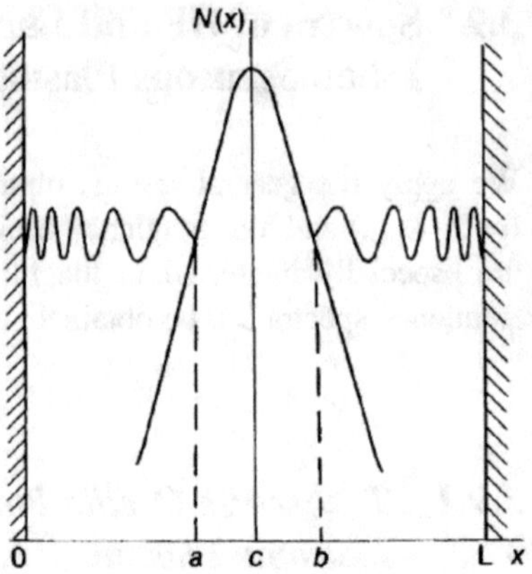

Fig. 3.3 Transparence range of transverse waves in isotropic plasma with density decreasing towards the periphery

It can be seen from Eq. (3.236) that the frequency of the transverse waves always exceeds the local value of the electron plasma frequency $\omega^2 > \omega_{\text{pe}}^2(x)$ in weakly inhomogeneous plasma. The range of transparency is determined by the condition that $\omega_{\text{pe}}^2(x)$ has to be smaller than the local value $\omega_{\text{pe}}^2(b)$at the turning point $x = b$:

$$\left(k_y^2 + k_z^2\right)c^2 + \omega_{\text{pe}}^2(b) = \omega^2. \tag{3.237}$$

In the majority of real experiments the distribution of the charged particle density achieves a maximum at some point along the direction of the inhomogeneity and then falls off smoothly. Examples are the radial distribution of the charged particle density in plasma of a gas-discharge, in thermonuclear fusion plasma (Fig. 3.3) and also in ionospheric plasma.

It is easily seen that there exist two turning points a and b for frequencies $\omega^2 < \omega_{\text{pe}}^2(c) \approx \omega_{\text{pe, max}}$, the range of plasma opacity lying between them. For frequencies $\omega^2 > \omega_{\text{pe, max}}^2$ plasma is totally transparent, i.e., the electromagnetic waves propagate freely (the oscillations are untrapped) and their frequency spectrum is not discrete. The number of oscillations is unbounded, too, in the ranges of transparency lying to the left and to the right from the turning points a and b, respectively. Note that those assumptions are valid for spatially infinite plasma with a density which smoothly goes to zero at infinity. The points 0 and L (metallic surfaces bounding plasma) where the electromagnetic waves are reflected cannot be described within the model. Due to the presence of such points, the oscillations become trapped between the points 0 and a, and between b and L. Therefore, the quantized spectra apply to these regions. Electromagnetic oscillations of longer wavelengths can be bound strongly due to the spectrum quantization.

3.9.2 Langmuir Oscillations. Tonks-Dattner Resonances

Next we consider the case of longitudinal waves, in particular high-frequency ($\omega \gg \nu_e, kv_{\mathrm{Te}}$) longitudinal plasma oscillations. Using relations (2.126) and (2.127) in the high-frequency limit, we obtain from the general eikonal equation (1.276)

$$1 - \frac{\omega_{\mathrm{pe}}^2}{\omega^2}\left(1 + 3\frac{k^2 v_{\mathrm{Te}}^2}{\omega^2} - \imath\frac{\nu_e}{\omega}\right) + \imath\sqrt{\frac{\pi}{2}}\frac{\omega\omega_{\mathrm{pe}}^2}{k^3 v_{\mathrm{Te}}^3}\exp\left(-\frac{\omega^2}{2k^2 v_{\mathrm{Te}}^2}\right) = 0. \qquad (3.238)$$

Solving this equation for $\mathrm{Re}\{k_x\}$ and $\mathrm{Im}\{k_x\}$, and substituting the result into relations (3.236) gives the dispersion equation for high-frequency longitudinal waves in inhomogeneous plasma:

$$\int dx\,\mathrm{Re}\{k_x\} = \int dx\left(-k_y^2 - k_z^2 + \frac{\omega^2 - \omega_{\mathrm{pe}}^2}{3v_{\mathrm{Te}}^2}\right)^{1/2} = \pi n,$$

$$\delta = -\int \frac{dx}{\mathrm{Re}\{k_x\}}\frac{1}{v_{\mathrm{Te}}^2}\left[\sqrt{\frac{\pi}{2}}\nu_e + \frac{\omega_{\mathrm{pe}}^4}{k^3 v_{\mathrm{Te}}^3}\exp\left(-\frac{3}{2} - \frac{1}{2k^2 r_{\mathrm{De}}^2}\right)\right] \times \left(\int \frac{dx}{\mathrm{Re}\{k_x\}}\frac{1}{v_{\mathrm{Te}}^2}\right)^{-1}. \qquad (3.239)$$

In the expression for δ the quantity k^2 means $k^2 = \mathrm{Re}\{k_x^2\} + k_y^2 + k_z^2$. The integration is performed over the range of transparency, i.e., over the range with $\mathrm{Re}\{k_x^2\} > 0$ and, consequently, $\omega^2 > \omega_{\mathrm{pe}}^2$. As for the transverse modes, there exist two turning points, a and b, determined by $\omega^2 = 3\left(k_y^2 + k_z^2\right)v_{\mathrm{Te}}^2 + \omega_{\mathrm{pe}}^2(x)$ when the spatial density distribution is of the type shown in Fig. 3.3. Of course the range of plasma opacity lies between these points.

Thus, longitudinal plasma waves can exist only in the peripheral range of plasma. However, in contrast to the transverse waves, the range of transparency is rather narrow, since a strong collisionless damping occurs when ω^2 differs slightly from ω_{pe}^2. As a result, the range of transparency is limited by the turning point on the one hand, and by the range of strong absorption, on the other hand. Thus, the oscillation spectrum is not quantized. The only exception is a particular plasma configuration where the plasma density changes smoothly first and then falls steeply near the boundaries. The plasma waves can reach the boundaries before they are damped in this case, and they are reflected from the wall. Then, we have to integrate in Eq. (3.239) over the domain lying between the plasma boundaries and the turning points a and b. As a result, the oscillation spectrum becomes quantized. In gas-discharge devices, these oscillations are called the Tonks-Dattner resonances when they are resonantly excited by external electric fields.

3.9.3 Ion-Acoustic Oscillations of Inhomogeneous Isotropic Plasma

Another longitudinal mode is the low-frequency ion-acoustic mode existing in inhomogeneous isotropic plasma with $T_e \gg T_i$. When the space charge depends on the coordinates as shown in Fig. 3.3, they are trapped in plasma. The eikonal equation (1.276) for these waves (in the frequency range $kv_{\mathrm{Ti}} \ll \omega \ll kv_{\mathrm{Te}}$, [see Sect. 2.7]) takes the form

$$1 - \frac{\omega_{\mathrm{pi}}^2}{\omega^2}\left(1 - \imath\alpha\frac{\nu_i}{\omega}\right) + \frac{\omega_{\mathrm{pe}}^2}{k^2 v_{\mathrm{Te}}^2}\left(1 + \imath\sqrt{\frac{\pi}{2}}\frac{\omega}{kv_{\mathrm{Te}}}\right) = 0, \tag{3.240}$$

where we have $\alpha = 1, \nu_i = \nu_{in}$ or $\alpha = 8k^2 v_{\mathrm{Ti}}^2/(5\omega^2), \nu_i = \nu_{ii}$ for weakly or completely ionized plasma, respectively. We obtain from Eq. (3.240) for the ion-acoustic waves of inhomogeneous plasma the following dispersion equation:

$$\int dx \mathrm{Re}\{k_x\} = \int dx \left(-k_y^2 - k_z^2 + \frac{\omega^2}{v_s^2}\frac{\omega_{\mathrm{pi}}^2}{\omega_{\mathrm{pi}}^2 - \omega^2}\right)^{1/2} = \pi n,$$

$$\delta = -\frac{\omega^2}{2}\left[\int dx \frac{\omega_{\mathrm{pi}}^2}{\mathrm{Re}\{k_x\}\left(\omega_{\mathrm{pi}}^2 - \omega^2\right)}\left(\sqrt{\frac{\pi}{2}}\frac{1}{kv_{\mathrm{Te}}v_s^2} + a\frac{k^2\nu_i}{\omega^4}\right)\right]\left[\int \frac{dx}{\mathrm{Re}\{k_x\}}\frac{\omega_{\mathrm{Li}}^4}{v_s^2\left(\omega_{\mathrm{pi}}^2 - \omega^2\right)^2}\right]^{-1}. \tag{3.241}$$

According to the condition of weak damping, $k^2 = \mathrm{Re}\left\{k_x^2\right\} + k_y^2 + k_z^2$ has to be taken in the expression for δ. From Eq. (3.241) it follows that these oscillations exist in the frequency range $\omega^2 \ll \omega_{\mathrm{pi}}^2(x)$. In the case of a bell-shaped density distribution (Fig. 3.4) they are trapped in plasma between the points a and b given by $\omega_{\mathrm{pi}}^2(a) = \omega_{\mathrm{pi}}^2(b) = \omega^2$. However, these points are not the turning points of the system. Near them $\mathrm{Re}\left\{k_x^2\right\} \to \infty$ holds, and the applicability conditions of geometrical optics are not violated. On the contrary, they are fulfilled even better since the wavelength sharply decreases when the wave approaches these points. Consequently, they are called the clustering points. The ion-acoustic waves are strongly

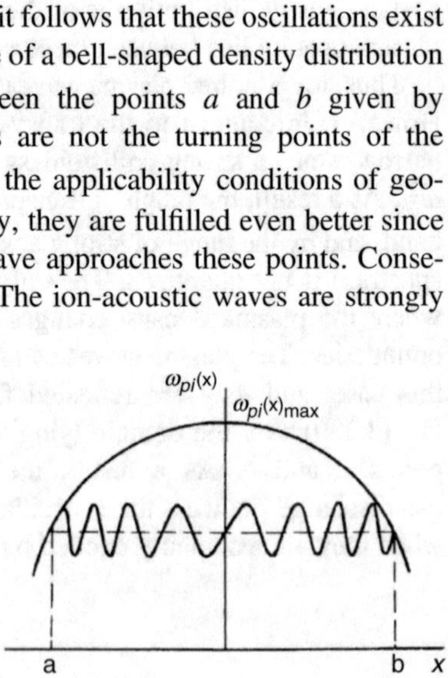

Fig. 3.4 Range of transparency of low-frequency oscillations of isotropic plasma with bell-shaped density distribution

damped near these points and, therefore, they are non-quantized. Only if there are walls of the plasma limiting container between a and b reflecting the waves they become quantized.

3.9.4 Oscillations of Weakly Inhomogeneous Magneto-Active Plasma

We confine our investigation to cold magneto-active plasma when spatial dispersion is absent. Since the oscillation spectra do not depend on the thermal velocity in this limit, the results obtained above remain valid both for non-degenerate and degenerate plasmas. The eikonal equation (1.275) is conveniently written as

$$\begin{aligned} &k_\perp^4\varepsilon_{xx} + k_\perp^2\left[\left(k_z^2 - \frac{\omega^2}{c^2}\varepsilon_{xx}\right)(\varepsilon_{xx}+\varepsilon_{zz}) - \frac{\omega^2}{c^2}\varepsilon_{xy}^2 - \frac{\omega^2}{c^2}\left(\varepsilon_{xx}\varepsilon_{yy} - \varepsilon_{xx}^2\right)\right] \\ &+\varepsilon_{zz}\left[\left(k_z^2 - \frac{\omega^2}{c^2}\varepsilon_{xx}\right)\left(k_z^2 - \frac{\omega^2}{c^2}\varepsilon_{yy}\right) + \frac{\omega^4}{c^4}\varepsilon_{xy}^2\right] = 0, \end{aligned} \tag{3.242}$$

where $k_\perp^2 = k_x^2 + k_y^2$, and the components of tensor $\varepsilon_{ij}\left(\omega, \vec{k}, x\right)$ are given by expressions (3.73)–(3.74) and (3.105)–(3.106). It is important that these components do not depend on the perpendicular projection of the wave vector $k_\perp$. Thus, from Eq. (3.242) we find the two solutions $k_{x1,2}^2(\omega, x)$, which correspond to the ordinary and extraordinary waves in cold magneto-active plasma. Neglecting the small dissipation associated with the anti-Hermitian part of tensor $\varepsilon_{ij}\left(\omega, \vec{k}, x\right)$and determining $\mathrm{Re}\{k_x(\omega,x)\}$from Eq. (3.242), using the quantization rule (1.278), we obtain the dispersion equation:

$$\int dx\,\mathrm{Re}\,\{k_x\} = \int dx\left(-k_y^2 - p \pm \sqrt{p^2 - q}\right)^{1/2} = \pi n. \tag{3.243}$$

Here, the following notations are introduced:

$$\begin{aligned} p &= \frac{1}{2\varepsilon_\perp^H}\left[\left(k_z^2 - \frac{\omega^2}{c^2}\varepsilon_\perp^H\right)\left(\varepsilon_\perp^H + \varepsilon_\parallel^H\right) + \frac{\omega^2}{c^2}g^{H2}\right], \\ q &= \frac{\varepsilon_\parallel^H}{\varepsilon_\perp^H}\left[\left(k_z^2 - \frac{\omega^2}{c^2}\varepsilon_\perp^H\right)^2 - \frac{\omega^4}{c^4}g^{H2}\right]. \end{aligned} \tag{3.244}$$

The frequency dependence of these oscillation branches can be determined from Eq. (3.243) (see Sects. 3.2, 3.4, and 3.5). We consider in detail the spectra of the Alfven, the fast magnetosonic, and the helical waves only.

To analyze the Alfven and the fast magnetosonic waves, we have to study the frequency range $\omega \ll \Omega_i$.It is easily seen that Eq. (3.243) goes over to

$$\int dx \operatorname{Re}\{k_x\} = \int dx \left[-k_y^2 - k_z^2 + \frac{\omega^2}{c^2}\left(1 + \frac{c^2}{v_A^2}\right)\right]^{1/2} = \pi n, \tag{3.245}$$

$$\int dx \operatorname{Re}\{k_x\} = \int dx \left[-k_y^2 - \frac{\omega_{pe}^2}{c^2}\left(1 - \frac{k_z^2 v_A^2}{\omega^2 (1 + v_A^2/c^2)}\right)\right]^{1/2} = \pi n, \tag{3.246}$$

in this limit. In homogeneous plasma the first oscillation branch corresponds to the fast magnetosonic wave and the second one to the Alfven wave. The frequency spectra are given by relation (3.77) in this case. Further, it follows from Eqs. (3.245) and (3.246) that the turning points of fast magnetosonic waves are determined by $\omega^2 = \left(k_y^2 + k_z^2\right) v_A^2(x)$ and that the range of transparency is given by the inequality $\omega^2 > \left(k_y^2 + k_z^2\right) v_A^2(x)$. For a bell-shaped spatial distribution of the plasma density (Fig. 3.4) this implies that the fast magneto sonic waves are trapped inside plasma between the turning points, their spectra thus being quantized. The Alfven waves of inhomogeneous plasma can exist in the range of $\omega^2 < k_z^2 v_A^2(x)$, i.e., there are two ranges of transparency. These ranges are situated in the plasma periphery at values of the space coordinate x, smaller or larger than the respective coordinate of the turning points given by $\omega^2 = k_z^2 v_A^2(x)$. In plasma with a free surface, these ranges are of infinite extension and the oscillations are not trapped. Therefore, their spectra are not quantized. In fact, under laboratory conditions, there always exist walls confining plasma and limiting the peripheral range of transparency with respect to the propagation of the Alfven wave. Incidentally, with increasing distance from the center of the plasma layer, the condition of $\omega^2 < k_z^2 v_A^2(x)$ can be violated due to the growth of the Alfven velocity $v_A(x)$ and Eq. (3.246) may lose its sense.

Finally, we consider the helical waves, also called the helicons, for the weakly inhomogeneous cold plasma case. As shown in Sect. 3.2 [see relation (3.53)], these waves can exist only in the intermediate frequency range $\Omega_i \ll \omega \ll \Omega_e$ and if plasma is sufficiently dense. Under the condition $\omega_{\mathrm{pe}}^2 \gg \omega \Omega_e$ it is easy to obtain from Eq. (3.242) an approximation for $\operatorname{Re}\{k_x(\omega, x)\}$ and to determine the frequency spectrum of the helical waves from the dispersion equation:

$$\int dx \operatorname{Re}\{k_x\} = \int dx \left(-k_y^2 - k_z^2 + \frac{\omega_{\mathrm{pe}}^4 \omega^2}{c^4 k_z^2 \Omega_e^2}\right)^{1/2} = \pi n. \tag{3.247}$$

This relation follows directly from Eq. (3.243) if we account for the explicit form of the dielectric tensor in the given frequency range. Equation (3.247) shows that the helical waves can propagate in inhomogeneous plasma only in the frequency ranges where

$$\omega_{\mathrm{pe}}^2(x) \geq \sqrt{c^4 k_z^2 \left(k_y^2 + k_z^2\right) \Omega_e^2 / \omega^2}.$$

When the spatial distribution of the plasma density is bell-shaped, the range of transparency of these waves lies inside plasma between the turning points a and b given by $\omega_{\mathrm{pe}}^2(a) = \omega_{\mathrm{pe}}^2(b) = \sqrt{c^4 k_z^2 \left(k_y^2 + k_z^2\right) \Omega_e^2 / \omega^2}$. Therefore, the helical waves are trapped in plasma and their spectra are quantized.

We completely neglected dissipative effects in our analysis of the spectra of the Alfven, magnetosonic and helical waves. The account of dissipation leads to the appearance of an imaginary term in Eq. (3.242) and thus to an imaginary part $\mathrm{Im}\{k_x(\omega, k)\}$ of the wave vector. Also a damping decrement δ of the complex wave spectrum appears. Since the explicit expressions for δ are rather complicated, we do not give them here. Moreover, there exist no principally new effects differing from those studied in Chap. 3 for the case of spatially homogeneous plasma.

3.9.5 Drift Oscillations of Weakly Inhomogeneous Collisionless Plasma

As shown above, the inhomogeneity of plasma does not give rise to the appearance of new oscillation spectra in the high-frequency range satisfying the condition of $\omega \gg \omega_{\mathrm{dr}\,\alpha}$, where $\omega_{\mathrm{dr}\,\alpha}$ is defined by relation (3.233). The reverse is true for the low-frequency range

$$\omega \leq \omega_{\mathrm{dr}\,\alpha} \approx \frac{k_y v_{T\alpha}^2}{\Omega_\alpha L_0} \tag{3.248}$$

which we investigate here. The new characteristic frequency appearing in the range (3.248) is called the Larmor drift frequency$\omega_{\mathrm{dr}\,\alpha}$. We will show that a new mode of oscillation can be excited here, in particular when the frequencies are close to$\omega_{\mathrm{dr}\,\alpha}$.

3.9.5.1 Larmor Drift in Inhomogeneous Plasma

To begin with, we discuss in detail the physical meaning of this drift frequency. Considering steady state inhomogeneous plasma immersed into a magnetic field, it follows from Eq. (3.219) that the plasma current differs from zero and that it flows perpendicular to the magnetic field $\vec{B}_0 \parallel oz$ and to the direction of the plasma inhomogeneity (to the oz-axis) as well. In general, the current density can be written as

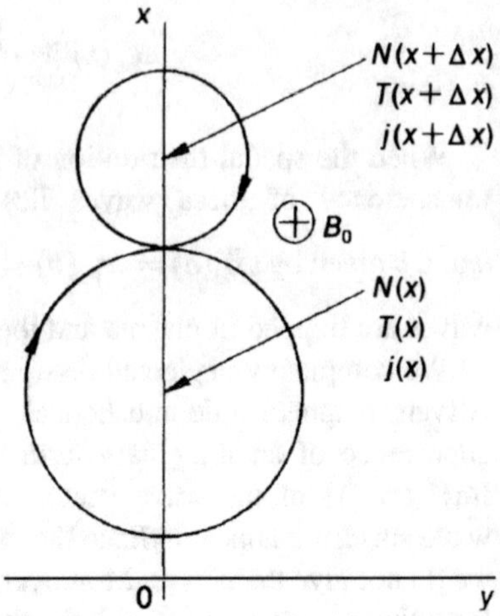

Fig. 3.5 Larmor drift of particles in magnetized inhomogeneous plasma

$$j_i = \sum_a j_{ia} = \frac{c\left(\vec{B}_0 \times \nabla p_0\right)_i}{B_0^2}. \tag{3.249}$$

This current can be compared with the effective drift velocity of the particles of type α oriented parallel to the oy-axis:

$$\vec{v}_{\mathrm{dr}\,\alpha} = \frac{\vec{j}_a}{e_\alpha N_\alpha} = \frac{c}{e_\alpha N_\alpha} \frac{\left(\vec{B}_0 \times \nabla p_{0\alpha}\right)}{B_0^2},$$

$$v_{\mathrm{dr}\,\alpha} \sim \frac{cT_\alpha}{e_\alpha B_0 L_0} \sim \frac{v_{T\alpha}^2}{\Omega_\alpha} \cdot \frac{1}{L_0}. \tag{3.250}$$

This drift motion is the well-known *Larmor drift* of the particles. We see that charged particles with different sign of the charge are drifting in opposite directions. Note that this drift does not correspond to a real motion of the guiding centers of the charged particles (the centers of the Larmor circles which are the particle orbits in the magnetic field). From Fig. 3.5 the nature of such a drift can be seen clearly. Here, we show two circular Larmor orbits and two elementary currents in inhomogeneous plasma. It results in a current perpendicular to the magnetic field (the oz-axis perpendicular to the plane of the drift) and perpendicular to the direction of the plasma inhomogeneity (the ox-axis) in plasma. The density of this current

$$j_{y\alpha} \sim [e_\alpha N_\alpha(x+\Delta x)v_{T\alpha}(x+\Delta x) - e_\alpha N_\alpha(x)v_{T\alpha}(x)] \sim$$
$$\sim e_\alpha \Delta x \frac{\partial N_\alpha v_{T\alpha}}{\partial x} \sim e_\alpha \rho_{\lambda\alpha} \frac{\partial N_\alpha v_{T\alpha}}{\partial x} \sim e_\alpha N_\alpha \frac{v_{T\alpha}^2}{\Omega_\alpha L_0} \quad (3.251)$$

coincides with the current density of the form of Eq. (3.249), $\rho_{\lambda\alpha} = v_{T\alpha}/\Omega_\alpha$ is the Larmor radius of the particles of the type α.

This description shows that this current exists in the steady state due to the diamagnetic effect and not due to a real motion of the charges. Then $j_{y\alpha}$ must be interpreted as a differential diamagnetic current in inhomogeneous plasma. Nevertheless, the Larmor drift can lead to the development of specific instabilities, as for example the beam instability. The instabilities associated with drift motions in plasma are called *drift instabilities*. They possess qualitatively new characteristics. In particular, they can develop even in plasma with Maxwellian velocity distributions of the particles. The density or temperature must be inhomogeneous, however.

The physical nature of the quantity $\omega_{\mathrm{dr}\,\alpha}$ becomes clear from Eqs. (3.248)–(3.251). It can be regarded as the Doppler shift of the frequency due to the Larmor drift of the particles. In thermonuclear plasma ($N \sim 10^{14} - 10^{15}\,\mathrm{cm}^{-3}$, $T \sim 10^8\,\mathrm{K}$, $L_0 \sim 10$ cm, $B_0 \sim 10^5$ Gauss), we have $v_{\mathrm{dr}} \sim 10^5\,\mathrm{cm/s}$and $\omega_{\mathrm{dr}} \sim 10^4\,\mathrm{s}^{-1}$ for $k_y \sim L_0^{-1}$. In plasma of a gas-discharge ($N \sim 10^{10} - 10^{12}\,\mathrm{cm}^{-3}$,$T \sim 10^4 - 10^5\,\mathrm{K}$, $L_0 \sim 1$ cm and $B_0 \sim 10^3 - 10^4$ Gauss), the order of magnitude is $v_{\mathrm{dr}} \sim 10^5 - 10^6\,\mathrm{cm/s}$ and $\omega_{\mathrm{dr}} \sim 10^5 - 10^6\,\mathrm{s}^{-1}$. In ionospheric plasma ($N \sim 10^7\,\mathrm{cm}^{-3}$, $T \sim 10^4\,\mathrm{K}$, $L_0 \sim 10 - 30\,\mathrm{km}$ and$B_0 \sim 1$ Gauss), we obtain $v_{\mathrm{dr}} \sim 10^2\,\mathrm{cm/s}$ and $\omega_{\mathrm{dr}} \sim 10^4\,\mathrm{s}^{-1}$. Finally, in degenerate solid-state plasma with a Fermi energy $\epsilon_F \sim 0.1 - 1\,\mathrm{eV}$ (typical for semiconductors and metals) placed in a magnetic field of $B_0 \sim 10^4$ Gauss and with an inhomogeneity scale $L_0 \sim 1$ cm, we have $v_{\mathrm{dr}} \sim 10^4 - 10^5\,\mathrm{cm/s}$and $\omega_{\mathrm{dr}} \sim 10^4 - 10^5\,\mathrm{s}^{-1}$. It follows from these estimates that the drift frequencies are much smaller than the electron and ion Larmor frequencies, which significantly simplifies the analysis of the drift oscillations of inhomogeneous magnetized plasma.

To simplify as much as possible, we make the following assumptions. First, we confine our interest to low-pressure plasma $\beta \ll 1$. In this limit, the low-frequency drift oscillations are longitudinal, with a high degree of accuracy. Our analysis of the oscillation spectra of homogeneous magneto-active plasma confirms this fact (Chap. 3). Further, for $\beta \ll 1$ plasma oscillations cannot significantly perturb a strong external magnetic field, which consequently remains constant. For longitudinal plasma oscillations the eikonal equation is of the form of Eq. (1.276), i.e.,

$$\varepsilon\left(\omega, \vec{k}, x\right) = 0, \quad (3.252)$$

where the longitudinal dielectric permittivity $\varepsilon\left(\omega, \vec{k}, x\right)$ is given by expression (3.232) for non-degenerate plasma.

Secondly, we are interested in the analogs of the local spectra only, which are directly defined by Eq. (3.252). It is important to obtain the qualitatively new results first, i.e., the spectra and the stability properties of inhomogeneous plasma. Exact quantitative results can be easily obtained later by means of the quantization rules. In the approximation of geometrical optics the use of the eikonal equation as a local dispersion equation is justified for very short-wavelengths compared to the scale of the plasma inhomogeneity. This approximate analysis corresponds to the evaluation of exact integral relations with the aid of the mean-value theorem.

3.9.5.2 Dispersion Equation for Drift Oscillations

Consequently, the local dispersion equation for the drift oscillations of inhomogeneous non-degenerate plasma is

$$\varepsilon\left(\omega, \vec{k}, x\right) = 1 + \sum_{\alpha} \frac{\omega_{p\alpha}^2}{k^2 v_{T\alpha}^2} \left\{1 - \sum_{\alpha} \frac{\omega}{\omega - n\Omega_\alpha} \left[1 - \frac{k_y v_{T\alpha}^2}{\omega \Omega_\alpha} \times \right.\right.$$
$$\left.\left. \times \left(\frac{\partial \ln N_\alpha}{\partial x} + \frac{\partial T_\alpha}{\partial x} \frac{\partial}{\partial T_\alpha}\right)\right] A_n\left(\frac{k_\perp^2 v_{T\alpha}^2}{\Omega_\alpha^2}\right) I_+\left(\frac{\omega - n\Omega_\alpha}{k_z v_{T\alpha}}\right)\right\} = 0. \tag{3.253}$$

The evaluation of this equation is rather complex (see Chap. 3). Therefore, the complete analysis is impossible. We assume that the frequency of the drift oscillations satisfies the inequality $\omega \ll \Omega_i$ and that the longitudinal wavelength is much larger than the Larmor radius $k_z v_{T\alpha} \ll \Omega_\alpha$. Under these conditions only the term $n = 0$ appears to be significant in Eq. (3.253). The contributions of the higher harmonics can be neglected and the eikonal equation for longitudinal oscillations becomes

$$\varepsilon\left(\omega, \vec{k}, x\right) = 1 + \sum_{\alpha} \frac{\omega_{p\alpha}^2}{k^2 v_{T\alpha}^2} \left\{1 - \left[1 - \frac{k_y v_{T\alpha}^2}{\omega \Omega_\alpha} \left(\frac{\partial \ln N_\alpha}{\partial x} + \frac{\partial T_\alpha}{\partial x} \frac{\partial}{\partial T_\alpha}\right)\right] \right.$$
$$\left. \times A_0\left(\frac{k_\perp^2 v_{T\alpha}^2}{\Omega_\alpha^2}\right) I_+\left(\frac{\omega}{k_z v_{T\alpha}}\right)\right\} = 0. \tag{3.254}$$

Note that Eq. (3.254) has no zero for $\omega \ll k_z v_{\mathrm{Ti}}$. Oscillations of plasma are impossible in this range. Instead, screening of a potential field by plasma occurs. In collisionless magnetized plasma, the quantity $1/(k_z v_{\mathrm{Ti}}) \sim \lambda_{\|}/v_{\mathrm{Ti}}$ characterizes the time interval during which the density and the temperature of the electrons as well as the ions relax towards equilibrium at the distance of the wavelength due to the free flight of the particles. Thus, the condition $\omega \ll k_z v_{\mathrm{Ti}}$ implies that the relaxation time of the density and temperature perturbations is short compared to the oscillation period. Obviously, longitudinal waves cannot exist under these conditions. Thus, it is sufficient to analyze the case $\omega \gg k_z v_{\mathrm{Ti}}$ and to neglect the ion Landau damping which is exponentially small here.

3.9.5.3 Spectra of Fast Long-Wavelength Drift Oscillations

Further, we consider only drift oscillations of long-wavelengths which satisfy the condition $\lambda_\perp \gg \rho_{\lambda i}(\lambda_\perp \sim 1/k_\perp)$. From the point of view of creating a stable plasma state, the long-wavelength oscillations are most dangerous since they can perturb rather large plasma regions in radial direction. Oscillations of short-wavelengths are less dangerous since they perturb comparatively small plasma regions.

Another important parameter in the analysis of drift oscillations is the ratio of their phase velocity to the thermal velocity of the particles. One distinguishes the fast, $\omega/k_z \gg v_{\mathrm{Te}}$ and the slow $v_{\mathrm{Ti}} \ll \omega/k_z \ll v_{\mathrm{Te}}$ drift oscillations. We begin our analysis with the fast long-wavelength oscillations $\lambda_\perp \gg \rho_{\lambda i}$, $\omega/k_z \gg v_{\mathrm{Te}}$. In this case, the eikonal equation (3.254) is of the form

$$1-\frac{\omega_{\mathrm{pe}}^2}{\omega^2}\frac{k_z^2}{k^2}\left[1-\frac{k_y v_{\mathrm{Te}}^2}{\omega\Omega_e}\frac{\partial\ln(NT_e)}{\partial x}\right]+\frac{k_\perp^2\omega_{\mathrm{pi}}^2}{k^2\Omega_i^2}\left[1-\frac{k_y v_{\mathrm{Ti}}^2}{\omega\Omega_i}\frac{\partial\ln(NT_i)}{\partial x}\right]=0. \quad (3.255)$$

Under the condition $c^2 \gg v_A^2\left(\omega_{\mathrm{pi}}^2 \gg \Omega_i^2\right)$ we have in the low-frequency range $\omega \ll \omega_{\mathrm{dr}\,\alpha}$

$$\frac{k_\perp^2 c^2}{v_A^2}\frac{k_y v_{\mathrm{Ti}}^2}{\Omega_i}\frac{\partial\ln(NT_i)}{\partial x}-k_z^2\frac{\omega_{\mathrm{pe}}^2}{\omega^2}\frac{k_y v_{\mathrm{Te}}^2}{\Omega_e}\frac{\partial\ln(NT_e)}{\partial x}=0. \quad (3.256)$$

Using this relation as local dispersion equation, we obtain the following local spectrum[16]:

$$\omega^2=-\frac{k_z^2}{k_\perp^2}\frac{T_e}{T_i}\frac{M}{m}\Omega_i^2\frac{\partial\ln(NT_e)}{\partial\ln(NT_i)}. \quad (3.257)$$

In Eqs. (3.256) and (3.257), $k_\perp^2$, as usual, denotes $k_\perp^2 = k_y^2 + \mathrm{Re}\left\{k_x^2\right\} \approx k_y^2 + \pi^2 n^2/L_\perp^2$, where $L_\perp$ is the extension of the device in the direction of the $0x$-axis since this is at the same time the order of the characteristic length of the plasma inhomogeneity L_0.

3.9.5.4 Universal Instability of Inhomogeneous Plasma

We have obtained a qualitatively new oscillation branch which does not occur in spatially homogeneous plasma. This branch is aperiodically unstable when the inequality

[16] Here and in the following we use the abbreviation: $\frac{\partial\ln A}{\partial\ln B}=\frac{\partial\ln A}{\partial x}\Big/\frac{\partial\ln B}{\partial x}=\frac{B}{A}\frac{\partial A}{\partial x}\Big/\frac{\partial B}{\partial x}$.

$$\frac{\partial \ln (NT_e)}{\partial \ln (NT_i)} > 0 \tag{3.258}$$

is satisfied. Almost all kinds of real plasmas are subject to this instability. Since the electron and ion pressure generally fall from the center to the plasma boundaries, and since the particle temperature T decreases on a longer scale than the density N, the derived instability is called *universal*. However, for the excitation of this instability the condition $\omega^2 \ll \Omega_i^2$ leads to

$$\frac{k_z^2}{k_\perp^2} \sim \frac{L_\perp^2}{L_\parallel^2} \ll \frac{m}{M}\frac{T_i}{T_e}\frac{\partial \ln (NT_i)}{\partial \ln (NT_e)} \sim \frac{m}{M}. \tag{3.259}$$

The instability can develop only in sufficiently long devices, with a longitudinal plasma extension at least $\sqrt{M/m} \geq 40$ times larger than the transverse one.

The universal instability is purely hydrodynamic and not related to the Cherenkov energy dissipation. Due to the fact that this instability is caused by the guiding center drift in plasma it is one special case of the drift instabilities.

3.9.5.5 Spectra of Slow Long-Wavelength Drift Oscillations

The requirement regarding the longitudinal plasma dimension appears less restrictive for the domain of existence of long-wavelength drift oscillations in the range of phase velocities $v_{\text{Ti}} \ll \omega/k_z \ll v_{\text{Te}}$. In this frequency range, the eikonal equation (3.254) takes the form

$$\begin{aligned} 1 + \frac{\omega_{\text{pe}}^2}{k^2 v_{\text{Te}}^2}\left[1 + \frac{k_y v_s^2}{\omega \Omega_i}\frac{\partial \ln N}{\partial x} - \frac{k_z^2 v_s^2}{\omega^2}\left(1 - \frac{k_y v_{\text{Ti}}^2}{\omega \Omega_i}\frac{\partial \ln (NT_i)}{\partial x}\right)\right] \\ + \imath \frac{\omega_{\text{pe}}^2}{k^2 v_{\text{Te}}^2}\sqrt{\frac{\pi}{2}}\frac{\omega}{|k_z| v_{\text{Te}}}\left(1 - \frac{k_y v_{\text{Te}}^2}{\omega \Omega_e}\frac{\partial}{\partial x}\ln\frac{N}{\sqrt{T_e}}\right) = 0. \end{aligned} \tag{3.260}$$

For isothermal plasma ($T_e \sim T_i$) the last term in the square brackets can be neglected. Then, it is easy to obtain the local spectrum for $\omega \gg k_z v_s$:

$$\begin{aligned} \omega_1 &= -\frac{k_y v_s^2}{\Omega_i}\frac{\partial \ln N}{\partial x}, \\ \delta_1 &= \sqrt{\frac{\pi}{2}}\frac{\omega_1^2}{|k_z| v_{\text{Te}}}\left(k^2 r_{\text{De}}^2 + \frac{k_\perp^2 v_s^2}{\Omega_i^2} - \frac{1}{2}\frac{\partial \ln T_e}{\partial \ln N}\right). \end{aligned} \tag{3.261}$$

3.9.5.6 Drift-Dissipative and Drift-Temperature Instabilities

We see that a kinetically unstable slow oscillation mode of long-wavelength can be excited if

$$\frac{\partial \ln T_e}{\partial \ln N} < 2\left(k^2 r_{De}^2 + \frac{k_\perp^2 v_s^2}{\Omega_i^2}\right) \tag{3.262}$$

These oscillations of inhomogeneous plasma are called the drift-dissipative oscillations. Here, the Cherenkov mechanism of dissipation by the electrons is responsible for the buildup of the oscillations. In the range of the drift frequencies, the Cherenkov term can have the opposite sign and produce a buildup of the oscillations. Note that these oscillations have a frequency larger than the ion-acoustic frequency: $\omega_1 \gg k_z v_s$. In homogeneous isothermal plasma ($T_e \sim T_i$), no oscillations can arise in this frequency range.

The last term in the square brackets of Eq. (3.260), which is proportional to $k_z v_s/\omega$, becomes significant in non-isothermal plasma with $T_e \gg T_i$. This term can drive oscillations in the frequency range $\omega^2 \ll k_z v_s^2$, i.e., at the frequencies smaller than the ion-acoustic frequency. Under these conditions, it is easy to obtain from Eq. (3.260) the local spectrum ($\omega \to \omega + \imath\delta$)

$$\omega_2 = \frac{k_z^2 \Omega_i}{k_y \partial \ln N/\partial x}, \qquad \delta_2 = -\sqrt{\frac{\pi}{2}} \frac{\omega_2^2}{|k_z| v_{\mathrm{Te}}} \left(1 - \frac{1}{2}\frac{\partial \ln T_e}{\partial \ln N}\right). \tag{3.263}$$

It can be interpreted as the continuation of the acoustic branch into the low-frequency range $\omega \ll k_z v_s$. From relations (3.263) we see that the electron Cherenkov dissipation drives these oscillations unstable if

$$\frac{\partial \ln T_e}{\partial \ln N} > 2. \tag{3.264}$$

As in the special case before, the instability is kinetic and can be called a drift-dissipative instability, too.

Finally, in the range of very low frequencies $\omega \ll \omega_{\mathrm{dr}\,\alpha}$, we obtain from Eq. (3.260) two more hydrodynamically unstable oscillation modes:

$$\omega_3^2 = -k_z^2 v_{\mathrm{Ti}}^2 \frac{\partial \ln T_i}{\partial \ln N}, \tag{3.265}$$

$$\omega_4^3 = -k_z^2 v_s^2 \frac{k_y v_{Ti}^2}{\Omega_i} \frac{\partial \ln T_i}{\partial x}. \tag{3.266}$$

They can exist when ($\partial \ln T_i/\partial \ln N) \gg 1$ and therefore are called *drift-temperature instabilities.*

We already mentioned that in the range of intermediate phase velocities $v_{\rm Ti} \ll \omega/k_z \ll v_{\rm Te}$ the long-wavelength drift oscillations can be excited in relatively short plasma devices. From $\omega \sim \omega_{{\rm dr}\,\alpha} \gg k_z v_{\rm Ti}$ it follows that the condition $L_{\|}/L_{\perp} \sim L_{\perp}/\rho_{\lambda i} > 1$is sufficient for their occurrence. On the other hand, the neglect of particle collisions $(\nu_a \ll k_z v_{\rm Ta})$ implies that the longitudinal extension of the system should be smaller than the mean free path of the particles, which gives $L_{\|} < v_{T\alpha}/\nu_\alpha \equiv l_\alpha$. Thus, the condition of validity for the collisionless description of long-wavelength drift oscillations can be written as

$$1 < \frac{L_{\perp}}{\rho_{\lambda i}} < \frac{L_{\|}}{L_{\perp}} < \frac{v_{T\alpha}}{\nu_\alpha}\frac{1}{L_{\perp}} \sim \frac{l_\alpha}{L_{\perp}}. \tag{3.267}$$

Strictly speaking, the following conditions must be satisfied to ensure the validity of the picture of this section. Either the frequencies and increments of the drift oscillations must greatly exceed the frequencies of the particle collisions, $|\omega| \gg \nu_\alpha$, or the mean free path of the particles must exceed the longitudinal wavelength of the drift oscillations, $\nu_\alpha \ll k_z v_{T\alpha}$. Only under these conditions it is possible to apply the Vlasov kinetic equation, which completely ignores particle collisions. Since in real devices the drift frequencies are of the order of $10^4 - 10^6\,{\rm s}^{-1}$, this limitation can be satisfied only in high-temperature non-degenerate plasmas at relatively small charged particle densities. In degenerate cold solid-state plasma, this condition is not fulfilled. However, it can be shown that the drift instabilities may develop in dense plasma with a large number of particle collisions, as well [131]. Even more, in inhomogeneous plasma, the collisional friction, especially due to the electron collisions, can become the driving force which excites drift instabilities.

3.10 Instability of Boundary of Magnetically Confined Plasma

We now study surface waves in magnetized plasma with a sharp boundary and with the characteristic dimension of the boundary inhomogeneity shorter than the Larmor radius. Under the condition of mirror reflection from the plasma surface the general dispersion equation for surface waves in magnetized plasma is analogous to that in the absence of the magnetic field. But the derivation is rather unwieldy and requires the calculation of the inverse dielectric tensor. Thus let us confine our consideration to quasi-longitudinal waves since their electric field is derived from a potential field to a good approximation. Here, we study them and the stability of surface waves in plasma confined by a strong magnetic field will be analyzed.

Let us analyze semi-bounded plasma confined by a strong magnetic field parallel to the plasma surface and oriented along the z-axis. The boundary inhomogeneity of such plasma has a characteristic dimension greatly exceeding the Larmor radii of particles. This boundary is assumed to be set near the plane $x = 0$ (Fig. 3.6). For

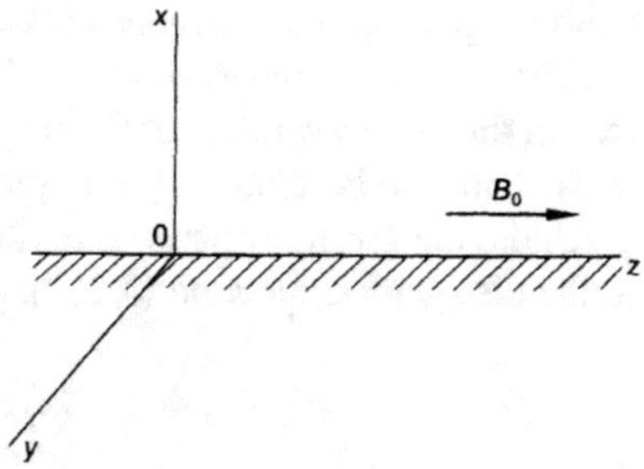

Fig. 3.6 Semi-bounded plasma confined by a strong magnetic field parallel to the plasma surface

simplicity, collisions of charged particles in plasma can be neglected, assuming $\Omega_\alpha \gg \nu_\alpha$, where $\alpha = e, i$. The distribution function of particles α that is unperturbed by an electromagnetic field of oscillations can be obtained analogously to that in Sect. 3.8:

$$f_{0\alpha} = \left(1 + \frac{v_\perp \sin\phi}{\Omega_\alpha} \frac{\partial}{\partial x}\right) F_{0\alpha}(\epsilon, x), \qquad (3.268)$$

where $F_{0\alpha}(\epsilon, x)$ is either Maxwellian (2.210) or Fermian (2.211) with x-dependent temperature and density.

In contrast to Sects. 3.8 and 3.9, where plasma has been considered to be smoothly inhomogeneous, here plasma is inhomogeneous within a thin layer near the plane $x = 0$ (plasma surface). The difference of the distribution function (3.268) from that in the thermodynamic equilibrium is manifested in this layer where the diamagnetic currents are localized due to the Larmor rotation of particles in spatially inhomogeneous plasma. As shown in Sect. 3.9, in the region of plasma inhomogeneity, diamagnetic currents may cause the excitation of short-wavelength (compared to the inhomogeneity dimension) drift oscillations described in the framework of geometrical optics approximation. Furthermore, it will be shown that diamagnetic currents can also excite oscillations with a wavelength being significantly greater than the dimension of the plasma boundary inhomogeneity, i.e., surface waves damping deep into plasma.

3.10.1 Poisson's Equation for Magnetically Confined Inhomogeneous Plasma

For surface waves, the plasma boundary can be regarded as infinitely thin and the diamagnetic currents can be considered as the boundary conditions for the electromagnetic field equations. Thus, the aim of our investigations is the derivation of the field equations and their boundary conditions accounting for the inhomogeneity of the plasma surface layer and diamagnetic currents. We solve this problem using longitudinal (potential) waves as an example, since in the case of magnetically confined plasmas the magnetic pressure greatly exceeds the gas pressure and plasma oscillations can be regarded as longitudinal with a high degree of accuracy. For this reason the inhomogeneity of the magnetic field, compared to the plasma inhomogeneity, will also be ignored (as in Sect. 3.8).

Under the given restrictions, the kinetic equation for the non-equilibrium addition to the distribution function (3.268), dependent on time and coordinates as

$$\delta f_\alpha = \delta f_\alpha(x) \exp\left(-\imath\omega t + \imath k_y y + \imath k_z z\right), \tag{3.269}$$

is of the form

$$\left(\omega - k_y v_y - k_z v_z\right)\delta f_\alpha + \imath v_x \frac{\partial \delta f_\alpha}{\partial x} - \imath\Omega_\alpha \frac{\partial \delta f_\alpha}{\partial \phi} = -e_\alpha \vec{E} \cdot \frac{\partial f_{0\alpha}}{\partial \vec{p}}, \tag{3.270}$$

where $\vec{E}$ is the potential of the perturbations $\vec{E} = -\nabla\Phi$.

Equation (3.270) is solved by integrating it over the characteristic (see Sect. 3.8)

$$v_\perp \sin\phi + \Omega_\alpha x = \text{const}. \tag{3.271}$$

Then

$$\delta f_\alpha(x) = -\frac{e_\alpha}{m_\alpha \Omega_\alpha} \int_\infty^\phi d\phi' \nabla\Phi(x') \frac{\partial f_{0\alpha}(x')}{\partial v}$$
$$\exp\left[\frac{\imath}{\Omega_\alpha} \int_\phi^{\phi'} d\phi'' \left(\omega - k_\perp v_\perp \sin\phi'' - k_z v_z\right)\right]. \tag{3.272}$$

Here, x', x and ϕ' are mutually related by relation (3.271). Substituting expression (3.272) into the formula for the charge density

$$\rho(x) = \sum_\alpha e_\alpha \int d\vec{p}\, \delta f_\alpha \tag{3.273}$$

and using Poisson's equation

$$\nabla^2\Phi = -4\pi\rho(x), \tag{3.274}$$

after rather unwieldy calculations (analogous to those given in Sect. 3.8), in the case of non-degenerate plasma, we finally obtain the equation for the potential of the oscillation field:

$$\nabla^2\Phi = \sum_\alpha \int dk_x \Phi(k_x)\, e^{ik_x x} \frac{\omega_{p\alpha}^2}{v_{T\alpha}^2} \left\{ 1 - \sum_s \frac{\omega}{\omega - s\Omega_\alpha} \left[1 - \frac{k_y v_{T\alpha}^2}{\omega\Omega_\alpha} \frac{\partial^0}{\partial x} \left(1 - \frac{s\omega}{z_\alpha \Omega_\alpha} \right) \right.\right.$$
$$\left.\left. -\imath \frac{k_x v_{T\alpha}^2}{\Omega_\alpha^2} \frac{\partial^0}{\partial x} \frac{A_s'(z_\alpha)}{A_s(z_\alpha)} \right] A_s(z_\alpha) I_+ \left(\frac{\omega - s\Omega_\alpha}{k_z v_{T\alpha}} \right) \right\}, \tag{3.275}$$

where

$$z_\alpha = \frac{k_\perp^2 v_{T\alpha}^2}{\Omega_\alpha^2}, \qquad \frac{\partial^0}{\partial x} = \frac{\partial \ln N_\alpha}{\partial x} + \frac{\partial T_\alpha}{\partial x} \frac{\partial}{\partial T_\alpha}.$$

The operator $\partial^0/\partial x$ acts on all the values to its right. Equation (3.275) is valid for the whole space both in plasma ($x \geq 0$) and in vacuum ($x < 0$). Thus, there is no need to set special boundary conditions which can be obtained by means of integrating Eq. (3.275) over a physically infinitely thin intermediate layer near the plasma surface. The thickness of this layer is small compared to the wavelength of surface waves.

3.10.2 Surface Oscillations of Cold Magneto-Active Plasma with a Sharp Boundary

We start the analysis of Eq. (3.275) with cold plasma when diamagnetic currents in the surface layer are totally neglected. In other words, we study oscillations with the phase velocity much higher than thermal velocities of particles and with the wavelength exceeding their Larmor radii. The limit $T \to 0$ should be taken in Eq. (3.275). Then the equation

$$\left(1 - \sum_\alpha \frac{\omega_{p\alpha}^2}{\omega^2 - \Omega_\alpha^2}\right)\left(\frac{\partial^2}{\partial x^2} - k_y^2\right)\Phi - \left(1 - \sum_\alpha \frac{\omega_{p\alpha}^2}{\omega^2}\right) k_z^2 \Phi$$
$$+ \frac{\partial \Phi}{\partial x}\frac{\partial}{\partial x}\left(1 - \sum_\alpha \frac{\omega_{p\alpha}^2}{\omega^2 - \Omega_\alpha^2}\right) + k_y \Phi \frac{\partial}{\partial x} \sum_\alpha \frac{\omega_{p\alpha}^2 \Omega_\alpha}{\omega\left(\omega^2 - \Omega_\alpha^2\right)} = 0, \tag{3.276}$$

being independent of the form of the distribution function of particles, is valid both for non-degenerate and degenerate plasmas. Besides, in the approximation of cold plasma, the diamagnetic currents are totally neglected in the surface layer, nothing to say this fact that here the surface layer together with the Larmor radius of particles vanishes. Therefore, Eq. (3.276) is valid for describing oscillations of plasmas with an arbitrarily sharp boundary, in particular, of plasma confined by the walls of a real dielectric vessel (glass).

In the plasma volume (for $x \geq 0$), where the density may be regarded as homogeneous from Eq. (3.276), we have

$$\left(1-\sum_{\alpha}\frac{\omega_{p\alpha}^2}{\omega^2-\omega_{\alpha}^2}\right)\left(\frac{\partial^2}{\partial x^2}-k_y^2\right)\Phi_1-\left(1-\sum_{\alpha}\frac{\omega_{p\alpha}^2}{\omega^2}\right)k_z^2\Phi_1=0. \qquad (3.277)$$

In vacuum ($x < 0$), Eq. (3.276) is reduced to the Laplace equation

$$\nabla^2\Phi_2=0. \qquad (3.278)$$

Finally, by integrating Eq. (3.276) we obtain the boundary conditions relating Φ_1 to Φ_2 on the surface of the plasma-vacuum partition ($x = 0$)

$$\begin{aligned} &\{\Phi\}_{x=0}=0, \\ &\left\{\left(1-\sum_{\alpha}\frac{\omega_{p\alpha}^2}{\omega^2-\Omega_{\alpha}^2}\right)\frac{\partial\Phi}{\partial x}+k_y\Phi\sum_{\alpha}\frac{\omega_{p\alpha}^2\Omega_{\alpha}}{\omega\left(\omega^2-\Omega_{\alpha}^2\right)}\right\}_{x=0}=0. \end{aligned} \qquad (3.279)$$

In the ranges $x \geq 0$ and $x < 0$, the field equations and their boundary conditions are known. They can be solved and their solutions can be jointed. Then

$$\begin{aligned} &\Phi_1(x)=C_1\exp\left(-\sqrt{k_y^2+k_z^2\frac{\varepsilon_{\parallel}}{\varepsilon_{\perp}}}x\right) \quad \text{for} \quad x\geq 0, \\ &\Phi_2(x)=C_2\exp\left(-\sqrt{k_y^2+k_z^2}x\right) \quad \text{for} \quad x<0, \end{aligned} \qquad (3.280)$$

where

$$\varepsilon_{\parallel}=1-\sum_{\alpha}\frac{\omega_{p\alpha}^2}{\omega^2}, \quad \varepsilon_{\perp}=1-\sum_{\alpha}\frac{\omega_{p\alpha}^2}{\omega^2-\Omega_{\alpha}^2}. \qquad (3.281)$$

On substituting these solutions into the boundary conditions (3.279), we obtain the system of homogeneous algebraic equations for the constants C_1 and C_2. Its solvability condition yields the dispersion equation for surface waves in semi-bounded plasma with a sharp boundary:

$$\varepsilon_{\perp}\sqrt{k_y^2+k_z^2\frac{\varepsilon_{\parallel}}{\varepsilon_{\perp}}}+k_y g+\sqrt{k_y^2+k_z^2}=0. \qquad (3.282)$$

Here,

$$g = \sum_{\alpha} \frac{\omega_{p\alpha}^2 \Omega_\alpha}{\omega\left(\omega^2 - \Omega_\alpha^2\right)}. \tag{3.283}$$

In the absence of the magnetic field, when $\Omega_\alpha \to 0$, from Eq. (3.282) we obtain dispersion equation (2.240) for longitudinal surface waves in semi-bounded isotropic plasma. The frequency spectrum of these waves is given by the second expression (2.239). The external magnetic field, parallel to the plasma surface, essentially modifies the derived frequency spectrum if $\Omega_e \geq \omega = \omega_{\rm pe}/\sqrt{2}$. For the modes with $k_y = 0$, according to Eq. (3.282), surface waves exist in purely electron plasma only for $\omega_{\rm pe}^2 > \Omega_e^2$ and in the frequency range $\omega_{\rm pe}^2 > \omega^2 > \Omega_e^2$. Here, the frequency spectrum of these waves is determined by

$$\omega^2 = \frac{1}{2}\left(\omega_{\rm pe}^2 + \Omega_e^2\right). \tag{3.284}$$

For $k_y = 0$ surface waves can also exist in the frequency range $\omega < \Omega_e$. For example, for the flute modes with $k_z = 0$ in the case of purely electron plasma from Eq. (3.282) we obtain

$$\omega = \frac{k_y}{2|k_y|}\Omega_e \pm \frac{1}{2}\sqrt{\Omega_e^2 + 2\omega_{\rm pe}^2}. \tag{3.285}$$

Hence, low-frequency surface waves are possible in strong magnetic fields when $\Omega_e^2 \gg \omega_{\rm pe}^2$ and $\omega \approx \omega_{\rm pe}^2/\Omega_e$.

It is straightforward to analyze Eq. (3.282) also in the low-frequency range $\omega \ll \omega_{\rm pi}$, when the ion motion becomes significant. In the range of the lowest frequencies$\omega \ll \Omega_i$, both plasma electrons and ions are strongly magnetized, and $g \to 0$, $\varepsilon_\perp \to 0$. As a result, Eq. (3.282) has no solutions, i.e., in cold strongly magnetized plasma with a sharp boundary in the frequency range $\omega \ll \Omega_i$ surface waves are non-existent. They are possible only in the frequency range $\omega \geq \Omega_i$. Actually, in the low-frequency range $\varepsilon_\parallel \gg \varepsilon_\perp$, from Eq. (3.282) we obtain

$$\omega^2 = \left(\omega_{\rm pi}^2 + \Omega_i^2\right) \tag{3.286}$$

for the waves propagating not strictly perpendicular to the magnetic field ($k_z \neq 0$). For $k_z = 0$ (flute modes), the spectrum of low-frequency waves is determined by

$$\omega = \frac{k_y}{|k_y|}\left(\Omega_i + \frac{\omega_{\rm pi}^2}{2\Omega_i}\right). \tag{3.287}$$

The above analysis shows that the surface waves of plasmas with sharp boundaries are always stable in collisionless magnetized plasmas when the thermal motion of particles is ignored. Moreover, accounting for particle collisions results in their damping. This is not surprising since in the approximation discussed the diamagnetic currents, which lead to oscillations buildup, are totally ignored in the inhomogeneous surface layer of plasma.

3.10.3 Instability of the Surface of Magnetically Confined Plasma

Accounting for the finite plasma temperature and the finite Larmor radius of the particles, the situation is qualitatively altered. Besides dissipative effects, specified by the collisionless Cherenkov absorption and wave radiation by plasma particles, the diamagnetic currents which can cause the instability of plasma surface waves become considerable in the inhomogeneous surface layer of plasma. Let us consider Eq. (3.275) in the range of low-frequency $\left(\omega^2 \ll \Omega_\alpha^2\right)$ and short-wavelength $\left(k_\perp^2 v_{T\alpha}^2 \ll \Omega_\alpha^2\right)$ oscillations. Expanding the function $A_s(z_\alpha)$ in a power series of z_α and keeping only the term with $s = 0$, in the sum over cyclotron harmonics, from the integro-differential equation (3.275), we obtain the second-order differential equation[17]

$$\nabla^2\Phi = \sum_\alpha \frac{\omega_{p\alpha}^2}{v_{T\alpha}^2}\left\{\left(1 - I_+\left(\frac{\omega}{k_z v_{T\alpha}}\right) + \frac{k_y v_{T\alpha}^2}{\omega\Omega_\alpha}\frac{\partial^0}{\partial x} I_+\left(\frac{\omega}{k_z v_{T\alpha}}\right)\right)\Phi + \right.$$
$$\left. + \frac{v_{T\alpha}^2}{\Omega_\alpha^2} I_+\left(\frac{\omega}{k_z v_{T\alpha}}\right)\left(k_y^2 - \frac{\partial^2}{\partial x^2}\right)\Phi - \frac{\partial\Phi}{\partial x}\frac{1}{\Omega_\alpha^2}\frac{\partial^0}{\partial x} v_{T\alpha}^2 I_+\left(\frac{\omega}{k_z v_{T\alpha}}\right)\right\}. \tag{3.288}$$

When deriving this equation, we also assumed $\omega^2 {k_\perp}^2 \ll {k_z}^2\Omega_i^2$. Integrating Eq. (3.288) over the intermediate layer near the plasma boundary gives the boundary conditions for the potential. As a result, we have

$$\{\Phi\}_{x=0} = 0,$$
$$\left\{\frac{\partial\Phi}{\partial x} + \sum_\alpha \frac{\omega_{p\alpha}^2}{\Omega_\alpha^2} I_+\left(\frac{\omega}{k_z v_{T\alpha}}\right)\left(\frac{\partial\Phi}{\partial x} - \frac{\Omega_\alpha}{\omega} k_y \Phi\right)\right\}_{x=0} = 0. \tag{3.289}$$

With these boundary conditions, Eq. (3.288) can now be written as:
for plasma ($x \geq 0, \;\; \Phi = \Phi_1$)

[17] In the approximation of geometrical optics from Eq. (3.275) we obtain the eikonal equation which is reduced to Eq. (3.255) in the low-frequency limit.

$$\nabla^2 \Phi_1 = \sum_\alpha \frac{\omega_{p\alpha}^2}{v_{T\alpha}^2} \left[1 - I_+ \left(\frac{\omega}{k_z v_{T\alpha}} \right) + \frac{v_{T\alpha}^2}{\Omega_\alpha^2} I_+ \left(\frac{\omega}{k_z v_{T\alpha}} \right) \left(k_y^2 - \frac{\partial^2}{\partial x^2} \right) \right] \Phi_1, \quad (3.290)$$

and for vacuum ($x < 0, \quad \Phi = \Phi_2$):

$$\nabla^2 \Phi_2 = 0. \quad (3.291)$$

Their solutions, damped for $x \rightarrow \pm \infty$, are

$$\Phi_1 = C_1 e^{-\kappa x}, \qquad \Phi_2 = C_2 \exp \left(\sqrt{k_y^2 + k_z^2} x \right) , \quad (3.292)$$

where

$$\kappa^2 = k_y^2 + \frac{k_z^2 + \sum_\alpha \frac{\omega_{p\alpha}^2}{v_{T\alpha}^2} \left[1 - I_+ \left(\frac{\omega}{k_z v_{T\alpha}} \right) \right]}{1 + \sum_\alpha \frac{\omega_{p\alpha}^2}{\Omega_\alpha^2} I_+ \left(\frac{\omega}{k_z v_{T\alpha}} \right)}. \quad (3.293)$$

Substitution of the solutions (3.292) into the boundary conditions (3.289) yields a system of homogeneous algebraic equations for the constants C_1 and C_2. The solvability condition of this system is the dispersion equation for surface waves in semi-bounded plasma accounting for the diamagnetic current on its surface:

$$\left[1 + \sum_\alpha \frac{\omega_{p\alpha}^2}{\Omega_\alpha^2} I_+ \left(\frac{\omega}{k_z v_{T\alpha}} \right) \right] \kappa + \sum_\alpha \frac{\omega_{p\alpha}^2}{\omega \Omega_\alpha} k_y I_+ \left(\frac{\omega}{k_z v_{T\alpha}} \right) + \sqrt{k_y^2 + k_z^2} = 0. \quad (3.294)$$

We note that surface waves are possible only in the frequency range $\kappa^2 > 0$. In the frequency range $\omega > k_z v_{\text{Te}}$, the thermal motion of the particles can be neglected, and Eq. (3.294) reduces to Eq. (3.282). In this limit, the diamagnetic currents on the plasma surface are also neglected, and the oscillations do not grow with time. As mentioned above, accounting for diamagnetic currents can result in the buildup of surface oscillations. To verify this, we study the solutions of Eq. (3.294) in the frequency range $k_z v_{\text{Ti}} \ll \omega \ll k_z v_{\text{Te}}$ where the effects of the thermal motion of the electrons are significant. Here, we obtain from Eq. (3.294)

$$\left(1 + \frac{\omega_{\text{pi}}^2}{\Omega_i^2} \right) \kappa + \frac{\omega_{\text{pi}}^2 k_y}{\omega \Omega_i} \left(1 + \imath \sqrt{\frac{\pi}{2}} \frac{\omega}{|k_z| v_{\text{Te}}} \right) + \sqrt{k_y^2 + k_z^2} = 0, \quad (3.295)$$

and

$$\kappa^2 = k_y^2 + \frac{k_z^2\left(1 - \frac{\omega_{\rm pi}^2}{\omega^2}\right) + \frac{\omega_{\rm pi}^2}{v_s^2}\left(1 + \imath\sqrt{\frac{\pi}{2}}\frac{\omega}{|k_z|v_{\rm Te}}\right)}{1 + \omega_{\rm pi}^2/\Omega_i^2}. \tag{3.296}$$

For the modes with $k_y^2 \gg k_z^2$ under the condition $\omega_{\rm pi}^2 \gg \Omega_i^2$ (usually satisfied with ample reserve in real plasma confined by a magnetic field) from Eq. (3.295) we obtain the spectrum of slowly increasing oscillations ($\omega \to \omega + \imath\delta$):

$$\omega = k_y v_s, \qquad \delta = \sqrt{\frac{\pi}{8}}\frac{\omega^2}{|k_z|v_{\rm Te}}. \tag{3.297}$$

Hence, it follows that under the influence of diamagnetic currents in the inhomogeneous surface layer the buildup of surface ion-acoustic waves occurs in semi-bounded plasma confined by a magnetic field. These waves propagate along the plasma boundary at a large angle to the magnetic field and subside deep into plasma with the characteristic space scale of the order v_s/Ω_i. In strongly non-isothermal plasmas with $T_e \gg T_i$, this scale is many times greater than the ion Larmor radius and the dimension of the plasma inhomogeneity.

Finally, we note that the instabilities of surface waves may be more dangerous for the problem of magnetic confinement than drift instabilities, since the latter lead to the excitation of short-wavelength oscillations localized in the inhomogeneous layer near the plasma surface while unstable surface waves can extend much deeper into plasma. This is particularly manifested when either a current or gravitational drift, caused by curvature of field lines of the confining magnetic field, is present in plasma.

3.11 Problems

3.11.1. Along a cylindrical column of hot plasma of radius a there flows a current J distributed over the cross section with the density $\vec{j}(r)$ (z-pinch). What is the r-dependence of the plasma pressure in a stationary state provided it is balanced by the magnetic pressure exerted by the current flowing along the column. Find the relation between the total current and the pressure in the pinch, integrated over the cross section. Let plasma be isothermal and let it satisfy the perfect gas equation of state. Express the current strength J in terms of the plasma temperature T and the total number N of particles of one charge sign per unit length of the plasma column. Calculate the current strength taking $N \approx 10^{15}$ particles/cm and $T \approx 10^8 K$ (these values are typical for thermonuclear studies) [132].

Solution The magnetic field has a single projection

$$B_\varphi \equiv B(r) = \frac{4\pi}{cr}\int_0^r rj(r)\,\mathrm{dr},$$

Integrating the equilibrium equation which follows from Eq. (3.136) at $V = 0$ and taking into account the boundary condition$p|_{r\geq a} = 0$, we have

$$p\,(r) = \frac{1}{8\pi}\int_r^a \frac{1}{r^2}\frac{d}{\mathrm{dr}}\left(r^2B^2\right)\mathrm{dr}, \tag{3.298}$$

where $B = (4\pi/\mathrm{cr})\int_0^r rj\,(r)\,\mathrm{dr}$ at $r < a$, and $B = 2J/\mathrm{cr}$ at $r > a$. In order to connect the total current with the pressure, we integrate Eq. (3.298) over the entire cross section and use the relation $a\,B(a) = 2J/c$. We then get

$$\int_0^a p(r)\,2\pi r\,\mathrm{dr} = \frac{J^2}{2c^2}. \tag{3.299}$$

Assuming that plasma is an equilibrated perfect gas with a given temperature T (in energy units) and that $p = 2n(r)\,T$, from Eq. (3.299) we have

$$J = 2c\,\sqrt{NT}, \tag{3.300}$$

Substituting numerical values into Eq. (3.300), we have $J = 7.5 \times 10^4$ A. In practice, plasma is usually non-isothermal, and the temperature of electrons is higher than that of ions. To maintain the equilibrium, the current according to Eq. (3.300) should increase, because it increases the temperature and pressure of plasma. In addition, equilibrium is unstable with respect to bending and constrictions of plasma filaments. When the current flows in a thin surface layer, then the pressure inside the plasma column is constant:

$$p = \frac{J^2}{2\pi c^2a^2}.$$

3.11.2. Find the equilibrium condition for a cylindrical plasma column of radius in which the current has only the azimuthal component $j_\varphi(r)$ (theta pinch). The pressure of the medium outside the column can be neglected. Is it possible to maintain the equilibrium using a magnetic field of external sources [132]?

Solution The magnetic field inside the cylinder has a single component $B_z(r) \equiv B(r) = (4\pi/c)\int_r^a j_\varphi(r)\,\mathrm{dr}$. The equilibrium inside the cylinder requires constant total pressure,

$$p(r) + \frac{B^2(r)}{8\pi} = \text{const.}$$

Outside the cylinder, in the absence of matter, $p = 0$ and the magnetic field of azimuthal currents is zero, $B = 0$. Hence, the internal pressure can be balanced only by an external magnetic field, which is parallel to the axis of the cylinder and is equal to $B_0 = \sqrt{8\pi p(a)}$ at the boundary. The magnetic field inside the plasma column is always smaller than the external field,

$$\frac{B^2}{8\pi} = \frac{B_0^2}{8\pi} - p.$$

Therefore, plasma is diamagnetic.

3.11.3. The velocity field of a conducting medium in the spherical coordinates is given by the vector $\vec{V} = \nabla \times \vec{e}_r \psi\left(\vec{r}\right)$, while the electric conductivity $\sigma(r)$depends only on the distance from the center. Show that the magnetic field decays during a finite time, regardless of the initial state (Elsasser's antidynamo-theorem) [132].

Solution Let us write down $\vec{V} = \nabla\psi \times \vec{e}_r$and project the induction equation

$$\frac{\partial \vec{B}}{\partial t} = \nabla \times \left[\vec{V} \times \vec{B}\right] + \nu_m \nabla^2 \vec{B},$$

on to the direction $\vec{e}_r$. We get $\vec{e}_r \cdot \left(\nabla \times \left[\vec{V} \times \vec{B}\right]\right) = -\left(\vec{V} \cdot \nabla\right) \vec{B}_r;$

$$-\vec{e}_r \cdot \left(\nabla \times \nu_m \left[\nabla \times \vec{B}\right]\right) = \nabla^2 B_r + \frac{2}{r}\frac{\partial B_r}{\partial r} + \frac{2}{r^2} B_r.$$

In the last equation, we took into account that $\nabla \nu_m$ is directed along $\vec{e}_r$, and that $\nabla \cdot \vec{B} = 0$. Hence, we obtain the equation for B_r,

$$\frac{\partial B_r}{\partial t} + \left(\vec{V} \cdot \nabla\right) B_r = \nu_m \left(\nabla^2 B_r + \frac{2}{r}\frac{\partial B_r}{\partial r} + \frac{2}{r^2} B_r\right). \tag{3.301}$$

Next, we multiply Eq. (3.298) by $B_r/\nu_m r^2$ and integrate over the entire space. Rearranging the integral which includes $\nabla^2 B_r$ in accordance with the Gauss theorem, we finally obtain

$$\frac{d}{dt}\int \frac{B_r^2}{\nu_m r^2}\, dV = -2\int \frac{(\nabla B_r)^2}{r^2}\, dV - \frac{8\pi}{r} \overline{B_r^2}\Big|_{r\to 0} \le 0, \tag{3.302}$$

where the upper line denotes the averaging over orientations of r. For a finite system, the field at large distances must be zero. Hence, the right-hand side will be negative if B_r differs from zero somewhere in space.

3.11.4. Study the longitudinal waves of collisionless non-degenerate electron plasma which propagate strictly across the magnetic field (Bernstein modes) [131].

Solution According to (3.17), the dispersion equation of these waves is

$$1 = 2\sum_{n=1}^{\infty} \frac{\omega_{pe}^2 n^2 \Omega_e^2}{k^2 v_{Te}^2 \left(\omega^2 - n^2\Omega_e^2\right)} A_n \left(\frac{k^2 v_{Te}^2}{\Omega_e^2}\right) \tag{3.303}$$

or in the long-wave range $k^2 v_{\mathrm{Te}}^2 \ll \Omega_e^2$,

$$1 - \frac{\omega_{\mathrm{pe}}^2}{\omega^2 - \Omega_e^2} - \sum_{n=2}^{\infty} \frac{\omega_{\mathrm{pe}}^2 n^2}{\omega^2 - n^2\Omega_e^2} \frac{1}{n!} \left(\frac{k^2 v_{\mathrm{Te}}^2}{2\Omega_e^2}\right)^{n-1} = 0. \tag{3.304}$$

In the limit $k \to 0$ the solutions of this equation are

$$\omega = \sqrt{\omega_{\mathrm{pe}}^2 + \Omega_e^2}, \quad \omega = n\Omega_e. \tag{3.305}$$

On the other hand, in the short-wave range $k^2 v_{\mathrm{Te}}^2 \ll \Omega_e^2$, we have

$$1 = \sum_{n=1}^{\infty} \frac{\omega_{\mathrm{pe}}^2 n^2}{\omega^2 - n^2\Omega_e^2} \sqrt{\frac{2}{\pi}} \left(\frac{\Omega_e^2}{k^2 v_{\mathrm{Te}}^2}\right)^{3/2}. \tag{3.306}$$

In the limit $k \to \infty$ the solutions of the dispersion equation are $\omega \to n\Omega_e$. The dispersion laws are shown in Fig. 3.7.

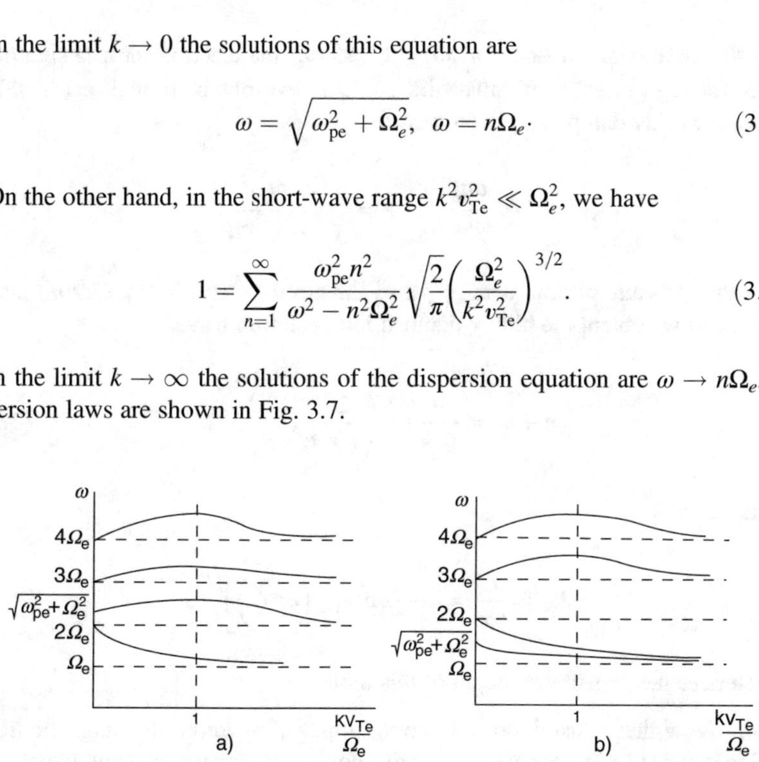

Fig. 3.7 (**a**, **b**) Bernstein modes in non-degenerate plasma: (**a**) $\omega_{\mathrm{pe}} > \Omega_e$, (**b**) $\omega_{\mathrm{pe}} < \Omega_e$

3.11.5. Derive the refractive index and the damping coefficient of the electromagnetic waves propagating along the external magnetic field for frequencies near the ion cyclotron frequency in collisionless non-degenerate plasma [131].

Solution Near the frequency $\omega \approx \Omega_i$ the general dispersion equation (3.48) reduces to

$$k^2c^2 = \omega^2\left[1 + \frac{\omega_{\rm pe}^2}{\omega\Omega_e} - \frac{\omega_{\rm pi}^2}{\omega(\omega-\Omega_i)}I_+\left(\frac{\omega-\Omega_i}{kv_{\rm Ti}}\right)\right]. \tag{3.307}$$

Far from the resonance absorption line $\omega \gg |\omega - \Omega_i| \gg kv_{\rm Ti}$, the contribution of the electron term may be ignored in Eq. (3.298). Under these conditions the refractive index and the damping coefficient of the ion cyclotron wave are

$$n^2 = -\frac{\omega_{pi}^2}{\omega(\omega-\Omega_i)}, \quad \chi = \sqrt{\frac{\pi}{8}}\frac{\omega_{pi}^2 c}{n^2\omega^2 v_{Ti}}\exp\left(-\frac{c^2(\omega-\Omega_i)^2}{2n^2\omega^2 v_{Ti}^2}\right). \tag{3.308}$$

Inside the absorption line, for $|\omega - \Omega_i| \ll kv_{\rm Ti}$ the electron term is significant. Hence, for high-pressure plasma with $v_{\rm Ti}^2 \gg v_A^2$ we obtain from Eq. (3.298) the following weakly damped electron wave:

$$n^2 \approx \frac{\omega_{\rm pe}^2}{\omega\Omega_e}, \quad \chi = \sqrt{\frac{\pi}{8}}\frac{\omega_{\rm pi}^2 c}{n^2\omega^2 v_{\rm Ti}}. \tag{3.309}$$

For low-pressure plasma with $v_{\rm Ti}^2 \ll v_A^2$ the electron term in Eq. (3.298) may be ignored and we obtain the highly damped ion cyclotron wave

$$n + \imath\chi = \frac{\imath + \sqrt{3}}{2}\left(\frac{2}{\sqrt{\pi}}\frac{\omega_{\rm pi}^2 c}{\omega^2 v_{\rm Ti}}\right)^{1/3}. \tag{3.310}$$

The quantity

$$\lambda_{\rm sk} = \frac{c}{\omega\chi} = 2\left[\frac{2}{\sqrt{\pi}}c^2 v_{\rm Ti}/\left(\omega\omega_{\rm pi}^2\right)\right]^{1/3}$$

characterizes the penetration depth of this mode.

3.11.6. Show that transmission of waves propagating across the magnetic field is possible in the collisionless magneto-active degenerate gas in semiconductors, under the condition $\omega_{\rm pe}^2 \gg \Omega_e^2 \gg \omega^2$, and find the transmission condition [131].

Solution Dispersion equation for electromagnetic (ordinary) wave is

$$k^2c^2 - \omega^2\varepsilon_{zz} = 0. \tag{3.311}$$

Under the mentioned conditions, the problem is strongly simplified since the cyclotron harmonics with $n = 0$ have the main contribution in ε_{zz}:

$$\varepsilon_{zz} = 1 - \frac{3}{2}\frac{\omega_{\rm pe}^2}{\omega^2}\int\limits_0^\pi d\theta \sin\theta \cos^2\theta J_0^2\left(\frac{kv_{\rm Fe}}{\Omega_e}\sin\theta\right). \tag{3.312}$$

Taking into account the relations

$$J_0(z) = \frac{1}{\pi}\int\limits_0^\pi \cos\left(z\sin\theta\right)d\theta,$$

$$J_0''(z) = -\frac{1}{\pi}\int\limits_0^\pi \sin^2\theta\cos\left(\sin\theta\right)d\theta,$$

in the limit $kv_{\rm Fe} \gg \Omega_e$, we find

$$\varepsilon_{zz} = 1 - \frac{3\omega_{\rm pe}^2}{\omega^2}\frac{\Omega_e}{kv_{\rm Fe}}\left[J_0\left(\frac{kv_{\rm Fe}}{\Omega_e}\right) + J_0''\left(\frac{kv_{\rm Fe}}{\Omega_e}\right)\right] = 0. \tag{3.313}$$

In this case, for $\omega \ll \omega_{\rm Le}$, from Eq. (3.298), we obtain

$$k^2c^2 + 3\omega_{\rm pe}^2\frac{\Omega_e}{kv_{\rm Fe}}\left[J_0\left(\frac{kv_{\rm Fe}}{\Omega_e}\right) + J_0''\left(\frac{kv_{\rm Fe}}{\Omega_e}\right)\right] = 0. \tag{3.314}$$

The solution of this equation exists only when

$$J_0''(z) + J_0(z) = -\frac{1}{z}J_0'(z) = \frac{J_1(z)}{z} \approx 0.$$

In this case, we find

$$k^2c^2 + 3\frac{\omega_{\rm pe}^2\Omega_e^2}{k^2v_{\rm Fe}^2}J_1\left(\frac{kv_{\rm Fe}}{\Omega_e}\right) = 0.$$

This equation has solutions with $k^2 > 0$ (region of gas transparency) only if $\frac{kv_{\rm Fe}}{\Omega_e} \approx \mu_{1s}$ where μ_{1s} are the zeros of the Bessel function $J_1(\mu_{1s}) = 0$. This implies that the degenerate dense gas is transparent for low-frequency oscillations when the wavelength is a multiple of the electron Larmor radius. Such transmission of dense magneto-active plasma was observed in metallic films at low temperatures and was called size effect [131].

3.11.7. Study the interaction of a low-density relativistic straight electron beam with the high-frequency electrostatic oscillations of cold magneto-active plasma [131].

Solution The dispersion equation pertaining to the described system is written from Eq. (3.204) as

$$\frac{k_i k_j}{k^2}\varepsilon_{ij}\left(\omega,\vec{k}\right)=\frac{k_z^2}{k^2}\left(1-\frac{\omega_{\rm pe}^2}{\omega^2}\right)+\frac{k_\perp^2}{k^2}\left(1-\frac{\omega_{\rm pe}^2}{\omega^2-\Omega_e^2}\right)-\frac{k_z^2}{k^2}\frac{\omega_b^2\gamma^{-3}}{(\omega-k_z u)^2}-\frac{k_\perp^2\omega_b^2\gamma^{-1}}{(\omega-k_z u)^2-\Omega_e^2\gamma^{-2}}=0. \tag{3.315}$$

In the absence of the beam we obtain the following longitudinal oscillations of plasma:

$$\omega_{1,2}=\frac{1}{2}\left[\omega_{\rm pe}^2+\Omega_e^2\pm\sqrt{\left(\omega_{\rm pe}^2+\Omega_e^2\right)^2-4\frac{k_z^2}{k_\perp^2+k_z^2}\omega_{\rm pe}^2\Omega_e^2}\right]. \tag{3.316}$$

Figure 3.8 shows the spectra $\omega_{1,\,2}(k_z)$ and the straight lines

$$\omega=k_z u,\qquad \omega=k_z u\pm\frac{\Omega_e}{\gamma}, \tag{3.317}$$

which correspond to the Cherenkov and cyclotron resonance (with the normal and anomalous Doppler effects) of the beam interaction with plasma oscillations. The

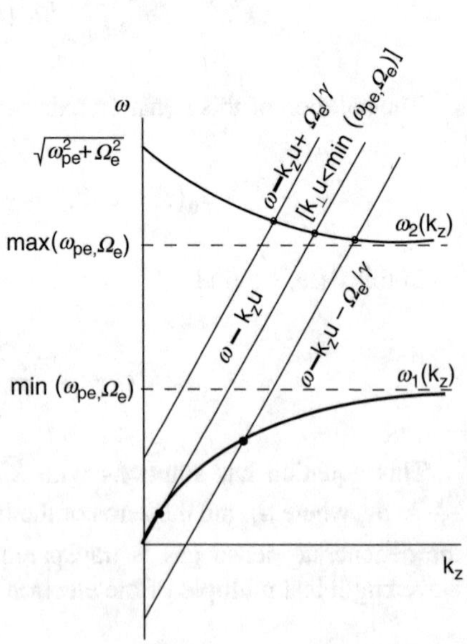

Fig. 3.8 Cherenkov and cyclotron resonance (the normal and anomalous Doppler effects) of the beam interaction with plasma oscillations

strongest interaction arises at the intersections of these straight lines with the curves $\omega_{1,2}(k_z)$ where the development of the Cherenkov and cyclotron instabilities is possible. It is shown in Fig. 3.8 that the straight line corresponding to the anomalous Doppler effect$\omega = k_z u - \Omega_e/\gamma$ intersects both branches of the longitudinal waves. The Cherenkov line $\omega = k_z u$ always intersects the upper branch; for $k_\perp u <$ min $\{\omega_{\text{pe}}, \Omega_e\}$ it also intersects the lower branch. The straight line corresponding to the normal Doppler effect $\omega = k_z u + \Omega_e/\gamma$ intersects the upper branch only. Further, Fig. 3.8 shows that the group velocity of the upper oscillation branch is always negative, whereas that of the lower one is positive. Therefore, the instability of the upper mode is always absolute and that of the lower one is convective.

In the limiting cases of dense $\left(\omega_{\text{pe}}^2 \gg \Omega_e^2\right)$ and rarefied $\left(\Omega_e^2 \gg \omega_{\text{pe}}^2\right)$ plasmas, one can obtain simple formulas for the oscillation frequencies and the increments.

In dense plasma $\left(\omega_{\text{pe}}^2 \gg \Omega_e^2\right)$ the Cherenkov instability $\omega = k_z u + \delta$ predominantly excites the upper Langmuir oscillations with the spectrum

$$\omega \approx \omega_{\text{pe}}, \qquad \frac{\delta}{\omega} = \frac{-1 + \imath\sqrt{3}}{2\gamma}\left(\frac{N_b}{2N_p}\right)^{1/3}. \tag{3.318}$$

Under these conditions the cyclotron instability is possible with the anomalous Doppler ($\omega = k_z u - \Omega/\gamma + \delta$) only and it primarily excites the upper oscillation branch

$$\omega \approx \omega_{\text{pe}}, \qquad \frac{\delta}{\omega} \simeq \frac{\imath}{2}\left(\frac{N_b}{N_p}\frac{\omega_{\text{pe}}}{\Omega_e}\right)^{1/2}. \tag{3.319}$$

We see that the increment of the Cherenkov instability is $\text{Im}\{\delta\} \sim \omega\gamma^{-1}(N_b/N_p)^{1/3}$, compared with the increment of the cyclotron instability $\text{Im}\{\delta\} \sim \omega(N_b/N_p)^{1/2}$. Nevertheless, the cyclotron instability can prevail if

$$\frac{1}{\gamma^{1/3}} > \left(\frac{\omega_{\text{pe}}}{\Omega_e}\right)^{1/2}\left(\frac{N_b}{N_p}\right)^{1/6} > \frac{1}{\gamma}. \tag{3.320}$$

This is possible in the case of ultra-relativistic electron beams only. In rarefied plasma $\left(\omega_{\text{pe}}^2 \ll \Omega_e^2\right)$, the Cherenkov instability predominantly excites the lower branch of the Langmuir oscillations with the spectrum

$$\omega \approx \sqrt{\omega_{\text{pe}}^2 - k_\perp^2 u^2}, \qquad \frac{\delta}{\omega} \approx \frac{-1 + \imath\sqrt{3}}{2\gamma}\left(\frac{N_b}{2N_p}\right)^{1/3}. \tag{3.321}$$

For the cyclotron instability with the anomalous Doppler effect we have

$$\omega \approx \omega_{\text{pe}}, \qquad \frac{\delta}{\omega} \approx \frac{1}{3}\left(\frac{N_b}{6N_p}\right)^{1/2}\left(\frac{\omega_{\text{pe}}}{\Omega_e}\right)^{1/2}. \tag{3.322}$$

This increment is always much less than the increment (3.321). Therefore, the cyclotron instability cannot develop in rarefied plasma except for the case $k_\perp u > \omega_{\text{pe}}$ where the Cherenkov instability is impossible.

3.11.8. Study the interaction between two counter-streaming identical plasma beams moving parallel to the external magnetic field with velocities much smaller than the thermal velocity of the electrons [131].

Solution Since the velocities of the beams are much smaller than the velocity of light, we can confine our analysis to electrostatic perturbations. The electrostatic dispersion equation for the system of two colliding plasma beams is written as [compare to Eq. (3.204)]

$$1 + \sum_{e,i} \frac{\omega_{p\alpha}^2}{k^2 v_{T\alpha}^2}\left[2 - \sum_s \frac{\omega - \vec{k}\cdot\vec{u}}{\omega - \vec{k}\cdot\vec{u} - s\Omega_\alpha} A_s\left(\frac{k_\perp^2 v_{T\alpha}^2}{\Omega_\alpha^2}\right) I_+\left(\frac{\omega - \vec{k}\cdot\vec{u} - s\Omega_\alpha}{k_z v_{T\alpha}}\right)\right.$$
$$\left. - \sum_s \frac{\omega + \vec{k}\cdot\vec{u}}{\omega + \vec{k}\cdot\vec{u} - s\Omega_\alpha} A_s\left(\frac{k_\perp^2 v_{T\alpha}^2}{\Omega_\alpha^2}\right) I_+\left(\frac{\omega + \vec{k}\cdot\vec{u} - s\Omega_\alpha}{k_z v_{T\alpha}}\right)\right] = 0. \tag{3.323}$$

We investigate the two limiting cases without an external magnetic field and with an infinitely strong one. When there is no magnetic field, Eq. (3.298) simplifies to

$$1 + \sum_{e,i} \frac{\omega_{p\alpha}^2}{k^2 v_{T\alpha}^2}\left[2 - I_+\left(\frac{\omega - \vec{k}\cdot\vec{u}}{k v_{T\alpha}}\right) - I_+\left(\frac{\omega + \vec{k}\cdot\vec{u}}{k v_{T\alpha}}\right)\right] = 0. \tag{3.324}$$

Under the condition $u \ll v_{\text{Ti}}$ the oscillation obeying this equation are stable, moreover, they are damping with time. This implies that there is no interaction between the beams.

For $v_{\text{Ti}} \ll u \ll v_{\text{Te}}$ we have from Eq. (3.299)

$$1+2\frac{\omega_{\text{pe}}^2}{k^2v_{\text{Te}}^2}\left(1-\iota\sqrt{\frac{\pi}{2}}\frac{\omega}{kv_{\text{Te}}}\right)-\frac{\omega_{\text{pi}}^2}{\left(\omega-\vec{k}\cdot\vec{u}\right)^2}-\frac{\omega_{\text{pi}}^2}{\left(\omega+\vec{k}\cdot\vec{u}\right)^2}=0. \tag{3.325}$$

If the small imaginary term describing the Cherenkov absorption by the electrons is neglected we obtain the spectrum

$$\omega_{1,2}^2=\alpha^{-1}\left\{\omega_{\text{pi}}^2+\left(\vec{k}\cdot\vec{u}\right)^2\alpha\pm\sqrt{\left[\omega_{\text{pi}}^2+\left(\vec{k}\cdot\vec{u}\right)^2\alpha\right]^2+4\left(\vec{k}\cdot\vec{u}\right)^2\alpha\left[2\omega_{\text{pi}}^2-\left(\vec{k}\cdot\vec{u}\right)^2\alpha\right]}\right\}, \tag{3.326}$$

where $\alpha = 1 + 2\omega_{\text{pe}}^2/k^2v_{\text{Te}}^2$. Under the condition

$$2\omega_{\text{pi}}^2 > \left(\vec{k}\cdot\vec{u}\right)^2\alpha, \tag{3.327}$$

the root $\omega_2^2 \approx -\left(\vec{k}\cdot\vec{u}\right)^2 < 0$ which corresponds to aperiodically unstable oscillations appears. This indicates a strong interaction of the colliding beams. According to condition (3.319), this interaction is possible for velocities $u < v_s = \sqrt{T_e/M}$. It occurs only in non-isothermal plasma with $T_e \gg T_i$, because condition $u \gg v_{\text{Ti}}$ was assumed. This hydrodynamic instability also exists in the presence of a strong longitudinal magnetic field. In the limit $B_0 \to \infty$, assuming $v_{\text{Ti}} \ll u \ll v_{\text{Te}}$, we obtain from Eq. (3.298)

$$1+2\frac{\omega_{pe}^2}{k^2v_{Te}^2}\left(1-\iota\sqrt{\frac{\pi}{2}}\frac{\omega}{|k_z|v_{Te}}\right)-\frac{k_z^2}{k^2}\left(\frac{\omega_{pi}^2}{\left(\omega-\vec{k}.\vec{u}\right)^2}+\frac{\omega_{pi}^2}{\left(\omega+\vec{k}.\vec{u}\right)^2}\right)=0. \tag{3.328}$$

Comparing with Eq. (3.300), we conclude that the instability is of the same nature as in the case of non-magnetized plasma. The modifications are slight. Condition (3.319) becomes

$$2\omega_{\text{pi}}^2 > k^2u^2\alpha,$$

and the instability increment becomes larger than $\omega_2^2 \approx -k^2u^2$.

3.11.9. Show that even for $u < v_{\text{Ti}}$ collisionless non-magnetized plasma with a current is unstable with respect to transverse perturbations [131].

Solution Dispersion equation (3.170) for arbitrary oscillations of collisionless non-magnetized plasma with a current is written as ($u \ll c$):

$$D\left(\omega,\vec{k}\right)\equiv\left\{k^2-\frac{\omega^2}{c_2}\left[\delta\varepsilon_i^{\text{tr}}(\omega,k)+1+\frac{\omega'^2}{\omega^2}\delta\varepsilon_e^{\text{tr}}(\omega',k)\right]\right\}\left[\delta\varepsilon_e^l(\omega',k)+\delta\varepsilon_i^l(\omega,k)+1\right]-$$
$$-\frac{k^2u^2-\left(\vec{k}\cdot\vec{u}\right)^2}{c^2}\left\{\delta\varepsilon_e^l(\omega',k)+\frac{\omega'^2}{k^2c^2}\left[\delta\varepsilon_e^{\text{tr}}(\omega',k)-\delta\varepsilon_e^l(\omega',k)\right]\right\}\times$$
$$\times\left\{\delta\varepsilon_i^l(\omega,k)+\frac{\omega^2}{k^2c^2}\left[\delta\varepsilon_i^{\text{tr}}(\omega,k)-\delta\varepsilon_i^l(\omega,k)\right]\right\}=0. \tag{3.329}$$

Here, $\delta\varepsilon_\alpha^l(\omega,k)$ and $\delta\varepsilon_\alpha^{\text{tr}}(\omega,k)$ for $\alpha=e,\,i$ are the contributions of electrons and ions to the longitudinal and the transverse permittivities. In the limit of very low frequencies, $\omega\ll k_z v_{\text{Ti}}$, and for wave propagation transverse to the current, $\vec{u}\cdot\vec{k}=0$, this equation is of the form

$$D(0,k)+\omega\frac{\partial D(0,k)}{\partial\omega}=0, \tag{3.330}$$

where

$$\begin{aligned}D(0,k)&=k^2\left(1+\sum_{\alpha=e,\,i}\frac{\omega_{p\alpha}^2}{k^2v_{T\alpha}^2}\right)-\frac{u^2}{c^2}\frac{\omega_{\text{pe}}^2\omega_{\text{pi}}^2}{k^2v_{\text{Te}}^2v_{\text{Ti}}^2},\\ \frac{\partial D(0,k)}{\partial\omega}&=\imath\sqrt{\frac{\pi}{2}}\left(\frac{\omega_{\text{pi}}^2}{v_{\text{Ti}}^2}-\frac{u^2}{c^2}\frac{\omega_{\text{pe}}^2\omega_{\text{pi}}^2}{k^2v_{\text{Te}}^2v_{\text{Ti}}^2}\right)\frac{1}{kv_{\text{Ti}}}.\end{aligned} \tag{3.331}$$

Perturbations with the wave vector k satisfying

$$k\leq k_0\approx\frac{u}{c}\frac{\omega_{\text{pe}}\omega_{\text{pi}}}{\sqrt{\omega_{\text{pe}}^2v_{\text{Ti}}^2+\omega_{\text{pi}}^2v_{\text{Te}}^2}} \tag{3.332}$$

are unstable for arbitrarily small velocities $u<v_{\text{Ti}}$. They increase aperiodically with the increment

$$Im\{\omega\}=(k_0-k)v_{\text{Ti}}=k_0v_{\text{Ti}}\left(1-\frac{k}{k_0}\right). \tag{3.333}$$

Note that a strong external magnetic field aligned with the current stabilizes this instability.

3.11.10 Find the limiting current of the monoenergetic relativistic electron beam traveling through the equipotential vacuum drift space in the strong external longitudinal magnetic field. Analyze cases of plane and cylindrical geometry [131].

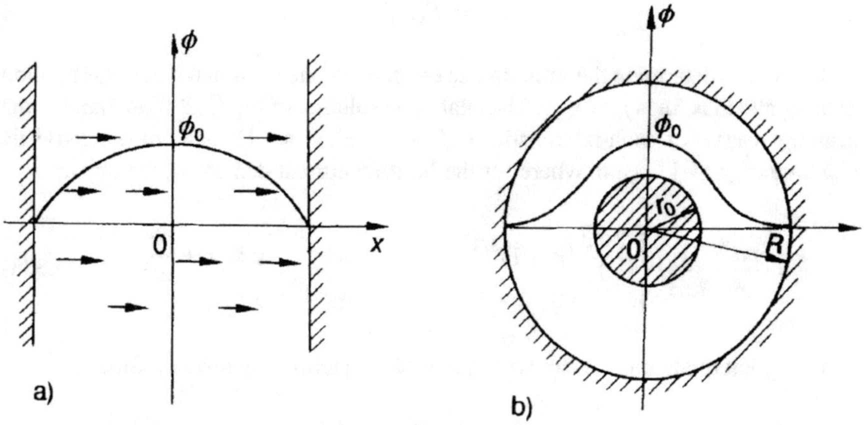

Fig. 3.9 Distribution of the beam potential: (**a**) in plane drift space; (**b**) in cylindrical drift space

Solution When traveling in the equipotential drift space, beam electrons produce a spatially distributed charge which causes a spatial modification of the potential and therefore brake the beam. As a result, the distribution of the potential with a maximum in the center will be established in the plane drift space (Fig. 3.9a). Poisson's equation for the potential distribution $\varphi(x)$ can be written as

$$\frac{d^2\phi}{dx^2} = -4\pi\rho = -\frac{4\pi j}{v} = -\frac{4\pi j}{c}\left[1 - \frac{1}{\left(1 + e\phi/mc^2\right)^2}\right]^{-1/2}. \tag{3.334}$$

Here, $j = \rho v = \text{const}$ is the current density of the electron beam, $\gamma = \left(1 - v^2/c^2\right)^{-1/2}$ is the relativistic factor, and v is the velocity of the injected electrons. When deriving Eq. (3.298), the integral of motion

$$\frac{mc^2}{\sqrt{1 - v^2/c^2}} - mc^2 = e\phi \tag{3.335}$$

is taken into account. Equation (3.298) must be supplemented with the boundary conditions (Fig. 3.9a)

$$\phi|_{x=\pm d} = 0, \quad \phi|_{x=0} = \phi_0. \tag{3.336}$$

The given problem is overdetermined: the potential on the axis is determined uniquely for the assigned current density j. Therefore, the solution permits us to relate j to ϕ_0:

$$j = F(\phi_0). \tag{3.337}$$

In order to determine the limiting current density one must maximize this relation over ϕ_0 and thus finds $j_0 = j_{\max}$. The analytical solution of Eq. (3.298) is found in the limiting cases of non-relativistic ($e\phi \ll mc^2, \gamma \approx 1$) and ultra-relativistic ($e\phi \gg mc^2, \gamma \gg 1$) beam, where for the limiting current density j_0, we obtain

$$j_0 = \frac{mc^3}{e}\frac{1}{2\pi d^2}\begin{cases} \frac{2\sqrt{2}}{9}(\gamma-1)^{3/2}, & \text{for} \quad \gamma = 1 + \frac{v^2}{2c^2}, \\ \gamma, & \text{for} \quad \gamma \gg 1. \end{cases} \tag{3.338}$$

Using interpolation, we can write these two equations by a single formula

$$j_0 = \frac{mc^3}{e}\frac{\left(\gamma^{3/2}-1\right)^{3/2}}{2\pi d^2}. \tag{3.339}$$

For $\gamma \gg 1$, this expression coincides with expression (3.319), and for $\gamma = 1 + v^2/(2c^2)$, it differs from the latter by the factor $\sqrt{3}$.

In case of the beam traveling along the axis of the cylindrical drift space (Fig. 3.9b), the problem is formulated and solved analogously:

$$\frac{1}{r}\frac{d}{\mathrm{d}r}r\frac{d\phi}{\mathrm{d}r} = -\frac{4\pi j}{c}\left[1 - \frac{1}{\left(1 + e\phi/mc^2\right)^2}\right]^{-1/2} \tag{3.340}$$

$$\phi|_{r=R} = 0, \qquad \phi_{r=0} = \phi_0. \tag{3.341}$$

Then the current distribution over a radius is

$$j(r) = \begin{cases} 0, & \mathit{for} \quad r > r_0, \\ \text{const}, & \mathit{for} \quad r \le r_0. \end{cases} \tag{3.342}$$

When solving this problem, we must take account of the continuity of the potential and its derivative at the beam boundary for $r = r_0$. Hence

$$I_0 = \pi r_0^2 j_0 = \frac{mc^3}{e}\frac{\left(\gamma^{2/3}-1\right)^{3/2}}{1 + 2\ln\left(R/r_0\right)}. \tag{3.343}$$

Finally, note that such an estimate for a thin hollow beam with an average radius r_0 and thickness $a \ll r_0$ leads to the following expression for the limiting current:

$$I_0 = \frac{mc^3}{e} \frac{\left(\gamma^{2/3} - 1\right)^{3/2}}{\frac{a}{r_0} + 2\ln\left(R/r_0\right)}. \tag{3.344}$$

Here, $mc^3/e \approx 17\ kA$.

3.11.11. Show that the surface of plasma, confined by a magnetic field with a positive curvature of the field lines (i.e., with a positive normal oriented away from the plasma surface), is unstable with respect to low-frequency flute perturbations [131].

Solution The curvature of the magnetic field lines can be accounted for by introducing a centrifugal acceleration of the particles of type α. The acceleration is oriented along the positive normal to the plasma surface and is equal to $v_{T\alpha}^2/R$, R being the radius of the curvature of the magnetic field lines. This results in a particle drift along the plasma surface (oy-axis) with the velocity $u_\alpha = -v_{T\alpha}^2/(\Omega_\alpha R)$. Finally, in Eq. (3.282) $\omega - k_y u_\alpha$ must be substituted for the value ω. For flute oscillation modes in the low-frequency $|\omega - k_y u_\alpha| \ll \Omega_\alpha$, we obtain

$$2 + \frac{\omega_{\mathrm{pi}}^2}{\omega_i^2} + |k_y| \frac{\omega_{\mathrm{pi}}^2}{\Omega_i^2} \frac{v_s^2 + v_{\mathrm{Ti}}^2}{R\omega^2} = 0, \tag{3.345}$$

where $v_s^2 = T_e/M$ is the ion-acoustic velocity. Hence

$$\omega^2 = -\frac{|k_y| g_{\mathrm{eff}}}{1 + 2v_A^2/c^2}. \tag{3.346}$$

Here, $g_{\mathrm{eff}} = \left(v_s^2 + v_{\mathrm{Ti}}^2\right)/R$ is an effective gravitational field accounting for the curvature of field lines of the magnetic field, confining plasma. Since$\omega^2 < 0$, the surface of plasma, confined by the magnetic field with the positive curvature of field lines, is always unstable. This instability is analogous to the convective (flute) instability of volumetric waves in inhomogeneous magnetized plasma.

3.11.12. Study the low-frequency quasi-longitudinal modes of oscillations in thin layers of magneto-active dispersive plasma.

Solution Now, we consider the low-frequency quasi-longitudinal modes of oscillations in a thin layer of a magneto-active plasma medium. Since in magneto-active plasma, the dielectric permittivity tensor with spatial dispersion $\varepsilon_{ij}\left(\omega, \vec{k}\right)$ has six independent components with complex forms, the problem is essentially simplified if we restrict our analysis to quasi-potential oscillations or to potential fields [133]. It is possible to show that in this case in the model of mirror-like reflecting boundaries, dispersion equation (2.266) retains its form, if we use $\varepsilon\left(\omega, \vec{k}\right)$ instead of $\varepsilon^l(\omega, k)$.

The validity of such an equation is simply verified by the direct generalization of this equation to the anisotropic medium. We start from the analysis of ion-acoustic waves, assuming the magnetic field to be along the medium surface. In the ion-acoustic frequency range,

$$kv_{\mathrm{Ti}}, \nu_i \ll \begin{cases} \nu_e \ll \omega \ll k_z v_{\mathrm{Te}}, \\ \omega \ll \dfrac{k_z^2 v_{\mathrm{Te}}^2}{\nu_e} \ll \nu_e, \end{cases} \tag{3.347}$$

we have

$$\varepsilon\left(\omega, \vec{k}\right) = 1 - \frac{\omega_{\mathrm{pi}}^2}{\omega^2}\frac{k_z^2}{k^2}\left(1 - \imath\frac{\nu_i}{\omega}\right) + \frac{\omega_{\mathrm{pe}}^2}{k^2 v_{\mathrm{Te}}^2}\left(1 + \imath\begin{cases} \sqrt{\dfrac{\pi}{2}}\dfrac{\omega}{|k_z| v_{\mathrm{Te}}} \\ \dfrac{\omega\nu_e}{k_z^2 v_{\mathrm{Te}}^2} \end{cases}\right). \tag{3.348}$$

Therefore, substituting Eq. (3.299) into Eq. (2.266), we obtain

$$1 + \frac{2}{a\mid k_{\|}\mid}\sum_{n=0}^{\prime}\frac{\left[1 \pm (-1)^n\right] k_{\|}^2}{k_{\|}^2 + \frac{n^2\pi^2}{a^2} - \frac{\omega_{\mathrm{pi}}^2}{\omega^2}k_{\|}^2\left(1 - \frac{\imath\nu_i}{\omega}\right) + \frac{\omega_{\mathrm{pi}}^2}{v_s^2}\left(1 + \imath\alpha\left(k_{\|}\right)\right)} = 0, \tag{3.349}$$

where $\alpha(k_{\|})$ is given by Eq. (2.272). To solve Eq. (3.300), we assume that $a \mid k_{\|} \mid \ll 1$ because in the opposite case, this equation reduces to an equation for the semi-bounded case. If $a \mid k_{\|} \mid \ll 1$, then from Eq. (3.300), we obtain

$$1 + \frac{2a\mid k_{\|}\mid}{\pi}\phi^{\pm}(\eta) = 0, \tag{3.350}$$

where

$$\eta^2 = \frac{k_{\|}^2 a^2}{\pi^2} - \frac{a^2 k_{\|}^2 \omega_{pi}^2}{\pi^2\omega^2}\left(1 - \frac{\imath\nu_i}{\omega}\right) + \frac{a^2\omega_{pi}^2}{\pi^2 v_s^2}\left(1 + \imath\alpha\left(k_{\|}\right)\right). \tag{3.351}$$

The function $\phi^{\pm}(\eta) = \sum_{n=0}[1 \pm (-1)^n/(n^2 + \eta^2)$ has the following asymptotic forms:

$$\phi^{-}(\eta) = \begin{cases} \dfrac{\pi^2}{2} & \eta \ll 1, \\ \dfrac{\pi}{2\eta} & \eta \gg 1, \end{cases} \qquad \phi^{+}(\eta) = \begin{cases} \dfrac{2}{\eta^2} & \eta \ll 1, \\ \dfrac{\pi}{2\eta} & \eta \gg 1. \end{cases} \tag{3.352}$$

It is obvious that the non-trivial solution of Eq. (3.306) exists only if $\phi^{+}(\eta) = \frac{2}{\eta^2} \gg 1$ (symmetric modes with even n when $\eta \ll 1$). Therefore, in this limit, from Eq. (3.306), we find

$$1 + \frac{2}{a \mid k_{\parallel} \mid} \frac{1}{1 - \frac{\omega_{\rm pi}^2}{\omega^2}\left(1 - \frac{\imath\nu_i}{\omega}\right) + \frac{\omega_{\rm pi}^2}{k_{\parallel}^2 v_s^2}\left(1 + \imath\alpha(k_{\parallel})\right)} = 0. \tag{3.353}$$

The solution of this equation has the form of

$$\omega^2 = \frac{\omega_{pi}^2}{\frac{2}{a|k_{\parallel}|} + \frac{\omega_{pi}^2}{k_{\parallel}^2 v_s^2}}, \qquad \frac{\delta}{\omega} = -\frac{\nu_i}{2\omega} - \alpha(k_{\parallel})\frac{\omega^2}{k_{\parallel}^2 v_s^2}. \tag{3.354}$$

In the long-wavelength limit, when $k_{\parallel}^2 v_s^2 \ll \omega_{\rm pi}^2 a k_{\parallel}/4$, the frequency spectrum (3.322) coincides with the oscillation spectrum of ion-acoustic surface waves, but in the short-wavelength limit it leads to the frequency spectrum of deep water $\sim \omega_{\rm pi}\sqrt{ak_{\parallel}/2}$. On the whole, spectrum (3.322) coincides with spectrum (2.273) for isotropic plasma.

Now, we consider the diffusion spectrum for a thin layer of a magneto-active plasma medium under the condition that the magnetic field is parallel to the layer surface. In the diffusion frequency range,

$$k_z v_{T\alpha}, \omega \ll \nu_\alpha, \qquad k_\perp v_{T\alpha} \ll \Omega_\alpha, \tag{3.355}$$

we have

$$\varepsilon\left(\omega, \vec{k}\right) = 1 + \sum_\alpha \frac{\omega_{p\alpha}^2}{k^2 v_{T\alpha}^2} \frac{\frac{\imath k_\perp^2 v_{T\alpha}^2}{\nu_\alpha} + \frac{\imath k_\perp^2 v_{T\alpha}^2 \nu_\alpha}{\Omega_\alpha^2 + \nu_\alpha^2}}{\omega + \imath\nu_\alpha\left[\frac{k_z^2 v_{T\alpha}^2}{\nu_\alpha^2} + \frac{k_\perp^2 v_{T\alpha}^2}{\Omega_\alpha^2 + \nu_\alpha^2}\right]}, \tag{3.356}$$

where $k_\perp = n\pi/a$. Therefore, substituting Eq. (3.343) into Eq. (2.266), in one-component plasma, in the limit of $a^2 \ll r_D^2$, we obtain the monopolar characteristics of symmetric diffusion modes

$$\omega = -\imath D_{\parallel\alpha}\left(k_{\parallel}^2 + \frac{a \mid k_{\parallel} \mid}{2r_{D_\alpha}^2}\right). \tag{3.357}$$

Asymmetric modes, in the same limit, do not have any solution.

If $a^2 \gg r_D^2$, diffusion has ambipolar characteristics and for both symmetric and asymmetric modes, we find

$$\omega = -\imath k_{\parallel}^2 D_{\parallel e}\left(1 + \frac{T_i}{T_e}\right) = -\imath k_{\parallel}^2 D_{\parallel a}. \tag{3.358}$$

Finally, we consider low-frequency oscillation modes in a magneto-active plasma layer under the condition that the magnetic field is normal to the layer. We start our analysis from ion-acoustic oscillations. Again, the z axis is oriented along the magnetic field, but the x axis is parallel to the layer. It is simply verified that when $a \mid k_{\perp} \mid \ll 1$, ion-acoustic waves, propagating across the magnetic field and damping inside layer, cannot exist. Therefore, we consider diffusion modes in the similar geometry of the field and layer. Substituting Eq. (3.343) in Eq. (2.266) for rarefied plasma, we find

$$\omega = -\imath D_{\perp a}\left(k_{\perp}^2 + \frac{a \mid k_{\perp} \mid}{2r_{D_a}^2}\right). \tag{3.359}$$

Here, as above, the transverse diffusion coefficient $D_{\perp \alpha}$ and the transverse wave vector $\vec{k}_{\perp}$ appear. We note that in rarefied plasma the second term in expression (3.359) is a small correction. But, in deriving this formula, we did not assume it to be small. In addition, when it is determined, Eq. (3.359) describes the Maxwell relaxation of charge carriers. In this case, Eq. (3.359) is valid only for one-component electron plasma.

If plasma is dense, the monopolar diffusion is replaced by ambipolar diffusion and the solution of Eq. (2.266) is written in the form of

$$\omega = -\imath k_{\perp}^2 D_{\perp \mathrm{i}}\left(1 + \frac{T_e}{T_i}\right) = -\imath k_{\perp}^2 D_{\perp \mathrm{a}}. \tag{3.360}$$

From the above analysis in a plasma layer with finite thickness, we may conclude two qualitative results. Firstly, in the frequency range of ion-acoustic oscillations, layer thickness appears in the short-wavelength region and essentially affects the frequency of such oscillations. Secondly, in the diffusion region of oscillation modes, the thickness of the layer appears in the long-wavelength region. The thickness of the layer is important only in the monopolar diffusion of perturbations in a rarefied plasma medium.

References

1. B.A. Trubnikov, *Plasma Physics and Problem of Controlled Thermonuclear Reactions*, vol 3 (Izd-vo Academy Science of USSR, Moscow, 1958)
2. B.A. Trubnikov, Dissertation, MIFI, 1958
3. V.L. Ginzburg, *Theory of Radio Wave Propagation in Ionosphere* (Gostechizdat, Moscow, 1949)

4. V.L. Ginzburg, *Electromagnetic Waves in Plasma* (Fizmatgiz, Moscow, 1960)
5. V.M. Galitskii, A.B. Migdal, *Plasma Physics and Problem of Thermonuclear Fusion*, vol 1 (Izd-vo Academy Science of USSR, Moscow, 1958)
6. M.E. Gertsenshtein, J. Exp. Theor. Phys. **27**, 180 (1954)
7. Y.A. Alpert, V.L. Ginzburg, E.L. Feinberg, *Radio Wave Propagation* (Gostechizdat, Moscow, 1953)
8. V.D. Shafranov, *Plasma Physics and Problem of Controlled Thermonuclear Reactions*, vol 4 (Izd-vo Academy Science of USSR, Moscow, 1958)
9. B.N. Gershman, V.L. Ginzburg, N.G. Denisov, Usp. Fiz. Nauk **61**, 561 (1957)
10. V.P. Silin, Bull. Lebedev Inst. Acad. Sci. USSR **6**, 200 (1955)
11. A.G. Sitenko, K.N. Stepanov, J. Exp. Theor. Phys. **31**, 642 (1956)
12. B.N. Gershman, J. Exp. Theor. Phys. **24**, 659 (1953)
13. B.A. Trubnikov, V.S. Kudryatsev, *Proceeding of Second International Conference on Peaceful Uses of Atomic Energy*, Geneva, vol 31 (Reports of the Soviet Delegation (Nuclear Physics), 1958), p. 93
14. W.E. Drummond, M.N. Rosenbluth, Phys. Fluids **3**, 45 (1960)
15. V.D. Shafranov, J. Exp. Theor. Phys. **34**, 1475 (1958)
16. S.T. Beliaev, *Plasma Physics and Problem of Controlled Thermonuclear Reactions*, vol 3 (Izd-vo Academy Science of USSR, Moscow, 1958)
17. G.V. Gordeev, J. Exp. Theor. Phys. **24**, 445 (1953)
18. V.P. Silin, J. Exp. Theor. Phys. **23**, 649 (1959)
19. A.I. Akhiezer, L.E. Pargamanik, Kharkov Univ. A.M. Gorki **27**, 75 (1948)
20. G.V. Gordeev, J. Exp. Theor. Phys. **28**, 660 (1952)
21. B.N. Gershman, *Collections: Memory of A. A. Andronov* (Izd-vo Academy Science of USSR, Moscow, 1955)
22. L.M. Pyatigorsky, Lect. Note Kharkov Univ. A.M. Gorki **64**, 23 (1955)
23. M. Lozzi, R. Jancel, T. Kanen, Appl. Sci. Res. B **5**, 327 (1956)
24. M. Lozzi, R. Jancel, T. Kanan, C. R. Acad. Sci. **240**, 162 (1955)
25. B.N. Gershman, J. Exp. Theor. Phys. **31**, 707 (1956)
26. S.I. Braginskii, Dokl. Acad. Sci. USSR **115**, 475 (1957)
27. J.B. Bernstein, Bull. Am. Phys. Soc. **2**, 85 (1957)
28. T. Pradhan, Phys. Rev. **107**, 1222 (1957)
29. J.B. Bernstein, Phys. Rev. **109**, 10 (1958)
30. K.N. Stepanov, V.S. Tkalich, J. Tech. Phys. **28**, 1789 (1958)
31. K.N. Stepanov, J. Exp. Theor. Phys. **35**, 283 (1958)
32. V.L. Ginzburg, V.V. Zhelezniakov, Radiofizika **1**, 59 (1958)
33. V.V. Zhelezniakov, Radiofizika **2**, 14 (1959)
34. W.E. Drummond, Phys. Rev. **112**, 1460 (1958)
35. D. Beard, Phys. Rev. Lett. **2**, 81 (1959)
36. B.N. Gershman, J. Exp. Theor. Phys. **37**, 695 (1959)
37. B.N. Gershman, M.S. Kovner, Radiofizika **2**, 28 (1958)
38. B.N. Gershman, J. Exp. Theor. Phys. **38**, 912 (1960)
39. L. Oster, Rev. Mod. Phys. **32**, 141 (1960)
40. B.N. Gershman, Radiofizika **3**, 146 (1960)
41. A.I. Akhiezer, Ya.B. Fainberg, A.G. Sitenko, et al, *Proceeding of Second International Conference on Peaceful Uses of Atomic Energy*, Geneva, vol 31 (Reports of the Soviet Delegation (Nuclear Physics), 1959), p. 99
42. B.N. Gershman, Dissertation, FIAN, USSR, 1960
43. R.Z. Sagdeev, V.D. Shafranov, *Proceeding of Second International Conference on Peaceful Uses of Atomic Energy*, Geneva, vol 31 (Reports of the Soviet Delegation (Nuclear Physics), 1958), p. 118
44. E.P. Gross, Phys. Rev. **82**, 232 (1951)
45. T.H. Stix, Phys. Fluids **3**, 19 (1960)

46. S.I. Buchsbaum, L. Mower, S.C. Brown, Phys. Fluids **3**, 806 (1960)
47. K.N. Stepanov, J. Exp. Theor. Phys. **38**, 265 (1960)
48. K.N. Stepanov, J. Exp. Theor. Phys. **38**, 1564 (1960)
49. A.A. Rukhadze, V.P. Silin, J. Tech. Phys. **32**(4), 423 (1962)
50. Y.L. Klimontovich, Dokl. Acad. Sci. USSR **87**, 927 (1952)
51. V.P. Silin, J. Exp. Theor. Phys. **38**, 1577 (1960)
52. A.I. Akhiezer, R.V. Polovin, J. Exp. Theor. Phys. **30**, 915 (1956)
53. S.T. Beliaev, G.I. Budker, Dokl. Acad. Sci. USSR **107**, 807 (1956)
54. B.A. Trubnikov, Dokl. Acad. Sci. USSR **118**, 913 (1958)
55. Y.L. Klimontovich, J. Exp. Theor. Phys. **37**, 735 (1959)
56. D.B. Beard, Phys. Fluids **3**, 324 (1960)
57. B.N. Gershman, Radiofizika **3**, 534 (1960)
58. B.A. Trubnikov, *Plasma Physics and Problem of Controlled Thermonuclear Reactions*, vol 4 (Izd-vo Academy Science of USSR, Moscow, 1958)
59. V.L. Ginzburg, G.P. Motulevich, Usp. Fiz. Nauk **55**, 469 (1955)
60. S.I. Braginskii, A.P. Kazantsev, *Plasma Physics and Problem of Controlled Thermonuclear Reactions* (Izd-vo Academy Science of USSR, Moscow, 1958)
61. K.N. Stepanov, J. Exp. Theor. Phys. **34**, 1292 (1958)
62. K.N. Stepanov, J. Exp. Theor. Phys. **35**, 1155 (1958)
63. V.L. Ginzburg, J. Exp. Theor. Phys. **21**, 788 (1951)
64. B.N. Gershman, J. Exp. Theor. Phys. **24**, 453 (1953)
65. B.N. Gershman, Radiofizika **1**, 3 (1958)
66. B.N. Gershman, M.S. Kovner, Radiofizika **3**, 19 (1958)
67. T.F. Volkov, J. Exp. Theor. Phys. **37**, 422 (1959)
68. J. Paddington, Nature **176**, 875 (1955).; Philos. Mag. **46**, 1037 (1955)
69. E. Astrom, Ark. Fys. **2**, 443 (1951)
70. K.N. Stepanov, Ukr. Phys. J. **4**, 678 (1959)
71. B.N. Gershman, Radiofizika **1**, 49 (1958)
72. P.H. Doyle, J. Neufeld, Phys. Fluids **2**, 39 (1959)
73. L.S. Bogdankevich, A.A. Rukhadze, J. Tech. Phys. **32**(3), 322 (1962)
74. V.L. Ginzburg, A.V. Gurevich, Usp. Fiz. Nauk **71**, 201 (1960)
75. A.V. Gurevich, J. Exp. Theor. Phys. **35**, 392 (1958)
76. V.P. Silin, J. Exp. Theor. Phys. **38**, 1771 (1960)
77. A.V. Gurevich, J. Exp. Theor. Phys. **30**, 112 (1956)
78. R.C. Hwa, Phys. Rev. **110**, 307 (1958)
79. V.V. Dolgopolov, K.N. Stepanov, Ukr. Fiz. Zh. **5**, 59 (1960)
80. L.D. Landau, E.M. Lifshitz, *Electrodynamics of Continuous Media*, 2nd edn. (Pergamon, New York, 1984)
81. K.M. Watson, Phys. Rev. **102**, 12 (1956)
82. G. Chew, M. Goldberger, F. Low, Proc. Roy. Soc. A **236**, 112 (1956)
83. S. Chandrasekhar, A. Kaufman, K. Watson, Ann. Phys. **2**, 443 (1957)
84. B.B. Kadomtsev, *Plasma Physics and Problem of Controlled Thermonuclear Reactions*, vol 4 (Izd-vo Academy Science of USSR, Moscow, 1958)
85. L.I. Rudakov, R.Z. Sagdeev, *Plasma Physics and Problem of Controlled Thermonuclear Reactions*, vol 3 (Izd-vo Academy Science of USSR, Moscow, 1958)
86. Y.L. Klimontovich, V.P. Silin, J. Exp. Theor. Phys. **40**(4), 1213 (1961)
87. N.N. Bogolubov, D.N. Zubarev, Ukr. Math. J. **7**(1), 5 (1955)
88. A.B. Bernstein, E.A. Frieman, M.D. Kruskal, R.M. Kulsrud, Proc. Roy. Soc. Lond. A **244**, 17 (1958)
89. A.A. Vedenov, R.Z. Sagdeev, *Plasma Physics and Problem of Controlled Thermonuclear Reactions* (Izd-vo Academy Science of USSR, Moscow, 1958)
90. A.B. Kitsenko, K.N. Stepanov, J. Exp. Theor. Phys. **38**, 1841 (1960)
91. S.I. Syrovatskii, Usp. Fiz. Nauk **62**, 247 (1957)

92. A.I. Akhiezer, Y.B. Fainberg, Dokl. Acad. Sci. USSR **69**, 555 (1949)
93. A.I. Akhiezer, Y.B. Fainberg, J. Exp. Theor. Phys. **21**, 1262 (1951)
94. A.I. Akhiezer, Y.B. Fainberg, Usp. Fiz. Nauk **44**, 321 (1951)
95. D. Bohm, E.P. Gross, Phys. Rev **75**, 1851 (1949)
96. V.V. Zhelezniakov, Radiofizika **2**, 1 (1959)
97. V.L. Feinberg, Usp. Fiz. Nauk **69**, 537 (1959)
98. M.E. Gertsenshtein, J. Exp. Theor. Phys. **23**, 669 (1953)
99. G.V. Gordeev, J. Exp. Theor. Phys. **27**, 24 (1954)
100. A.I. Akhiezer, G.Y. Lubarskii, Y.B. Fainberg, Lect. Note Kharkov Univ. A.M. Gorki **64**, 73 (1955)
101. A.I. Akhiezer, G.Y. Lubarskii, Lect. Note Kharkov Univ. A.M. Gorkii **64**, 13 (1955)
102. A.I. Akhiezer, Nuovo Cimento **3**(4), 591 (1956)
103. P.L. Auer, Phys. Rev. Lett. **1**, 411 (1958)
104. G.G. Getmantsev, J. Exp. Theor. Phys. **37**, 843 (1959)
105. R.V. Polovin, N.L. Tsintsadze, J. Tech. Phys. **29**, 831 (1959)
106. A.G. Sitenko, Y.L. Kirochkin, J. Tech. Phys. **31**, 801 (1960)
107. V.O. Rapoport, Radiofizika **3**, 148 (1960)
108. L.M. Kovrizhnikh, A.A. Rukhadze, J. Exp. Theor. Phys. **38**, 850 (1960)
109. L.M. Kovrizhnikh, Dissertation, FIAN, USSR, 1960
110. Ya.B. Fainberg, Dissertation, FIAN, USSR, 1960
111. J. Pierce, Proc. IRE **35**, 111 (1947)
112. V.L. Ginzburg, V.V. Zhelezniakov, Radiofizika **1**, 9 (1959)
113. V.V. Zhelezniakov, Astron. Zh. **35**, 230 (1958)
114. V.V. Zhelezniakov, Usp. Fiz. Nauk **64**, 113 (1958)
115. P.I. Kellogg, H. Ziemolh, Phys. Fluids **3**, 40 (1960)
116. O. Penrose, Phys. Fluids **3**, 258 (1960)
117. V.O. Rapoport, Radiofizika **3**, 737 (1960)
118. M.S. Kovner, Radiofizika **2**, 631 (1960)., 746
119. P.A. Sturrok, Phys. Rev. **117**, 1426 (1960)
120. W.E. Nexsen Jr., W.F. Cummins, F.H. Coensgen, A.E. Sherman, Phys. Rev. **119**, 1457 (1960)
121. K.M. Watson, S.A. Bludman, M.N. Rosenbluth, Phys. Fluids **3**(741), 747 (1960)
122. M.S. Kovner, J. Exp. Theor. Phys. **40**, 527 (1961)
123. K.N. Stepanov, A.B. Kitsenko, J. Tech. Phys. **31**(167), 176 (1961)
124. J. Neufield, P.H. Doyle, Phys. Rev. **121**, 654 (1961)
125. B. Shokri, B. Jazi, Phys. Plasmas **10**, 4622 (2003)
126. M.V. Kuzelev, A.A. Rukhadze, *Basics of Plasma Free Electron Lasers* (Editions Frontiers, Paris, 1995)
127. M.V. Kuzelev, A.A. Rukhadze, *Methods of Wave Theory in Dispersive Media* (World Scientific Publishing, Hackensack, 2009)
128. B. Shokri, M. Khorashadizadeh, Phys. Lett. A **352**, 520 (2006)
129. B. Jazi, M. Nejati, B. Shokri, Phys. Lett. A **370**, 319 (2007)
130. V.M. Bolotovskii, A.A. Rukhadze, J. Exp. Theor. Phys. **37**, 1346 (1959)
131. A.F. Alexandrov, L.S. Bogdankevich, A.A. Rukhadze, *Principles of Plasma Electrodynamics* (Springer, Heidelberg, 1984)
132. I.N. Toptygin, *Electromagnetic Phenomena in Matter: Statistical and Quantum Approaches*, 1st edn. (Wiley-VCH, Weinheim, 2015)
133. B. Shokri, Phys. Plasmas **7**, 3867 (2000)

Chapter 4
Quantum Plasma: Influence of Spatial Dispersion on Some Phenomena in Metals

4.1 Quantum Kinetic Equation with Self-consistent Fields and Magnetic Permittivity of an Electron Gas

There exist conditions in some types of plasmas (for example, for electrons in metals) under which the classic description is incorrect. In this case, it is necessary to use the quantum version of the classic kinetic equation. Such an equation can be easily obtained for an electron gas with weak interactions between its particles when the kinetic energy of particles is much higher than their average energy of the Coulomb interaction $eN_e^{1/3}$. The electrons in metals usually are in the Fermi-degenerate state and, therefore, their average kinetic energy per electron is of the order of $\epsilon_0 \sim p_0^2/2m$, where $p_0 = (3\pi^2)^{1/3}\hbar N_e^{1/3}$ is the Fermi momentum, and $\hbar$ is plank's constant [1]. In this sense, we may talk about the electron gas when its particle density is sufficiently high

$$N_e^{\frac{1}{3}} \gg \frac{e^2 m}{\hbar^2} \equiv \frac{1}{a_B}, \tag{4.1}$$

where $a_B = 0.529 \times 10^{-8}$ cm is the so-called Bohr radius.

For real metals, inequality (4.1) is only weakly satisfied or, more correctly, the quantities $N_e^{1/3}$ and a_B^{-1} are of the same order. However, the investigations of Fermi-liquid effects of electrons in metals show that these effects are not essential for qualitative analysis [2]. Therefore, below we will follow the quantum theory of the electron gas in metals, having in mind that sometimes the results may be quantitatively slightly different from the results of the quantum liquid theory [3, 4].

In both of quantum and classical plasmas, one can talk about the gas approximation when particles correlation is neglected, compared to their interaction stipulated

B. Shokri, A. A. Rukhadze, *Electrodynamics of Conducting Dispersive Media*, Springer Series on Atomic, Optical, and Plasma Physics 111,
https://doi.org/10.1007/978-3-030-28968-3_4

by the self-consistent fields. Then, particles may be described by their own wave functions $\psi\left(\vec{r},t\right)$ or their own density matrix $\rho\left(\vec{r},\vec{r}',t\right)$.[1]

To be as much similar as possible to the classic consideration, we will use the well-known Wigner representation of the density matrix [6–13]:

$$f\left(\vec{P},\vec{r},t\right)=\frac{1}{(2\pi)^3}\int d\vec{\tau}\exp\left(-\imath\vec{\tau}\cdot\vec{P}\right)\rho\left(\vec{r}-\frac{\hbar\vec{\tau}}{2},\vec{r}+\frac{\hbar\vec{\tau}}{2},t\right). \tag{4.2}$$

This function will be called the quantum distribution function.[2]

The density matrix satisfies the Schrodinger equation

$$\imath\hbar\frac{\partial}{\partial t}\rho\left(\vec{r}',\vec{r},t\right)=\left(\hat{H}-\widehat{H'}^*\right)\rho\left(\vec{r}',\vec{r},t\right), \tag{4.3}$$

where the Hamiltonian $\hat{H}$ acts on $\vec{r}$, whereas the Hamiltonian $\widehat{H'}^*$ acts on $\vec{r}'$. Taking the latter into account, one can obtain the following equation for the quantum distribution function [12]:

$$\begin{aligned}\frac{\partial f\left(\vec{P},\vec{r},t\right)}{\partial t}=&\frac{1}{(2\pi)^6}\frac{\imath}{\hbar}\int d\vec{\tau}d\vec{\eta}d\vec{p}'d\vec{r}'\exp\left\{\imath\left[\vec{\tau}.\left(\vec{p}'-\vec{P}\right)+\vec{\eta}.\left(\vec{r}'-\vec{r}\right)\right]\right\}f\left(\vec{p}',\vec{r}',t\right)\times\\&\times\left[H\left(\vec{p}'+\frac{\hbar\vec{\eta}}{2},\vec{r}'-\frac{\hbar\vec{\tau}}{2}\right)-H\left(\vec{p}'-\frac{\hbar\vec{\eta}}{2},\vec{r}'+\frac{\hbar\vec{\tau}}{2}\right)\right].\end{aligned} \tag{4.4}$$

Here, $H\left(\vec{p},\vec{r}\right)$ is the well-known Hamilton function for the charged particle in the electromagnetic fields:

---

[1]For the pure states with the wave function $\psi\left(\vec{r},t\right)$ the density matrix can be represented as $\rho\left(\vec{r},\vec{r}',t\right)=\psi^*\left(\vec{r},t\right)\psi\left(\vec{r}',t\right)$ (for more details, see [5–7]).

[2]The function $f\left(\vec{P},\vec{r},t\right)$ is not a positively definite quantity and, therefore, it does not present the probability distribution of particles' coordinates and momenta, which cannot be determined simultaneously by the measurements. On the other hand, the quantity $\rho\left(\vec{r},t\right)=\int d\vec{P}f\left(\vec{P},\vec{r},t\right)$ presents the probability distribution of particles' coordinates and the quantity $\int d\vec{r}f\left(\vec{P},\vec{r},t\right)$ is the probability distribution of momenta. To find the average of quantities by making use of the quantum distribution function, one can use $\left\langle\hat{A}\left(\widehat{\vec{P}},\widehat{\vec{r}}\right)\right\rangle=\int d\vec{P}d\vec{r}A\left(\vec{P},\vec{r}\right)f\left(\vec{P},\vec{r},t\right)$, where A is the classic function, which corresponds to the quantum operator $\hat{A}\left(\widehat{\vec{P}},\widehat{\vec{r}}\right)$, and depends on the canonical conjugate operators $\widehat{\vec{P}}$ and $\widehat{\vec{r}}$.

$$H\left(\vec{P},\vec{r}\right)=\frac{1}{2m}\left(\vec{P}-\frac{e}{c}\vec{A}\right)^2+e\phi, \qquad (4.5)$$

where ϕ and $\vec{A}$ are the scalar and vector field potentials, respectively; $\vec{P}$ is the canonical momentum of the particle.

According to the above-mentioned prescription for the current density of electrons, we can write the following expression:

$$\vec{j}\left(\vec{r},t\right)=\frac{e}{m}\int d\vec{P}f\left(\vec{P},\vec{r},t\right)\left(\vec{P}-\frac{e}{c}\vec{A}\right). \qquad (4.6)$$

If we introduce the usual kinetic momentum

$$\vec{p}=\vec{P}-\frac{e}{c}\vec{A},$$

then the expression (4.6) in its form coincides with classic expression (2.4). In this case, one can obtain the following linear equation for the small deviation of the quantum distribution function from the equilibrium distribution $f_0\left(\vec{p}\right)$ [14, 15]:

$$\frac{\partial\delta f}{\partial t}+\vec{v}\cdot\frac{\partial\delta f}{\partial\vec{r}}+e\vec{E}\cdot\frac{\partial f_0(p)}{\partial\vec{p}}=\frac{e}{(2\pi)^3}\frac{\imath}{\hbar}\int d\vec{\tau}d\vec{p'}\exp\left[\imath\vec{\tau}\cdot\left(\vec{p'}-\vec{p}\right)\right]f_0\left(\vec{p'}\right)\times$$
$$\times\left\{\left[\frac{\partial\phi}{\partial\vec{r}}\cdot\hbar\vec{\tau}-\phi\left(\vec{r}+\frac{\hbar\vec{\tau}}{2}\right)+\phi\left(\vec{r}-\frac{\hbar\vec{\tau}}{2}\right)\right]-\frac{\vec{v'}}{c}\cdot\left[\left(\hbar\vec{\tau}\cdot\frac{\partial}{\partial\vec{r}}\right)\vec{A}-\vec{A}\left(\vec{r}+\frac{\hbar\vec{\tau}}{2}\right)+\vec{A}\left(\vec{r}-\frac{\hbar\vec{\tau}}{2}\right)\right]\right\}. \qquad (4.7)$$

Here, there is no external magnetic field. Moreover, Coulomb gauge has been applied:

$$\nabla\cdot\vec{A}=0. \qquad (4.8)$$

In the classical limit, when $\hbar\to 0$, Eq. (4.7) tends to Eq. (2.2.2).

When the function $\delta f\left(\vec{p},\vec{r},t\right)$ at $t=t_0$ is known, the solution of Eq. (4.7) can be written as

$$\delta f\left(\vec{p},\vec{r},t\right)=\delta f\left(\vec{p},\vec{r}-\vec{v}(t-t_0),t_0\right)-e\frac{\partial f_0}{\partial\vec{p}}\cdot\int_{t_0}^{t}dt'\vec{E}\left(\vec{r}-\vec{v}(t-t'),t'\right)+$$
$$+\frac{e}{(2\pi)^3}\frac{\imath}{\hbar}\int_{t_0}^{t}dt'\int d\vec{\tau}d\vec{p'}\exp\left[\imath\vec{\tau}\cdot\left(\vec{p}-\vec{p'}\right)\right]f_0\left(\vec{p'}\right)\left\{\left[\hbar\vec{\tau}\cdot\frac{\partial}{\partial\vec{r}}\phi\left(\vec{r}-\vec{v}(t-t'),t'\right)-\right.\right.$$

$$-\phi\left(\vec{r}-\vec{v}(t-t')+\frac{\hbar\vec{\tau}}{2},t\right)+\phi\left(\vec{r}-\vec{v}(t-t')-\frac{\hbar\vec{\tau}}{2},t'\right)\Big]-$$

$$-\frac{\vec{v}}{c}\cdot\left[\left(\hbar\vec{\tau}\cdot\frac{\partial}{\partial\vec{r}}\right)\vec{A}\left(\vec{r}-\vec{v}(t-t'),t'\right)-\vec{A}\left(\vec{r}-\vec{v}(t-t')+\frac{\hbar\vec{\tau}}{2},t'\right)+\vec{A}\left(\vec{r}-\vec{v}(t-t')-\frac{\hbar\vec{\tau}}{2},t'\right)\right]\Bigg\}. \tag{4.9}$$

Assuming that the perturbations were adiabatically switched on in the infinite past (or $\delta f \to 0$ when $t \to -\infty$), from Eq. (4.9), we can calculate the current density induced by the electromagnetic field in quantum plasma. But first it is necessary to express the potentials ϕ and $\vec{A}$ in terms of the electric field. It is easy to show that

$$\phi\left(\vec{k},\omega\right)=\frac{\imath}{k^2}\left(\vec{k}\cdot\vec{E}^{l}\right),$$
$$\vec{A}\left(\vec{k},\omega\right)=-\frac{\imath c}{\omega}\vec{E}^{tr}\left(\vec{k},\omega\right).$$

Considering these relations, we may present the current density as

$$\vec{j}\left(\vec{k},\omega\right)=\sigma^{l}(\omega,k)\vec{E}^{l}(k,\omega)+\sigma^{\mathrm{tr}}(\omega,k)\vec{E}^{\mathrm{tr}}\left(\vec{k},\omega\right), \tag{4.10}$$

where

$$\sigma^{l}(\omega,k)=-\frac{\imath e^2}{\hbar k^2}\int\frac{\left(\vec{k}\cdot\vec{v}\right)d\vec{p}}{\left(\omega-\vec{k}\cdot\vec{v}\right)}\left[f_0\left(\vec{p}+\frac{\hbar\vec{k}}{2}\right)-f_0\left(\vec{p}-\frac{\hbar\vec{k}}{2}\right)\right], \tag{4.11}$$

$$\sigma^{\mathrm{tr}}(\omega,k)=-\frac{\imath e^2}{2\omega k^2}\int\frac{\left(\vec{k}\times\vec{v}\right)^2 d\vec{p}}{\left(\omega-\vec{k}\cdot\vec{v}\right)}$$
$$\times\left\{\left(\omega-\vec{k}\cdot\vec{v}\right)f_0'\left(\vec{p}\right)+\frac{1}{\hbar}\left[f_0\left(\vec{p}+\frac{\hbar\vec{k}}{2}\right)-f_0\left(\vec{p}-\frac{\hbar\vec{k}}{2}\right)\right]\right\}. \tag{4.12}$$

Here, $f_0'(p)=\partial f_0(p)/\partial\epsilon(p)$, and the integration over momentum must be performed by considering a small imaginary positive part of ω, or (see Fig. 2.2)

$$\frac{1}{\omega-\vec{k}\cdot\vec{v}}=2\pi\imath\delta_{+}\left(\omega-\vec{k}\cdot\vec{v}\right).$$

Then, for $\varepsilon^l(\omega, k)$ and $\varepsilon^{\mathrm{tr}}(\omega, k)$, by making use of Cauchy integral formula, we obtain[3]

$$\varepsilon^l(\omega,k) = 1 + \frac{4\pi e^2}{\omega\hbar k^2}\int d\vec{p}\,\frac{\vec{k}\cdot\vec{v}}{\omega-\vec{k}\cdot\vec{v}}\left[f_0\left(\vec{p}+\frac{\hbar\vec{k}}{2}\right)-f_0\left(\vec{p}-\frac{\hbar\vec{k}}{2}\right)\right], \quad (4.13)$$

$$\varepsilon^{\mathrm{tr}}(\omega,k) = 1 + \frac{2\pi e^2}{\omega^2 k^2}\int d\vec{p}\left(\vec{k}\times\vec{v}\right)^2\left\{f_0'(p) + \frac{f_0\left(\vec{p}+\frac{\hbar\vec{k}}{2}\right)-f_0\left(\vec{p}-\frac{\hbar\vec{k}}{2}\right)}{\hbar\left(\omega-\vec{k}\cdot\vec{v}\right)}\right\}. \quad (4.14)$$

Let us use these expressions for calculating the magnetic permittivity of the quantum electron gas. In this case, we have (see Sect. 1.3)

$$1-\frac{1}{\mu(\omega,k)} = \frac{\omega^2}{c^2k^2}\left[\varepsilon^{\mathrm{tr}}(\omega,k)-\varepsilon^l(\omega,k)\right].$$

In the static limit, when $\omega \to 0$, from Eqs. (4.13) and (4.14), it follows that

$$1-\frac{1}{\mu(0,k)} = \frac{2\pi e^2}{c^2k^4}\int d\vec{p}\left(\vec{k}\times\vec{v}\right)^2\left\{f_0'\left(\vec{p}\right) - \frac{f_0\left(\vec{p}+\frac{\hbar\vec{k}}{2}\right)-f_0\left(\vec{p}-\frac{\hbar\vec{k}}{2}\right)}{\hbar\vec{k}\cdot\vec{v}}\right\}. \quad (4.15)$$

In the sufficiently homogeneous field limit, when $\hbar\vec{k} \ll \vec{p}$, from Eq. (4.15) by expanding in powers of $\hbar\vec{k}$, one can easily obtain

$$1-\frac{1}{\mu(0,0)} = \frac{\pi e^2\hbar^2}{3m^2c^2}\int d\vec{p}\,f_0'\left(\vec{p}\right). \quad (4.16)$$

Noting that the right-hand side of this relation is much less than unity and using the definition given by Eq. (1.38) for the magnetic susceptibility, we obtain[4]

---

[3] The dielectric permittivity of the quantum electron gas was studied sufficiently in more details in [16].

[4] The formula for the weak diamagnetism of an electron gas, given by Eq. (4.17), was first obtained by L.D. Landau [17].

$$\chi_D = \frac{e^2\hbar^2}{12m^2c^2}\int d\vec{p}\, f_0'\left(\vec{p}\right). \tag{4.17}$$

For degenerate electrons, when

$$f_0\left(\vec{p}\right) = \begin{cases} \dfrac{2}{(2\pi\hbar)^3} & \text{for} \quad p < p_0 = (3\pi^2)^{\frac{1}{3}}\hbar N_e^{\frac{1}{3}}, \\ 0 & \text{for} \quad p > p_0, \end{cases} \tag{4.18}$$

from Eq. (4.17), it follows

$$\chi_D = -\left(\frac{e\hbar}{2mc}\right)^2 \frac{4mN_e^{1/3}}{(2\pi\hbar)^2}\left(\frac{\pi}{3}\right)^{2/3}. \tag{4.19}$$

For non-degenerate Maxwellian electrons with non-relativistic temperature $\kappa T_e \ll mc^2$, we have [see Eq. (2.47)]:

$$\chi_D = -\left(\frac{e\hbar}{2mc}\right)^2 \frac{N_e}{3\kappa T_e}. \tag{4.20}$$

The paramagnetic effects stipulated by the electron's spin were not considered above. For this reason, expressions $\varepsilon^l(\omega, k)$ and $\varepsilon^{\text{tr}}\left(\omega, \vec{k}\right)$, obtained above, strictly speaking, are incorrect. For describing the effects of the electron's spin, the density matrix must consider spin variables as well. As a result, the matrix $f_{\alpha\beta}\left(\vec{p}, \vec{r}, t\right)$ works as the quantum distribution function. Then, the distribution function discussed above is

$$f\left(\vec{p}, \vec{r}, t\right) = \sum_\alpha f_{\alpha\alpha}\left(\vec{p}, \vec{r}, t\right). \tag{4.21}$$

Beside $f\left(\vec{p}, \vec{r}, t\right)$, it is convenient to introduce the distribution function of the electron's spin

$$\vec{\sigma}\left(\vec{p}, \vec{r}, t\right) = \frac{1}{(2\pi)^3}\sum_{\alpha\beta}\int d\vec{\tau}\, \exp\left(-\iota\vec{\tau}\cdot\vec{p}\right)\widehat{\sigma}_{\alpha\beta}\,\rho_{\beta\alpha}\left(\vec{r} - \frac{\hbar\vec{\tau}}{2}, \vec{r} + \frac{\hbar\vec{\tau}}{2}, t\right). \tag{4.22}$$

Here, $\widehat{\sigma}_{\alpha\beta}$ are Pauli matrices:

$$\widehat{\sigma}_x = \begin{pmatrix} 0 & 1 \\ 1 & 0 \end{pmatrix}, \ \widehat{\sigma}_y = \begin{pmatrix} 0 & -\imath \\ \imath & 0 \end{pmatrix}, \ \widehat{\sigma}_z = \begin{pmatrix} 1 & 0 \\ 0 & -1 \end{pmatrix}.$$

Since the Hamiltonian of an electron with spin in the electromagnetic field is equal to[5]

$$\widehat{H}_{\alpha\beta} = \frac{1}{2m}\left(\vec{P} - \frac{e}{c}\vec{A}\right)^2 \delta_{\alpha\beta} + e\phi\delta_{\alpha\beta} - \frac{e\hbar}{2mc}\vec{B}\widehat{\sigma}_{\alpha\beta}, \tag{4.23}$$

then the equation of the electron spin distribution function can be written as

$$\frac{\partial\vec{\sigma}\left(\vec{P},\vec{r},t\right)}{\partial t} = \frac{1}{(2\pi)^6}\frac{\imath}{\hbar}\int d\vec{\tau}d\vec{\eta}d\vec{r}'d\vec{p}'\exp\left\{\imath\left[\vec{\tau}\cdot\left(\vec{p}'-\vec{P}\right)+\vec{\eta}\cdot\left(\vec{r}-\vec{r}'\right)\right]\right\}\times$$
$$\times\vec{\sigma}\left(\vec{p}',\vec{r}',t\right)\left[H\left(\vec{p}'+\frac{\hbar\vec{\eta}}{2},\vec{r}'-\frac{\hbar\vec{\tau}}{2}\right)-H\left(\vec{p}'-\frac{\hbar\vec{\eta}}{2},\vec{r}'+\frac{\hbar\vec{\tau}}{2}\right)\right]+\frac{1}{(2\pi)^3}\frac{\mu}{\hbar}\int d\vec{\tau}d\vec{p}'\exp\left[\imath\vec{\tau}\cdot\left(\vec{p}'-\vec{P}\right)\right]\times$$
$$\times\left\{2\left(\vec{B}\left(\vec{r}-\frac{\hbar\vec{\tau}}{2}\right)+\vec{B}\left(\vec{r}+\frac{\hbar\vec{\tau}}{2}\right)\right)\times\vec{\sigma}\left(\vec{p}',\vec{r},t\right)-\imath f\left(\vec{P}',\vec{r},t\right)\left[\vec{B}\left(\vec{r}-\frac{\hbar\vec{\tau}}{2}\right)-\vec{B}\left(\vec{r}+\frac{\hbar\vec{\tau}}{2}\right)\right]\right\}, \tag{4.24}$$

where $\mu = e\hbar/2mc$, and H coincides with Eq. (4.5). It must be noted that, accounting spin modifies Eq. (4.4) and in its right-hand side, an additional term

$$-\frac{1}{(2\pi)^3}\frac{\imath\mu}{\hbar}\int d\vec{\tau}d\vec{p}'\exp\left[\imath\vec{\tau}\cdot\left(\vec{p}'-\vec{P}\right)\right]\vec{\sigma}\left(\vec{p}',\vec{r},t\right)\cdot\left[\vec{B}\left(\vec{r}-\frac{\hbar\vec{\tau}}{2}\right)-\vec{B}\left(\vec{r}+\frac{\hbar\vec{\tau}}{2}\right)\right]$$

appears. But, this term in the linear approximation for the small perturbations of the equilibrium state with $\vec{\sigma}_0 = 0$ can be neglected and, therefore, Eq. (4.7) remains unchanged. Then, for the perturbation of the spin distribution function, we have [16]

$$\frac{\partial\delta\vec{\sigma}}{\partial t} + \left(\vec{v}\cdot\nabla\right)\delta\vec{\sigma} = -\frac{1}{(2\pi)^3}\frac{\imath\mu}{\hbar}\int d\vec{\tau}d\vec{p}'\exp\left[\imath\vec{\tau}\cdot\left(\vec{p}'-\vec{P}\right)\right]f_0\left(\vec{p}'\right)$$
$$\left[\vec{B}\left(\vec{r}-\frac{\hbar\vec{\tau}}{2}\right)-\vec{B}\left(\vec{r}+\frac{\hbar\vec{\tau}}{2}\right)\right]. \tag{4.25}$$

The solution of this equation is

[5]The Schrodinger equation for the density matrix, in this case, looks as [compare to Eq. (4.3)] [5, 6]:

$$\widehat{H}_{\alpha\beta} = \frac{1}{2m}\left(\vec{P} - \frac{e}{c}\vec{A}\right)^2 \delta_{\alpha\beta} + e\phi\delta_{\alpha\beta} - \frac{e\hbar}{2me}\vec{B}\widehat{\sigma}_{\alpha\beta},$$

$$\delta\vec{\sigma}\left(\vec{p},\vec{r},t\right)=\delta\vec{\sigma}\left(\vec{p},\vec{r}-\vec{v}(t-t_0),t_0\right)-\frac{1}{(2\pi)^3}\frac{\imath\mu}{\hbar}\int\limits_{t_0}^{t}dt'\int d\vec{\tau}d\vec{p'}\exp\left[\imath\vec{\tau}\cdot\left(\vec{p'}-\vec{p}\right)\right]$$
$$\times f_0\left(\vec{p'}\right)\left[\vec{B}\left(\vec{r}-\vec{v}(t-t')-\frac{\hbar\vec{\tau}}{2},t'\right)-\vec{B}\left(\vec{r}-\vec{v}(t-t')+\frac{\hbar\vec{\tau}}{2},t'\right)\right].\tag{4.26}$$

Using this solution, we can determine the paramagnetic part of the current density as follows:

$$\delta\vec{j}=\mu c\nabla\times\sum_{\alpha\beta}\widehat{\sigma}_{\alpha\beta}\,\rho_{\alpha\beta}=\mu c\nabla\times\int d\vec{p}\;\vec{\sigma}\left(\vec{p},\vec{r},t\right).\tag{4.27}$$

Substituting Eq. (4.26) into Eq. (4.27), we obtain

$$\delta\vec{j}\left(\vec{k},\omega\right)=\frac{\imath\mu^2c}{\hbar}\left[\vec{k}\times\vec{B}\left(\vec{k},\omega\right)\right]\int d\vec{p}\frac{f_0\left(\vec{p}+\frac{\hbar\vec{k}}{2}\right)-f_0\left(\vec{p}-\frac{\hbar\vec{k}}{2}\right)}{\omega-\vec{k}\cdot\vec{v}}.\tag{4.28}$$

Considering the Maxwell's equation $\vec{B}\left(\vec{k},\omega\right)=(c/\omega)\left[\vec{k}\times\vec{E}\left(\vec{k},\omega\right)\right]$, one can show that $\delta\vec{j}\left(\vec{k},\omega\right)$ gives a contribution only in the transverse dielectric permittivity

$$\delta\varepsilon^{\mathrm{tr}}(\omega,k)=\frac{4\pi c^2\mu^2k^2}{\hbar\omega^2}\int d\vec{p}\frac{f_0\left(\vec{p}+\frac{\hbar\vec{k}}{2}\right)-f_0\left(\vec{p}-\frac{\hbar\vec{k}}{2}\right)}{\omega-\vec{k}\cdot\vec{v}}.\tag{4.29}$$

As a result,

$$\delta\left[\frac{1}{\mu(0,k)}\right]=4\pi\mu^2\int d\vec{p}\frac{f_0\left(\vec{p}+\frac{\hbar\vec{k}}{2}\right)-f_0\left(\vec{p}-\frac{\hbar\vec{k}}{2}\right)}{\hbar\vec{k}\cdot\vec{v}},$$

or, in the limit of $\hbar\vec{k}\ll\overline{p}$, we have

$$\delta\left[\frac{1}{\mu(0,0)}\right]=\frac{\pi e^2\hbar^2}{c^2m^2}\int d\vec{p}\;f_0'\left(\vec{p}\right).\tag{4.30}$$

Comparing this expression to Eq. (4.17), we can conclude that the paramagnetic part of the magnetic susceptibility is three times larger than the diamagnetic part.[6]

[6]For the relativistic electron gas, this result was generalized in [18].

4.2 Longitudinal Oscillations of a Degenerate Electron Gas and Characteristic Energy Loss of the Fast Moving Electrons

Let us use the above results to investigate the longitudinal oscillations of a degenerate electron gas at $T_e = 0$ [14–16, 19–30]. Such oscillations in isotropic media are described by

$$\varepsilon^l(\omega, k) = 0,$$

where $\varepsilon^l(\omega, k)$ is determined by Eq. (4.13). As a result, we obtain[7]

$$1 = \frac{4\pi e^2}{\hbar k^2} \int d\vec{p}\, \frac{f_0\left(\vec{p} + \frac{\hbar \vec{k}}{2}\right) - f_0\left(\vec{p} - \frac{\hbar \vec{k}}{2}\right)}{\vec{k} \cdot \vec{v} - \omega}. \tag{4.31}$$

When $T_e = 0$ the distribution function is given by Eq. (4.18). In this case, Eq. (4.31) takes the form [15]

$$\frac{2}{3}\frac{v_0^2\, k^2}{\omega_{\text{pe}}^2} = -1 + \frac{m}{2\hbar k^3 v_0}\left\{\left[\left(\omega + \frac{\hbar k^2}{2m}\right)^2 - v_0^2 k^2\right] \ln\left(\frac{\omega + \frac{\hbar k^2}{2m} + v_0 k}{\omega + \frac{\hbar k^2}{2m} - v_0 k}\right)\right.$$
$$\left. - \left[\left(\omega - \frac{\hbar k^2}{2m}\right)^2 - v_0^2 k^2\right] \ln\left(\frac{\omega - \frac{\hbar k^2}{2m} + v_0 k}{\omega - \frac{\hbar k^2}{2m} - v_0 k}\right)\right\}, \tag{4.32}$$

where

$$v_0 = \frac{p_0}{m} = \left(3\pi^2\right)^{1/3} \hbar N_e^{1/3}/m\,, \quad \omega_{\text{pe}} = \left(4\pi e^2 N_e/m\right)^{1/2}.$$

In the limit of $\omega \to \infty$, all logarithms in Eq. (4.32) are real and, therefore, the high-frequency oscillations become undamped. Only such oscillations will be investigated below. In the classical limit of $\hbar k \ll p_0$, from Eq. (4.32), it follows [20]

$$\frac{k^2 v_0^2}{3\omega_{\text{pe}}^2} = -1 + \frac{\omega}{2k v_0} \ln\left(\frac{\omega + k v_0}{\omega - k v_0}\right). \tag{4.33}$$

[7] About this equation, see also [16, 22–30].

This equation has the following undamped solution in the long-wave limit $(\omega_{pe} \gg kv_0)$ [19]

$$\omega^2 = \omega_{pe}^2 + \frac{3}{5}\, v_0^2 k^2. \tag{4.34}$$

Using Eq. (4.32), one can find the quantum correction to this solution [15]:

$$\omega^2 = \omega_{pe}^2 + \frac{3}{5}\, v_0^2 k^2 + \left(\frac{\hbar k^2}{2m}\right)^2. \tag{4.35}$$

When the wavelength decreases, the spectrum of longitudinal field oscillations will be similar to the single-particle excitation spectrum of electrons. Therefore, when $\hbar k^2/2m \ll \omega_{pe} \ll kv_0$, we have

$$\hbar\omega = \frac{(p_0 + \hbar k)^2 - p_0^2}{2m}\left\{1 - \exp\left(-2 - \frac{2}{3}\frac{k^2 v_0^2}{\omega_{pe}^2}\right)\right\}. \tag{4.36}$$

In the classical limit ($\hbar \to 0$), this formula was obtained in [20]. On the other hand, omitting the small exponential term in Eq. (4.36) and noting that the quantity $p = p_0 + \hbar k$ corresponds to the momentum of the excited particle, we find out that this formula takes the form

$$\hbar\omega = v_0(p - p_0), \tag{4.37}$$

which is typical for Fermi spectra and corresponds to the excitation of individual electrons above the Fermi surface.

The problem of the connection between the so-called characteristic or the discrete energy loss of the fast electrons passing through the thin films and the longitudinal oscillations of solid-state plasmas was studied in [31]. The energy loss has a quite definitely discrete character, leading to appearance of relatively distinct energy loss lines. The connection of such energy loss with the excitation of plasma oscillations in the solid state was first noted in [31–33]. Indeed, assuming the medium to be transparent, from Eq. (1.163), one can obtain an expression for the scattering probability of a fast particle emitting a longitudinal quanta with the frequency ω per unit time at the scattering angle $\theta \ll 1$

$$\frac{v dW^l}{\hbar\omega d\omega\theta d\theta} = \frac{2e^2}{\hbar v}\frac{1}{\theta^2 + \left(\frac{\hbar\omega}{pv}\right)^2}\delta\left[\varepsilon^l\left(\omega, \sqrt{\left(\frac{p\theta}{\hbar}\right)^2 + \frac{\omega^2}{v^2}}\right)\right]. \tag{4.38}$$

Here, p and v are the momentum and velocity of the fast electron. The existence of the δ-function in Eq. (4.38) indicates the discrete character of the energy loss spectra when a wave with the definite frequency ω is emitted at an angle θ.

The influence of spatial dispersion for the fast particles with velocity $v \gg v_0$ is determined by

$$\frac{kv_0}{\omega} \sim \frac{v_0 p\theta}{\hbar\omega}.$$

If $\theta \ll \hbar\omega_{\text{pe}}/pv_0$, then the influence of spatial dispersion in the dielectric permittivity is small. Thus, at the given scattering angle, the spectrum of the longitudinal oscillation, determined by Eq. (4.35), can be written as

$$\hbar^2\omega^2 \simeq \hbar^2\omega_{\text{pe}}^2 + \frac{3}{5}\; v_0^2 p^2\theta^2 + \left(\frac{p^2\theta^2}{2m}\right)^2. \tag{4.39}$$

According to this relation, the frequency of an emitted quantum (or energy loss $\Delta E = \hbar\omega$) increases with the scattering angle θ. Such dependence was really observed in experiments, showing also that, for real metals, the magnitude order of the coefficients in front of θ^2 and θ^4 in Eq. (4.39) is correct [34, 35].

For real metals, the theory of plasma oscillations and discrete energy loss, of course, is more complicated than the one discussed above for the free electron gas. Moreover, the existence of the periodic lattice potential leads to the zone structure in the spectra and, consequently, to the appearance of absorption bands determined by the dielectric permittivity.[8] Particularly, the zeros of the dielectric permittivity essentially depend on the zone structure and the lattice constant. Another circumstance arises from the difference of the properties of the electron liquid in metals from those of the electron gas. The theory of plasma oscillations based on the Landau-Fermi liquid theory was discussed in [3].

It must be noted that, in contrast to the theory, which does not take into account the dependence of ε_{ij} on k (see, for example, [38]), in the theory, which considers the spatial dispersion of the dielectric permittivity, any difference between the so-called short and long distance energy loss does not appear. Indeed, from Eq. (4.31), it follows that for large k the spectrum looks as [15]

$$\omega^2 = \omega_{\text{pe}}^2 + \left(\frac{\hbar k^2}{2m}\right)^2, \tag{4.40}$$

which corresponds to the dielectric permittivity [16]

[8] About the influence of the periodic lattice potential on the longitudinal oscillations and discrete energy loss of electrons, see [36, 37].

$$\varepsilon^l(\omega, k) = 1 - \frac{\omega_{\text{pe}}^2}{\omega^2 - \left(\frac{\hbar k^2}{2m}\right)^2}. \tag{4.41}$$

Then, if one calculates the energy loss of a fast particle using this expression, it will be clear that at the large scattering angles θ, the energy loss practically corresponds to the energy transmitted to the free particles of plasma.

Substituting Eq. (4.41) into Eq. (4.38), one can obtain the following expression for the scattering probability of a fast particle per unit length at the angle $\theta \gg v_0/v$ when it emits the quantum energy $\hbar\omega$:

$$\frac{v dW^l}{\hbar\omega\, d\hbar\omega\, d\Omega} = \delta\left(\hbar\omega - \frac{p^2\theta^2}{2m}\right) N_e \frac{d\sigma_{\text{res}}}{d\Omega}. \tag{4.42}$$

Here, $d\sigma_{\text{res}} = \left(2e^2/vp\theta^2\right)^2 d\Omega$ is the well-known Rutherford formula for the electron–electron scattering cross section. Therefore, we see that the spatial dispersion of the dielectric permittivity takes account of the Rutherford scattering, or short distance collisions of particles too.

4.3 Anomalous Skin-Effect in Metals

The phenomenon of the anomalous skin-effect was discovered when the absorption of radio-frequency radiation in metals was studied [39, 40]. It was observed that at the sufficiently low temperatures the field penetration depth is less than the mean free path of conduction electrons in metals and less than the average length covered by a conduction electron during one field period. Moreover, in accordance to measurements, the dependence of the field penetration depth on the frequency, under these conditions, is proportional to $\omega^{-1/3}$ compared to $\omega^{-1/2}$ taking place for the ordinary skin-effect. Moreover, the field perturbation depth does not depend on the length of the electron mean free path.

The first qualitative explanation of observed field behaviors was done in [41], where it was noted that the anomalous skin-effect is stipulated by a small part of electrons moving toward the metallic surface under the very small angle with a small ratio of the field penetration depth to the electron mean free path. Then, field penetration depth dependence on the field frequency turns out to be as $\omega^{-1/3}$ and is independent of the mean free path (see Sect. 2.9)

The quantitative theory of the anomalous skin-effect was developed in [42] and was based on the gaseous model of conduction electrons in metals. This theory differs from the theory discussed above in Sects. 2.8 and 2.9 only by the equilibrium distribution function taken as the degenerate Fermi distribution (4.18). Since the anomalous skin-effect mainly is determined under the conditions when $p_0 \gg \hbar k$, then for the electrons we can use the kinetic equation (2.139) with the equilibrium function (4.18). This leads to the following expression for the transverse dielectric permittivity [compare to Eq. (4.14)]:

$$\varepsilon^{tr}(\omega,k) = 1 - \frac{3\omega_{pe}^2}{2\omega(\omega+\imath\nu)}\left\{1+\left[\left(\frac{\omega+\imath\nu}{kv_0}\right)^2-1\right]\left[1-\frac{\omega+\imath\nu}{2kv_0}\ln\frac{\omega+\imath\nu+kv_0}{\omega+\imath\nu-kv_0}\right]\right\} \quad (4.43)$$

Here, $v_0 = p_0/m$ is the electron's velocity on the Fermi level, and ν is the electron collision frequency.

Under the conditions of the anomalous skin-effect, when $kv_0 \gg \omega, \nu$, from Eq. (4.43), it follows

$$\varepsilon^{tr}(\omega,k) \approx \imath\frac{3\pi\omega_{pe}^2}{4\omega \mid k \mid v_0}, \quad (4.44)$$

which corresponds to Eq. (2.195) if we take

$$C = \frac{3}{16}\frac{\omega_{pe}^2}{v_0}. \quad (4.45)$$

As a result, we obtain the effective field penetration depth of the mirror and diffusion cases of the electron reflection from the surface, respectively [compare to Eq. (2.198)]

$$\lambda_M = \frac{2}{3}\left(1+\frac{\imath}{\sqrt{3}}\right)\left(\frac{4v_0c^2}{3\pi\omega\omega_{pe}^2}\right)^{1/3},$$
$$\lambda_D = \frac{3}{4}\left(1+\frac{\imath}{\sqrt{3}}\right)\left(\frac{4v_0c^2}{3\pi\omega\omega_{pe}^2}\right)^{1/3}. \quad (4.46)$$

Using Eqs. (2.177) and (4.46) for the surface impedance of metals, we obtain [42]

$$\begin{pmatrix} Z_M \\ Z_D \end{pmatrix} = \left(1-\imath\sqrt{3}\right)\left(\frac{4\sqrt{3}\pi^2 v_0\omega^2}{c^4\omega_{pe}^2}\right)^{1/3} \times \begin{cases} 8/9, \\ 1. \end{cases} \quad (4.47)$$

In real metals, in the anomalous skin-effect region, usually $\nu \gg \omega$ and, therefore, the validity of Eqs. (4.46)–(4.47) is determined by the inequality

$$l = \frac{v_0}{\nu} \gg \left(\frac{v_0c^2}{\omega\omega_{pe}^2}\right)^{1/3} \sim \left(\frac{p_0c^2}{4\pi e^2 N_e\omega}\right)^{1/3} \sim 0,1\omega^{-1/3}. \quad (4.48)$$

Here, we assume that for most metals $v_0 \sim 10^8 \mathrm{cm/s}$ and $\omega_{pe} \sim 10^{16} \mathrm{s}^{-1}$, and ω is measured in s^{-1}. The best experimental conditions for observing the anomalous skin-effect and the difference between mirror and diffusion reflections may be realized in the infrared range of field frequency, where the length covered by the thermal electrons during one field period is less than the penetration depth. But, for most experimental conditions, the latter is of the order of the electron mean free path in metals.[9] Then, besides the electron's collisions with the lattice and with each other, it is necessary to take into account its collisions with the surface as well. Since in the infrared range of frequency the following inequality holds usually [compare to Eq. (2.189)]

$$| \varepsilon(\omega) | \ll \frac{c^2}{v_0^2}, \tag{4.49}$$

then we can obtain expressions for the surface impedance, following the calculations of Sect. 2.9. But, it must be noticed that in contrast to non-degenerate electrons (see Sect. 2.9) the dielectric permittivity $\varepsilon^{tr}(\omega, k)$ of degenerate electrons has branch points at $k = \pm(\omega + \imath\nu)/v_0$. This leads to slightly different expressions for the effective field penetration depth [42, 44]:

$$\delta\lambda_M = -2\imath \frac{1}{\left(\frac{\omega^2}{c^2}\frac{\partial\varepsilon^{tr}}{\partial k} - 2k\right)_{k_0=k}} - \frac{2\imath}{\pi}\frac{v_0^2}{\omega c^2}\left(1 + \imath\frac{\nu}{\omega}\right)$$
$$\times \int_1^\infty \frac{dx \mathrm{Im}\varepsilon_+^{tr}\left(\omega, \frac{\omega + \imath\nu}{v_0}x\right)}{\left[\frac{v_0^2}{c^2}\mathrm{Re}\,\varepsilon_+^{tr}\left(\omega, \frac{\omega+\imath\nu}{v_0}x\right) - x^2\left(1 + \imath\frac{\nu}{\omega}\right)^2\right]^2 + \left[\frac{v_0^2}{c^2}\mathrm{Im}\varepsilon_+^{tr}\left(\omega, \frac{\omega+\imath\nu}{v_0}x\right)\right]^2}, \tag{4.50}$$

$$\delta\{\lambda_D\}^{-1} = \int_0^1 da \frac{\imath\omega^2}{c^2}\varepsilon^{tr}(\omega, k_0)\left\{a\frac{\omega^2}{c^2}\frac{\partial\varepsilon^{tr}}{\partial k} - 2k\right\}^{-1}_{k=k_0} + \frac{\imath}{\pi}\frac{v_0}{c}\frac{\omega}{c}\left(1 + \frac{\imath\nu}{\omega}\right)^3 \int_0^1 \frac{da}{a^2} \times$$
$$\times \int_1^\infty \frac{dx x^2 \mathrm{Im}\varepsilon_+^{tr}\left(\omega, \frac{\omega + \imath\nu}{v_0}x\right)}{\left[\frac{v_0^2}{c^2}\mathrm{Re}\,\varepsilon_+^{tr}\left(\omega, \frac{\omega+\imath\nu}{v_0}x\right) - \left(1 + \imath\frac{\nu}{\omega}\right)^2\frac{x^2}{a}\right]^2 + \left[\frac{v_0^2}{c^2}\mathrm{Im}\varepsilon_+^{tr}\left(\omega, \frac{\omega+\imath\nu}{v_0}x\right)\right]^2}. \tag{4.51}$$

Here, k_0 is determined by (in Eq. (4.50), $a = 1$)

[9] Very often in optics of metals, the anomalous skin-effect is realized in the experiments just under this condition [43].

$$a\frac{\omega^2}{c^2}\varepsilon^{\text{tr}}(\omega, k_0) - k_0^2 c^2 = 0, \tag{4.52}$$

and

$$\operatorname{Re}\varepsilon_+^{\text{tr}}(\omega, k) = 1 - \frac{3\omega_{\text{pe}}^2}{2\omega(\omega + \imath\nu)}\left\{1 - \left[1 - \left(\frac{\omega + \imath\nu}{kv_0}\right)^2\right]\left[1 - \frac{\omega + \imath\nu}{2kv_0}\ln\left|\frac{\omega + \imath\nu + kv_0}{\omega + \imath\nu - kv_0}\right|\right]\right\}, \tag{4.53}$$

$$\operatorname{Im}\varepsilon_+^{\text{tr}}(\omega, k) = \frac{3\pi\omega_{\text{pe}}^2}{4\omega \mid k \mid v_0}\left[1 - \left(\frac{\omega + \imath\nu}{kv_0}\right)^2\right]. \tag{4.54}$$

One must take the solution of Eq. (4.52), which corresponds to the wave damped in metals. In the case of our interest, when$\omega^2 \gg \omega_{\text{pe}}^2 v_0^2/c^2$, this solution practically does not depend on the characteristics of the electron reflection from the surface and looks as ($\omega \gg \nu$):

$$\frac{1}{k_0} = \lambda = \frac{c}{\omega_{\text{pe}}}\left(1 + \imath\frac{\nu}{2\omega}\right). \tag{4.55}$$

Strictly speaking, this formula is valid for metals only if $\omega \ll \omega_{\text{pe}}$. Otherwise, in metals when $\omega \sim \omega_{\text{pe}}$, the quantum absorption effects connected to the zone–zone transitions become important and, as a result, the model of free electron plasma is irrelevant.

Considering the small parameter $\left(v_0^2/c^2\right)\left(\omega_{pe}^2/\omega^2\right) \ll 1$, one can approximately calculate the integrals in Eqs. (4.50) and (4.51) and obtain the following expressions [44]:

$$\begin{aligned} \delta\lambda_M &= \frac{\imath}{8}\frac{v_0}{\omega}\frac{\omega_{\text{pe}}^2}{\omega^2}\frac{v_0^2}{c^2}, \\ \delta(\lambda_D)^{-1} &= -\frac{3\imath}{16}\frac{\omega}{v_0}\frac{\omega_{\text{pe}}^2}{\omega^2}\frac{v_0^2}{c^2}. \end{aligned} \tag{4.56}$$

Using Eqs. (2.178)–(2.181), (4.43) and (4.56), one can obtain the following expressions for the ratio of the absorbed energy per unit area of surface and per unit time to the average flux of the energy incident on the metal[10]

---

[10] It must be noted that the collision frequencies of electrons with the crystal lattice and with each other in general depend on the field frequency ω. Thus, the electron mean free path appearing in the static conductivity increases when the temperature decreases, i.e., the collision frequency decreases. But, the effective electron–phonon collision frequency in the infrared range does not decrease when the temperature decreases [45, 46]. At the same time, the electron–electron collision frequency increases when the field frequency ω increases [47–49].

$$A_M = \frac{2\nu}{\omega_{\text{pe}}} + \frac{\omega_{\text{pe}}^2}{2\omega^2} \frac{v_0^3}{c^3}, \tag{4.57}$$

$$A_D = \frac{2\nu}{\omega_{\text{pe}}} + \frac{3}{4} \frac{v_0}{c}. \tag{4.58}$$

The second terms of these expressions can be obtained by direct calculations of the field energy absorbed by electrons due to their collisions with the metallic surface, as done in Sect. 2.9 [50]. These calculations differ from the ones carried out in Sect. 2.9 only because of using the Fermi distribution (4.18).

The second term on the right-hand side of Eq. (4.57) increases when the field frequency ω decreases, but the range of its validity is less than that of the second term of Eq. (4.58). It must be noted that the both terms on the right-hand side of Eq. (4.58) are of the same order for good conductors. Using this formula allows us to define the electron's velocity from the experimental data [43, 51].

For real metals, there are essential reasons which lead to incorrectness of the gaseous model of free electrons to the quantitative description of the phenomena stipulated by the conduction electrons in metals. The reason is the periodic potential of lattice under the action of which the momentum dependence of electron energy may significantly differ from $p^2/2m$ corresponding to the free electron.[11] Then, electron's energy $\epsilon\left(\vec{p}\right)$ depends on the mutual orientation of $\vec{p}$ and the crystallographic axes. Of course, all these effects lead to the complication of the theory of real conduction electrons in metals [55]. But, for the anomalous skin-effect in radio-frequency range it is possible to develop a relatively simple theory, which allows us to get some information about the Fermi surface from the experimental data [56–58].[12] Below, we will discuss the theory of the anomalous skin-effect, which accounts the arbitrary dependence of $\epsilon\left(\vec{p}\right)$.

To describe the electrons, it is possible to use Eq. (2.139). All the effects stipulated by the anisotropic distribution of electrons are accounted for the equilibrium distribution function, which depends on $\epsilon\left(\vec{p}\right)$. Since this distribution function is the Fermi distribution, then

[11] More exactly, one must talk about quasi-momentum [6, 52]. Moreover, there also exist the effects stipulated by the electron–electron interaction accounted in the Fermi-liquid theory [53, 54].

[12] In the phenomena related to the conduction electrons, only the electrons with energy near the Fermi energy ϵ_F are essential. Below this energy, all levels are occupied, and above this energy, all levels are unoccupied (at $T = 0$). Equation

$$\epsilon\left(\vec{p}\right) = \epsilon_F,$$

describes the so-called Fermi surface in the space of $\vec{p}$. In the isotropic case, this surface is spherical, whereas, in the anisotropic case, it presents a relatively complicated surface.

$$f_0'\left[\epsilon\left(\vec{p}\right)\right] = -\frac{2}{(2\pi\hbar)^3}\delta\left[\epsilon\left(\vec{p}\right) - \epsilon_F\right]. \tag{4.59}$$

Using the results of Sect. 2.2, it is easy to show that Eq. (2.139) leads to the following expression for the dielectric permittivity[13]

$$\varepsilon_{ij}\left(\omega, \vec{k}\right) = \delta_{ij} + \frac{4\pi e^2}{\omega}\int d\vec{p}\,\frac{v_i v_j}{\omega + \imath\nu - \vec{k}\cdot\vec{v}} f_0'(\epsilon). \tag{4.60}$$

In the anomalous skin-effect range [compare to Eqs. 2.195 and (4.44)], we have

$$\varepsilon_{ij}\left(\omega, \vec{k}\right) = \frac{4\pi\imath}{\omega \mid k \mid} C_{ij}, \tag{4.61}$$

where

$$C_{ij} = -\pi e^2 \int d\vec{p} v_i v_j f_0'(\epsilon)\delta\left(\frac{\vec{k}.\vec{v}}{k}\right). \tag{4.62}$$

Therefore, it is seen that in the anomalous skin-effect range the conductivity tensor does not depend on frequency. Thus, considering Eqs. (1.46) and (4.61), one can obtain

$$\sigma_{ij}\left(\omega, \vec{k}\right) = \frac{C_{ij}}{\mid k \mid}. \tag{4.63}$$

It is noticeable that this tensor, similar to C_{ij}, is purely transverse, or $k_i\sigma_{ij} = 0$. For this reason, it follows that in the anomalous skin-effect range one can account for only the parallel field component with respect to the metal surface. Moreover, for the polarization of the electric field $\vec{E}$ parallel to one of the main axes of C_{ij}, the field equations are separated. As a result, it is possible to use the relations of Sect. 2.9. Thus, quite analogically to Eq. (2.198), we obtain

$$\begin{aligned}\lambda_M &= \frac{2}{3}\left(1 + \frac{\imath}{\sqrt{3}}\right)\left(\frac{c^2}{4\pi\omega C_a}\right)^{1/3}, \\ \lambda_D &= \frac{3}{4}\left(1 + \frac{\imath}{\sqrt{3}}\right)\left(\frac{c^2}{4\pi\omega C_a}\right)^{1/3}.\end{aligned} \tag{4.64}$$

Here, C_a is the principal value of the tensor C_{ij}.

[13] We suppose that the metal is not optically active.

The experimental measurements of tensor C_{ij} give the information about the curvature of the Fermi surface [57–59]. To clarify this, let us slightly change the right-hand side of Eq. (4.62) by using Eq. (4.59) to

$$C_{ij} = \frac{2\pi e^2}{(2\pi\hbar)^3} \int \frac{dS}{v} v_i v_j \delta\left(\frac{\vec{k}\cdot\vec{v}}{k}\right), \tag{4.65}$$

where S is the surface of the constant Fermi energy. Then, introducing the Gauss curvature of the Fermi surface $K(\vartheta, \varphi)$ and supposing $\vec{k}\cdot\vec{v} = kv\cos\vartheta$, we obtain [58]

$$C_{ij} = \frac{2\pi e^2}{(2\pi\hbar)^3} \int\limits_0^{2\pi} \frac{n_i n_j d\varphi}{K\left(\frac{\pi}{2}, \varphi\right)}, \tag{4.66}$$

where $n_i = v_i/v$. This formula estimates the relation between the curvature of the Fermi surface and the field penetration depth or, consequently, the surface impedance of the metal [see Eq. (2.178)].

4.4 Paramagnetic Resonance Absorption of Metals by Conduction Electrons

When the spin direction of a free electron changes in the opposite direction in the presence of the external field $\vec{B}_0$, its energy changes in the amount of $2\mu B_0 = (e\hbar/mc)$ B_0. This means that the resonance absorption of electromagnetic waves with the frequency $2\mu B_0/\hbar = \Omega = eB_0/mc$ must take place.[14] The absorption width is determined by the characteristic time of the spin relaxation U and the resonance really takes place if $\Omega U \gg 1$. Considering this case, one must consider another extremely important phenomenon. The point is that the field penetrating depth due to the skin-effect is smaller than the distance traversed by electrons during the spin relaxation time. Since this time is much longer than the electron mean free time, $U\nu \gg 1$, electrons undergoes a lot of collisions during the relaxation time U. Therefore, the diffusion of electrons from the skin layer into the volume of the metal takes place and this process determines the character of the resonance absorption.[15]

[14] For real metals, the resonance frequency slightly differs from the resonance frequency Ω of free electrons.

[15] The experimental investigation of such a diffusion process was performed in [60]. The theory of this phenomenon was firstly developed in [61] (see also [62, 63]).

The theory of the paramagnetic resonance absorption can be developed with the help of equations obtained in Sect. 4.1. But before this, let us obtain an approximate equation for the electron's spin diffusion using the kinetic equation [61, 62, 64].

For the fields and distributions weakly varying in the distances of the order of the average distance between conduction electrons, Eq. (4.24) can be written as

$$\frac{\partial \vec{\sigma}}{\partial t} + \left(\vec{v} \cdot \nabla\right)\vec{\sigma} + e\left[\left(\vec{E} + \frac{1}{c}\left[\vec{v} \times \vec{B}\right]\right) \cdot \frac{\partial}{\partial \vec{p}}\right]\vec{\sigma} - \frac{2\mu}{\hbar}\left[\vec{B} \times \vec{\sigma}\right] + \mu\left(\frac{\partial f}{\partial \vec{p}} \cdot \nabla\right)\vec{B} = \vec{J}. \tag{4.67}$$

Here, $\vec{J}$ is the particle's collision integral which is absent in Eq. (4.24).

Now, we suppose that in the equilibrium state, there exists a constant magnetic field $\vec{B}_0$. Assuming the quantity μB_0 to be much less than the Fermi energy, we can take the Fermi distribution for the degenerate electron gas. It is obvious that the vector distribution function of the spin $\vec{\sigma}$ is parallel to $\vec{B}_0$. Moreover, since the distribution function of particles with spin projection parallel to $\vec{B}_0$ is equal to $1/2 f_0(\epsilon - \mu B_0)$, whereas for the particles with the opposite direction of spins it is equal to $(-1/2) f_0(\epsilon + \mu B_0)$, then

$$\vec{\sigma}_0 = -\mu \vec{B}_0 \frac{\partial f_0}{\partial \epsilon}. \tag{4.68}$$

From Eq. (4.67) for the small perturbations of the equilibrium state, it follows

$$\begin{aligned} &\frac{\partial \delta\vec{\sigma}}{\partial t} + \left(\vec{v} \cdot \nabla\right)\delta\vec{\sigma} + \frac{e}{c}\left(\left[\vec{v} \times \vec{B}_0\right] \cdot \frac{\partial}{\partial \vec{p}}\right)\delta\vec{\sigma} - \frac{2\mu}{\hbar}\left[\vec{B}_0 \times \delta\vec{\sigma}\right] \\ &+ e\left(\vec{E} \cdot \frac{\partial}{\partial \vec{p}}\right)\vec{\sigma}_0 - \frac{2\mu}{\hbar}\left[\delta\vec{B} \times \vec{\sigma}_0\right] + \mu f_0'\left(\vec{v} \cdot \nabla\right)\delta\vec{B} = \vec{J}. \end{aligned} \tag{4.69}$$

Here, $\delta\vec{\sigma}$ and $\delta\vec{B}$ are the small perturbation of the spin distribution function and magnetic induction, respectively.

Now, let us derive an equation for the magnetization density

$$\vec{M}\left(\vec{r}, t\right) = \mu \int d\vec{p}\,\vec{\sigma}. \tag{4.70}$$

From Eq. (4.68), it follows that in the equilibrium state

$$\vec{M}_0 = -\mu^2 \vec{B}_0 \int d\vec{p} f_0' = \chi_p \vec{B}_0, \tag{4.71}$$

where χ_p is the paramagnetic part of the static magnetic susceptibility [see Eq. (4.30)]. To derive an equation for the small perturbation $\delta\vec{M}\left(\vec{M}=\vec{M}_0+\delta\vec{M}\right)$, let us multiply Eq. (4.69) by μ and integrate it over momentum. As a result, we obtain[16]

$$\frac{\partial\delta\vec{M}}{\partial t}-\frac{2\mu}{\hbar}\left[\vec{B}_0\times\delta\vec{M}\right]-\frac{2\mu}{\hbar}\left[\delta\vec{B}\times\vec{M}_0\right]+\int d\vec{p}\left(\vec{v}\cdot\nabla\right)\mu\delta\vec{\sigma}=\mu\int\vec{J}d\vec{p}. \quad (4.72)$$

Now, it is necessary to specify the dependence of the right side of Eq. (4.72) on $\delta\vec{M}$. For this aim, let us present the quantity $\vec{J}$ as a sum of two terms $\vec{J}_U$ and $\vec{J}_\tau$. The first term corresponds to the collisions with spin transfer and the second term to the usual particles' collisions without spin-flipping. It is obvious that

$$\int\vec{J}_\tau d\vec{p}=0, \quad (4.73)$$

because the usual collisions cannot change the particle's spin. For spin transfer collisions, we suppose that

$$\vec{J}_U=-\frac{1}{U}\left(\vec{\sigma}-\vec{\sigma}_0\right). \quad (4.74)$$

Then, the right-hand side of Eq. (4.72) takes the form

$$-\frac{1}{U}\delta\vec{M}. \quad (4.75)$$

For the spatially homogeneous distributions, Eq. (4.72) with account of Eq. (4.75) corresponds to the well-known Bloch equation [65]. But, we are interested in the inhomogeneous distribution of spins. Therefore, suppose that

$$\delta\vec{\sigma}=\frac{1}{\mu}F(\epsilon)\delta\vec{M}\left(\vec{r},t\right)+\vec{\Sigma}, \quad (4.76)$$

where

[16]By this way, from Eq. (4.67), it follows

$$\frac{\partial}{\partial t}\vec{M}-\frac{2\mu}{\hbar}\left[\vec{B}\times\vec{M}\right]+\int d\vec{p}\left(\vec{v}\cdot\nabla\right)\mu\vec{\sigma}=\mu\int\vec{J}d\vec{p}.$$

$$\int d\vec{p}F(\epsilon) = 1, \qquad \int d\vec{p}\vec{\Sigma} = 0. \tag{4.77}$$

If the spatial gradients are relatively small and the usual collision frequency is higher than other characteristic frequencies, then it is easy to show that $\vec{\sum} \ll \delta\vec{\sigma}$. From Eqs. (4.72) and (4.69), it follows

$$\frac{2}{\hbar}\left[\delta\vec{B}\times M_0\right]F(\epsilon) - \frac{\delta\vec{M}}{U}\frac{1}{\mu}F(\epsilon) + e\left(\vec{E}\cdot\frac{\partial}{\partial\vec{p}}\right)\vec{\sigma}_0 - \frac{2\mu}{\hbar}\left[\delta\vec{B}\times\vec{\sigma}_0\right] + \mu f_0'\left(\vec{v}\cdot\nabla\right)\delta\vec{B}$$
$$+\left(\vec{v}\cdot\nabla\right)\frac{1}{\mu}\delta\vec{M}F(\epsilon) = \vec{J}\left(\delta\vec{\sigma}\right) = \vec{J}\left(\frac{1}{\mu}\delta\vec{M}F(\epsilon)\right) + \vec{J}\left(\vec{\Sigma}\right).$$

This equation is an integral equation. But, in our case, one can suppose that $\vec{J}\left(\vec{\Sigma}\right) = -\vec{\Sigma}/\tau$ and determine $\vec{\Sigma}$.[17] Then, for the last term on the left-hand side of Eq. (4.72), we obtain

$$\int d\vec{p}\left(\vec{v}\cdot\nabla\right)\mu\delta\vec{\sigma} = -D_{ik}\frac{\partial^2\delta\vec{M}}{\partial r_i\partial r_k} + C_{ik}\frac{\partial^2\delta\vec{B}}{\partial r_i\partial r_k}, \tag{4.78}$$

where

$$D_{ik} = \tau\int d\vec{p}F(\epsilon)v_iv_k; \qquad C_{ik} = \tau\frac{2\mu^2}{(2\pi\hbar)^3}\int d\vec{p}\delta(\epsilon-\epsilon_F)v_iv_k. \tag{4.79}$$

Considering Eq. (4.77) and having in mind that for the degenerate gas the distribution function is varied only near the Fermi surface, we obtain:

$$F(\epsilon) = \frac{2}{(2\pi\hbar)^3}\delta(\epsilon-\epsilon_F)\left\{-\int d\vec{p}f_0'\right\}^{-1}.$$

Therefore,

$$\chi_p D_{ik} = C_{ik}. \tag{4.80}$$

In this case, one can represent Eq. (4.72) in the following form:

$$\frac{\partial\delta\vec{M}}{\partial t} - \frac{2\mu}{\hbar}\left[\vec{B}_0\times\delta\vec{M}\right] - \frac{2\mu}{\hbar}\left[\delta\vec{B}\times\vec{M}_0\right] - D_{ik}\frac{\partial^2}{\partial r_i\partial r_k}\left(\delta\vec{M} - \chi_p\delta\vec{B}\right) = -\frac{\delta\vec{M}}{U}. \tag{4.81}$$

For metals with cubic symmetry, we have

[17] Relation (4.73) holds when Eq. (4.77) holds.

$$D_{ik} = D\delta_{ik}. \tag{4.82}$$

Quantity D can be easily estimated in the case of the isotropic Fermi surface [61]:

$$D = \frac{v_0^2 \tau}{3}. \tag{4.83}$$

Now, we can calculate the paramagnetic absorption of electromagnetic radiation normally incident on a plate surface of such an isotropic metallic medium, which occupies the semi-space $z \geq 0$, using Eq. (4.81). But here, it is necessary to supplement this equation by boundary conditions. If we suppose that electron's collisions with the metallic surface do not change their spin, then such boundary conditions can be written as[18]

$$\left(\vec{n} \cdot \nabla\right)\left(\delta\vec{M} - \chi_p \delta\vec{B}\right) = 0\cdot \tag{4.84}$$

The electron's collision with the metallic surface with spin transfer is not essential and, therefore, we neglect them [61].

In the considered case, one can suppose that the electromagnetic field in a metal has the form of $f(z)$ exp $(-\iota\omega t)$ and, therefore, Eq. (4.81) can be written as

$$\begin{aligned} &D\frac{d^2}{dz^2}\left\{\delta\vec{M} - \chi_p\delta\vec{B}\right\} + \left(\iota\omega - \frac{1}{U}\right)\left\{\delta\vec{M} - \chi_p\delta\vec{B}\right\} \\ &+\frac{2\mu}{\hbar}\left[\vec{B} \times \left(\delta\vec{M} - \chi_p\delta\vec{B}\right)\right] = -\chi_p\left(\iota\omega - \frac{1}{U}\right)\delta\vec{B}. \end{aligned} \tag{4.85}$$

[18]It is easy to prove this. Let us denote $\delta\vec{\sigma} = \sigma_1$ for $\vec{n} \cdot \vec{v} > 0$ and $\delta\vec{\sigma} = \sigma_2$ for $\vec{n} \cdot \vec{v} < 0$. If

$$\vec{\sigma}_1\left(-\vec{v} \cdot \vec{n};0\right) = \vec{\sigma}_2\left(\vec{v} \cdot \vec{n};0\right),$$

then the collisions of electrons with the surface does not change their magnetic moments. Therefore, it follows

$$\int d\vec{p}\left(\vec{v} \cdot \vec{n}\right)\delta\vec{\sigma}\left(\vec{p},0\right) = 0\cdot$$

Substituting the quantity $\vec{\Sigma}$ in this expression and noting that $\vec{J}\left(\vec{\Sigma}\right) = -(1/\tau)\vec{\Sigma}$, we can easily see that on the metallic surface

$$n_i D_{ik} \frac{\partial}{\partial r_k}\{\delta M_k - \chi_p \delta\beta_k\} = 0.$$

In the isotropic diffusion, when Eq. (4.82) is valid, this equation coincides with Eq. (4.84).

We will investigate the solution of this equation only in the range of the resonance frequencies, $\omega \simeq \Omega$ and, therefore, only the terms which are important in this range will be considered. Moreover, we also suppose that the field penetration depth is much smaller than $\sqrt{UD}$ being the particles displacement due to the diffusion during the spin relaxation time U. Under such a restriction, the approximate solution of Eq. (4.85) takes the form

$$\delta\vec{M} = \frac{\chi_p \Omega}{2}\sqrt{\frac{U}{D}}\frac{1}{\sqrt{1 - \imath U(\omega - \Omega)}}\int_{\infty}^{0} dz' \left\{ \frac{\left[\vec{B}_0 \times \delta\vec{B}\right]}{B_0} + \imath \frac{\left[\vec{B}_0 \times \left(\vec{B}_0 \times \delta\vec{B}\right)\right]}{B_0^2} \right\}, \tag{4.86}$$

which does not depend on coordinates. Taking into account the field equation

$$\imath \frac{\omega}{c}\delta\vec{B} = \frac{d}{dz}\left[\vec{n} \times \vec{E}\right], \tag{4.87}$$

we find the following expression for $\delta\vec{M}$:

$$\delta\vec{M} = \frac{c\chi_p}{2\imath}\sqrt{\frac{U}{D}}\frac{1}{\sqrt{1 - \imath U(\omega - \Omega)}}\left\{ \frac{\left(\vec{B}_0 \times \left(\vec{n} \times \vec{E}(0)\right)\right)}{B_0} + \imath \frac{\left[\vec{B}_0 \times \left(\vec{B}_0 \times \left(\vec{n} \times \vec{E}(0)\right)\right)\right]}{B_0^2} \right\}, \tag{4.88}$$

where $\vec{E}(0)$ is the electric field on the metallic surface.

For describing the paramagnetic resonance absorption, it is more convenient to use the field equation in the form of Eq. (1.4). Let us neglect the displacement current in this system and denote $(1/c)\partial\vec{D}/\partial t = (4\pi/c)\vec{j}$, where the current $\vec{j}$ is not connected with the electron paramagnetic properties (see Sect. 4.1). Then, according to Eq. (4.88), we can present the system of field equations (1.4):

$$\nabla \times \delta\vec{B} = \frac{4\pi}{c}\vec{j}; \qquad \nabla \times \delta\vec{E} = -\frac{1}{c}\frac{\partial \delta\vec{B}}{\partial t}; \qquad \nabla \cdot \delta\vec{B} = 0. \tag{4.89}$$

These equations allow us to express the electromagnetic field in a metal in terms of the fields being obtained when the paramagnetic part of magnetization is neglected. For the normal skin-effect, $\vec{E}_0$ and $\delta\vec{B}_0$ are determined very easily. For the anomalous skin-effect, these quantities can be determined by using the results of Sects. 2.9 and 4.3.

When one determines the electromagnetic fields in the presence of the paramagnetic magnetization $\delta\vec{M}$, it must be noted that, in this case, the tangential components of the electric and magnetic fields (not magnetic induction) are continuous. Then, $\vec{E}_0$ remains unchanged on the metallic surface, whereas

$$\delta\vec{H}(0) = \delta\vec{B}_0(0) - 4\pi\delta\vec{M}_\perp(0), \tag{4.90}$$

where $\delta\vec{M}_\perp$ is the tangential component of $\delta\vec{M}$ [see Eq. (4.88)].

To calculate the paramagnetic absorption, it is necessary to find out the average flux of energy input in the metal

$$\langle S\rangle = \frac{c}{2\pi}\,\mathrm{Re}\left(\vec{E}(0)\times\delta\vec{H}^*(0)\right)\cdot\vec{n}. \tag{4.91}$$

According to Eq. (2.177), when paramagnetic behavior is neglected, this quantity looks as

$$\frac{c^2}{8\pi^2}\left|\delta\vec{B}_0(0)\right|^2\,\mathrm{Re}\,Z_0, \tag{4.92}$$

where Z_0 is the surface impedance in the absence of magnetization given by Eq. (4.88). Now, we must calculate the paramagnetic correction $\sim\chi_p$ to Eq. (4.92). Introducing the complex vector quantity $\vec{b} = \delta\vec{B}_0/\left|\delta\vec{B}_0\right|$, it is easily shown that

$$\left(\vec{n}\times\vec{E}(0)\right) = \frac{c}{4\pi}Z_0\vec{b}\mid\delta\vec{B}_0\mid. \tag{4.93}$$

Having this relation in mind and substituting Eqs. (4.88) and (4.90) into the right-hand side of Eq. (4.91), we find

$$\langle S\rangle = \frac{c^2}{8\pi^2}\left|\delta\vec{H}(0)\right|^2\,\mathrm{Re}\,Z, \tag{4.94}$$

where

$$Z = Z_0\left\{1 - \frac{c^2\chi_p}{2}\sqrt{\frac{U}{D}}Z_0\frac{f}{\sqrt{1-\imath U(\omega-\Omega)}}\right\}, \tag{4.95}$$

$$f = \frac{\left(\vec{b}\times\vec{B}_0\right)\cdot\left(\vec{b}^*\times\vec{B}_0\right)}{B_0^2} + \imath\frac{\vec{B}_0}{B_0}\cdot\left(\vec{b}\times\vec{b}^*\right). \tag{4.96}$$

The most essential frequency dependence of absorption is determined by

$$\frac{1}{\sqrt{1-\imath U(\omega-\Omega)}}=\frac{\sqrt{1+\sqrt{1+U^2(\omega-\Omega)^2}}+\imath\frac{\omega-\Omega}{|\omega-\Omega|}\sqrt{-1+\sqrt{1+U^2(\omega-\Omega)^2}}}{\sqrt{2}\sqrt{1+U^2(\omega-\Omega)^2}}. \tag{4.97}$$

For example, in the case of the normal skin-effect, we have

$$Z_0=(1-\imath)\sqrt{\frac{2\pi\omega}{\sigma_0 c^2}}\equiv(1-\imath)(\sigma_0\delta)^{-1}, \tag{4.98}$$

where $\delta=c/\sqrt{2\pi\sigma_0\omega}$ is the field penetration depth. Then, from Eq. (4.94), it follows[19]

$$\mathrm{Re}\,Z=\sqrt{\frac{2\pi\omega}{\sigma_0 c^2}}-\frac{\sqrt{2}\pi\Omega}{\sigma_0}\chi_p\sqrt{\frac{U}{D}}f\frac{\omega-\Omega}{\mid\omega-\Omega\mid}\frac{\sqrt{-1+\sqrt{1+U^2(\omega-\Omega)^2}}}{\sqrt{1+U^2(\omega-\Omega)^2}}. \tag{4.99}$$

We see that the second term determining the paramagnetic absorption is antisymmetric due to dependence on $(\omega-\Omega)/\mid\omega-\Omega|$. The absorption frequency width is determined by the spin relaxation time, whereas the diffusion coefficient D, as seen from Eq. (4.99), determines only the absolute value of absorption. In fact, diffusion denotes not only the absolute value of absorption, but it also influences the width of absorption line quite significantly. Let us prove this point by neglecting the diffusion of magnetization. This is possible when the skin-depth is larger than $\sqrt{UD}$, i.e., the particles diffusion length during the spin relaxation time. Such conditions may be realized in the low-frequency range or sufficiently small U, which takes place in the presence of a large number of paramagnetic impurities in the metal.

Neglecting the spatial derivatives term in Eq. (4.85), we find the resonance part of magnetization (when ω~Ω)

$$\delta\vec{M}\simeq\chi(\omega)\delta\vec{B},\quad \chi(\omega)=\frac{\imath\chi_p\Omega U}{2\left(1+U^2(\omega-\Omega)^2\right)}. \tag{4.100}$$

For media with the dielectric permittivity $\varepsilon(\omega)$ and the magnetic permittivity $\mu(\omega)=1+4\pi\chi(\omega)$, the surface impedance looks as[20]

---

[19] Using Eqs. (4.47), one can obtain the analogous expression in the case of the anomalous skin-effect.

[20] The surface impedance was determined without $4\pi/c$ in [38].

$$Z = \frac{4\pi}{c}\sqrt{\frac{\mu}{\varepsilon}}.$$

Using this expression, one can obtain the resonance part of the real Z in the case of the normal skin-effect

$$\sqrt{\frac{2\pi\omega}{\sigma_0 c}}\frac{\pi\chi_p\Omega U}{1+U^2(\omega-\Omega)^2}. \tag{4.101}$$

Comparing this expression to Eq. (4.99) shows that the diffusion of magnetization qualitatively influences the absorption lines, particularly, when diffusion is strong. Namely, in contrast to Eq. (4.99), which describes the anti-symmetric form of the absorption line, the expression (4.101), which was obtained when diffusion is neglected, corresponds to the symmetric line. But, in both cases, the most essential point for the line width is the spin relaxation time.

4.5 Collisionless Absorption and Excitation of Sound Waves in Condensed Matters

Usually, in metals and semiconductors, the sound wave absorption in the low temperature range is determined by conduction electrons. When the wavelength is larger than the electron mean free path, the absorption, which has hydrodynamic character, is stipulated by the electron viscosity. Another quite similar situation takes place in the range of the short-wavelength, when the electron mean free path is infinitely large compared to the sound wavelength. In this case, sound wave absorption is similar to the collisionless Cherenkov mechanism of wave dissipation in the case of the anomalous skin-effect. Below, we will consider such a mechanism of sound wave absorption in the absence of the external magnetic field.[21] Besides, separately, the metals and the piezo-semiconductors will be investigated.

To describe these phenomena in metals, one must account for the energy change due to the lattice deformation under the action of sound waves [74]

$$\epsilon\left(\vec{p},\vec{r}\right) = \epsilon\left(\vec{p}\right) + \Lambda_{ij}\left(\vec{p}\right)\frac{\partial u_i}{\partial r_j}. \tag{4.102}$$

[21] Such a theory was developed for metals in [66–71], and for piezo-semiconductors in [72]. Review of experiments is given in [73].

Here, $\epsilon\left(\vec{p}\right)$ and $\epsilon\left(\vec{p},\vec{r}\right)$ are the electron's energy in the absence and presence of sound waves, respectively; $\vec{u}$ is the lattice deformation, and $\Lambda_{ij}\left(\vec{p}\right)$ is the tensor characterizing the energy change due to the lattice deformation.

Considering Eq. (4.102), one can present the kinetic equation for the collisionless electrons

$$\frac{\partial f}{\partial t}+\vec{v}\cdot\frac{\partial f}{\partial \vec{r}}+\left\{e\left(\vec{E}+\frac{1}{c}\left[\vec{v}\times\vec{B}\right]\right)-\Lambda_{ij}\frac{\partial^2 u_i}{\partial \vec{r}\,\partial r_j}\right\}\cdot\frac{\partial f}{\partial \vec{p}}=0. \qquad (4.103)$$

Here, we have taken into account the correction term to the Lorentz force (1.3) due to energy dependence of electrons on the lattice deformation given by Eq. (4.102). Besides, because of Eq. (4.102) we have

$$\vec{v}\equiv\frac{\partial\epsilon\left(\vec{p},\vec{r}\right)}{\partial\vec{r}}=\frac{\partial\epsilon\left(\vec{p}\right)}{\partial\vec{p}}+\frac{\partial\Lambda_{ij}}{\partial\vec{p}}\frac{\partial u_i}{\partial r_j}. \qquad (4.104)$$

The kinetic equation (4.103) must be supplemented by the Maxwell's equations. But, in the considered case, when the wavelength is larger than the Debye length, instead of equation $\nabla\cdot\vec{E}=4\pi\rho$, we can use the condition of plasma neutrality

$$\rho=\rho_{(e)}+\rho_{(i)}=e\int d\vec{p}f+\rho_{(i)}=0, \qquad (4.105)$$

where $\rho_{(i)}$ is the lattice charged density. Then, the field equation without the displacement current, can be written as

$$\nabla\times\nabla\times\vec{E}=-\frac{4\pi}{c^2}\frac{\partial}{\partial t}\left(\vec{j}_{(e)}+\vec{j}_{(i)}\right)=-\frac{4\pi}{c^2}\frac{\partial}{\partial t}\left[e\int d\vec{p}\,\vec{v}f_e+\vec{j}_{(i)}\right], \qquad (4.106)$$

where $\vec{j}_{(i)}=\vec{\dot{u}}\rho_{(i)}$ is the lattice current density. Finally, the equation of lattice motion (equation of elasticity) looks as

$$\rho_m\ddot{u}_i=\lambda^{(0)}_{ikjl}\frac{\partial^2 u_j}{\partial r_k\partial r_l}+\rho_{(i)}E_i+\frac{1}{c}\left[\vec{j}_{(i)}\times\vec{B}\right]_i+\frac{\partial}{\partial r_k}\int d\vec{p}\,\Lambda_{ik}(p)f. \qquad (4.107)$$

Here, ρ_m is the lattice mass density. The last term in Eq. (4.107) is the force stipulated by the energy dependence of Eq. (4.102); the first term is the usual elastic force and the second and third terms present the Lorentz force, acting on the lattice [75]. It should be recalled that considering the electric field, which enters into the interaction of electrons and the lattice, and taking into account the interaction, which arises from the change of electron's energy in the field of the acoustic wave determined by Eq. (4.102), results in additional elasticity. It is possible to talk that, in this case, the renormalization of the elasticity modulus tensor occurs. Therefore, in formula (4.107), one has to consider the renormalized value of the elasticity tensor. In this connection, the elasticity tensor $\lambda^{(0)}_{ikjl}$ refers to the elasticity tensor in the absence of the electromagnetic fields.

The system of Eqs. (4.102)–(4.107) is complete. Now, we can solve these equations in the linear approximation and investigate the problem of sound wave's propagation and absorption. In the equilibrium state, the distribution function $f_0(\epsilon(p))$ is supposed to be arbitrary (for example, the degenerate Fermi distribution). Then, for the small perturbation δf, from Eq. (4.103), we obtain

$$\frac{\partial \delta f}{\partial t} + \vec{v}_0 \cdot \frac{\partial \delta f}{\partial \vec{r}} + \left(e\vec{E} - \Lambda_{ij} \frac{\partial^2 u_i}{\partial \vec{r}\, \partial r_j} \right) \cdot \vec{v}_0 f_0' = 0. \tag{4.108}$$

Here, $\vec{v}_0 = \partial \epsilon(p)/\partial \vec{r}$ and $f_0' = \partial f_0/\partial \epsilon(p)$. Besides, we omitted the magnetic term because in the equilibrium state the external magnetic field is absent.

In the linear approximation, we can present all perturbations in space and time as $\exp\left(-\imath\omega t + \imath \vec{k} \cdot \vec{r}\right)$. Then, from Eq. (4.108), we obtain (see Sect. 2.2)

$$\delta f = -\imath f_0' \left(e\vec{v}_0 \cdot \vec{E} + \Lambda_{ij} k_j u_i \vec{k} \cdot \vec{v}_0 \right) \left[\mathcal{P} \frac{1}{\omega - \vec{k} \cdot \vec{v}_0} - \imath\pi\delta\left(\omega - \vec{k} \cdot \vec{v}_0\right) \right]. \tag{4.109}$$

The presence of the δ-function in the right-hand side of the latter equation corresponds to the adiabatic switching on of the interaction in the infinite past (see Sect. 2.2).

Using this fact that sound speed in metals is always less than the electron's velocity on the Fermi surface and substituting Eq. (4.109) into Eqs. (4.105) and (4.106), one can easily obtain expressions for the transverse and longitudinal components of the electric field:

$$E_i^{\mathrm{tr}} = \imath\omega e B_{ij}^{-1} \left\{ u_l k_s \left\langle \frac{v_{oj}^{\mathrm{tr}}}{\vec{k} \cdot \vec{v}_0} \left(\Lambda_{ls} - \frac{\langle \Lambda_{ls} \rangle}{\langle 1 \rangle} - \frac{N_e \delta_{ls}}{\langle 1 \rangle} \right) \right\rangle + N_e u_j^{\mathrm{tr}} \right\}, \tag{4.110}$$

$$\begin{aligned} \vec{E}^{l} = &- \frac{\vec{k} u_j k_l}{e\langle 1 \rangle} \langle \Lambda_{jl} + N_e \delta_{jl} \rangle - \frac{\vec{k}\vec{E}^{\mathrm{tr}}}{\langle 1 \rangle} \cdot \left\langle \frac{\vec{v}_0}{\vec{k} \cdot \vec{v}_0} \right\rangle - \frac{\imath\pi\omega \vec{k}}{e\langle 1 \rangle} \\ &\times \left\langle \delta\left(\vec{k} \cdot \vec{v}_0\right) \left(\Lambda_{il} - \frac{\langle \Lambda_{il} \rangle}{\langle 1 \rangle} - \frac{N_e \delta_{il}}{\langle 1 \rangle} \right) \right\rangle u_i k_l. \end{aligned} \tag{4.111}$$

Here, $\vec{v}_0^{\mathrm{tr}} = \vec{v} - \frac{\vec{k}\left(\vec{k} \cdot \vec{v}_0\right)}{k^2}$, $\langle A \rangle = -\int d\vec{p} f_0' A$, and

$$B_{ij} = -\sigma_{ij} - \imath \frac{c^2 k^2}{4\pi\omega} \delta_{ij}, \tag{4.112}$$

where $\sigma_{ij} = -\pi e^2 \left\langle v_{0i} v_{0j} \delta\left(\vec{k} \cdot \vec{v}_0\right)\right\rangle$ is the conductivity tensor of metals in the frequency range of the anomalous skin-effect [see Eq. (4.63)]. In addition, $\langle 1 \rangle$ is the normalization of the derivative of the distribution function, i.e., $\langle 1 \rangle = \int f'_0 d\vec{p}$.

Substituting Eqs. (4.109)–(4.111) into Eq. (4.107), in the absence of the magnetic field, in the linear approximation, we find

$$\rho_m \ddot{u}_i = \frac{\partial}{\partial r_j} \Xi_{ij}\left(\vec{r}, t\right) = \frac{\partial}{\partial r_j} \int_{-\infty}^{t} dt' \int d\vec{r}' \hat{\lambda}_{ijlm}\left(t - t', \vec{r} - \vec{r}'\right) \frac{\partial u_l\left(\vec{r}', t'\right)}{\partial r'_m}. \tag{4.113}$$

Assuming dependency $\exp\left(-\imath\omega t + \imath\vec{k} \cdot \vec{r}\right)$ for all quantities and following Sects. 1.2 and 1.4, we calculate the average energy loss of the sound waves per unit time and per unit volume of the medium

$$\frac{Q}{V} = \imath\omega\left(\lambda_{ijlm} - \lambda^*_{lmij}\right) k_j k_m u^*_i u_l, \tag{4.114}$$

where[22]

---

[22]From the Onsager symmetry relation, it follows that

$$k_j k_m \lambda_{ijlm}\left(\vec{B}_0, \omega, \vec{k}\right) = k_j k_m \lambda_{lmij}\left(-\vec{B}_0, \omega, -\vec{k}\right).$$

Moreover, in non-absorbing media, from Eq. (4.114), it follows that

$$k_j k_m \lambda_{ijlm} = k_j k_m \lambda^*_{lmij},$$

and

$$k_j k_m \lambda'_{ijlm} = k_j k_m \lambda'_{lmij}; \qquad k_j k_m \lambda''_{ijlm} = -k_j k_m \lambda''_{lmij}.$$

The anti-symmetric imaginary part can be expressed by

$$k_j k_r \lambda''_{ijlr} = e_{irl} G_r \rho_m k^2,$$

where ρ_m is the lattice mass density, and the vector $\vec{G}$ is determined by the acoustic properties [71, 76].

$$\lambda_{ijlm}\left(\omega,\vec{k}\right)=\int d\vec{r}\exp\left(-\imath\vec{k}.\vec{r}\right)\int_0^\infty dt\exp\left(\imath\omega t\right)\widehat{\lambda}_{ijlm}\left(t,\vec{r}\right)=\lambda'_{ijlm}+\imath\lambda''_{ijlm}. \quad (4.115)$$

According to Eqs. (4.109)–(4.111), we have:

$$k_m k_l \lambda'_{imjl}=k_m k_l\left\{\lambda^{(0)}_{ijml}+\langle\Lambda_{im}\Lambda_{jl}\rangle-\frac{1}{\langle 1\rangle}\left(\langle\Lambda_{im}\rangle+N_e\delta_{im}\right)\left(\langle\Lambda_{jl}\rangle+N_e\delta_{jl}\right)\right\}, \quad (4.116)$$

$$k_m k_l \lambda''_{imjl}=\pi\omega\left\langle\delta\left(\vec{k}\cdot\vec{v}_0\right)L_iL_j\right\rangle+\omega e^2\left(\left\langle\frac{v^{\text{tr}}_{0s}}{\vec{k}\cdot\vec{v}_0}L_i\right\rangle+N_e\delta_{is}\right)\text{Im}\left(\imath B^{-1}_{sq}\right)\left(\left\langle\frac{v^{\text{tr}}_{0q}}{\vec{k}\cdot\vec{v}_0}L_j\right\rangle+N_e\delta_{jq}\right), \quad (4.117)$$

where

$$L_i=k_m\left(\Lambda_{im}-\frac{\langle\Lambda_{im}\rangle}{\langle 1\rangle}-\frac{N_e\delta_{im}}{\langle 1\rangle}\right).$$

Then, considering $-4\pi\imath\omega B_{ij}\simeq\omega^2\varepsilon_{ij}-c^2k^2\delta_{ij}$, one can easily show that the second term in Eq. (4.117) is determined by the work of the transverse field. For example, when the Fermi surface is spherical and $\left\langle v^{\text{tr}}_{0q}L_j/\vec{k}\cdot\vec{v}_0\right\rangle=0$, from the second term of Eq. (4.117), one can obtain the following expression for the heat delivered in the metal due to the transverse electromagnetic field:

$$8\pi\omega j^*_{(i)s}j_{(i)q}\ \text{Im}\left(c^2k^2\delta_{sq}-\omega^2\varepsilon_{sq}\right)^{-1}=-2\,\text{Re}\,\vec{j}^{\,*}_{(i)}.\vec{E}=\frac{\omega}{2\pi}\varepsilon''_{sq}E^*_sE_q.$$

The main difference between Eq. (4.117) describing collisionless sound wave absorption and usual absorption stipulated by the hydrodynamic viscosity consists of the frequency dependence. In hydrodynamics, the quantity

$$q^t=\frac{Q/V}{2\rho_m\omega^2\vec{u}\cdot\vec{u}^*}$$

is proportional to ω^2, whereas, in the considered collisionless case, it is proportional to k [75]. But, strictly speaking, this statement is correct when the second term in Eq. (4.112) is negligible, or when

$$\lambda\sim\frac{1}{k}\gg\left(p_0c^2/\omega e^2N_e\right)^{1/3}.$$

In other words, sound wavelength is larger than the anomalous skin-depth. In the opposite case, collisionless sound wave absorption is relatively small [71, 77, 78].

4.5.1 Acoustic Wave Absorption and Excitation in Piezo-Semiconductors

Piezoelectric materials are the kinds of materials in which a link between an electrical field and a deformation field takes place. In this way, the phenomenon of piezoelectricity can be described as the generation of an electric polarization due to the applied pressure. Piezoelectric materials have found many applications among engineers and scientists who are looking for the ways to create smart material structures. Researchers and engineers have worked with piezoelectric materials in order to make piezoelectric materials act as actuators, sensors, or both [72, 79–81]. Moreover, the rapid development of micro-electro-mechanical systems (MEMs) in industrial applications has attracted a surge interest in modeling of piezoelectric materials' behavior [72, 79–81]. Beside many dielectric materials, there is a strong piezoelectric effect in some semiconductors, such as the wurtzite family with hexagonal structure and in piezoelectric semiconductors such as *ZnO*, *CdS*, *ZnS*, *CdSe*, *CdTe,* and *ZnTe* [72, 79–81]. In the 1960s, the properties of piezoelectric semiconductors led to the acousto-electronic device development. Bulk acoustic wave (BAW), delay lines, traveling wave amplifiers, surface acoustic wave devices (SAW), and oscillators were developed based on such materials. Later, it was found that the interaction of piezoelectric and semiconducting properties can affect the velocity and attenuation of acoustic waves [72, 79–81]. Usually, acoustic wave's propagation in a piezoelectric material is accompanied by an electric field. In a piezoelectric semiconductor, because of charge carriers, these fields produce currents and space charge, which can cause the dispersion and acoustic loss. Furthermore, the interaction of traveling acoustic waves with charge carriers, called the acoustoelectric effect, takes place in such materials. It was found that applying a DC electric field can cause the acoustoelectric amplification of acoustic waves [79–81]. It seems that all these special effects appear because of the coupling of piezoelectric and semiconducting properties existing in piezoelectric semiconductor materials. Actually, a DC bias in a piezoelectric semiconductor may derive a beam of electrons which can cause some phenomena [79–81]. In this part, we study the possibility of elastic acoustic wave excitation in a piezoelectric semiconductor in the presence of an external electric field. The reason for this excitation is thought to be the electric drift of charge carriers. By considering a piezoelectric semiconductor sample with hexagonal symmetry, we may approximately model it as an isotropic plasma-like medium [72, 79–81]. As it was shown, the coupling between the elastic waves and the plasma properties of the piezoelectric semiconductor medium may exist in this case [72, 79–81]. Hence, an equation describing such a coupling between the elasticity wave (longitudinal and transverse acoustic waves) and the quasi-electromagnetic (longitudinal) oscillations of charge carriers is obtained. In other words, the oscillation spectra of the piezoelectric semiconductor as a plasma-like medium, showing the coupling between elastic wave and the oscillation of charge carriers, were obtained and then such spectra for the surface wave in the semi-bounded piezoelectric semiconductor material were investigated [72, 79–81]. It is well-known that in plasma media a beam may excite the wave emerging from an initial fluctuation in the medium by transferring its energy into the wave. Based on such a fact, in the present part by taking into account the electric drift of charge carriers due to an external DC electric field, we investigate the elasticity-acoustic excitation as an instability which will grow in time by the injection of the beam energy into the lattice vibrations via the piezoelectric effect.

Let us, now, consider sound wave absorption in the collisional limit in piezo-semiconductors. In such media the coupling between lattice oscillations and plasma oscillations is very strong. Therefore, the charge and current densities in the lattice can be neglected. In this case, the so-called press elastic force which is proportional to the self-consistent electric field is very important.

Here, we consider a piezoelectric semiconductor medium with hexagonal symmetry whose main symmetry axis is along the z axis. Moreover, we suppose that in this stage there are no external electric and magnetic fields, i.e., $E_0 = 0$ and $B_0 = 0$ for simplicity. It should be noted that for a homogeneous piezoelectric plasma-like medium in an equilibrium and stationary state the velocity distribution of charge carriers will be Maxwellian if charge carriers are non-degenerate and will be the Fermi distribution if they are degenerate [72, 79]. Hence, by neglecting collisions, the charge carrier's motion can be described by the Vlasov equation,

$$\frac{\partial f_{0\alpha}}{\partial t} + \left(\vec{v} \cdot \nabla\right) f_{0\alpha} + \frac{e}{m_\alpha}\left\{\vec{E} + \frac{1}{c}\vec{v} \times \vec{B}\right\} \cdot \frac{\partial f_{0\alpha}}{\partial \vec{v}} = 0, \tag{4.118}$$

where $f_{0\alpha}$ is the equilibrium distribution function for α species of charge carriers. Furthermore, the elastic equation in the absence of the charges induced in the crystal is written as [see Eq. (4.107)] [38]

$$\rho_m \frac{\partial^2 u_i}{\partial t^2} = \lambda_{iklm} \frac{\partial^2 u_l}{\partial r_k \partial r_m} + \frac{\partial \beta_{ilk} E_l}{\partial r_k}, \tag{4.119}$$

where, ρ_m is the lattice density, u_i is the ith component of the lattice displacement, λ_{iklm} with dimension Pa is the elastic modulus tensor determining the elasticity force in the deformed lattice, and β_{ilk} with dimension cm^2 is the piezoelectric tensor of the medium [38]. Now, in the perturbed medium, the small perturbations of the equilibrium values of the fields and current arise in the medium. Linearizing aforementioned equations with respect to these perturbations and using the Maxwell's equations, we find

$$\frac{\partial \delta f_\alpha}{\partial t} + \left(\vec{v} \cdot \nabla\right) \delta f_\alpha + \frac{e}{m_\alpha} \vec{E} \cdot \frac{\partial f_{0\alpha}}{\partial \vec{v}} = 0, \quad \nabla \times \nabla \times \vec{E} + \frac{1}{c^2}\frac{\partial^2 \vec{E}}{\partial t^2} + \frac{4\pi}{c}\frac{\partial \vec{j}}{\partial t} = 0,$$

$$\rho_m \frac{\partial^2 u_i}{\partial t^2} = \lambda_{iklm} \frac{\partial^2 u_l}{\partial r_k \partial r_m} + \frac{\partial \beta_{ilk} E_l}{\partial r_k},$$

$$\vec{j} = \sum_\alpha e_\alpha \int \vec{v} \delta f_\alpha d\vec{p} + j^{(p)} = \sum_\alpha e_\alpha \int \vec{v} \delta f_\alpha d\vec{p} + \beta_{ikl} \frac{\partial^2 u_k}{\partial t \partial r_l}, \quad (4.120)$$

where ∂f_α is the perturbation of the equilibrium distribution function $f_{0\alpha}$ for α species of charge carriers. Considering a plane wave as $\exp\left(-\imath\omega t + \imath\vec{k}\cdot\vec{r}\right)$ for the solution of these equations and using the dielectric permittivity tensor of plasma, $\varepsilon_{ij}\left(\omega, \vec{k}\right)$, for charge carriers, we find a system of two coupling equations

$$\begin{aligned} &\omega^2 \rho_m u_i - \lambda_{ijls} k_j k_s u_l + \imath k_l \beta_{ijl} E_j = 0, \\ &\left\{ k^2 \delta_{ij} - k_i k_j - \frac{\omega^2}{c^2} \varepsilon_{ij}(\omega, k) \right\} E_j - \frac{4\pi\imath\omega^2}{c^2} \beta_{ijl} k_l u_j = 0. \end{aligned} \quad (4.121)$$

The solvability condition of this equation system gives the dispersion equation for the coupled quasi-elasto-electromagnetic wave in piezoelectric semiconductor plasma as follows:

$$\left| \omega^2 \rho_m \delta_{ij} - \lambda_{ikjl} k_k k_l - \frac{4\pi\omega^2 k_k \beta_{lik}}{k^2} \left\{ \frac{k_l k_s}{\omega^2 \varepsilon^l} - \frac{k^2 \delta_{ls} - k_l k_s}{k^2 c^2 - \omega^2 \varepsilon^{tr}} \right\} \beta_{sjm} k_m \right| = 0. \quad (4.122)$$

This equation is quite general and describes the coupled sound-electromagnetic waves in arbitrary solid-state isotropic plasmas with elasto-electrical properties of the crystal lattice. Below, we will consider this equation in the limit $c \to \infty$ when potential approximation for the electromagnetic field is valid. Since acoustic crystal vibrations are relatively slow, they may strongly be coupled only with slow oscillations in plasma of charge carriers. Moreover, as it has been mentioned before, we consider a crystal lattice with hexagonal symmetry whose main axis coincides with the z axis and its four axes of second-order symmetry lie on the xy plane. In such a crystal, the piezoelectric tensor β_{ilk} can be determined by three quantities β_i, i.e., the piezoelectric coupling constants ($\imath = 1, 2, 3$) [38]

$$\beta_1 = \beta_{xxz} = \beta_{xzx} = \beta_{yyz} = \beta_{yzy}, \quad \beta_2 = \beta_{zxx} = \beta_{zyy}, \quad \beta_3 = \beta_{zzz}. \quad (4.123)$$

In this case, the crystal is very similar to an isotropic medium. Furthermore, the elastic moduli can be approximated by an isotropic tensor with two non-zero components as follows [1, 72, 79]:

$$\lambda_{ijkl}\frac{k_k k_l}{k^2} = \left(\delta_{ij} - \frac{k_i k_j}{k^2}\right)\lambda^{\text{tr}} + \frac{k_i k_j}{k^2}\lambda^l. \tag{4.124}$$

Here, λ^{tr} and λ^l are the transverse and the longitudinal components of the elasticity modulus, respectively. In this case, the quantities λ^l and λ^{tr} are related to the elasticity coefficient of the medium, the Young module ξ, and the Poisson coefficient σ [1, 72]:

$$\lambda^{\text{tr}} = \frac{\xi}{2(1+\sigma)}, \quad \lambda^l = \frac{\xi(1-\sigma)}{(1+\sigma)(1-2\sigma)}.$$

It is clear that $\lambda^{\text{tr}} < \lambda^l < 1/2$. In fact, a crystal with hexagonal symmetry is characterized by five numbers

$$\lambda_{xxxx} = \lambda_{yyyy} = a + b, \ \lambda_{xxyy} = \lambda_{yyxx} = a - b, \ \lambda_{xxzz} = \lambda_{zzxx} = \lambda_{yyzz} = \lambda_{zzyy} = c,$$
$$\lambda_{xyxy} = \lambda_{yxxy} = \lambda_{yxyx} = \lambda_{xyyx} = b, \quad \lambda_{zzzz} = f,$$
$$\lambda_{xzxz} = \lambda_{yzyz} = \lambda_{zxzx} = \lambda_{xzzx} = \lambda_{zxzy} = \lambda_{yzzy} = \lambda_{zyyz} = \lambda_{zyzy} = d.$$

Transition to an isotropic medium takes place through $a + b = f$, $a - b = c$, $b = d$, which in a real hexagonal crystal is always satisfied. Therefore, $b = \lambda^{\text{tr}}$ and $a = \lambda^l - \lambda^{\text{tr}}$.

From relations (4.122), using the above assumptions, we obtain

$$\begin{aligned} &\varepsilon^l(\omega)\left[\left(\omega^2\rho_m - k^2\lambda^{\text{tr}}\right)\left(\omega^2\rho_m - k^2\lambda^l\right) + k_z^2 k_\perp^2\left(\lambda^{\text{tr}} - \lambda^l\right)^2\right] = \\ &= 4\pi k^2\left[(\beta_1+\beta_2)^2\frac{k_\perp^2 k_z^2}{k^4}\left(\omega^2\rho_m - k_\perp^2\lambda^{\text{tr}} - k_z^2\lambda^l\right) + \right. \\ &+2(\beta_1+\beta_2)\left(\beta_1\frac{k_\perp^2}{k^2} + \beta_3\frac{k_z^2}{k^2}\right)\left(\frac{k_\perp^2 k_z^2}{k^2}\right)\left(\lambda^l - \lambda^{\text{tr}}\right) \\ &\left. +\left(\beta_1^2\frac{k_\perp^4}{k^4} + \beta_3^2\frac{k_z^4}{k^4} + 2\beta_1\beta_3\frac{k_\perp^2 k_z^2}{k^4}\right)\left(\omega^2\rho_m - k_z^2\lambda^{\text{tr}} - k_\perp^2\lambda^l\right)\right]. \end{aligned} \tag{4.125}$$

This equation describes the coupling between the elasticity wave (longitudinal and transverse acoustic wave) and the electromagnetic (longitudinal) oscillations of charge carriers. Now, we analyze this coupling for purely longitudinal ($k_\perp = 0$) and purely transverse ($k_z = 0$) propagation of waves, respectively. In this case, we find

$$\begin{cases} \varepsilon^l(\omega,k)\left(\omega^2\rho_m - k^2\lambda^l\right) = 4\pi k^2\beta_3^2, \\ \varepsilon^l(\omega,k)\left(\omega^2\rho_m - k^2\lambda^{\text{tr}}\right) = 4\pi k^2\beta_1^2. \end{cases} \tag{4.126}$$

These equations imply that there is a coupling between plasma properties of charge carriers, the first terms on the left-hand side of Eqs. (4.126), and the elastic properties of the lattice, the second terms, via the piezoelectric effect appearing on the right-hand side of Eqs. (4.126). The first equation shows a coupling between the longitudinal plasma wave and the longitudinal elasticity. The second equation means that the longitudinal plasma wave is coupled to a transverse acoustic wave.

Equations (4.126) have a similar structure and, therefore, we analyze both of them. It is clear that a strong interaction between plasma and elasticity waves occurs when the following equations

$$\varepsilon^l(\omega, k) = 0, \quad \omega^2 - k^2 \frac{\lambda^{l,\mathrm{tr}}}{\rho_m} \equiv \omega^2 - k^2 v_{l,\mathrm{tr}}^2 = 0 \tag{4.127}$$

are simultaneously satisfied. Here, $v_{l,\mathrm{tr}} = \sqrt{\lambda^{l,\mathrm{tr}}/\rho_m}$ is the longitudinal (transverse) acousto-elastic wave's velocity. This means that the frequencies of acousto-elastic and plasma waves are close to each other. It is clear that, in this case, plasma oscillations should be low-frequency oscillations. Such low-frequency oscillations exist only in non-isothermal isotropic plasmas when $T_e \gg T_i$ (or $E_F \gg T_i$ where T_α is the thermal energy of α species, and E_F is the Fermi energy) and their frequency spectrum is given by

$$\omega^2 = \omega_s^2 = \frac{k^2 v_s^2}{1 + k^2 r_{\mathrm{De}}^2}, \quad \delta_s = -\sqrt{\frac{\pi}{8}}\kappa \frac{\omega^2}{k v_0}, \tag{4.128}$$

where v_s is the ion-acoustic frequency; $\kappa = 1$ when $v_0 = v_{\mathrm{Te}}$, or $\kappa = \pi/2$ when $v_0 = v_{\mathrm{Fe}}$ is the Fermi velocity; $r_{\mathrm{De}}^2 = v_{\mathrm{Te}}/\omega_{\mathrm{pe}}$ is the Debye length for non-degenerate plasma and $r_{\mathrm{De}} = v_{\mathrm{Fe}}/\sqrt{3}\omega_{\mathrm{pe}}$ for degenerate plasma. In the absence of the piezo effect (when $\beta_i = 0$), elasticity and plasma waves are not coupled and, in our approximation, elasticity oscillations are not damped, but ion-acoustic oscillation is strongly damped due to the Cherenkov dissipation by electrons. Taking into account the piezoelectric coupling of these waves, acousto-elastic oscillations are damped due to their absorption by plasma electrons. In fact, in the collisionless limit for $k = k_\perp$, and in the ion-acoustic frequency region for $k = k_z$, Eq. (4.125) for charge carriers can be written as follows [72, 82]:

$$\left(\omega^2 - k^2 v_{l,\mathrm{tr}}^2\right)\left(\omega^2 - \omega_s^2 + \imath\sqrt{\frac{\pi}{2}}\kappa \frac{\omega^3}{k v_0}\right) = \frac{4\pi \beta_{3,1}^2 \omega^2 k^4 v_s^2}{\rho_m \omega_{pi}^2}, \tag{4.129}$$

where $\omega_{p\alpha}^2 = 4\pi e_\alpha^2 n_0/m_\alpha$ is the plasma frequency, e_α and m_α are charge and mass of α species, respectively. In the crossing region of the dispersion curves $\omega = \omega_s$ and $\omega = k v_{l,\mathrm{tr}}$, the solution of Eq. (4.129) may have the form of

$$\omega = k v_{l,\mathrm{tr}} + \imath\delta = \omega_s + \imath\delta, \tag{4.130}$$

where

$$\delta_{1,2} = -\sqrt{\frac{\pi}{8}}\kappa\frac{\omega^2}{2kv_0} \pm \sqrt{\frac{\pi\kappa\omega^4}{32k^2v_0^2} - \frac{\pi\beta_{3,1}^2k^4v_s^2}{\rho_m\omega_{pi}^2}}. \tag{4.131}$$

The positive sign is associated to the lattice elasticity oscillation and the negative sign is related to the charge carrier oscillation. When the piezoelectric coupling constants β_1, β_3 are large, we have

$$\delta_{1,2} = \pm\imath\sqrt{\frac{\pi}{\rho_m}\frac{k^2v_s\beta_{3,1}}{\omega_{pi}}} - \sqrt{\frac{\pi}{2}}\kappa\frac{\omega^2}{2kv_0}. \tag{4.132}$$

It is clear that, in this case, the region of oscillation spectra is extended. Taking into account the dissipation effects, we are led to a weak damping rate for both elasticity and ion-acoustic waves. However, ion-acoustic wave is damped in a weaker way than the case in which the piezo effects are absent. When β_1, β_3 are small, from Eq. (4.131), we find

$$\delta_1 = -\frac{2\pi\beta_3^2k^4v_s^2}{\rho_m\omega_{pi}^2\kappa}\frac{kv_0}{\omega^2}\sqrt{\frac{2}{\pi}}, \quad \delta_2 = -\sqrt{\pi}\frac{\kappa\omega^2}{4kv_0}. \tag{4.133}$$

Here, the damping rate of the elasticity wave, stipulated by the piezo effect, is weaker than that of plasma waves.

Now, we consider the damping rate of acoustic waves due to the collision with electrons in the non-resonant interaction of charged particles with lattice vibrations. The piezo effect does not appear only when the dispersion curves of plasma waves and elasticity waves intersect with each other. In this case, the piezo effect has resonance character and, as a result, strongly appears. In order to verify the aforementioned, we consider how the high-frequency longitudinal field affects the damping of elasticity waves due to the collisions with electrons. It is simply seen that the dispersion equation (4.122) is valid when particles' collisions are taken into account as well. In fact, when collisions are taken into account, Eq. (4.118) and the first equation of the set (4.110) are changed. However, the aforementioned method for obtaining coupled waves is not changed. Thus, Eqs. (4.121)–(4.122) completely keep their forms if particles' collisions are taken into account. Therefore, Eq. (4.125), in the high-frequency limit, when electrons collide, is written [72]

$$\left(\omega^2 - k^2v_{l,\mathrm{tr}}^2\right)\left(1 + \imath\frac{\omega_{\mathrm{pe}}^2}{\omega\nu_e}\right) = \frac{4\pi\beta_{3,1}^2k^2}{\rho_m} \tag{4.134}$$

for $k = k_\perp$ and $k = k_z$, respectively. Thus, in the acoustic frequency region $\omega \ll \omega_{\mathrm{pe}}^2/\nu_e$, we can obtain the frequency spectrum of elasticity waves and their damping rate stipulated by the electron friction ($\omega \to \omega + \imath\delta$)

$$\omega^2 \simeq k^2 v_{l,\mathrm{tr}}^2, \quad \delta_{l,\mathrm{tr}} = -\frac{2\pi\beta_{3,1}^2 k^2 \nu_e}{\rho^{(m)}\omega_{\mathrm{pe}}^2} \tag{4.135}$$

In the range of sufficiently low temperatures, this absorption is much higher than the hydrodynamic one stipulated by viscosity.

Let us investigate the possibility of acousto-elastic wave excitation in the bulk of a piezoelectric semiconductor medium in the presence of an external electric field. In this case, an electron current flows through the bulk of the medium. It is well-known that, in plasma media, when the electron beam velocity is greater than the wave's velocity, the beam electrons overtake the wave and transfer part of their energy to the wave [82–86]. Noting this fact, it should be expected that in the aforementioned piezoelectric semiconductor, such an excitation takes place when the directed velocity of charge carriers exceeds the velocity of acousto-elastic waves. Furthermore, the acousto-elastic wave's velocity is much smaller than the electron's thermal velocity. Moreover, we confine our study on the weak electric field limit where the directed velocity of electrons is assumed to be sufficiently small to avoid the acceleration of electrons and reaching to the runaway electrons domain. By taking into account the carrier drift in this case, one may consider the equilibrium distribution of electrons to be Maxwellian or Fermi distribution. Here, there is no external magnetic field and the excitation of the acoustic oscillation in piezo-semiconductors is investigated when charge carriers drift in the bulk of the medium. Since the piezo-semiconductor has hexagonal symmetry with the main symmetry axis along the z axis, the wave's propagation may take place either along ($k_\perp = 0$) or across ($k_\| = 0$) the symmetry axis. Moreover, here, it is assumed that the electron's directed velocity u is much smaller than electron's thermal velocity v_{Te}. The dispersion relation of coupled elasto-electromagnetic waves given by Eqs. (4.126), can be written as

$$\left(\omega^2 - k^2 v_l^2\right)\varepsilon\left(\omega - \vec{k}\cdot\vec{u}, \vec{k}\right) = 4\pi\frac{\beta_3^2 k^2}{\rho_m}, \tag{4.136}$$

when $k_\perp = 0$, $k = k_z$, and

$$\left(\omega^2 - k^2 v_{\mathrm{tr}}^2\right)\varepsilon\left(\omega - \vec{k}\cdot\vec{u}, \vec{k}\right) = 4\pi\frac{\beta_1^2 k^2}{\rho_m}, \tag{4.137}$$

when $k_z = 0$, $k = k_\perp$. Here, $\varepsilon\left(\omega - \vec{k}\cdot\vec{u}, \vec{k}\right)$ is the longitudinal dielectric permittivity taking into account the electric drift of charge carriers. It should be noted that the direction of the directed velocity $\vec{u}$ is arbitrary. In order to investigate the possibility of acousto-elastic wave excitation in such a system, we consider collisional electron plasma with the following dielectric permittivity for charge carriers [82]

$$\varepsilon\left(\omega - \vec{k}\cdot\vec{u}, \vec{k}\right) = 1 - \frac{\omega_{\rm pe}^2}{\left(\omega - \vec{k}\cdot\vec{u}\right)\left(\omega - \vec{k}\cdot\vec{u} + \imath\nu_e\right)} \approx 1 + \imath\frac{\omega_{\rm pe}^2}{\nu_e\left(\omega - \vec{k}\cdot\vec{u}\right)}, \tag{4.138}$$

where $\omega_{\rm pe}$ is the electron Langmuir frequency and ν_e is the electron-hole collision frequency. It should be noted that since all the involved frequencies in the piezoelectric system are smaller than ν_e, plasma is considered to be collisional. Substituting the latter equation into Eqs. (4.136) or (4.137) for $k = k_\perp$ and $k_z = 0$, or $k = k_z$ and $k_\perp = 0$, we find

$$\left(\omega^2 - k^2 v_{{\rm tr},l}^2\right)\left[1 + \imath\frac{\omega_{\rm pe}^2}{\nu_e\left(\omega - \vec{k}\cdot\vec{u}\right)}\right] = \frac{4\pi\beta_{1,3}^2 k^2}{\rho_m}. \tag{4.139}$$

Assuming the piezo effect to be small, i.e., $\beta_{1,3}^2/\rho_m \ll v_{{\rm tr},l}^2$, we find the frequency spectrum of the acousto-elastic wave. In this way, we consider $\omega \to \omega + \imath\delta$ which is the common method for finding the decrement or growth rate of the instability in plasma media, depending on the sign of δ. Hence, we have

$$\delta = -\frac{2\pi\beta_1^2 k^2 \nu_e}{\rho_m \omega_{\rm pe}^2}\left(1 - \frac{u\cos\theta}{v_{{\rm tr},l}}\right). \tag{4.140}$$

It is clear that under the condition

$$u > \frac{v_{{\rm tr},l}}{\cos\theta}, \tag{4.141}$$

the sign of dissipation defined by Eq. (4.140) is changed. This means that acousto-elastic wave grows due to its excitation by the electric drift of carriers. The nature of the aforementioned instability is determined by expressions (4.140) and (4.141). It is attributed to the Cherenkov radiation of the electrons during their scattering in the piezo-semiconductor. It should be noted that Eqs. (4.136)–(4.140) and their consequences are valid for the both strongly and weakly ionized plasmas. Furthermore, charge carriers are assumed to be non-degenerate or degenerate. In each case, it is only sufficient to choose the relevant collision frequency ν_e.

It has been shown earlier that an electron beam can excite the acousto-elastic oscillations in piezo-semiconductors [80]. In piezo-semiconductors, it is better to excite acousto-elastic waves by using the charge carrier current in the sample itself. However, in a piezo-dielectric, due to the absence of charge carriers in the sample the excitation of acousto-elastic waves is possible only by external sources. Moreover, since acousto-elastic waves have small phase velocity, their excitation will be more effective by making use of ion beams. Besides, it is well-known that by Cherenkov mechanism, it is possible to excite slow waves, such as ion-acoustic waves, Alfven waves, helicon-waves or coupled elasto-electromagnetic waves of semiconductors in plasma-like media. Their phase velocity is smaller than the thermal velocity of electrons and is greater than the thermal velocity of ions. Therefore, it is clear that the excitation of such waves by fast electron beams is weakly effective because of the large difference between beam velocity and wave's velocity. However, the excitation of such waves by ion beams is more effective. Here, we investigate the stimulated Cherenkov radiation of slow waves by injecting the relatively fast ion beams, not reaching the relativistic regime, on the surface of a semi-bounded piezo-dielectric. We indicate how the ion beam can excite the surface acousto-elastic oscillation in semi-bounded piezo-dielectric media. Moreover, we obtain the growing increment of such waves. For this goal, we consider a semi-bounded piezo-dielectric with hexagonal symmetry with the main symmetry axis being along the z axis. Hence, the elastic properties of the piezoelectric crystal in the plane perpendicular to the hexagonal axis are isotropic and the crystal can be considered as an isotropic medium. The ion beam flowing on the surface of such a piezoelectric can excite the surface elasticity waves due to the piezoelectric effect. On the other hand, the piezoelectric effect transfers the beam energy into the surface oscillation on the piezoelectric surface. It seems that such excitation can be used as a method for beam diagnostic and it should also be considered as a dissipative mechanism in the piezoelectric waveguide in the presence of the ion beam.

We assume that a semi-bounded piezo-dielectric has hexagonal symmetry with the main symmetry axis along the z axis. By this assumption, the piezoelectric medium can approximately be considered as an isotropic medium. Moreover, we consider that the piezo-dielectric surface is placed on the plane $x = 0$ such that the piezo-dielectric occupies the region $x > 0$. In this case, we again use Eqs. (4.123) and (4.124). Moreover, in the region $x < 0$, parallel to the piezo-dielectric surface, a monoenergetic ion beam with the dielectric permittivity

$$\varepsilon_{\text{bi}} = 1 - \frac{\omega_{\text{bi}}^2}{\left(\omega - k_y u\right)^2}\,, \tag{4.142}$$

flows along the y axis. Here, u is the ion beam velocity, and $\omega_{\text{bi}} = 4\pi n_{\text{bi}} e^2/m_i$ where n_{bi} and m_i are the ion beam density and ion mass, respectively. It should be noted that there is no gap between ion beam and the piezo-dielectric surface. Later, we will find the dispersion relation of the surface wave excited on the piezo-dielectric surface and the frequency spectrum of such a system.

For simplicity, we assume that $k_z = 0$, without losing generality. It means that we restrict our consideration to the waves propagating perpendicular to the main symmetry axis (z). The formulation of this problem is the same as done for the propagation of electromagnetic surface waves on plasma-like media in [72]. However, comparing these two cases, it is easily seen that in the present case the plane $x = 0$ is the interface of two media and not the interface of vacuum. It should be noted that in the equilibrium, the force acting on a surface element in each medium should be balanced by the force of the internal stresses acting on that element. So, the stress tensors on the boundary interface of two piezoelectric media will be equal to zero. The general formalism to find the dispersion equation in such cases is to solve the field equations in two media and then to equate the surface impedances of two media [80, 87]. Therefore, by generalizing the old case for two media, we can find the dispersion relation by matching the surface impedances of two media (see Sect. 2.10.1)

$$\int_0^\infty \frac{dk_x}{k^2\varepsilon^{(0)}(\omega,k) - \frac{4\pi k^2 \beta_1^{(0)2}}{\omega^2\rho_m^{(0)} - \lambda^{\mathrm{tr}\,(0)}k^2}} + \int_0^\infty \frac{dk_x}{k^2\varepsilon^{(1)}(\omega,k) - \frac{4\pi k^2 \beta_1^{(1)2}}{\omega^2\rho_m^{(1)} - \lambda^{\mathrm{tr}\,(1)}k^2}} = 0\ , \tag{4.143}$$

where $\varepsilon(\omega, k)$ and ρ_m are the dielectric permittivity and lattice density, respectively. Moreover, indices (0) and (1) indicate the value of the considered quantity in the left and right side of the surface boundary. Actually, the above equation is the aforementioned dispersion relation for the spectrum of the surface waves propagating along the interface of two piezoelectric media. In our problem, an ion beam with the dielectric permittivity $\varepsilon_{\mathrm{bi}}$ [Eq. (4.142)] and the piezoelectric constant $\beta_1^{(0)} = 0$ flows on the left-hand side of the boundary surface. Moreover, there is a piezo-dielectric with the dielectric permittivity equal to zero (for simplicity) and the piezoelectric constant $\beta^{(1)} = \beta_1$ on the right-hand side of the boundary surface. After replacing these values, the dispersion equation will be

$$\int_0^\infty \frac{dk_x}{k^2\varepsilon_{\mathrm{bi}}} + \int_0^\infty \frac{dk_x}{k^2 - \frac{4\pi k^2 \beta_1^2}{\omega^2\rho_m^{(1)} - \lambda^{\mathrm{tr}\,(1)}k^2}} = 0, \tag{4.144}$$

It should be noted that the last term in the denominators of Eq. (4.143) corresponds to the piezo effect. Hence, in the absence of the piezo effect ($\beta_1^{(0)}, \beta_1^{(1)} \to 0$), the aforementioned dispersion equation reduces to the dispersion equation for the surface wave excited on the surface of the dielectric by the ion beam flowing through the surface. Assuming the piezo effect to be weak, i.e., $4\pi\beta_1^2/\left(\rho_m v_{\mathrm{tr}}^2\right) \ll 1$ and omitting index (1) for simplicity, from Eq. (4.144), we find

$$\left(\omega^2 - k_y^2 v_{\mathrm{tr}}^2\right)(1+\varepsilon_{\mathrm{bi}})^2 = \left(\frac{4\pi\beta_1^2}{\rho_m v_{\mathrm{tr}}^2}\varepsilon_{\mathrm{bi}}\right)^2 k_y^2 v_{\mathrm{tr}}^2. \tag{4.145}$$

The latter equation has a solution, which corresponds to a growing in time surface wave. Such a solution exists when

$$\omega = k_y v_{tr} + \delta = k_y u + \frac{\omega_{bi}}{\sqrt{2}} + \delta. \tag{4.146}$$

Considering $\omega \to \omega + \delta$ with the condition $|\delta/\omega| \ll 1$ as a solution for the dispersion equation is a typical method to find the growth or damping rate of the excited oscillation. The plus or minus sign of the imaginary part of δ shows the increment or decrement of the corresponding frequency. Here, it should be noted that δ should be much less than ω in order to find the propagating waves in the system. Otherwise, there are only evanescent waves. Thus, the growth increment obtained from Eq. (4.145) is

$$\frac{\delta}{\omega} = \frac{-1 + \imath\sqrt{3}}{8} \left(\frac{4\pi\beta_1^2}{\rho_m v_{tr}^2} \right)^{2/3} \left(\frac{\omega_{bi}^2}{\omega^2} \right)^{1/3}. \tag{4.147}$$

Regarding the positive imaginary part of the damping decrement obtained in Eq. (4.147), the surface wave can be excited in the interface of piezo-dielectric and the ion beam. From Eq. (4.146), it is clear that such an instability has a Raman-type character which means a collective behavior can be created in the medium. In such instability development, the beam energy is partially transformed into the elasticity energy due to the piezo effect. Hence, the surface elasticity waves will grow in time. In other words, the surface wave is excited on the interface due to the piezoelectric effect.

4.5.2 *Excitation of Coupled Quasi Elasto-Electromagnetic Surface Waves*

Because of the analysis of the coupling between elasticity and electromagnetic characteristics of plasma-like media, surface waves cannot be purely electromagnetic.

At the boundary of a piezo-semiconductor sample, the electromagnetic field of surface wave penetrates into the sample. This causes the elements of the crystal lattice move and surface waves become elasto-electromagnetic waves.

It is clear that the analogous phenomena take place in non-piezo-semiconductors, if one takes into account the field effect of the surface waves on the deformation of the crystal lattice through potential deformation. This effect, however, is very small and we do not consider it.

The dispersion of elasto-electromagnetic surface waves in plasma media is very complicated, especially when the thermal effect of charge carries is present and it is needed to make use of the kinetic equation. Therefore, hereafter, we restrict our consideration only on the simple case, i.e., an isotropic plasma medium in the absence of external fields. In addition, the electromagnetic fields of surface waves are assumed to be potential, i.e., $\vec{E} = -\vec{\nabla}\phi$, $\vec{B}_0 = 0$. Thus, we refer to these waves as quasi elasto-electromagnetic waves. The equation of elasticity for lattice vibrations can be written as [compare to Eq. (4.113)]

$$\rho_m \frac{\partial^2 u_i}{\partial t^2} = \frac{\partial}{\partial x_j} \Xi_{ij}, \tag{4.148}$$

where

$$\Xi_{ij}\left(\vec{x}\right) = \lambda_{ijlm}\left(\vec{x}\right) U_{lm} - \beta_{lij} \frac{\partial \phi}{\partial x_l}, \tag{4.149}$$

Here, Ξ_{ij} and U_{lm} are the stress and strain tensors, respectively; ρ_m is the lattice density, λ_{iklm} is the elastic modulus tensor determining the elasticity force in the deformed lattice, u_i is the ith component of the displacement of the lattice element with respect to the equilibrium state, and β_{ilk} is the piezoelectric tensor of the medium [38]. Furthermore, the field equations reduce to the Poisson's equation (4.170)

$$\frac{\partial}{\partial x_i}\left[\frac{\partial \phi}{\partial x_i} + 4\pi \beta_{ikl}\left(\vec{x}\right) U_{kl}\right] = -4\pi \rho\left(\vec{x}\right), \tag{4.150}$$

where $\rho(x)$ is the charge carrier density, which is determined by

$$\rho(x) = \sum_\alpha e_\alpha \int \delta f_\alpha d\vec{p}, \tag{4.151}$$

and $\delta f_\alpha(x)$, in the linear approximation, can be found by the Vlasov equation

$$\frac{\partial \delta f_\alpha}{\partial t} + \left(\vec{v} \cdot \nabla\right) \delta f_\alpha + \frac{e_\alpha}{m_\alpha} \frac{\partial \phi}{\partial \vec{r}} \cdot \frac{\partial f_{0\alpha}}{\partial \vec{v}} = 0, \tag{4.152}$$

where δf_α is the perturbation of the equilibrium distribution function $f_{0\alpha}$ for α species of charge carries. It is clear that the surface waves essentially depend on the boundary conditions. So, we should complete the obtained equations by the boundary conditions. In the equilibrium, in view of the balance between the force acting on a surface element and the force of the internal stresses acting on it, we can obtain the boundary condition on the surface of the piezo-semiconductor:

$$\Xi_{xj}|_{x=0} = 0. \tag{4.153}$$

Here, it is assumed that the z axis is along the surface medium and the x axis is across it.

To determine the boundary conditions for the Vlasov equation, we make use of the mirror reflection model

$$\delta f_\alpha(0, v_x > 0)\big|_{x=0} = \delta f_\alpha(0, v_x < 0)\big|_{x=0}. \tag{4.154}$$

which means that charge carriers have mirror reflection from the surface of the piezo-semiconductor [87]. In other words, such a condition contains all information about the surface characteristics, which determine the interaction of the charged particles with the surface confining solid-state plasma.

As already mentioned, for simplicity we assume that the piezo-crystal has hexagonal symmetry, its main symmetry axis is parallel to the z axis, and tensor λ_{ijlm} is approximated by an isotropic tensor. Moreover, we restrict our consideration only on the waves propagating perpendicular to the main symmetry axis and assume that all perturbed quantities have the form $A(x)$ exp $(-\imath\omega t + k_y y)$. In this case, Eq. (4.148) for the component u_z is written as

$$-\omega^2\rho_m u_z = \frac{\partial}{\partial x}\lambda^{\text{tr}}\frac{\partial u_z}{\partial x} - k_y^2\lambda^{\text{tr}}u_z - \frac{\partial \beta_1}{\partial x}\frac{\partial \phi}{\partial x} - k_y^2\beta_1\phi. \tag{4.155}$$

The equations for the components u_x and u_y are split off and do not contain any electromagnetic fields, i.e., they describe only elastic crystal lattice vibration and, as a result, we do not study them.From Eqs. (4.153) and (4.155), we obtain the following boundary condition:

$$\left(\lambda^{\text{tr}}\frac{\partial u_z}{\partial x} - \beta_1\frac{\partial \phi}{\partial x}\right)|_{x=0} = 0. \tag{4.156}$$

Furthermore, Poisson's Eq. (4.150) for the considered waves finds the form

$$\frac{\partial^2\phi}{\partial x^2} - k_y^2\phi - 4\pi k_y^2\beta_1 u_z + 4\pi\frac{\partial \beta_1}{\partial x}\frac{\partial u_z}{\partial x} + 4\pi\rho(x) = 0, \tag{4.157}$$

where $\rho(x)$ is the induced charge carrier density, which is obtained from Eq. (4.151) by making use of the solution of the Vlasov Eq. (4.152) with boundary condition (4.154). Finally, the dispersion equation for the longitudinal surface wave on the interface of a piezo-semiconductor with vacuum may be written as [72, 79, 87]

$$1 + \frac{2}{\pi}\int_0^{+\infty} dk_x \frac{|\, k_y \,|}{k^2\varepsilon^l(\omega, k) - \frac{4\pi k^4\beta_1^2}{\omega^2\rho_m - \lambda^{\text{tr}}k^2}} = 0, \tag{4.158}$$

where $\varepsilon^l\left(\omega, \vec{k}\right) = \varepsilon_{ij}\left(\omega, \vec{k}\right)k_i k_j/k^2$ is the longitudinal dielectric permittivity of the plasma-like medium.

In the absence of the piezo effect, i.e., in the limit $\beta \to 0$, Eq. (4.158) reduces to the dispersion equation for the longitudinal wave on a semi-bounded isotropic plasma medium. From Eqs. (4.148) and (4.149), one can directly show that in the considered geometry, purely elastic surface waves cannot exist. Therefore, the appearance of elasto-electrostatic surface waves described by Eq. (4.158), in fact, is the appearance of surface electrostatic waves on the piezo-semiconductor.

Due to the piezo-coupling of the electromagnetic and elastic properties, electromagnetic surface waves stimulate elastic crystal lattice vibrations which are damped in the bulk of the piezo-semiconductor medium, as happened for the electromagnetic waves.

To study the aforementioned effect, we consider high-frequency surface waves. In this case, neglecting spatial dispersion, from Eq. (4.158), we find

$$\omega^2 \rho_m = k_y^2 \left(\lambda^{\mathrm{tr}} + \frac{4\pi\beta_1^2}{\varepsilon(\omega)}\right) \left\{ 1 - \left(\frac{4\pi\beta_1^2}{[1+\varepsilon(\omega)]\left[\lambda^{\mathrm{tr}}\varepsilon(\omega) + 4\pi\beta_1^2\right]} \right)^2 \right\}. \tag{4.159}$$

When we have a piezo-dielectric and $\varepsilon(\omega) = \text{const}$, Eq. (4.159) describes the dispersion of the so-called Gulyaev-Blustein surface waves [38]. It should be noted that this wave has widespread applications in acousto-electronics as an effective tool for producing a periodic structure on the surface of piezo-dielectrics.

We note that Eq. (4.159) could be derived not from the Vlasov Eq. (4.152). Starting with Eqs. (4.155) and (4.157),

$$\frac{\partial}{\partial x_i}\left[\varepsilon(\omega)\frac{\partial \phi}{\partial x_i} + 4\pi\beta_{ikl}\left(\vec{x}\right)U_{kl}\right] = 0. \tag{4.160}$$

and making use of boundary condition (4.156) and continuity equation for potential ϕ on the surface $x = 0$, we find

$$\left(\varepsilon(\omega)\frac{\partial \phi}{\partial x} + 4\pi\beta_1 \frac{\partial u_z}{\partial x}\right)\Big|_{x=0} = 0, \tag{4.161}$$

which is obtained from Eq. (4.157) by integration over an infinitely small surface layer near $x = 0$.

The solution of these equations inside and outside of the piezo-semiconductor along with their matching on the boundary conditions leads to the dispersion equation (4.159).

As already mentioned, the considered surface wave is a quasi-longitudinal elastic wave because when $\beta_1 \to 0$, i.e., in the absence of the piezo effect, this wave does not exist and it is transformed into the bulk elastic waves. The existence of the piezo effect slows down the velocity of bulk waves and stipulates the surface elastic waves. Beside deceleration, the piezo effect leads to a qualitatively new effect, i.e., plasma damping of elastic oscillations. This is formally an evidence for the presence of an imaginary part in $\varepsilon(\omega)$ which has the form [82]

$$\varepsilon(\omega) = 1 - \frac{\omega_{\mathrm{pe}}^2}{\omega^2} + \imath\frac{\omega_{\mathrm{pe}}^2 \nu_e}{\omega^3}. \tag{4.162}$$

in the high-frequency region. Here, ν_e is the electric collision frequency and, as a result, the dissipation of acoustic waves is brought about by the friction of electrons in piezo-semiconductors.

We now pay close attention to the special behavior of the coupled quasi elasto-electromagnetic wave near the eigen frequency of electromagnetic surface oscillations [72, 87]:

$$\varepsilon(\omega) + 1 = 0, \quad \omega = \omega_l. \tag{4.163}$$

In this frequency region, taking into account a small piezo effect, Eq. (4.159) can be written as

$$\left(\omega^2 - k_y^2 v_{\rm tr}^2\right)[\varepsilon(\omega) + 1]^2 = -k_y^2 v_{\rm tr}^2 \left(\frac{4\pi\beta_1^2}{\rho_m v_{\rm tr}^2}\right)^2. \tag{4.164}$$

The solution of this equation has the form of

$$\omega = k_y v_{\rm tr} + \delta = \frac{\omega_{\rm pe}}{\sqrt{2}} + \delta. \tag{4.165}$$

Therefore,

$$2\delta^3 \left(\frac{\partial \varepsilon}{\partial \omega}\right)^2 = -\omega \left(\frac{4\pi\beta_1^2}{\rho_m v_{\rm tr}^2}\right)^2. \tag{4.166}$$

This equation has three solutions

$$\delta_1 = -\left(\frac{\omega}{2}\right)^{\frac{1}{3}} \left(\frac{4\pi\beta_1^2}{\rho_m v_{\rm tr}^2 \frac{\partial \varepsilon}{\partial \omega}}\right)^{\frac{2}{3}},$$

$$\delta_{2,3} = \frac{1 \pm \imath\sqrt{3}}{2} \left(\frac{\omega}{2}\right)^{\frac{1}{3}} \left(\frac{4\pi\beta_1^2}{\rho_m v_{\rm tr}^2 \frac{\partial \varepsilon}{\partial \omega}}\right)^{\frac{2}{3}}. \tag{4.167}$$

However, the first solution deals with the localization of the surface wave whose spatial damping is

$$L^{-1} = \frac{4\pi\beta_1^2}{\rho_m v_{\rm tr}^2} \frac{1}{v_{\rm tr} \frac{\partial \varepsilon}{\partial \omega}}. \tag{4.168}$$

This expression is positive when $\partial\varepsilon(\omega)/\partial\omega > 0$. In fact, such a situation occurs in a medium under thermal equilibrium which is described by $\varepsilon(\omega)$ given by Eq. (4.162).

Finally, it should be noted that all results obtained above are valid for degenerate plasmas with corresponding substitutions.

As a conclusion, in this part, at the end of the present section by making use of the mirror reflection model when charge carriers have mirror reflection from the surface of the piezo-semiconductor, we studied the excitation of surface quasi elasto-electromagnetic waves. We found that beside deceleration, the piezo effect leads to a qualitatively new effect, i.e., plasma damping of elastic oscillations. In the ion-acoustic frequency region, the coupling of ion-acoustic and elastic waves was investigated.

4.6 Problems

4.6.1 Starting from the Schrodinger equation for an electron without spin in the field of an electromagnetic wave, drive quantum hydrodynamic equation for a collisionless electron gas and on its basis obtain an expression for dielectric permittivity.

Solution Here, Schrodinger equation is of the form

$$\imath\hbar\frac{\partial\psi}{\partial t}=\widehat{H}\psi=\left[-\frac{\hbar^2}{2m}\nabla^2+\imath\hbar\frac{e}{mc}\vec{A}\cdot\nabla+\frac{e^2}{2mc^2}\vec{A}^2+e\varphi\right]\psi, \tag{4.169}$$

where, $\vec{A}$ and φ are vector and scalar potentials of fields $\vec{E}$ and $\vec{B}$:

$$\vec{E}=-\frac{1}{c}\frac{\partial\vec{A}}{\partial t}-\nabla\varphi,\quad \vec{B}=\left[\nabla\times\vec{A}\right],\quad \left(\nabla\cdot\vec{A}\right)=0. \tag{4.170}$$

Presenting wave function as

$$\psi=a\left(\vec{r},t\right)\exp\left[\frac{\imath}{\hbar}S\left(\vec{r},t\right)\right], \tag{4.171}$$

and making use of charge and current densities

$$\begin{gathered}\rho=en=e|\psi|^2=ea^2,\\ \vec{j}=en\vec{V}=\frac{\imath e\hbar}{2m}(\psi\nabla\psi*-\psi*\nabla\psi)-\frac{e^2}{mc}\vec{A}\psi\psi*=\frac{ea^2}{m}\left(\nabla S-\frac{e}{c}\vec{A}\right),\end{gathered} \tag{4.172}$$

from Eq. (4.169), we find

$$\frac{\partial n}{\partial t} + \nabla \cdot \left(n\vec{V} \right) = 0,$$
$$\frac{\partial \vec{V}}{\partial t} + \left(\vec{V} \cdot \nabla \right) \vec{V} = \frac{e}{m} \left\{ \vec{E} + \frac{1}{c} \left[\vec{V} \times \vec{B} \right] \right\} + \frac{\hbar^2}{4m^2} \nabla \left\{ \frac{1}{n} \left[\nabla^2 n - \frac{1}{2n} (\nabla n)^2 \right] \right\}. \tag{4.173}$$

The first equation coincides with the continuity equation, and the second equation is Euler equation. Therefore, in analogy with the hydrodynamic equation, we will call the equation system (4.173) as quantum hydrodynamic equations of cold plasma.

Equations (4.173) are different from hydrodynamic equations by the presence of the quantum force in the Euler equation due to the uncertainty principle of Heisenberg. This is easily seen by considering small perturbations of the homogeneous state with $n = \text{const}$ and $\vec{V} = 0$. In the limit $n \to 0$, when it is possible to neglect the self-consistent fields $\vec{E}$ and $\vec{B}$, from linearized Eq. (4.173) for the solution of the form $\exp\left(\imath\omega t + \imath\vec{k} \cdot \vec{r} \right)$, we find the following dispersion equation

$$\omega = \frac{\hbar k^2}{2m} \equiv \omega_q, \tag{4.174}$$

which describes the oscillations of one electron. This relation connects temporal (proportional to $1/\omega$) and spatial (proportional to $1/k$) regions of localization of a free electron, or energy $\hbar\omega$ and momentum $\hbar\vec{k}$ to each other. Quantity (4.174) is the frequency of the quantum oscillations of a free electron.

Now we proceed to find dielectric permittivity of a quantum electron gas. To do this, first we find dielectric permittivity of particles with density n and velocity $\vec{V}$ based on Eq. (4.173). Linearizing this equation system with respect to the perturbation of the form $\exp\left(-\imath\omega t + \imath\vec{k} \cdot \vec{r} \right)$, in the absence of the external magnetic field, ($\vec{B}_0 = 0$), we find

$$\varepsilon_{ij}^{a}\left(\omega, \vec{k} \right) = \varepsilon_{ij}^{cl}\left(\omega, \vec{k} \right) - \frac{\omega_q^2}{\omega_{\text{Le}}^2} \delta\varepsilon_{i\mu}^{cl} \frac{k_\mu k_\nu}{k^2} \delta\varepsilon_{\nu j}^{cl}\left(\omega, \vec{k} \right) \left[1 + \frac{\omega_q^2}{\omega_{\text{Le}}^2} \frac{k_\mu k_\nu}{k^2} \delta\varepsilon_{\mu\nu}^{cl} \right]^{-1}. \tag{4.175}$$

Here,

$$\varepsilon_{ij}^{cl} = \delta_{ij} + \delta\varepsilon_{ij}^{cl}\left(\omega, \vec{k} \right) = \delta_{ij} - \frac{\omega_{\text{pe}}^2}{\omega^2} \left(\delta_{ij} + \frac{k^2 V_i V_j}{\left(\omega - \vec{k} \cdot \vec{V} \right)^2} + \frac{k_i V_j + V_i k_j}{\omega - \vec{k} \cdot \vec{V}} \right). \tag{4.176}$$

In order to take into account the velocity distribution, we should pass to the kinetic consideration by the following replacement

$$n \to \int d\vec{p} f_0\left(\vec{p}\right)\ldots, \tag{4.177}$$

where $f_0\left(\vec{p}\right)$ is the equilibrium distribution function normalized to density. As a result, from expressions (4.175) and (4.176), we find

$$\begin{aligned}\varepsilon^l\left(\omega,\vec{k}\right) &= 1 + \frac{4\pi e^2}{\hbar k^2}\int \frac{d\vec{p}}{\omega - \vec{k}\cdot\vec{V}}\left[f_0\left(\vec{p}+\frac{\hbar\vec{k}}{2}\right) - f_0\left(\vec{p}-\frac{\hbar\vec{k}}{2}\right)\right],\\ \varepsilon^{\mathrm{tr}}\left(\omega,\vec{k}\right) &= 1 - \frac{\omega_{\mathrm{pe}}^2}{\omega^2} + \frac{2\pi e^2}{\hbar\omega^2}\int \frac{d\vec{p}}{\omega - \vec{k}\cdot\vec{V}}\vec{V}_\perp^2\left[f_0\left(\vec{p}+\frac{\hbar\vec{k}}{2}\right) - f_0\left(\vec{p}-\frac{\hbar\vec{k}}{2}\right)\right].\end{aligned} \tag{4.178}$$

These expressions can coincide with expression (4.13) and (4.14) obtained by solving the quantum kinetic equation.

4.6.2 Show that the magnetic moment of a system of charged particles, which are moving in a magnetic field according to the *laws of classical mechanics* is zero in the stationary state (the *Bohr–Van Leeuwen theorem*). For this purpose, write down the energy of the system, averaged over Gibbs ensemble, in the presence and in the absence of the magnetic field and show that the energy is independent of the external field [88].

Solution Consider a system of particles whose Hamiltonian function in the absence of any external magnetic field is given by

$$H_0 = \sum_a \frac{p_a^2}{2m_a} + U, \tag{4.179}$$

where the potential energy U is the function of coordinates. In the presence of a magnetic field, the Hamiltonian function takes the form

$$H = \sum_a \frac{1}{2m_a}\left(\vec{P}_a - \frac{e_a}{c}\vec{A}_a\right)^2 + U, \tag{4.180}$$

where $\vec{P}_a$ is the generalized momentum of the a-th particle, and $\vec{A}_a$ is the vector potential of the external field at the point where the particle is localized. The energy of the system averaged over the Gibbs ensemble (its internal energy in the thermodynamic approach) is expressed as an integral over the phase space

$$\overline{E} = \frac{1}{Z}\int H \exp\left(-\frac{H}{T}\right) d\Gamma, \qquad (4.181)$$

where $d\Gamma = \prod_i dP_i dx_i$ is an element of phase space. Let us change momentum variables in Eq. (4.181)

$$\vec{P}_a - \frac{e_a}{c}\vec{A}_a \to \vec{p}_a, \qquad (4.182)$$

keeping the same coordinates. With such a replacement, H becomes H_0, and $d\Gamma = \prod_i dP_i dx_i$ because the Jacobian of transition to the new variables equals 1. As a result, the internal energy of the system in a magnetic field is expressed just as it is expressed in the absence of the field, that is, the energy is field-independent. A body that does not possess the magnetic moment in the absence of the field will not acquire it in the presence of the field. However, such a result holds only in the classical case and fails when particles move according to quantum mechanical laws.

4.6.3 A rarefied electron gas at temperature T is in a weak uniform magnetic field and obeys the Maxwell–Boltzmann statistics. Calculate the magnetic susceptibility of the electron gas and separate its part, which is due to the orientation of spin magnetic moments and the contribution associated with the effect of the magnetic field on the orbital motion of particles. Make use of the quantum mechanical expression for the electron energy in the magnetic field [88].

Hint An electron in a homogeneous magnetic field has the energy

$$E_n = \left(n + \frac{1}{2}\right)\hbar\omega_B + \frac{p_z^2}{2m} - \mu_B m_s B,$$

Here, $\omega_B = |e|B/mc$ is the cyclotron frequency, $n = 0, 1\ldots, m_s = \pm 1/2$. The values of energy degenerate due to the position uncertainty of Larmor circle. The number of quantum states in volume V per dp_z interval is

$$dQ = \frac{eBV}{(2\pi\hbar)^2 c} dp_z.$$

Solution In order to calculate the magnetization, we use the formulas of statistical physics for the free energy: $F = -T\ln z$ and $M = -\partial F/\partial B$. Since the gas is rarefied and homogeneous, we ignore the electron–electron interaction and calculate the free energy per unit volume from the formula $F = -NT\ln z$, where z is the statistical sum of an individual electron regarded as a quasi-independent equilibrium subsystem. With the aid of the data given in the statement of the problem, we find

$$z = z_s z_{\text{orb}}, \qquad z_s = \sum_{m_s=-1/2}^{m_s=1/2} \exp\left(-\frac{2\mu_B B m_s}{T}\right) = 2\cosh\left(\frac{\mu_B B}{T}\right), \tag{4.183}$$

$$\begin{aligned} z_{\text{orb}} &= \sum_{n=0}^{\infty} \exp\left(-\frac{\hbar\omega_B}{2T}(2n+1)\right) \int \exp\left(-\frac{p_z^2}{2mT}\right) \frac{eBV\,dp_z}{(2\pi\hbar)^2 c} = \\ &= \sqrt{2\pi m T}\,\frac{eBV}{(2\pi\hbar)^2 c}\,\frac{1}{2\sinh(\hbar\omega_B/2T)}, \end{aligned} \tag{4.184}$$

where z_s relates to the spin states and z_{orb} to the states of orbital motion. Further on, we take into consideration that $\hbar\omega_B/2 = \mu_B B$ and $\mu_B B \ll T$. Carrying out the small-argument expansion of the hyperbolic functions, we find

$$M = NT\left(\frac{\partial z_s}{\partial B} + \frac{\partial z_{\text{orb}}}{\partial B}\right) = \frac{N\mu_B^2 B}{T} - \frac{N\mu_B^2 B}{3T}\,. \tag{4.185}$$

The first and second terms in the right-hand side describe the paramagnetic and diamagnetic effects, respectively. Now the corresponding susceptibilities depend on temperature; however, the relation between them remains the same as for the degenerate gas:

$$\chi_{\text{para}} = \frac{N\mu_B^2}{T}, \qquad \chi_{\text{dia}} = -\frac{1}{3}\chi_{\text{para}}. \tag{4.186}$$

4.6.4 Study magnetic permeability of non-degenerate and degenerate isotropic equilibrium electron gases in different frequency ranges.

Solution Static magnetic permeability of every classical equilibrium medium is equal unity, i.e., an equilibrium medium in the classical limit has no magnetic permeability. However, dynamic (i.e., $\omega \neq 0$) magnetic permeability of an equilibrium medium can be different from unity. This can be seen, for example, for the classical equilibrium electron gas.

1. Non-degenerate electron gas. We begin with a collisionless non-degenerate electron gas.

   Making use of Eq. (2.63), in the general case, we find

$$1 - \frac{1}{\mu(\omega,k)} = -\frac{\omega^2}{k^2c^2}\left[\frac{\omega_{\text{pe}}^2}{\omega^2} I_+\left(\frac{\omega}{kv_{\text{Te}}}\right) + \frac{\omega_{\text{pe}}^2}{k^2 v_{\text{Te}}^2}\left(1 - I_+\left(\frac{\omega}{kv_{\text{Te}}}\right)\right)\right]. \tag{4.187}$$

In the limit $\omega^2 \ll k^2c^2$, we can assume magnetic permeability close to unity and relation (4.187) may be represented in the form of

$$\mu(\omega, k) = 1 - \frac{\omega^2}{k^2c^2}\frac{\omega_{pe}^2}{k^2v_{Te}^2}\left[1 - \left(1 - \frac{k^2v_{Te}^2}{\omega^2}\right)I_+\left(\frac{\omega}{kv_{Te}}\right)\right]. \tag{4.188}$$

In the high-frequency range, $\omega \gg kv_{Te}$, from Eq. (4.188) we find

$$\mu(\omega, k) = 1 + \frac{2\omega_{pe}^2 v_{Te}^2}{\omega^2 c^2} - \imath\sqrt{\frac{\pi}{2}}\frac{\omega_{pe}^2\omega^3}{k^5c^2v_{Te}^3}\exp\left(-\omega^2/2k^2v_{Te}^2\right) \tag{4.189}$$

It follows that an electron gas in the high-frequency range is weakly paramagnetic, and the imaginary part of permeability is exponentially small.

In the low-frequency range, $\omega \ll kv_{Te}$, from Eq. (4.188) we find

$$\mu(\omega, k) = 1 - \frac{2\omega_{pe}^2\omega^2}{k^4c^2v_{Te}^2} + \imath\sqrt{\frac{\pi}{2}}\frac{\omega_{pe}^2\omega}{k^3c^2v_{Te}} \tag{4.190}$$

It follows that a non-degenerate electron gas in the low-frequency range is weakly diamagnetic, and the imaginary part of permeability is a positive value greatly exceeding the diamagnetic correction to the real part of magnetic permeability.

2. Degenerate electron gas. The general relation in all frequency ranges is obtained by making use of Eqs. (2.63), (4.13) and (4.14):

$$1 - \frac{1}{\mu(\omega, k)} = -\frac{\omega^2}{k^2c^2}\left[\frac{3\omega_{pe}^2}{2\omega^2}\left(2 + \frac{\omega^2}{k^2v_{Fe}^2}\right) + \frac{3\omega_{pe}^2\omega}{4k^3v_{Fe}^3}\left(1 + \frac{\omega^2}{k^2v_{Fe}^2}\right)\ln\frac{\omega + kv_{Fe}}{\omega - kv_{Fe}}\right] \tag{4.191}$$

As above, the correction is assumed small. Then, in the low-frequency limit, $\omega \gg kv_{Fe}$, the imaginary part is equal to zero, $\mathrm{Im}\mu = 0$, while the real part is

$$\mu(\omega, k) = 1 - \frac{3\omega_{pe}^2\omega^4}{2k^6c^2v_{Fe}^4}. \tag{4.192}$$

This means that the degenerate electron gas is weakly diamagnetic at the high-frequency range.

In the low-frequency range, $\omega \ll kv_{Fe}$, from Eq. (4.191), we obtain

$$\mu(\omega, k) = 1 - \frac{3\omega_{pe}^2}{k^2c^2} + \imath\pi\frac{\omega_{pe}^2\omega^3}{k^5c^2v_{Fe}^3} \tag{4.193}$$

It is evident that the degenerate electron gas is weakly diamagnetic in the low-frequency range, and the imaginary part of magnetic permeability is positive.

4.6.5 In a semiconductor with $m_+ \gg m_-$ in which collisionless sound may be present find the interaction potential of two electrons which is stipulated by the sound waves exchange.

Solution For degenerate light carriers (electrons), we have

$$v_{\mathrm{Fe}} = \sqrt{\frac{\epsilon_{\mathrm{Fe}}}{m_e}} \gg v_s \simeq \sqrt{\frac{\epsilon_{\mathrm{Fe}}}{m_+}} \gg \begin{cases} \sqrt{T_i/m_+} \ \ for \ \ T_+ > \epsilon_{F+}, \\ \sqrt{\epsilon_{F+}/m_+} \ \ for \ \ \epsilon_{A+} \gg T_+. \end{cases} \tag{4.194}$$

Sound waves are described by the dispersion equation

$$\varepsilon^l(\omega, k) = 1 + \frac{3\omega_{\mathrm{pe}}^2}{k^2 v_{\mathrm{Fe}}^2}\left(1 + \imath \frac{\pi\omega}{2kv_{\mathrm{Fe}}}\right) - \frac{\omega_{p+}^2}{\omega^2} = 0. \tag{4.195}$$

The spectrum of eigen frequencies of this equation describes the weakly damped sound waves ($\omega \to \omega + \imath\delta$):

$$\omega^2 = \frac{\omega_{p+}^2}{1 + 3\omega_{\mathrm{pe}}^2/k^2 v_{\mathrm{Fe}}^2}, \qquad \frac{\delta}{\omega} = -\frac{3\pi}{4}\frac{m_+}{m_e}\frac{\omega^3}{k^3 v_{\mathrm{Fe}}^3} \ll 1. \tag{4.196}$$

Now let us write the exchange interaction potential of two electrons which are placed at a distance r from each other:

$$U\left(\vec{r}\right) = e\varphi\left(\vec{r}\right) = \frac{4\pi e^2}{(2\pi)^3}\int d\vec{k}\frac{\exp\left(\imath\vec{k}.\vec{r}\right)}{k^2\varepsilon^l\left(\vec{k}\cdot\vec{V}, k\right)}, \tag{4.197}$$

where $\vec{V}$ is the relative velocity of electron's motion which is of the order of Fermi velocity.

Substituting expression (4.197) into Eq. (4.195), integrating over $\vec{k}$, and taking into account a small positive imaginary part for dielectric permittivity (4.195) which allows us to use

$$\frac{1}{x+0} = \mathcal{P}\frac{1}{x + \imath 0} - \imath\pi\delta(x),$$

finally we obtain

$$\begin{aligned} U\left(\vec{r}\right) &= \frac{e^2 r_D^2}{\pi}\int_0^{\infty}\frac{k^2 dk}{1 + k^2 r_D^2}\int_{-1}^{+1}\frac{x^2 dx \exp(\imath krx)}{x^2 - x_0^2 + \imath 0} \\ &= -\imath\frac{e^2 r_D^2}{2}\int_0^{\infty}\frac{k^2 dk}{1 + k^2 r_D^2}\int_{-1}^{+1} x^2 dx[\delta(x - x_0) - \delta(x + x_0)], \end{aligned} \tag{4.198}$$

where $r_D^2 = v_{\text{Fe}}^2/3\omega_{\text{pe}}^2$, $x_0 = a/\left(1 + k^2 r_D^2\right)$, $a = v_{\text{Fe}}^2 \omega_{p+}^2/3V^2\omega_{\text{pe}}^2 \simeq 0.001$. After simple calculations, from (4.198), we obtain

$$U(r) = \frac{e^2 r_D^2 \sqrt{a}}{r^3} \int_0^\infty \frac{y^2 dy}{\left(1 + y^2 r_D^2/r^2\right)^{3/2}} \sin\left(\frac{y\sqrt{a}}{\sqrt{1 + y^2 r_D^2/r^2}}\right). \tag{4.199}$$

This integral is taken only numerically. However, as we are interested in its form at large distances $r \gg r_D$, where it differs from Coulomb interaction potential, we find

$$U(r) = \frac{e^2 r_D}{ar^3} \int_0^\infty z^2 dz \sin z = -\frac{2e^2 r_D^2}{ar^3}. \tag{4.200}$$

This interaction potential, firstly has negative sign which corresponds to an attraction between electrons. Secondly, at large distances $r^2 \sim r_D^2/a \gg r_D^2$, it is quite large and can be compared to the Coulomb potential. Thus, when $r \approx 100\, r_D$, potential (4.200) is only two orders of magnitude smaller than the Coulomb potential.

4.6.6 Investigate the possibility of superconductivity in semiconductors and electron-ion plasma at cryogenic temperatures (compare with the previous problem).

Solution First, we consider the condition of existence of ion-acoustic waves in semiconductors at cryogenic temperatures when electrons (light negative charge carriers) are degenerate and the following inequality holds

$$v_{\text{Fe}} = \sqrt{\epsilon_{\text{Fe}}/m^*} \gg \omega/k \gg \sqrt{\epsilon_+/m_+}. \tag{4.201}$$

Here, v_{Fe} and ϵ_{Fe} are Fermi velocity and Fermi energy of electrons with effective mass $m^* \approx 0.03m$ which is much less than the mass of free electron; m_+ and ϵ_+ are mass and energy of the positive charge carriers (holes); indeed $m_+ \approx m$, and ω/k is the phase wave velocity. Holes can be either non-degenerate or degenerate, while electrons are degenerate, i.e.,

$$\epsilon_{\text{Fe}} = \frac{\left(3\pi^2\right)^{2/3} \hbar^2 n_e^{2/3}}{2m^*} \gg \kappa T. \tag{4.202}$$

Here, n_e is the electron density. At $T_e < 100$ K, for density $n_e > 10^{17}\text{cm}^{-3}$, which is typical for most semiconductors, this inequality is well satisfied. In this way, at $T_e \leq 10$ K the degeneracy condition of electrons in semiconductors is easily realized. As about the positive charge carriers, their degeneracy requires much more stringent conditions. Therefore, we consider them arbitrary.

When inequality (4.201) holds, the frequency spectrum of ion-acoustic waves is determined by zeros of longitudinal dielectric permittivity:

$$\varepsilon^l(\omega, k) = 1 + \frac{1}{k^2 r_{\mathrm{De}}^2}\left(1 + \imath \frac{\pi}{2}\frac{\omega}{k v_{\mathrm{Fe}}}\right) - \frac{\omega_{p_+}^2}{\omega^2} = 0, \tag{4.203}$$

where $r_{\mathrm{De}} = \sqrt{v_{\mathrm{Fe}}^2/3\omega_{\mathrm{pe}}^2}$ is the electron Debye radius, $\omega_{\mathrm{pe}} = \sqrt{4\pi e^2 n_e/m^*}$ is the electron Langmuir frequency, $\omega_{p_+} = \sqrt{4\pi e^2 n_+/m_+}$ is the hole Langmuir frequency, $n_+ = n_e$, and ω and k are the frequency and wave vector of ion-acoustic waves, respectively.

From Eq. (4.203), we find the frequency spectrum of weakly damped wave $(\omega \to \omega + \imath\delta)$

$$\omega^2 = \frac{\omega_{p_+}^2}{1 + 1/k^2/r_{\mathrm{De}}^2}, \qquad \frac{\delta}{\omega} = -\frac{3\pi m_+}{4m}\frac{\omega^3}{k^3 v_{\mathrm{Fe}}^3} \ll 1. \tag{4.204}$$

In the long-wave range, the frequency spectrum (4.204) has the acoustic form $\omega = k v_s = k v_{\mathrm{Fe}}\sqrt{m^*/m}$, while in the short-wave range, it gets the oscillating form $\omega^2 = \omega_{p_+}^2$.

Now we consider the field produced by a test electron moving with velocity $\vec{v}$, where $v = v_{\mathrm{Fe}}$ in the considered medium. Field potential at a distance $\vec{r} = \mathrm{const}$ from the test electron is given by

$$\varphi\left(\vec{r}\right) = \frac{4\pi e}{(2\pi)^3}\int d\vec{k}\,\frac{\exp\left(\imath\vec{k}.\vec{r}\right)}{k^2\varepsilon^l\left(\vec{k}.\vec{v}, k\right)}. \tag{4.205}$$

Here, $\varepsilon^l(\omega, k)$ is determined by Eq. (4.203). Now let us suppose that at the same distance from the test electron the second electron moves with the same velocity $\vec{v}$. The interaction potential energy of this electron with the test electron is equal to

$$U\left(\vec{r}\right) = e\varphi\left(\vec{r}\right) = \frac{4\pi e^2}{(2\pi)^3}\int d\vec{k}\,\frac{\exp\left(\imath\vec{k}\cdot\vec{r}\right)}{k^2\varepsilon^l\left(\vec{k}\cdot\vec{v}, k\right)}. \tag{4.206}$$

Substituting expression (4.203) into Eq. (4.206), after simple calculations, we find

$$U\left(\vec{r}\right) = \frac{e^2}{\pi}\int_0^\infty dk\int_{-1}^{1} dx\,\frac{\exp(\imath krx)}{1 + \frac{1}{k^2 r_{\mathrm{De}}^2} - \frac{a^2}{k^2 r_{\mathrm{De}}^2}\frac{1}{x^2} + \imath\beta}. \tag{4.207}$$

Here, $a^2 = m/3m_+ \approx 0.01$, $\beta = \pi\omega/2k^2 r_{\text{De}}^2 k v_{\text{Fe}} \to +0$. Taking into account the latter condition allows us to write Eq. (4.207) in the form

$$U\left(\vec{r}\right) = \frac{e^2}{\pi}\int_0^\infty dk \int_{-1}^{1} dx \frac{k^2 r_{\text{De}}^2}{1 + k^2 r_{\text{De}}^2} \frac{x^2 \exp\left(\imath krx\right)}{x^2 - x_0^2 + \imath 0}$$

$$= \frac{e^2 r_{\text{De}}^2 a}{r^3} \int_0^\infty \frac{y^2 dy \sin\left(\frac{ay}{\sqrt{1+y^2 r_{\text{De}}^2/r^2}}\right)}{\left(1 + y^2 r_{\text{De}}^2/r^2\right)^{3/2}}, \qquad (4.208)$$

where $x_0^2 = a^2/\left(1 + k^2 r_{\text{De}}^2\right)^2$. The integral in Eq. (4.208) is taken only numerically. However, it is possible to find it analytically in two opposite limits:

1. At small distances $r \leq r_{\text{De}}$, or $kr_{\text{De}} \leq 1$, we find

$$U\left(\vec{r}\right) = \frac{e^2}{r} \exp\left(-r/r_{\text{De}}\right), \qquad (4.209)$$

which corresponds to the well-known Debye screening of the interaction at the distances exceeding electron Debye radius. In this limit, the exchange interaction of electrons is neglected.

2. At the large distances greatly exceeding the Debye radius, the exchange interaction between electrons becomes dominant. When $kr_{\text{De}} \ll 1$, from Eq. (4.208), we find

$$U(r) = \frac{e^2 r_D}{ar^3} \int_0^\infty z^2 dz \sin z = -\frac{2e^2 r_D^2}{ar^3} \qquad (4.210)$$

Here, it is clear that the exchange interaction potential at large distances leads to the attraction of two electrons in the considered medium. If we write the interpolating formula for the interaction potential of electrons as the sum of expressions (4.209) and (4.210),

$$U\left(\vec{r}\right) = \frac{e^2}{r} \exp\left(-r/r_{\text{De}}\right) - \frac{2e^2 r_{\text{De}}^2}{a^2 r^3}, \qquad (4.211)$$

then we can simply show that the attraction of electrons predominates over repulsion at all distances, $r > r_{\text{De}}$, for which formula (4.209) is applicable. At such a distance, the damping of ion-acoustic waves, providing the exchange interaction of electrons, is negligibly small.

Now we will transfer the above theoretical results to low temperature (cryogenic) plasma with degenerate electrons. Ion mass is much greater than the electron mass in plasma, $M \gg m$. Electron mass in plasma is also much greater than the mass of light carriers in semiconductors and therefore, electrons degeneracy in cryogenic plasma occurs at much lower temperatures. Then, at $T \approx 10\,\mathrm{K}$, electrons are degenerate when $n_e > 10^{17}\mathrm{cm}^{-3}$ which will be used below in estimations.

All calculations within the notation coincide with those carried out above. As a result, we find Eq. (4.207) for the interaction potential, but we should change only one notation in it: $a^2 = m/3M \leq 1.6 \times 10^{-4}$. In fact, formulas (4.208)–(4.210) remains unchanged. However, the role of the exchange interaction in plasma is manifested much larger, because quantity a^2 is much smaller. Attraction predominates over Coulomb repulsion at every $r > r_{\mathrm{De}}$. From the above analysis it follows that in semiconductors and in gaseous electron-ion plasmas with degenerate electrons, in which weakly damped ion-acoustic wave propagation is possible, the exchange interaction between electrons may lead to the attraction of electrons. In fact, electrons exchange interaction, stipulated by electron–phonon (crystal sound) exchange, explains superconductivity in metals. In metals, the exchange interaction is responsible for Cooper's pair formation which couples an electron pair. In this case, however, in metals, independent experimental (insufficiently defined) data are used to determine the electron–phonon interaction constant. In the example of the semiconductor under consideration, the constant of such an interaction is exactly defined: it is the electron charge. In this sense, formulas (4.207) and (4.209)–(4.210) are more justified. Therefore, it is reasonable to raise the question of whether a superconducting state with Cooper's pairs can exist in a semiconductor and electron-ion plasma with degenerate electrons?

According to the estimation carried out above, for $n_t \geq 10^{17}\mathrm{cm}^{-3}$ (which is the typical density for both gaseous plasmas and semiconductors), electrons are degenerate at $T_e \leq 10\,\mathrm{K}$. Therefore, we can assume that the conditions for the appearance of superconductivity both in semiconductors and cryogenic electron-ion plasmas are quite realistic. At the same time, to fulfill the gas approximation condition, when inequalities (4.201), ensuring the existence of ion-acoustic waves, hold, the following condition should be satisfied

$$e^2 n_e^{1/3} < \epsilon_{\mathrm{Fe}}. \tag{4.212}$$

This condition along with the above-mentioned inequalities for realization of exchange attraction and formation of coupling state of electrons in semiconductors holds with a small margin. There are only a few electrons in the Debye sphere. But precisely because of this, the effective mass of an electron is much less than the mass of a free electron $m^* \approx 0/03\,m$. Possibly, under these conditions, the electron component of semiconductors forms an electron Fermi liquid. In this case, it is necessary to develop another theory.

4.6.7 Study the excitation of the zero-point sound in degenerate isotropic plasma by a low-density non-relativistic monoenergetic electron beam.

Solution The dispersion equation for the longitudinal waves of the system can be written as

$$1+\frac{3\omega_{\text{pe}}^2}{k^2 v_{\text{Fe}}^2}\left(1-\frac{\omega}{2kv_{\text{Fe}}}\ln\frac{\omega+kv_{\text{Fe}}}{\omega-kv_{\text{Fe}}}\right)-\frac{\omega_b^2}{\left(\omega-\vec{k}\cdot\vec{u}\right)^2}=0 \tag{4.213}$$

Assuming the beam to be a small perturbation, we obtain

$$\begin{aligned}
&\omega=\omega_0+\delta=kv_{\text{Fe}}+\delta\approx\vec{k}\cdot\vec{u}+\delta,\\
&\omega_0=kv_{\text{Fe}}\left[1+2\exp\left(-\frac{2}{3}\frac{\omega_0{}^2}{\omega_{\text{pe}}{}^2}-2\right)\right],\\
&\frac{\delta}{\omega_0}=\frac{-1+\imath\sqrt{3}}{2}\left[\frac{4}{3}\frac{N_b}{N_p}\exp\left(-\frac{2}{3}\frac{\omega_0^2}{\omega_{\text{pe}}^2}-2\right)\right]^{1/3}.
\end{aligned} \tag{4.214}$$

These formulas are applicable under the condition

$$\frac{N_b}{6N_p}\ll\exp\left[-4\left(1+\frac{\omega_0^2}{3\omega_{\text{pe}}^2}\right)\right]. \tag{4.215}$$

Along with the inequality Im $\{\delta\} > \nu_{\text{eff}}$, where ν_{eff} is the electron collision frequency in degenerate plasma, this condition is easily satisfied for metals and degenerate semiconductors.

4.6.8 Find the spectral density of the energy lost by a relativistic particle per unit path length, using the permittivity. Use the Fermi method, that is, calculate the flux of the Poynting vector through the cylindrical surface around the trajectory of the particle. Interpret the result in terms of the equivalent photons (see Problem 1.11.14) [88].

Solution From Maxwell's equations, we determine the monochromatic components of the electromagnetic field of a particle moving with a constant velocity $v=\text{const}$ in the cylindrical reference frame:

$$\begin{aligned}
&E_{\omega z}\left(\vec{r},t\right)=\frac{\imath q\omega}{\pi c^2}\left(1-\frac{1}{\beta^2\varepsilon}\right)K_0(s\rho)\,e^{\imath\omega(z/v-t)},\\
&B_{\omega\varphi}\left(\vec{r},t\right)=\frac{\imath q\omega}{\pi c^2}K_1(s\rho)e^{\imath\omega(z/v-t)},\quad s^2=\frac{\omega^2}{v^2}-\frac{\omega^2}{c^2\varepsilon(\omega)},\quad \text{Re } s>0,
\end{aligned} \tag{4.216}$$

where $K_n(x)$ is the MacDonald function of the order of n, and $\varepsilon(\omega)$ is given by

$$\varepsilon(\omega) = 1 - \frac{\omega_p^2}{\omega^2} + \imath \frac{\Gamma_p + \Gamma_C}{\omega},$$

where

$$\Gamma_p = cn_a\sigma_p, \qquad \Gamma_C = cn_a Z\sigma_C,$$

are the probabilities of absorption of the quantum per unit time due to the formation of the pair by this quantum or due to Compton scattering. With these formulas we calculate the energy lost by the particle per unit path length:

$$-\frac{dE}{dz} = 2\pi b \int_{-\infty}^{\infty} \frac{c}{4\pi}\left[\vec{E} \times \vec{B}\right]_\rho dt = -2\pi cb\,\mathrm{Re} \int_0^{\infty} B^*_{\omega\varphi} E_{\omega z} d\omega. \tag{4.217}$$

From the uncertainty relations, it follows that the radius b must be of the order of the Compton length Λ_C; and $|bs| \ll 1$. Substituting the components (4.216) in Eq. (4.217), we obtain the spectral power of the losses per unit path length:

$$-\frac{dE_\omega}{dz\,d(\hbar\omega)} = \hbar\omega\, n(\omega)\frac{\Gamma(\omega)}{c}, \tag{4.218}$$

where

$$n(\omega) = \frac{2}{\pi}\frac{q^2}{\hbar c}\frac{1}{\omega} \ln \frac{m_e c^2}{\hbar\omega\left[\Gamma^2/\omega^2 + \left(\omega_p^2/\omega^2 + \gamma^{-2}\right)^2\right]^{1/4}} \tag{4.219}$$

is the spectral power of photons, which are equivalent of the self-field of the particle. In derivation of this formula, we assume that the logarithm is much larger than unity; the value under the logarithm is determined up to the factor of the order of unity. In the absence of the medium ($\omega_p = \Gamma = 0$), the quantity (4.219) changes to the spectral power of the equivalent photons of the relativistic particle in vacuum. The result (4.218) has a simple meaning: the product $\Gamma_p n(\omega) d\omega$ is the number of quanta absorbed per unit time from the self-field of the particle due to formation of pairs. In other words, it is the number of pairs with the total energy of the particles $\hbar\omega$, which are formed by the field of the particle. The product $\Gamma_p n(\omega) d\omega$ gives the number of quanta absorbed due to the Compton effect. Along with this process, other quanta of lower energies are also produced, that is, bremsstrahlung of quanta by medium electrons takes place.

4.6.9 Investigate the electromagnetic surface waves on single-layered metal-like materials of the borophene type.

Solution Borophene is a fairly broadband metal. The energy spectrum of light carriers (electrons) in borophene, for the strongly degenerate case, is given by

$$\epsilon = v_F p. \tag{4.220}$$

Here, v_F is the Fermi velocity of carriers with charge e and with zero effective mass. It is assumed that the Fermi energy of charge carriers ϵ_F is much less than the work function of borophene layer and as a result the carriers can perform only two-dimensional motion on yz plane. Moreover, ox-axis is normal to the layer surface. In this case, the equilibrium distribution function of the degenerate carriers can be written in the form

$$f_0(v_x, v_y) = \begin{cases} \dfrac{2}{(2\pi\hbar)^2} & \text{for} \quad p \leq p_F, \\ 0 & \text{for} \quad p > p_F, \end{cases} \tag{4.221}$$

where $p_F = \epsilon_F / v_F$ is the Fermi momentum of the carriers determined by the surface density n (cm^{-2}) as

$$\int d\vec{p}_\perp f_0 = \frac{2\pi p_F^2}{(2\pi\hbar)^2} = n. \tag{4.222}$$

Finally, from expression (4.221) we introduce an important relation for further analysis

$$\frac{\partial f_0}{\partial \epsilon} = -\frac{2}{(2\pi\hbar)^2}\delta(\epsilon - \epsilon_F) = -\frac{n v_F^2}{\pi \epsilon_F^2}\delta(\epsilon - \epsilon_F) \tag{4.223}$$

Below we study the electrodynamic properties of planar and cylindrical single-layered metal-like materials when spatial dispersion is negligible. It should be remarked that cylindrical single layers (nanotubes) are important in nanotechnology. Here, we focus on the surface waves damping on the both sides of the surface. Naturally, volumetric waves localized on the material surface are discussed as well. The most interesting case is the slow waves ($v_{\text{ph}} \ll c$) which are known as surface plasmons.

To describe the two-dimensional motion of electrons in a planar single-layered material we begin from the linearized Vlasov equation:

$$\frac{\partial \delta f}{\partial t} + \vec{v}_\perp \cdot \frac{\partial \delta f}{\partial \vec{r}_\perp} + e\vec{E} \cdot \frac{\partial f_0}{\partial \vec{p}_\perp} = 0. \tag{4.224}$$

Here, $\delta f\left(\vec{r}_\perp, \vec{v}_\perp, t\right)$ is a small deviation of the electron distribution function which is caused by a small electric field $\vec{E}$. For the solution in the form of $\delta f \sim \exp\left(-\imath\omega t + \imath\vec{k}_\perp \cdot \vec{r}_\perp\right)$, we find

$$\delta f = \frac{2\imath e}{(2\pi\hbar)^2}\frac{\vec{E}\cdot\vec{v}_\perp}{\left(\omega - \vec{k}_\perp\cdot\vec{v}_\perp\right)}\delta(\epsilon - \epsilon_F). \tag{4.225}$$

Substituting this expression into the current density

$$j_i = e\int v_i \delta f\, d\vec{p}_\perp = \sigma_{ij}\left(\omega, \vec{k}_\perp\right)E_j, \tag{4.226}$$

we find the conductivity and dielectric permittivity tensors which are diagonal in the two-dimensional coordinate system:

$$\varepsilon_{ij} = \varepsilon(\omega, k_\perp)\delta_{ij}, \qquad \varepsilon = 1 + \frac{4\pi\imath}{\omega}\sigma = 1 + \frac{8\pi e^2\epsilon_F}{(2\pi\hbar)^2\omega v_F^2}\int_0^{2\pi}\frac{v_{Fi}v_{Fj}d\varphi}{\omega - k_\perp v_F\cos\varphi}. \tag{4.227}$$

Here, in the integral term of Eq. (4.227), we should have in mind that

$$\frac{1}{\omega - k_\perp v_F\cos\varphi} = \mathcal{P}\frac{1}{\omega - k_\perp v_F\cos\varphi} - \imath\pi\delta(\omega - k_\perp v_F\cos\varphi). \tag{4.228}$$

Therefore, dielectric permittivity generally is a complex quantity and its imaginary part corresponds to the Cherenkov absorption of waves in the sample. Such an absorption occurs in the frequency range $\omega \leq kv_F$. However, we are interested in the high-frequency range $\omega \gg kv_F$. In this range, wave absorption is absent and dielectric permittivity is a purely real quantity

$$\sigma = \frac{\imath e^2 v_F^2 n}{\omega\epsilon_F}\delta(x), \qquad \varepsilon = 1 - \frac{4\pi e^2 v_F^2 n}{\epsilon_F\omega^2}\delta(x). \tag{4.229}$$

The presence of delta functions in expressions (4.229) means that they describe the surface responses in the yz plane, while density n corresponds to the surface density and is measured in 1/cm^2. In real estimations of the surface density of single-layered materials, the finite size of an atom (10^{-8}cm) is taken into account. Therefore, in borophene, we have $n \leq 10^{14}\text{cm}^{-2}$.

1. Planar geometry. We begin the analysis of the oscillation spectra from the planar case and consider the E-mode of surface waves. Non-zero field components of

this mode are E_x, E_z, B_y, and oz-axis is along the wave propagation. From the Maxwell's equations, for the field E_z, we obtain

$$\varepsilon\left\{\frac{\partial^2 E_z}{\partial x^2} - \left(k_z^2 - \frac{\omega^2}{c^2}\right)E_z\right\} = 0. \tag{4.230}$$

This equation splits in two equations. The first equation is $\varepsilon = 0$ which describes the localized (in the layer) longitudinal (potential) volumetric waves with spectrum

$$\omega^2 = \omega_p^2, \qquad \omega_p^2 = \frac{4\pi e^2 v_F^2 n\delta(x)}{\epsilon_f} = \widetilde{\omega}_p^2\delta(x). \tag{4.231}$$

The second equation describes the longitudinal-transverse wave which is valid outside the layer, both below and above the layer. The solutions of this equation in these regions are

$$E_z = \begin{cases} C_1 e^{-\imath\kappa x}, & x > 0, \\ C_2 e^{\imath\kappa x}, & x < 0, \end{cases} \tag{4.232}$$

where $\kappa = \sqrt{k_z^2 - \omega^2/c^2}$.

These solutions satisfy the boundary conditions on the layer surface:

$$\{E_z\}_{x=0} = 0, \quad \left\{\frac{-\imath\omega}{c\kappa^2}\frac{\partial E_z}{\partial x}\right\}_{x=0} = \frac{4\pi}{c}\int dx\sigma E_z. \tag{4.233}$$

Substituting solutions (4.232) into boundary conditions (4.233) leads to the following dispersion equation for surface waves on the planar single-layer sample:

$$\frac{\widetilde{\omega}_p^2}{\omega^2}\sqrt{k_z^2 - \omega^2/c^2} = 2. \tag{4.234}$$

Hence, we find the frequency spectrum of surface wave in this case:

$$\omega = \begin{cases} k_z c, & \text{at} \quad k_z c \ll \widetilde{\omega}_p\sqrt{|k_z|}, \\ \widetilde{\omega}_p\sqrt{|k_z|}\,, & \text{at} \quad \widetilde{\omega}_p\sqrt{|k_z|} \ll k_z c. \end{cases} \tag{4.235}$$

This spectrum is different from the spectrum of the surface wave on the surface of the semi-bounded conducting medium at the large wave numbers (see Sect. 2.10). It is the spectrum that continues to grow at large wave vectors, although growth slows down and is proportional to the square root of the wave vector.

2. Cylindrical geometry. Now let us consider the spectrum of surface wave on the surface of a single-layer cylindrical tube of radius r_0. The first factor of Eq. (4.230) which leads to $\varepsilon = 0$ is not changed in the present case and as a result it keeps its form and spectrum of volumetric plasmon (4.231).

The second factor of Eq. (4.211) for the cylindrical case can be written as

$$\nabla_\perp^2 E_z - \left(k_z^2 - \omega^2/c^2\right) E_z = 0. \tag{4.236}$$

Here, again we assume that the wave propagates along the oz-axis. For axially symmetric modes, we write the solution of Eq. (4.236) in the form of

$$E_z = \begin{cases} C_1 I_0(\kappa r), & \text{at} \quad r < r_0, \\ C_2 K_0(\kappa r), & \text{at} \quad r > r_0. \end{cases} \tag{4.237}$$

Taking into account the boundary conditions

$$\{E_z\}_{r=r_0} = 0, \;\; \{B_\varphi\}_{r=r_0} = \frac{4\pi}{c} \int dx \sigma E_z, \tag{4.238}$$

we find the dispersion equation of surface waves:

$$\frac{K_0'(\kappa r_0)}{K_0(\kappa r_0)} - \frac{I_0'(\kappa r_0)}{I_0(\kappa r_0)} = -\frac{\widetilde{\omega}_p^2 \kappa}{\omega^2}. \tag{4.239}$$

In the limit $\kappa r_0 \gg 1$, Eq. (4.239) turns into Eq. (4.234) for the planar case, and as a result all the above results are valid in this limit. In the opposite limit of the small radius cylinder, when $\kappa r_0 \ll 1$, Eq. (4.239) reduces to

$$-\frac{\widetilde{\omega}_p^2}{\omega^2} \kappa^2 r_0 \ln \kappa r_0 = 1. \tag{4.240}$$

Solution of the latter equation is

$$\omega = \begin{cases} k_z c, & \text{at} \quad k_z c \ll \widetilde{\omega}_p \sqrt{k_z^2 r_0}, \\ \widetilde{\omega}_p \sqrt{k_z^2 r_0 |\ln k_z r_0|}, & \text{at} \quad \widetilde{\omega}_p \sqrt{k_z^2 r_0} \ll k_z c. \end{cases} \tag{4.241}$$

This spectrum is more different from the spectrum of the surface wave on the massive semiconductor sample. In fact, this spectrum not only does not reach the limiting value at large longitudinal wave numbers, but also continues to grow linearly with a slowed phase velocity.

4.6.10 Study the excitation of surface elasto-electrostatic waves on thin piezo plasma-like layers.

Solution To study the electromagnetic oscillations in a thin piezo-semiconductor layer, we use the mirror reflection model. Now, we consider a piezo-semiconductor layer with thickness a so that the z axis is along the surface of the layer and the x axis is across it. Furthermore, by taking into account the dispersion equation (4.158) for a semi-infinite sample and comparing to what obtained for a thin plasma-like layer in Sect. 2.10.2, the dispersion equation of the potential surface waves of a thin piezo plasma-like medium with thickness a along the x axis takes the form [79, 87]

$$1+\frac{2}{a\mid k_z\mid}\sum_{n=0}^{\infty'}[1\pm(-1)^n]\frac{k_z^2}{k^2\varepsilon^l(\omega,k)-\frac{4\pi k^4\beta_1^2}{\omega^2\rho_m-\lambda^{\rm tr}k^2}}=0, \tag{4.242}$$

where $\vec{k}=(n\pi/a,0,k_z)$ is the wave number, $n=$ even and $n=$ odd correspond to the symmetric and anti-symmetric modes, the prime over the sum means that for $n=0$ it should be divided by 2. For magneto-active plasma-like media, when $\vec{B}_0 \parallel oz$, we should replace $k_z^2\to k_z^2+k_y^2$, $k^2\to k_z^2+k_y^2+n^2\pi^2/a^2$ and $\varepsilon^l(\omega,k)=\varepsilon(\omega,k)$ where $\varepsilon(\omega,k)$ is the quasi-longitudinal dielectric permittivity of the medium.

In the limit $a\mid k_z\mid\gg 1$, we go from the thin layer case to the semi-bounded case, where we replace the sum over n to the integral and obtain Eq. (4.158). In the limit $a\mid k_z\mid\ll 1$, i.e., a thin layer, in the absence of spatial dispersion, we obtain the following dispersion equation:

$$\left(\varepsilon^l(\omega,k)+\frac{2}{a\mid k_z\mid}\right)(\omega^2-k_z^2v_{\rm tr}^2)=\frac{4\pi k_z^2\beta_1^2}{\rho_m}. \tag{4.243}$$

From this equation, it is clear that two kinds of oscillations are related to each other. The first factor in the left side of Eq. (4.243) corresponds to the longitudinal oscillation of a thin plasma-like layer and the second factor shows the transverse volume acoustic wave in the crystal lattice of the piezo-semiconductor.

The relation between these two oscillations is provided by the small term on the right-hand side of Eq. (4.243) which contains the piezo effects. Therefore, in the absence of spatial dispersion, having

$$\varepsilon(\omega)=1-\frac{\omega_{\rm pe}^2}{\omega(\omega+\imath\nu_e)}\simeq 1-\frac{\omega_{\rm pe}^2}{\omega^2}\left(1-\imath\frac{\nu_e}{\omega}\right), \tag{4.244}$$

and neglecting collisions, we find the maximum correction $\delta_{\max}$ to the frequency spectrum of the both vibrational systems. As a result, we find

$$\omega = \sqrt{\frac{\omega_{pe}^2 a \mid k_z \mid}{2}} + \delta_{max} = \mid k_z \mid v_{tr} + \delta_{max},$$

$$\delta_{max}^2 = \frac{\pi a |k_z|^3 \beta_1^2}{\rho_m}. \tag{4.245}$$

To consider the collision effects, we should use Eq. (4.244) and substitute it into Eq. (4.243). Therefore, we find

$$\left(\delta_{max} + \imath \frac{\nu_e}{2}\right) \delta_{max} = \frac{\pi a |k_z|^3 \beta_1^2}{\rho_m}. \tag{4.246}$$

When $\delta_{max} \gg \nu_e$, we find Eq. (4.245). In the opposite limit, we find

$$\delta_{max} = -\imath \frac{2\pi a |k_z|^3 \beta_1^2}{\rho_m \nu_e}. \tag{4.247}$$

which corresponds to the damping rate of shallow water and acousto-elastic waves.

In the ion-acoustic frequency region, $k v_{Ti} \ll \omega \ll k v_{Te}$, the longitudinal dielectric permittivity can be written as [82]

$$\varepsilon^l(k, \omega) = 1 - \frac{\omega_{pi}^2}{\omega^2} \left(1 - \imath \frac{\nu_i}{\omega}\right) + \frac{\omega_{pe}^2}{k^2 r_{De}^2} \left(1 + \imath \begin{cases} \sqrt{\frac{\pi}{2}} \frac{\omega}{k v_{Te}} \\ \frac{\pi \omega}{2 k v_{Fe}} \end{cases}\right), \tag{4.248}$$

where the upper expression in Eq. (4.248) is related to non-degenerate plasma in which $r_{De}^2 = v_{Te}^2/\omega_{pe}^2$ and the lower expression is related to degenerate plasma in which $r_{De}^2 = v_{Fe}^2/3\omega_{pe}^2$. By substituting the latter expression into Eq. (4.243), and by neglecting dissipation terms, we find

$$\left(1 - \frac{\omega_{pi}^2}{\omega^2} + \frac{\omega_{pe}^2}{k_z^2 v_{Te}^2} + \frac{2}{a \mid k_z \mid}\right) \left(\omega^2 - k_z^2 v_{tr}^2\right) = \frac{4\pi k_z^2 \beta_1^2}{\rho_m}, \tag{4.249}$$

for the coupling of ion-acoustic and elastic waves in a piezoelectric. Here, we have found a relation between oscillations in two systems. Therefore, from Eq. (4.249), we find the frequency spectrum

$$\omega = \sqrt{\frac{\omega_{pi}^2}{\frac{2}{a|k_z|} + \frac{1}{k_z^2 r_{De}^2}}} + \delta_{max} = \mid k_z \mid v_{tr} + \delta_{max}, \tag{4.250}$$

with

$$\delta_{\max}^2 = \frac{\omega^2}{\omega_{\rm pi}^2}\frac{2\pi k_z^2\beta_1^2}{\rho_m}, \tag{4.251}$$

which corresponds to the coupling between ion-acoustic and elastic waves caused by the piezoelectric effect. If we take into account spatial dispersion, starting with Eq. (4.249), we find

$$\left(\delta_{\max} + \imath\frac{\nu_e}{2} + \imath\begin{Bmatrix} \sqrt{\frac{\pi}{8}}\frac{\omega^2}{|k_z| v_{\rm Te}} \\ \frac{2\pi\omega^2}{|k_z| v_{\rm Fe}} \end{Bmatrix}\right)\delta_{\max} = \frac{\omega^2}{\omega_{\rm pi}^2}\frac{2\pi k_z^2\beta_1^2}{\rho_m}, \tag{4.252}$$

instead of Eq. (4.246). Thus, when the imaginary terms are neglected in Eq. (4.252), we find Eq. (4.251). In the opposite limit, we find the damping rate of ion-acoustic and elastic waves in these layers:

$$\delta_{\max} = -\imath\frac{\omega^2}{\omega_{\rm pi}^2}\frac{2\pi k_z^2\beta_1^2}{\rho_m}\left(\frac{\nu_i}{2} + \imath\begin{Bmatrix} \sqrt{\frac{\pi}{2}}\frac{\omega^2}{|k_z| v_{\rm Te}} \\ \frac{2\pi\omega^2}{|k_z| v_{\rm Fe}} \end{Bmatrix}\right)^{-1}. \tag{4.253}$$

References

1. L.D. Landau, E.M. Lifshitz, *Course of Theoretical Physics, Vol. 5: Statistical Physics*, 3rd edn. (Nauka, Moscow, 1976).; Pergamon Press, Oxford, 1980), Part 1
2. V.P. Silin, J. Exp. Theor. Phys. **33**, 1282 (1957)
3. V.P. Silin, J. Exp. Theor. Phys. **37**, 873 (1959)
4. Y.L. Klimontovich, V.P. Silin, Usp. Fiz. Nauk **70**, 247 (1960)
5. D.I. Blokhintsev, *Principles of Quantum Mechanics* (Gostechizdat, Moscow, 1949)
6. L.D. Landau, E.M. Lifshitz, *Course of Theoretical Physics, Vol. 3: Quantum Mechanics: Non-Relativistic Theory*, 3rd edn. (Nauka, Moscow, 1974; Pergamon, New York, 1977)
7. P.A.M. Dirac, *Principles of Quantum Mechanics* (Oxford University Press, Oxford, 1935)
8. E. Wigner, Phys. Rev. **40**, 749 (1932)
9. Y.P. Terletskii, J. Exp. Theor. Phys. **7**, 1290 (1937)
10. D.I. Blokhintsev, J. Phys. **2**, 71 (1940)
11. D.I. Blokhintsev, P.E. Nemirovsky, J. Phys. **3**, 191 (1940)
12. J.E. Moyal, Proc. Camb. Philol. Soc. **45**, 99 (1949)
13. M.A. Mokulskii, J. Exp. Theor. Phys. **20**, 688 (1950)
14. V.P. Silin, Bull. Lebedev Inst. Acad. Sci. USSR **6**, 200 (1955)
15. Y.L. Klimontovich, V.P. Silin, J. Exp. Theor. Phys. **23**, 151 (1952)
16. J. Lindhard, Det. Kong. Danske vid. Selskab. Dan. Mat. Fys. Med. **28**, 2 (1954)
17. L.D. Landau, Z. Phys. **64**, 629 (1930)
18. A.A. Rukhadze, V.P. Silin, J. Exp. Theor. Phys. **38**, 645 (1960)
19. A.A. Vlasov, J. Exp. Theor. Phys. **8**, 291 (1938)

20. I.I. Goldman, J. Exp. Theor. Phys. **17**, 681 (1947)
21. S. Tomonaga, Prog. Theor. Phys. **5**, 544 (1950)
22. D. Bohm, D. Pines, Phys. Rev. **92**, 609 (1953)
23. D.N. Zubarev, J. Exp. Theor. Phys. **25**, 548 (1953)
24. P.S. Zyrianov, V.M. Eleonskii, J. Exp. Theor. Phys. **30**, 592 (1956)
25. R.A. Ferrell, Phys. Rev. **107**, 450 (1957)
26. H. Sawada, K.A. Brueckner, N. Fukuda, R. Brout, Phys. Rev. **108**, 507 (1957)
27. A. Voloshinskii, L.Y. Kobelev, Fiz. Met. Metalloved **6**(2), 356 (1958)
28. V.L. Bonch-Bruevich, Fiz. Met. Metalloved. **6**, 590 (1958)
29. J. Hubbard, Proc. R. Soc. A **243**, 337 (1958)
30. E.S. Fradkin, Dissertation, FIAN, USSR, 1960
31. D. Pines, *Solid State Physics*, vol 1 (Academic Press, New York, 1955)
32. D. Pines, D. Bohm, Phys. Rev. **82**, 625 (1951).; Phys. Rev. **85**, 338 (1952)
33. D. Pines, Rev. Mod. Phys. **28**, 184 (1956).; D. Pincs, D. Nozieres, Nuovo Cimento, 9, 470 (1958)
34. H. Watanabe, J. Phys. Soc. Jpn. **11**, 112 (1956)
35. C. Fert, F. Pradal, C. R. Acad. Sci. **248**(5), 666 (1959)
36. E.L. Feinberg, J. Exp. Theor. Phys. **34**, 1125 (1958)
37. P.S. Zyrianov, J. Exp. Theor. Phys. **24**, 441 (1953)
38. L.D. Landau, E.M. Lifshitz, *Electrodynamics of Continuous Media*, 2nd edn. (Pergamon, New York, 1984)
39. H. London, Proc. R. Soc. A **176**, 522 (1940)
40. A.V. Pippard, Proc. R. Soc. A **191**, 385 (1947)
41. A.B. Pippard, Physica **15**, 45 (1949)
42. G.E.H. Reuter, E.H. Sondheimer, Proc. R. Soc. Lond. A Math. Phys. Sci. **195**, 336 (1948)
43. V.L. Ginzburg, G.P. Motulevich, Usp. Fiz. Nauk **55**, 469 (1955)
44. R.B. Dingle, Appl. Sci. Res. **B2**, 69 (1953)
45. T. Holstein, Phys. Rev. **96**, 535 (1954)
46. R.N. Gurzhi, J. Exp. Theor. Phys. **33**, 660 (1957)
47. L.D. Landau, J. Exp. Theor. Phys. **32**, 59 (1957)
48. L.D. Pitaevskii, J. Exp. Theor. Phys. **34**, 942 (1958)
49. R.N. Gurzhi, J. Exp. Theor. Phys. **35**, 965 (1958)
50. T. Holstein, Phys. Rev. **88**, 1427 (1952)
51. G.P. Motulevich, A.A. Shubin, Opt. Spektrosk. **2**, 633 (1957)
52. R. Peierls, *Elektrronnaya Teoriya Metallov* [Electron theory of metals] (IL, M, 1947)
53. L.D. Landau, J. Exp. Theor. Phys. **31**, 1058 (1956)
54. V.P. Silin, J. Exp. Theor. Phys. **33**, 495 (1957)
55. I.M. Lifshitz, M.I. Kaganov, Usp. Fiz. Nauk **69**, 419 (1959)
56. E.H. Sondheimer, Proc. R. Soc. A **224**, 260 (1954)
57. A.B. Pippard, Proc. R. Soc. A **224**, 273 (1945)
58. M.I. Kaganov, M.Y. Azbel, Dokl. Acad. Sci. USSR **102**, 49 (1955)
59. A.B. Pippard, Phil. Trans. R. Soc. A **250**, 325 (1957)
60. G. Feher, A.F. Kip, Phys. Rev. **98**, 337 (1955)
61. F.J. Dyson, Phys. Rev. **98**, 349 (1955)
62. V.P. Silin, J. Exp. Theor. Phys. **30**, 421 (1956)
63. M.Y. Azbel, V.I. Gerasimenko, I.M. Lifshitz, J. Exp. Theor. Phys. **31**, 357 (1956)
64. H.C. Torrey, Phys. Rev. **104**, 563 (1957)
65. F. Bloch, Phys. Rev. **70**, 460 (1946)
66. V.P. Silin, J. Exp. Theor. Phys. **23**, 649 (1959)
67. A.B. Pippard, Philos. Mag. **46**, 1104 (1955)
68. A.I. Akhiezer, M.I. Kaganov, J. Exp. Theor. Phys. **32**, 837 (1957)
69. Y.L. Klimontovich, S.V. Temko, J. Exp. Theor. Phys. **35**, 1141 (1958)
70. M.S. Steinberg, Phys. Rev. **3**, 425 (1958)

71. V.P. Silin, J. Exp. Theor. Phys. **38**, 977 (1960)
72. B. Shokri, S.K. Alavi, A.A. Rukhadze, Phys. Scr. **73**, 23 (2006)
73. W.P. Mason, H.E. Bommel, J. Acoust. Soc. Am. **28**, 430 (1956)
74. A.I. Akhiezer, J. Exp. Theor. Phys. **8**, 1318 (1938)
75. L.D. Landau, E.M. Lifshitz, *Course of Theoretical Physics, Vol. 1: Mechanics*, 3rd edn. (Nauka, Moscow, 1973; Pergamon Press, Oxford, 1976)
76. A.A. Andronov, Radiofizika **3**, 645 (1960)
77. V.L. Gurevich, J. Exp. Theor. Phys. **37**, 1680 (1959)
78. H. Stolz, Zur Theorie der Ultraaschallabsorption in Metallen. Z. Naturforsch., A **16**, 466 (1961)
79. B. Shokri, S.K. Alavi, A.A. Rukhadze, Waves Random Complex Media **7**, 87 (2006)
80. S.K. Alavi, B. Shokri, A.A. Rukhadze, Waves Random Complex Media **18**(4), 623 (2008)
81. S.K. Alavi, B. Shokri, Waves Random Complex Media **19**(2), 2783 (2009)
82. A.F. Alexandrov, L.S. Bogdankevich, A.A. Rukhadze, *Principles of Plasma Electrodynamics* (Springer, Heidelberg, 1984)
83. M.V. Kuzelev, A.A. Rukhadze, *Basics of Plasma Free Electron Lasers* (Editions Frontiers, Paris, 1995)
84. M.V. Kuzelev, A.A. Rukhadze, *Methods of Wave Theory in Dispersive Media* (World Scientific Publishing, Hackensack, 2009)
85. B. Shokri, M. Khorashadizadeh, Phys. Lett. A **352**, 520 (2006)
86. B. Jazi, M. Nejati, B. Shokri, Phys. Lett. A 370, **319** (2007)
87. B. Shokri, Phys. Plasmas **7**, 3867 (2000)
88. I.N. Toptygin, *Electromagnetic Phenomena in Matter: Statistical and Quantum Approaches*, 1st edn. (Wiley-VCH, Weinheim, 2015)

Chapter 5
Spatial Dispersion in Molecular Crystals

5.1 Dielectric Permittivity of Molecular Crystals

In this chapter, we shortly study the electromagnetic properties of the molecular crystals by taking account of spatial and temporal dispersions in the dielectric permittivity. It is well-known that neutral molecules and atoms are placed on the lattice points of the molecular crystals. The Van-der-Waals interaction between molecules and atoms is significantly smaller than the Coulomb interaction between charged particles in the molecules and atoms. This condition allows us to develop a relatively simple qualitative theory of such crystals, because the electromagnetic properties of the molecular crystals are mainly determined by the properties of the individual molecules and atoms [1–17]. Furthermore, the weak interaction between them plays a relatively small role.

It should be mentioned that the eigen frequencies of the Hamiltonian of the system consisting of the crystal and the electromagnetic field are found out by solving the quantum mechanics problem of determination of the electromagnetic oscillation spectra of the molecular crystals. However, quantum mechanics is used here only to derive the dielectric permittivity tensor $\varepsilon_{ij}\left(\omega, \vec{k}\right)$. On the contrary, the electromagnetic field in the medium in the long-wavelength limit is described classically.

It was shown that the spatial dispersion effects which are essential for optically active (gyrotropic) media are caused by the quantities of the order of the ratio of the molecule size to the radiation wavelength [13, 18]. Namely, for this reason, spatial dispersion is weak in such crystals even in optical spectra and the general theory of electromagnetic fields in the media with weak spatial dispersion, discussed in Sect. 1.7, is valid for their description. Below, we will present the quantum mechanical expression for the dielectric tensor $\varepsilon_{ij}\left(\omega, \vec{k}\right)$ and, in this sense, the expansion of $\varepsilon_{ij}\left(\omega, \vec{k}\right)$ in powers of $\vec{k}$ (see Sect. 1.7) will be justified.

B. Shokri, A. A. Rukhadze, *Electrodynamics of Conducting Dispersive Media*, Springer Series on Atomic, Optical, and Plasma Physics 111,
https://doi.org/10.1007/978-3-030-28968-3_5

An expression for the dielectric permittivity tensor is simply obtained by calculating the mean value of the current density induced in the medium under the action of the external field sources. In molecular crystals, in the absence of external electromagnetic fields, a microscopic field with the vector potential $\vec{A}_0\left(\vec{r}\right)$ exists. Under the action of external sources, an induced field with the potential $\vec{A}\left(\vec{r}\right)$ arises as well, which leads to the following induced current in crystals:

$$
\begin{aligned}
\vec{j}^{(n)}\left(\vec{r}\right) &= \left\langle \delta\psi_n^* \left| \sum_\alpha \frac{e_\alpha}{2m_\alpha}\left(\hat{\vec{p}}^{\alpha}\delta\left(\vec{r}-\vec{r}^{\alpha}\right)+\delta\left(\vec{r}-\vec{r}'\right)\hat{\vec{p}}^{\alpha}\right)\right| \psi_{n0}\right\rangle \\
&+\left\langle \psi_{n0}^* \left| \sum_\alpha \frac{e_\alpha}{2m_\alpha}\left(\hat{\vec{p}}^{\alpha}\delta\left(\vec{r}-\vec{r}^{\alpha}\right)+\delta\left(\vec{r}-\vec{r}^{\alpha}\right)\hat{\vec{p}}^{\alpha}\right)\right| \delta\psi_n^*\right\rangle \\
&-\left\langle \psi_n^* \left| \sum_\alpha \frac{e_\alpha^2}{m_\alpha c}\delta\left(\vec{r}-\vec{r}^{\alpha}\right)\vec{A}\left(\vec{r}^{\alpha}\right)\right| \psi_{n0}\right\rangle,
\end{aligned}
\tag{5.1}
$$

where $\hat{\vec{p}}^{\alpha} = (\hbar/\imath)\,\partial/\partial\vec{r}^{\alpha} - (e_\alpha/c)\vec{A}_0\left(\vec{r}^{\alpha}\right)$ is the operator of momentum and the summation is performed over all charged particles;[1] ψ_{n0} and $\delta\psi_n$ are the unperturbed wave function and its correction under the action of the field $\vec{A}\left(\vec{r}\right)$. If the medium is considered at a definite temperature, we have to average Eq. (5.1) over Gibbs distribution which is a simple procedure. Below, we consider only pure states and, as a result, averaging over Gibbs distribution is not considered.

We can calculate $\delta\psi_n$ by making use of the perturbation theory, assuming the field interaction operator (the perturbation energy operator) in the form of

$$
-\sum_\alpha \frac{e_\alpha}{2m_\alpha c}\left(\hat{\vec{p}}^{\alpha}\vec{A}\left(\vec{r}^{\alpha}\right)+\vec{A}\left(\vec{r}^{\alpha}\right)\hat{\vec{p}}^{\alpha}\right).
$$

As mentioned above, in the latter expression, we omitted the quadratic field terms and also we considered the gauge condition with zero scalar potential $\phi = 0$. Then, for the Fourier components $\vec{j}^{(n)}\left(\vec{k},\omega\right)$ and $\vec{E}\left(\vec{k},\omega\right)$, the following relations hold

$$
j_i^{(n)}\left(\vec{k},\omega\right) = \int d\vec{k}'\,\hat{\sigma}_{ij}^{(n)}\left(\omega,\vec{k},\vec{k}'\right)E_j\left(\vec{k}',\omega\right),
$$

where

[1] Since only the so-called π-electrons (weakly coupled) are important for the optical properties of the molecular crystals, the summation must be performed only over them.

$$\hat{\sigma}_{ij}^{(n)}\left(\omega,\vec{k},\vec{k'}\right)=\sum_{\alpha}\frac{\imath e_{\alpha}^{2}}{m_{\alpha}\omega}\delta_{ij}\delta\left(\vec{k}-\vec{k'}\right)-\sum_{\alpha,\beta}\sum_{m}\frac{\imath e_{\alpha}e_{\beta}}{(2\pi)^{3}4m_{\alpha}m_{\beta}\hbar\omega}$$
$$\times\left\{\left(p_i^{\alpha}\exp\left(-\imath\vec{k}\cdot\vec{r}^{\alpha}\right)+\exp\left(-\imath\vec{k}\cdot\vec{r}^{\alpha}\right)p_i^{\alpha}\right)\right)_{mn}\left(p_j^{\beta}\exp\left(\imath\vec{k'}\cdot\vec{r}^{\beta}\right)+\exp\left(\imath\vec{k'}\cdot\vec{r}^{\beta}\right)p_j^{\beta}\right)_{nm}\frac{1}{\omega+\omega_{mn}}$$
$$-\left(p_i^{\alpha}\exp\left(-\imath\vec{k}\cdot\vec{r}^{\alpha}\right)+\exp\left(-\imath\vec{k}\cdot\vec{r}^{\alpha}\right)p_i^{\alpha}\right)_{nm}\left(p_j^{\beta}\exp\left(\imath\vec{k'}\cdot\vec{r}^{\beta}\right)+\exp\left(\imath\vec{k'}\cdot\vec{r}^{\beta}\right)p_j^{\beta}\right)_{mn}\frac{1}{\omega-\omega_{mn}}\right\}.\tag{5.2}$$

Here, notation $(\ )_{mn}$ denotes the matrix element calculated by the unperturbed wave functions $\psi_{n0}\left(\vec{r}\right)$; $\hbar\omega_{mn}=E_m-E_n$, where E_n is the eigenvalue of energy in the absence of the electromagnetic field, and finally

$$\frac{1}{\omega\pm\omega_{mn}}=\mathcal{P}\frac{1}{\omega\pm\omega_{mn}}-\imath\pi\delta(\omega\pm\omega_{mn}).$$

For homogeneous media from Eq. (5.2), it follows [19, 20]

$$\hat{\sigma}_{ij}^{(n)}\left(\omega,\vec{k},\vec{k'}\right)=\delta\left(\vec{k}-\vec{k'}\right)\sigma^{(n)}\left(\omega,\vec{k}\right),$$

$$\hat{\sigma}_{ij}^{(n)}\left(\omega,\vec{k},\vec{k'}\right)=\sum_{\alpha}\frac{\imath e_{\alpha}^{2}}{m_{\alpha}\omega}\delta_{ij}\delta\left(\vec{k}-\vec{k'}\right)-\sum_{\alpha,\beta}\sum_{m}\frac{\imath e_{\alpha}e_{\beta}}{(2\pi)^{3}4m_{\alpha}m_{\beta}\hbar\omega}\times$$
$$\times\left\{\left(p_i^{\alpha}\exp\left(-\imath\vec{k}\cdot\vec{r}^{\alpha}\right)+\exp\left(-\imath\vec{k}\cdot\vec{r}^{\alpha}\right)p_i^{\alpha}\right)_{mn}\left(p_j^{\beta}\exp\left(\imath\vec{k'}\cdot\vec{r}^{\beta}\right)+\exp\left(\imath\vec{k'}\cdot\vec{r}^{\beta}\right)p_j^{\beta}\right)_{nm}\frac{1}{\omega+\omega_{mn}}-$$
$$-\left(p_i^{\alpha}\exp\left(-\imath\vec{k}\cdot\vec{r}^{\alpha}\right)+\exp\left(-\imath\vec{k}\cdot\vec{r}^{\alpha}\right)p_i^{\alpha}\right)_{nm}\left(p_j^{\beta}\exp\left(\imath\vec{k'}\cdot\vec{r}^{\beta}\right)+\exp\left(\imath\vec{k'}\cdot\vec{r}^{\beta}\right)p_j^{\beta}\right)_{mn}\frac{1}{\omega-\omega_{mn}}\right\}.\tag{5.3}$$

For the media without any magnetic structure and in the absence of an external magnetic field, from Eqs. (5.3) and (1.47) by considering the time reversal argument, one can find the following expression for the dielectric permittivity:

$$\varepsilon_{ij}^{(n)}\left(\omega,\vec{k}\right)=\left(1-\sum_{\alpha}\frac{4\pi e_{\alpha}^{2}}{\omega^{2}m_{\alpha}}\right)\delta_{ij}+\sum_{\alpha,\beta}\sum_{m}\frac{\pi e_{\alpha}e_{\beta}}{\hbar\omega^{2}m_{\alpha}m_{\beta}}\left\{\left(p_i^{\alpha}\exp\left(-\imath\vec{k}\cdot\vec{r}^{\alpha}\right)+\exp\left(-\imath\vec{k}\cdot\vec{r}^{\alpha}\right)p_i^{\alpha}\right)_{mn}\right.$$
$$\left.\times\left(p_j^{\beta}\exp\left(\imath\vec{k}\cdot\vec{r}^{\beta}\right)+\exp\left(\imath\vec{k}\cdot\vec{r}^{\beta}\right)p_j^{\beta}\right)_{nm}\left[\frac{1}{\omega+\omega_{mn}}-\frac{1}{\omega-\omega_{mn}}\right]\right\}.\tag{5.4}$$

In the long-wavelength range, we can expand this expression in powers of $\vec{k}$, which leads to an approximate expression as follows [compare to Eq. (1.132)]:

$$\varepsilon_{ij}^{(n)}\left(\omega,\vec{k}\right) = \varepsilon_{ij}^{(n)}(\omega) + \imath\gamma_{ijl}^{(n)}(\omega)\frac{c}{\omega}k_l + \alpha_{ijls}^{(n)}(\omega)\frac{c^2}{\omega^2}k_lk_s, \tag{5.5}$$

where, based on Eq. (5.3),

$$\begin{aligned}
\varepsilon_{ij}^{(n)}(\omega) &= \left(1 - \sum_{\alpha}\frac{4\pi e_\alpha^2}{m_\alpha\omega^2}\right)\delta_{ij} + \sum_{\alpha,\beta}\sum_{m}\frac{4\pi e_\alpha e_\beta}{m_\alpha m_\beta\hbar\omega^2}\left(p_i^\alpha\right)_{mn}\left(p_j^\beta\right)_{nm}\left[\frac{1}{\omega+\omega_{mn}} - \frac{1}{\omega-\omega_{mn}}\right],\\
\gamma_{ijl}^{(n)}(\omega) &= \sum_{\alpha\beta}\sum_{m}\frac{2\pi e_\alpha e_\beta}{m_\alpha m_\beta\omega c}\left\{\left(p_i^\alpha\right)_{mn}\left(p_j^\beta r_l^\beta + r_l^\beta p_j^\beta\right)_{nm}\right.\\
&\left. - \left(p_j^\beta\right)_{nm}\left(p_i^\alpha r_l^\alpha + r_l^\alpha p_i^\alpha\right)_{mn}\right\}\left[\frac{1}{\omega+\omega_{mn}} - \frac{1}{\omega-\omega_{mn}}\right],\\
\alpha_{ijls}^{(n)}(\omega) &= \sum_{\alpha\beta}\sum_{m}\frac{\pi e_\alpha e_\beta}{m_\alpha m_\beta\hbar c^2}\left\{\left(p_i^\alpha r_l^\alpha + r_l^\alpha p_i^\alpha\right)_{mn}\left(p_j^\beta r_s^\beta + r_s^\beta p_j^\beta\right)_{nm}\right.\\
&\left. - \left(p_i^\alpha\right)_{mn}\left(p_j^\beta r_l^\beta r_s^\beta + r_l^\beta r_s^\beta p_j^\beta\right)_{nm} - \left(p_j^\beta\right)_{nm}\left(p_i^\alpha r_l^\alpha r_s^\alpha + r_l^\alpha r_s^\alpha p_i^\alpha\right)_{mn}\right\}\left[\frac{1}{\omega+\omega_{mn}} - \frac{1}{\omega-\omega_{mn}}\right].
\end{aligned} \tag{5.6}$$

For gaseous media, when the wave function is the product of the single-particle wave functions, the expansion (5.5) coincides with the expansion in terms of the matrix elements of different multi-polarity.

The above-mentioned formulas of the complex dielectric permittivity tensor are admissible for every medium. However, when the energy spectrum of the medium is continuous, it is better to use other formulas. Moreover, translational invariance arises due to the presence of the periodic lattice in the crystals. This means that parallel translation transformation with the vector $\vec{a} = n_1\vec{a}_1 + n_2\vec{a}_2 + n_3\vec{a}_3$ keeps the Hamiltonian operator invariant where $n_{1,\,2,\,3}$ are integer numbers and $\vec{a}_i$ are three main lattice constants. As a result, the electron's wave function in the periodic field of the lattice is

$$\psi_n\left(\vec{k},\vec{r}\right) = \exp\left(\imath\vec{k}\cdot\vec{r}\right)u_n\left(\vec{k},\vec{r}\right),$$

where $\vec{k}$ is quasi-momentum and $u_n\left(\vec{k},\vec{r}+\vec{a}\right) = u_n\left(\vec{k},\vec{r}\right)$. Consequently, the electron's energy as a function of quasi-momentum is also periodic $E_n\left(\vec{k}+2\pi\vec{b}\right) = E_n\left(\vec{k}\right)$ where $\vec{b} = m_1\vec{b}_1 + m_2\vec{b}_2 + m_3\vec{b}_3$ (m_i are integer numbers, and $\vec{b}_i$ are the principal inverse lattice vectors, $\vec{a}_i\cdot\vec{b}_j = \delta_{ij}$). Considering aforementioned facts, relation (5.2) can be written as

$$\hat{\sigma}_{ij}^{(n)}\left(\omega, \vec{k}, \vec{k'}\right) = \sigma_{ij}^{(n)}\left(\omega, \vec{k}\right)\delta\left(\vec{k} - \vec{k'}\right) + \sum_{\vec{b}\neq 0}\sigma_{ij}^{(n)}\left(\omega, \vec{k}, \vec{k} + 2\pi\vec{b}\right)\delta\left(\vec{k'} - \vec{k} - 2\pi\vec{b}\right), \tag{5.7}$$

where

$$\sigma_{ij}^{(n)}\left(\omega, \vec{k}\right) = \sum_{\alpha}\frac{\imath e_\alpha^2}{\omega m_\alpha}\delta_{ij} - \sum_{\alpha,\beta}\sum_m \frac{\imath e_\alpha e_\beta}{4 m_\alpha m_\beta \hbar\omega}\times$$
$$\times\left\{\frac{\left(p_i^\alpha \exp\left(-\imath\vec{k}\cdot\vec{r}^\alpha\right) + \exp\left(-\imath\vec{k}\cdot\vec{r}^\alpha\right)p_i^\alpha\right)_{mn}\left(p_j^\beta \exp\left(\imath\vec{k}\cdot\vec{r}^\beta\right) + \exp\left(\imath\vec{k}\cdot\vec{r}^\beta\right)p_j^\beta\right)_{nm}}{\omega + \frac{1}{\hbar}\left[E_m\left(\vec{k}_n - \vec{k}\right) - E_n\left(\vec{k}_n\right)\right]} - \right.$$
$$\left. - \frac{\left(p_i^\alpha \exp\left(-\imath\vec{k}\cdot\vec{r}^\alpha\right) + \exp\left(-\imath\vec{k}\cdot\vec{r}^\alpha\right)p_i^\alpha\right)_{nm}\left(p_j^\beta \exp\left(\imath\vec{k}\cdot\vec{r}^\beta\right) + \exp\left(\imath\vec{k}\cdot\vec{r}^\beta\right)p_j^\beta\right)_{mn}}{\omega - \frac{1}{\hbar}\left[E_m\left(\vec{k}_n + \vec{k}\right) - E_n\left(\vec{k}_n\right)\right]}\right\},$$

$$\sigma_{ij}^{(n)}\left(\omega, \vec{k}, \vec{k} + 2\pi\vec{b}\right) = -\sum_{\alpha,\beta}\sum_m \frac{\imath e_\alpha e_\beta}{4 m_\alpha m_\beta \hbar\omega}\times$$
$$\times\left\{\frac{\left(p_i^\alpha \exp\left(-\imath\vec{k}\cdot\vec{r}^\alpha\right) + \exp\left(-\imath\vec{k}\cdot\vec{r}^\alpha\right)p_i^\alpha\right)_{mn}\left(p_j^\beta \exp\left[\imath\left(\vec{k} + 2\pi\vec{b}\right)\cdot\vec{r}^\beta\right] + \exp\left[\imath\left(\vec{k} + 2\pi\vec{b}\right)\cdot\vec{r}^\beta\right]p_j^\beta\right)_{nm}}{\omega + \frac{1}{\hbar}\left[E_m\left(\vec{k}_n - \vec{k}\right) - E_n\left(\vec{k}_n\right)\right]} - \right.$$
$$\left. - \frac{\left(p_i^\alpha \exp\left(-\imath\vec{k}\cdot\vec{r}^\alpha\right) + \exp\left(-\imath\vec{k}\cdot\vec{r}^\alpha\right)p_i^\alpha\right)_{nm}\left(p_j^\beta \exp\left[\imath\left(\vec{k} + 2\pi\vec{b}\right)\cdot\vec{r}^\beta\right] + \exp\left[\imath\left(\vec{k} + 2\pi\vec{b}\right)\cdot\vec{r}^\beta\right]p_j^\beta\right)_{mn}}{\omega - \frac{1}{\hbar}\left[E_m\left(\vec{k}_n + \vec{k}\right) - E_n\left(\vec{k}_n\right)\right]}\right\}. \tag{5.8}$$

It must be noted that, in a periodic and consequently inhomogeneous medium (a crystal), the functions of the type $\exp\left(\imath\vec{k}\cdot\vec{r}\right)$ (plane waves) are not the solution of the field equations. In fact, for example, for the electric field, we have $\vec{E}\left(\vec{r}\right) = \exp\left(\imath\vec{k}\cdot\vec{r}\right)\vec{E}_0\left(\vec{k}, \vec{r}\right)$, where $\vec{E}_0\left(\vec{k}, \vec{r}\right)$ is a periodic function, $\vec{E}_0\left(\vec{k}, \vec{r} + \vec{a}\right) = \vec{E}_0\left(\vec{k}, \vec{r}\right)$. Hence, it follows that the Fourier components of the electric field are periodic functions with the period $2\pi\vec{b}$. Namely, for this reason the diffraction of electromagnetic waves happens in the crystals. In the theory of the phenomena not related to the diffraction the summation over $\vec{b} \neq 0$ on the right-hand side of Eq. (5.7) can be neglected and, as a result, the crystal is considered as a homogeneous medium. In this case, the second term in Eq. (5.7) can be omitted and the following expression (in crystals and in the absence of the external magnetic field and magnetic properties) will be valid:

$$\varepsilon_{ij}^{(n)}\left(\omega,\vec{k}\right)=\left(1-\sum_{\alpha}\frac{4\pi e_{\alpha}^{2}}{m_{\alpha}\omega^{2}}\right)\delta_{ij}+\sum_{\alpha\beta}\sum_{m}\frac{\pi e_{\alpha}e_{\beta}}{m_{\alpha}m_{\beta}\hbar\omega^{2}}$$
$$\times\left(p_{i}^{\alpha}\exp\left(-\imath\vec{k}.\vec{r}^{\alpha}\right)+\exp\left(-\imath\vec{k}.\vec{r}^{\alpha}\right)p_{i}^{\alpha}\right)_{mn}\left(p_{j}^{\beta}\exp\left(\imath\vec{k}.\vec{r}^{\beta}\right)+\exp\left(\imath\vec{k}.\vec{r}^{\beta}\right)p_{j}^{\beta}\right)_{nm}$$
$$\times\left[\frac{1}{\omega+\frac{1}{\hbar}\left[E_{m}\left(\vec{k}_{n}-\vec{k}\right)-E_{n}\left(\vec{k}_{n}\right)\right]}-\frac{1}{\omega-\frac{1}{\hbar}\left[E_{m}\left(\vec{k}_{n}+\vec{k}\right)-E_{n}\left(\vec{k}_{n}\right)\right]}\right].\tag{5.9}$$

Matrix elements in Eqs. (5.4) and (5.9) are calculated by the unperturbed crystal wave functions. The advantage of Eq. (5.9) compared to Eq. (5.4) is that Eq. (5.9) explicitly considers the zone structure of the crystal energy spectrum. It must be noted that, quite analogically, relation (5.6) can be written.

The most interesting consequences of the zone characteristics of the energy spectrum arise when the reverse expansion of the dielectric tensor of the crystal is possible (see Sect. 1.7):

$$\frac{1}{\varepsilon_{ij}^{(n)}\left(\omega,\vec{k}\right)}=\frac{1}{\varepsilon_{ij}^{(n)}(\omega)}+\imath g_{ijl}^{(n)}(\omega)\frac{c}{\omega}k_{l}+\beta_{ijls}^{(n)}(\omega)\frac{c^{2}}{\omega^{2}}k_{l}k_{s}.\tag{5.10}$$

Usually, such an expansion is valid near one of the absorption lines (resonance frequency) when $\omega\approx 1/\hbar\left[E_{m}\left(\vec{k}_{n}\right)-E_{n}\left(\vec{k}_{n}\right)\right]$. Then, from Eq. (5.9), it follows

$$\frac{1}{\varepsilon_{ij}^{(n)}(\omega)}=\alpha_{ij}^{-1}(\omega)\left\{\omega-\frac{1}{\hbar}\left[E_{m}\left(\vec{k}_{n}\right)-E_{n}\left(\vec{k}_{n}\right)\right]\right\};$$
$$g_{ijl}^{(n)}(\omega)=-\imath\alpha_{ij}^{-1}(\omega)\frac{\omega}{\hbar c}\frac{\partial E_{m}\left(\vec{k}_{n}\right)}{\partial k_{nl}};\quad\beta_{ijls}^{(n)}(\omega)=\alpha_{ij}^{-1}(\omega)\frac{\omega^{2}}{\hbar c^{2}}\frac{\partial^{2}E_{m}\left(\vec{k}_{n}\right)}{\partial k_{nl}\partial k_{ns}}.\tag{5.11}$$

Here, we neglected the wave absorption and introduced

$$\alpha_{ij}(\omega)=-\sum_{\alpha,\beta}\frac{4\pi e_{\alpha}e_{\beta}}{m_{\alpha}m_{\beta}\hbar\omega^{2}}\left(p_{i}^{\alpha}\right)_{mn}\left(p_{j}^{\beta}\right)_{nm}\Bigg|_{\vec{k}=0}.$$

From Eq. (5.11), it follows that near the absorption band the expansion coefficients $g_{ijl}^{n}(\omega)$ and $\beta_{ijls}^{n}(\omega)$ are practically independent of the frequency. Equations (5.5), (5.6), (5.10), and (5.11) determine the coefficients of direct and inverse expansion of the dielectric permittivity tensor of the medium, which are necessary for describing the electromagnetic waves in the media with weak spatial dispersion (see Sect. 1.7).

Now, let us consider some consequences of the obtained formulas for the molecular crystals. As an example, we consider the Naphthalene crystal ($C_{10}H_8$). To determine the specific properties of this crystal, simultaneously, we consider the properties of Naphthalene molecules. Since the molecular Naphthalene spectrum is discrete, then, in this case, it is convenient to use the expression (5.4).

The Naphthalene molecule which has a planar structure belongs to the symmetry group D_{2h} [1, 21]. The symmetry elements determining this group are

1. E: identity symmetry element;
2. C_2^x, C_2^y, C_2^z: three axes of second-order rotation;
3. I: inversion;
4. σ^x, σ^y, σ^z: reflections in three mutually perpendicular planes.

The characteristics of irreducible representations of the group D_{2h} and the transformation properties of the quantities r_i and r_ir_j are listed in Table 5.1.

In the Naphthalene molecule, absorption, in the optical spectrum, is caused by π-electrons. It was shown that the excited state of π-electrons of the Naphthalene molecule may belong only to the following four irreducible representations: A_{1g}, A_{2g}, B_{1u}, and B_{2u} [14]. The normal state of this molecule is related to the irreducible totally symmetric representation A_{1g}. In the dipole approximation, the transition from the ground state A_{1g} to the states of B_{1u} (z-polarization) and B_{2u} (y-polarization) is permissible (see Table 5.1). The transition from the ground state A_{1g} to the excited states A_{1g} and A_{2g}, in the dipole approximation, is forbidden, but, in the quadrupole approximation, is permissible. The probability of multipole transitions is less than the probability of dipole transition by the quantity of the order of $(ak)^{2l}$~$(2\pi a/\lambda)^{2l}$ where a is the molecule size, and l is the multipole order. Therefore, in the spectrum of the Naphthalene molecule the lines corresponding to the transition to the states A_{1g} and A_{2g} should have much less intensity than the lines corresponding to the transition from the ground state to the states B_{1u}, and B_{2u}. In this case, the expansion (5.5) is an expansion in the power of the ratio $(2\pi a/\lambda)$. In spite of the smallness of the expansion parameter $(2\pi a/\lambda)$ in the expression (5.5), in the definite range of frequency, the terms, which correspond to the high multipole transition, can be essential. Thus, for example, near the line forbidden in the dipole approximation,

Table 5.1 Characteristics of irreducible representations of group D_{2h} and transformation properties of r_i and r_ir_j

D_{2h}	E	C^X2	C^Y2	C^Z2	I	σ^x	σ^y	σ^z	r_i	r_irj
A_{1g}	1	1	1	1	1	1	1	1	–	x^2; y^2; z^2
B_{1g}	1	−1	−1	1	1	−1	−1	1	–	xy
A_{2g}	1	1	−1	−1	1	1	−1	−1	–	yz
B_{2g}	1	−1	1	−1	1	−1	1	−1	–	xz
A_{1u}	1	1	1	1	−1	−1	−1	−1	–	–
B_{1u}	1	−1	−1	1	−1	1	1	−1	z	–
A_{2u}	1	1	−1	−1	−1	−1	1	1	x	–
B_{2u}	1	−1	1	−1	−1	1	−1	1	y	–

which corresponds to the transition to the excited state A_{1g} $(\lambda \approx 1960\,\text{Å})$ with the line half-width $\sim 10^{-2}$ Å, the value of the second coefficient in the expansion (5.5), $\alpha_{ijls}(\omega)$, may become of the order of unity.[2] The latter results in the substantial change of the electromagnetic wave's refractive index in such a frequency range [16]. It should be mentioned that the experimental observation of the above-mentioned effect of spatial dispersion is substantially facilitated due to the smallness of absorption near the line forbidden in the dipole approximation and permissible in the quadrupole approximation [23].

Now, let us consider the Naphthalene crystal which belongs to the symmetry group C_{2h} with two Naphthalene molecules in the unit cell. This group is determined by the following symmetry elements:

1. E: identity symmetry element;
2. C_2: the axis of second-order rotation;
3. I: inversion;
4. σ_h: reflection in the plane perpendicular to the rotation axis.

The characteristics of irreducible representations of the group C_{2h} and the transformation properties of the quantities r_i and r_ir_j are listed in Table 5.2.

The operation of the identity transformation E and inversion I coincides in the molecule and crystal. The operation C_2^y corresponds to the permutation of molecules 1 and 2 with successive rotations around its middle axis (y), i.e., the product of operations C_2^y and σ^y for the molecule. From Table 5.1, we have

Table 5.2 Characteristics of irreducible representations of group C_{2h} and transformation properties of r_i and r_ir_j

C_{2h}	E	C_2	I	σ_h	r_i	r_ir_j
A_g	1	1	1	1	–	x^2; y^2; z^2; xy
A_u	1	1	−1	−1	z	–
B_g	1	−1	1	−1	–	xz; yz
B_u	1	−1	−1	1	x, y	–

[2] In fact, from Eq. (5.8) it follows

$$\alpha(\omega) \sim \frac{4\pi N e^2}{m}\left(\frac{m\omega_{mn}a^2}{\hbar}\right)^2 \frac{\hbar\omega_{mn}}{mc^2}\frac{1}{\omega^2 - \omega_{mn}^2 - \iota\omega\nu}.$$

When $a \sim 10^{-8}$ cm, we have

$$\omega_0 = \left(\frac{4\pi e^2 N}{m}\right)^{1/2} \sim \omega_{mn} \sim \omega \sim 10^{16}\,\text{s}^{-1}, \qquad \frac{\Delta\lambda}{\lambda} \sim \frac{\nu}{\omega} \sim 10^{-5},$$

which corresponds to the line half-width $\Delta\lambda \sim 10^{-2}$ Å. Therefore, $\alpha(\omega) \sim 1$. The spectral line half-width of the molecule and crystal of Naphthalene at low temperatures is of the order of $\sim 10^{-2}$ to 10^{-1} Å [22, 23].

$$E\Phi_1^{B_{1u}} = \Phi_1^{B_{1u}}; \qquad C_2\Phi_1^{B_{1u}} = -\Phi_1^{B_{1u}};$$
$$I\Phi_1^{B_{1u}} = -\Phi_1^{B_{1u}}; \qquad \sigma_h\Phi_1^{B_{1u}} = \Phi_1^{B_{1u}}.$$

From here, it follows that the function $\Phi_1^{B_{1u}}$ is transformed by an irreducible representation of the crystal symmetry group B_u, i.e., $\Phi_1^{B_{1u}} \sim B_u$. Quite analogically, we find

$$\Phi_1^{B_{1u}}, \Phi_2^{A_{2g}} \sim A_g; \qquad \Phi_1^{A_{2g}}, \Phi_2^{A_{1g}} \sim B_g;$$
$$\Phi_1^{B_{2u}}, \Phi_2^{B_{1u}} \sim A_u; \qquad \Phi_1^{B_{1u}}, \Phi_2^{B_{2u}} \sim B_u.$$

From the latter relations, one can simply determine the selection rules for transition matrix elements in the crystals. In the dipole approximation (see Table 5.2), transitions from the crystal ground state A_{1g} to the state with the wave functions $\Phi_1^{B_{2u}}, \Phi_2^{B_{1u}}$ (~A_u, z polarization) and $\Phi_1^{B_{1u}}, \Phi_2^{B_{2u}}$ (~B_u, x or y-polarization) are permissible. Dipole transitions from the ground state to the states $\Phi_1^{A_{1g}}, \Phi_2^{A_{2g}}$ (~A_g) and $\Phi_1^{A_{2g}}, \Phi_2^{A_{1g}}$ (~B_g) are forbidden. However, these transitions are permissible in the quadrupole approximation.

Thus, each permissible transition in the molecule corresponds to the two permissible transitions in the crystal with the same multi-polarity. This means that the each energy term of the molecule in the crystal splits in two terms with different polarizations.[3] It should be mentioned that the above results, which have been obtained from the general symmetry properties of the crystal, are independent of the approximation used in the calculation of the crystal energy spectrum. The experimental values of the spectral lines of the Naphthalene molecule and the characteristics of the splitting of these lines in the crystal are listed in Table 5.3.

It must be noted that for molecular media discussed above, the expression (5.5) presents an expansion in terms of the matrix elements of different multipoles and, therefore, the parameter of such an expansion is the ratio of the molecule size to the radiation wavelength. At the same time, for crystals such an expansion (particularly, Eqs. (5.5) and (5.11)) corresponds to the expansion of $E_m\left(\vec{k}\right)$ in the

Table 5.3 Experimental values of the spectral lines of Naphthalene molecule and characteristics of the splitting of these lines in the crystal

Naphthalene pairs	Naphthalene crystal		
λ, Å	λ (*b*-component), Å	λ (*a*-component), Å	Splitting Δλ, Å
3100 (B_{2u})	3350	3355	~5
2700 (B_{1u})	3160	3200	~40
2800 (B_{1u})	3250	3255	~5
2650 (B_{1u})	3120	3130	~10

[3] For more details, see [1].

powers of $\vec{k}$ where $E_m\left(\vec{k}\right)$ are the eigenvalues of the crystal energy. Since, in molecular crystals, the interaction between molecules is weak, then the crystal energy spectrum slightly differs from the molecule spectrum $E_m\left(\vec{k}\right) = E_m + V_m\left(\vec{k}\right)$ where the correction $V_m\left(\vec{k}\right)$ arises due to the molecules interaction $V_m/E_m \sim 0.1$. Furthermore, the Van-der Waals attraction force, acting between molecules in crystals, is sufficiently short range: its range of action is of the order of the lattice constant. This means that the ratio of the lattice constant to the wavelength works as the expansion parameter of the function $E_m\left(\vec{k}\right)$ and, hence, the dielectric permittivity tensor of the crystal in terms of the wave vector.

In conclusion, let us notice that the dispersion of the dielectric permittivity of molecular crystals can be described phenomenologically using the Lagrangian of the medium, which interacts with the electromagnetic fields:

$$L = \hat{F}_{ij}^{(1)}\left(\frac{\partial}{\partial t}, \frac{\partial}{\partial \vec{r}}\right)\mathcal{P}_i\mathcal{P}_j + \frac{1}{c}A_i\frac{\partial}{\partial t}\hat{F}_{ij}^{(2)}\left(\frac{\partial}{\partial t}, \frac{\partial}{\partial \vec{r}}\right)\mathcal{P}_j. \tag{5.12}$$

Here, $\hat{F}_{ij}^{(1,2)}$ are the polynomial operators being invariant with respect to space and time reflections; the vector quantity $\vec{\mathcal{P}}$ characterizes the electromagnetic properties of the medium, and $\vec{A}$ is the vector potential of electromagnetic fields (for the gauge condition, we consider the scalar potential to be zero). Then, for the current density, one can write

$$j_i = c\frac{\partial L}{\partial A_i} = \frac{\partial}{\partial t}\hat{F}_{ij}^{(2)}\left(\frac{\partial}{\partial t}, \frac{\partial}{\partial \vec{r}}\right)\vec{\mathcal{P}}_j. \tag{5.13}$$

To determine the conductivity and dielectric permittivity tensors, $\vec{\mathcal{P}}$ should be expressed in terms of the electric field $\vec{E}$. Restricting on the first-order spatial and temporal derivatives of $\vec{\mathcal{P}}$ and making use of the Lagrange equation

$$\frac{\partial}{\partial t}\frac{\partial L}{\partial \frac{\partial \mathcal{P}_i}{\partial t}} + \frac{\partial}{\partial \vec{r}}\cdot\frac{\partial L}{\partial \frac{\partial \mathcal{P}_i}{\partial \vec{r}}} - \frac{\partial L}{\partial \mathcal{P}_i} = 0, \tag{5.14}$$

one can obtain the following expression for the conductivity and dielectric tensors [24] (for the field dependence of the type $\sim \exp\left(\imath\omega t + \imath\vec{k}\cdot\vec{r}\right)$):

$$\begin{aligned} \sigma_{ij}\left(\omega, \vec{k}\right) &= \imath\omega\left\{\omega^2\delta_{il} - \beta_{il} - \imath\gamma_{ilm}k_m - \alpha_{ilmr}k_rk_m\right\}^{-1} f_{lj}, \\ \varepsilon_{ij}\left(\omega, \vec{k}\right) &= \delta_{ij} - 4\pi\left\{\omega^2\delta_{il} - \beta_{il} - \imath\gamma_{ilm}k_m - \alpha_{ilmr}k_rk_m\right\}^{-1} f_{lj}, \end{aligned} \tag{5.15}$$

where f_{ij}, β_{ij}, γ_{ijl}, α_{ijlm} are the constants characterized by the electromagnetic properties of the medium. Tensor γ_{ijl} corresponding to the optical activity is zero for non-gyrotropic media.

For isotropic non-gyrotropic media, by taking into account

$$\beta_{ij} = \beta\delta_{ij}, \quad f_{ij} = f\delta_{ij}, \quad \gamma_{ijl} = 0, \quad \alpha_{ijlm}k_l k_m = \alpha_1 k^2 \delta_{ij} + \alpha_2 k_i k_j,$$

from Eq. (5.15), we find

$$\begin{aligned} \varepsilon^{tr}(\omega, k) &= 1 - \frac{\omega_0^2}{\omega^2 - (\beta + \alpha_1 k^2)}, \\ \varepsilon^{l}(\omega, k) &= 1 - \frac{\omega_0^2}{\omega^2 - (\beta + \alpha_1 k^2 + \alpha_2 k^2)}. \end{aligned} \tag{5.16}$$

Here, $\omega_0^2 = 4\pi f$. To find out the meaning of f, we consider this fact that in the limit of $\omega \to \infty$ [25]

$$\varepsilon(\omega) = 1 - \frac{4\pi e^2 N}{m\omega^2},$$

where N is the number of electrons in the unit volume. Therefore, we conclude that $\omega_0^2 = 4\pi f = 4\pi e^2 N/m$. In the derivation of Eqs. (5.15) and (5.16), we took into account only the first-order spatial and temporal derivatives of the quantity $\vec{\mathcal{P}}$ in the Lagrangian (5.12). In the general case of the derivatives of arbitrary order, from formula (5.12), we find the following expressions for transverse and longitudinal permittivities of the isotropic non-gyrotropic media:[4]

$$\begin{aligned} \varepsilon^{tr}(\omega, k) &= 1 - \sum_n \frac{4\pi f_n^{tr}\left(\dfrac{k}{\omega}\right)}{\omega^2 - \left[\omega_n^{tr}(k)\right]^2}, \\ \varepsilon^{l}(\omega, k) &= 1 - \sum_n \frac{4\pi f_n^{l}\left(\dfrac{k}{\omega}\right)}{\omega^2 - \left[\omega_n^{l}(k)\right]^2}, \end{aligned} \tag{5.17}$$

[4] It should be noted that the Lagrange equation, in this case, is written as

$$\sum_n \sum_m (-1)^{n+m} \frac{\partial^{n+m}}{\partial t^n \partial \vec{r}_1 \ldots \partial \vec{r}_m} \frac{\partial L}{\partial \frac{\partial^{n+m} \mathcal{P}_i}{\partial t^n \partial \vec{r}_1 \ldots \partial \vec{r}_m}} = 0.$$

When the Lagrangian depends on the first-order derivatives, this equation coincides with Eq. (5.14).

where $\omega_n^{tr}(k), f_n^{tr}(k/\omega), \omega_n^{tr}(k)$, and $f_n^l(k/\omega)$ are polynomials of the even powers of k. Furthermore, in the high-frequency limit, $\omega \to \infty$,

$$4\pi \sum_n f_n^{tr} = 4\pi \sum_n f_n^l = \frac{4\pi N e^2}{m}.$$

5.2 The Permittivity of a Monatomic Gas with Spatial Dispersion

In this section, permittivity of a monotonic gas is derived [26]. For this aim, first we note that determining the permittivity tensor $\varepsilon_{ij}\left(\omega, \vec{k}\right)$ reduces to calculating the density of the electric current induced by an electromagnetic field $\vec{E}\left(\vec{r}, t\right), \vec{B}\left(\vec{r}, t\right)$ in a medium [26–28]:

$$\vec{j}\left(\vec{r}, t\right) = \mathrm{Tr}\left(\hat{W}\hat{\vec{j}}\left(\vec{r}, t\right)\right)/\mathrm{Tr}\,\hat{W}. \tag{5.18}$$

Here, $\hat{\vec{j}}\left(\vec{r}, t\right)$ is the electric current density operator, and $\hat{W}$ is the density matrix that satisfies the equation [29, 30]

$$\dot{\hat{W}} = -\imath\left[\hat{H}, \hat{W}\right], \tag{5.19}$$

where $\hat{H}$ is the total Hamiltonian[5] of the "medium + electromagnetic field" system, including the interaction energy $\hat{U}$ between them.

Denote the complete set of quantum numbers that characterize the stationary states of the medium in the absence of a field by α (or β) and the corresponding energy levels by E_α)or E_β). In this notation, the solution of Eq. (5.19) in the linear (in $\hat{U}$) approximation is

$$W_{\alpha\beta}(t) = W_\alpha^{(0)}\delta_{\alpha\beta} + \left(W_\alpha^{(0)} - W_\beta^{(0)}\right)\int \frac{U_{\alpha\beta}(\omega)}{\omega_{\alpha\beta} - \omega} e^{-\imath\omega t}\, d\omega, \tag{5.20}$$

where $\omega_{\alpha\beta} = E_\alpha - E_\beta$, $W_{\alpha\beta}^{(0)} = W_\alpha^{(0)}\delta_{\alpha\beta}$ is the density matrix of the medium in the absence of a field and $U_{\alpha\beta}(\omega)$ is the Fourier component of the matrix element $U_{\alpha\beta}(t)$. We assume that the medium in the absence of a field is in thermodynamic equilibrium at temperature T. In this case [29],

[5]We use a system of units in which $\hbar = 1$.

$$W_\alpha^{(0)} = e^{(F-E_\alpha)/T}, \quad \mathrm{Tr}\hat{W}^{(0)} = \sum_\alpha W_\alpha^{(0)} = 1, \tag{5.21}$$

where F is the free energy of the medium.[6] We see from relation (5.20) and (5.21) that $\mathrm{Tr}\,\hat{W} = 1$.

The current density operator $\hat{\vec{j}}\left(\vec{r}, t\right)$ consists of two parts:

$$\hat{\vec{j}}\left(\vec{r}, t\right) = \hat{\vec{j}}^{(0)}\left(\vec{r}\right) + \delta\hat{\vec{j}}\left(\vec{r}, t\right), \tag{5.22}$$

where $\hat{\vec{j}}^{(0)}\left(\vec{r}\right)$ is the (time-independent) current density operator in the absence of a field, and the operator $\delta\hat{\vec{j}}\left(\vec{r}, t\right)$ is proportional to the field. Substituting relations (5.20) and (5.22) into Eq. (5.18), we obtain the following expression for the Fourier components of the current density in the linear (in field) approximation:

$$\vec{j}\left(\omega, \vec{k}\right) = \sum_\alpha W_\alpha^{(0)}\left[\delta\vec{j}_{\alpha\alpha}\left(\omega, \vec{k}\right) - \sum_\beta\left(\frac{U_{\alpha\beta}(\omega)\vec{j}^{(0)}_{\beta\alpha}\left(\vec{k}\right)}{\omega_{\beta\alpha} + \omega} + \frac{\vec{j}^{(0)}_{\alpha\beta}\left(\vec{k}\right)U_{\beta\alpha}(\omega)}{\omega_{\beta\alpha} - \omega}\right)\right]. \tag{5.23}$$

The permittivity tensor of the medium can be derived from Eq. (5.23), using (Sect. 1.2)

$$\begin{aligned} \varepsilon_{ij}\left(\omega, \vec{k}\right) &= \delta_{ij} + \frac{4\pi\imath}{\omega}\sigma_{ij}\left(\omega, \vec{k}\right), \\ j_i\left(\omega, \vec{k}\right) &= \sigma_{ij}\left(\omega, \vec{k}\right)E_j\left(\omega, \vec{k}\right). \end{aligned} \tag{5.24}$$

For reasons of symmetry, the permittivity tensor for an isotropic non-gyrotropic medium can be written as [see Eq. (1.24)]

$$\varepsilon_{ij}\left(\omega, \vec{k}\right) = \varepsilon^{\mathrm{tr}}(\omega, k)\left(\delta_{ij} - \frac{k_i k_j}{k^2}\right) + \varepsilon^{l}(\omega, k)\frac{k_i k_j}{k^2}. \tag{5.25}$$

The scalar potential of the electromagnetic field is assumed to be equal to zero. The Fourier components of the fields $\vec{E}\left(\omega, \vec{k}\right)$ and $\vec{B}\left(\omega, \vec{k}\right)$ and the vector potential $\vec{A}\left(\omega, \vec{k}\right)$ are then related by [27]

---

[6]The temperature is measured in energy units.

$$\vec{E}\left(\omega,\vec{k}\right)=\frac{\imath\omega}{c}\vec{A}\left(\omega,\vec{k}\right),\quad \vec{B}\left(\omega,\vec{k}\right)=\imath\vec{k}\times\vec{A}\left(\omega,\vec{k}\right). \tag{5.26}$$

Now let us calculate $\varepsilon^l(\omega,k)$ and $\varepsilon^{\text{tr}}(\omega,k)$. If the medium is an ideal gas, then to determine the current density $\vec{j}\left(\vec{r},t\right)$, it will suffice to calculate the current produced by one atom: multiplying it by the total number of atoms NV (N is the number of molecules per unit volume, and V is the volume of the gas) yields an expression for $\vec{j}\left(\vec{r},t\right)$ in the gas.

The coordinates of Z electrons and the atomic nucleus in the laboratory frame of reference in which the fields $\vec{E}\left(\vec{r},t\right)$, $\vec{B}\left(\vec{r},t\right)$ are denoted by $\vec{R}_a(a=1,2,\ldots,Z)$ and $\vec{R}_n$, respectively. If we disregard the terms of the order of m/M_n, where m and M_n are the electron and nuclear masses, respectively, then the atom-field interaction energy in the linear (in field) approximation is [21]

$$\hat{U}=\mu_B\sum_a\left[\hat{\vec{P}}_a\cdot\vec{A}\left(\vec{R}_a,t\right)+\vec{A}\left(\vec{R}_a,t\right)\cdot\hat{\vec{P}}_a+2\hat{\vec{s}}_a\cdot\vec{B}\left(\vec{R}_a,t\right)\right], \tag{5.27}$$

where $\mu_B=e/2mc$ is the Bohr magneton, $\hat{\vec{P}}_a=-\imath\partial/\partial\vec{R}_a$, and $\hat{\vec{s}}_a$ are the electron momentum and spin operators, respectively.

The expression for the current density operator is

$$\begin{aligned}\frac{\hat{\vec{j}}\left(\vec{r},t\right)}{NV}&=-\frac{e}{2m}\sum_a\left[\delta\left(\vec{R}_a-\vec{r}\right)\left(\hat{\vec{P}}_a+2\imath\hat{\vec{P}}_a\times\hat{\vec{s}}_a\right)+\left(\hat{\vec{P}}_a-2\imath\hat{\vec{P}}_a\times\hat{\vec{s}}_a\right)\delta\left(\vec{R}_a-\vec{r}\right)\right]\\&-\frac{e^2}{2mc}\vec{A}\left(\vec{r},t\right)\sum_a\left[\delta\left(\vec{R}_a-\vec{r}\right)\right].\end{aligned} \tag{5.28}$$

It differs from the standard expression for the current density only by the inclusion of spins.

To calculate the matrix elements in Eq. (5.23), we should pass from $\vec{R}_a$ and $\vec{R}_n$ to the coordinates of the electrons relative to the nucleus, $\vec{r}_a$, and to the coordinates of the atomic center of mass, $\vec{R}$ [31]:

$$\vec{r}_a=\vec{R}_a-\vec{R}_n,\quad \vec{R}=\frac{1}{M}\left(m\sum_a\vec{R}_a+M_n\vec{R}_n\right), \tag{5.29}$$

where $M=Zm+M_n$ is the atomic mass. The complete set of quantum numbers and the atomic energy can be represented as

$$\alpha = \left\{\vec{P}, J, M, n\right\}, \quad E_\alpha = \frac{1}{2M}P^2 + E_{Jn}. \tag{5.30}$$

Here, $\vec{P}$ is the momentum of the atomic center of mass, J is the angular momentum ($J = 0, 1/2, 3/2, \ldots$), $M = J, J = -1, \ldots, -J$ is the angular momentum component along the z axis, and E_{Jn} is the atomic energy in the frame of reference in which $\vec{P} = 0$; n numbers atomic states with equal J and M, but with different energies.

Simple calculations yield the following expressions for the matrix elements of the Fourier components of operators (5.27) and (5.28):

$$\begin{aligned}
\left\langle \vec{P}' J'M'n' \middle| \hat{U}(\omega) \middle| \vec{P}JMn \right\rangle &= -\frac{\imath e}{m\omega}\frac{(2\pi)^3}{V}\vec{E}\left(\omega, \vec{P}' - \vec{P}\right) \\
&\quad \times \left\langle J'M'n' \middle| \hat{\vec{P}}\left(\vec{P}' - \vec{P}\right) \middle| JMn \right\rangle, \\
\left\langle \vec{P}' J'M'n' \middle| \hat{\vec{j}}^{(0)}\left(\vec{k}\right) \middle| \vec{P}JMn \right\rangle &= -\frac{eNV}{m(2\pi)^3}\delta_{\boldsymbol{P}',\boldsymbol{P}-\boldsymbol{k}}\left\langle J'M'n' \middle| \hat{\vec{P}}\left(-\vec{k}\right) \middle| JMn \right\rangle, \\
&\left\langle \vec{P}' J'M'n' \middle| \delta\hat{\vec{j}}\left(\omega, \vec{k}\right) \middle| \vec{P}JMn \right\rangle \\
= \frac{\imath e^2 N}{m\omega}\vec{E}\left(\omega, \vec{k} - \vec{P} + \vec{P}'\right) &\times \left\langle J'M'n' \middle| \sum_a \exp\left[\imath\left(\vec{P} - \vec{P}'\right)\cdot \boldsymbol{r}_a\right] \middle| JMn \right\rangle.
\end{aligned} \tag{5.31}$$

We introduced the operators $\hat{\vec{p}}_a = -\imath\partial/\partial\vec{r}_a$ and

$$\hat{\vec{P}}\left(\vec{k}\right) = \sum_a \left\{\frac{1}{2}\left[\hat{\vec{p}}_a \exp\left(\imath\vec{k}\cdot\vec{r}_a\right) + \exp\left(\imath\vec{k}\cdot\vec{r}_a\right)\hat{\vec{p}}_a\right] + \imath\hat{\vec{s}}_a \times \vec{k}\exp\left(\imath\vec{k}\cdot\vec{r}_a\right)\right\}. \tag{5.32}$$

Given expressions (5.30), the normalization condition (5.21) after passing from summation to integration over $\vec{P}$ takes the form

$$V\left(\frac{MT}{2\pi}\right)^{3/2}\sum_{Jn}(2J+1)\exp\left(\frac{F - F_{Jn}}{T}\right) = 1. \tag{5.33}$$

Using formulas (5.30)–(5.33), we can determine the current density $\vec{j}\left(\omega, \vec{k}\right)$ [see Eq. (5.23)] and then the permittivity tensor [see Eq. (5.24)]:

$$\varepsilon_{ij}\left(\omega,\vec{k}\right)=\left(1-\frac{\omega_P^2}{\omega^2}\right)\delta_{ij}-\frac{4\pi e^2NV}{m^2\omega^2}\left(\frac{MT}{2\pi}\right)^{3/2}$$
$$\times\sum_{Jn}\exp\left(\frac{F-E_{Jn}}{T}\right)\times\sum_{J'n'}A^{ij}_{J'n',Jn}\left(\vec{k}\right)\Phi_{J'n',Jn}(\omega,k;T). \tag{5.34}$$

Here,$\omega_p=\sqrt{4\pi e^2N_e/m}$ is the plasma frequency of the atomic electrons ($N_e=ZN$),

$$\Phi_{J'n',Jn}(\omega,k;T)=\frac{1}{\sqrt{2}k\bar{\mathrm{v}}}\left[\frac{I_+\left(z^{(+)}_{J'n',Jn}\right)}{z^{(+)}_{J'n',Jn}}-\frac{I_+\left(z^{(-)}_{J'n',Jn}\right)}{z^{(-)}_{J'n',Jn}}\right],$$
$$z^{(\pm)}_{J'n',Jn}=\frac{\omega\pm\left(\omega_{J'n',Jn}+k^2/2M\right)}{\sqrt{2}k\bar{\mathrm{v}}}, \tag{5.35}$$

where $\bar{\mathrm{v}}=\sqrt{T/M}$ is the mean thermal velocity of the atoms,

$$I_+(z)=\frac{z}{\sqrt{\pi}}\int\limits_{-\infty}^{\infty}\frac{\exp\left(-t^2\right)}{t-z}dt, \tag{5.36}$$

$$A^{(ij)}_{J'n',Jn}\left(\vec{k}\right)=\sum_{MM'}\left\langle JMn\middle|\hat{P}_i\left(-\vec{k}\right)\middle|J'M'n'\right\rangle\left\langle J'M'n'\middle|\hat{P}_j\left(\vec{k}\right)\middle|JMn\right\rangle$$
$$=A^{ji}_{J'n',Jn}\left(-\vec{k}\right). \tag{5.37}$$

The last equality in Eq. (5.37) can be easily obtained by taking into account the fact that after time reversal [21],

$$|JMn\rangle\rightarrow(-1)^{J-M}|J,-M,n\rangle,$$
$$\vec{r}_a\rightarrow\vec{r}_a,\quad\hat{\vec{p}}_a\rightarrow-\hat{\vec{p}}_a,\quad\hat{\vec{s}}_a\rightarrow-\hat{\vec{s}}_a.$$

Since the states $|JMn\rangle$ with different M are transformed via the irreducible representation $D^{(J)}$ of the rotation group and since all of the states $|JMn\rangle$ (irrespective of M) are either even or odd, it is easy to show that (for given J and n) $A^{ij}_{J'n',Jn}\left(\vec{k}\right)$ are transformed through the representation $D^{(0)}+D^{(2)}$ of the rotation group. Consequently, they can be written as

$$A^{(ij)}_{J'n',Jn}\left(\vec{k}\right)=\left(\delta_{ij}-\frac{k_ik_j}{k^2}\right)A^{tr}_{J'n',Jn}(k)+\frac{k_ik_j}{k^2}A^{l}_{J'n',Jn}(k). \tag{5.38}$$

Finally, it follows from the equality $\hat{\vec{P}}^{(+)}\left(\vec{k}\right) = \hat{\vec{P}}\left(-\vec{k}\right)$ that $A^{l,\,tr}_{J'n',\,Jn}\left(\vec{k}\right)$ are real. They can be calculated, for example, by using the formulas

$$\begin{aligned} A^{l}_{J'n',\,Jn}(k) &= A^{zz}_{J'n',\,Jn}\left(k\vec{e}_z\right), \\ A^{tr}_{J'n',\,Jn}(k) &= A^{zz}_{J'n',\,Jn}\left(k\vec{e}_x\right), \end{aligned} \tag{5.39}$$

where $\vec{e}_i$ are the unit vectors along the coordinate axes.

As must be the case, substituting Eq. (5.38) reduces the permittivity tensor (5.34) to the form of Eq. (5.25), where the longitudinal and transverse permittivities are given by

$$\varepsilon^{l,\,tr}(\omega,k) = 1 - \frac{\omega_P^2}{\omega^2} - \frac{4\pi e^2 NV}{m^2\omega^2}\left(\frac{MT}{2\pi}\right)^{3/2}\sum_{Jn}\exp\left(\frac{F - E_{Jn}}{T}\right)\sum_{J'n'}\Phi_{J'n',\,Jn}(\omega,k;T)A^{l,\,tr}_{J'n',\,Jn}(k). \tag{5.40}$$

The integral in relation (5.36) for real z has no meaning. We assume that $\omega > 0$ in Eq. (5.40) (and below) and that the field proportional to $\exp(-\imath\omega t)$ adiabatically switches on for $t \to -\infty$ and make the corresponding substitution $\omega \to \omega + \imath\delta$, where $\delta \to +0$, (Landau bypassing rule [28]).

Below, we will need the following limiting expressions for the function $I_+(z)$, $z = z' + \imath z''$, $z'' > 0$ (Sect. 2.3):

$$I_+(z) \approx \begin{cases} \imath\sqrt{\pi}z - 2z^2, & |z| \ll 1, \\ -1 - \dfrac{1}{2z^2} + \imath\sqrt{\pi}ze^{-z^2}, & |z| \gg 1, \quad |z'| \gg z''. \end{cases} \tag{5.41}$$

We assume that the field is a moderately short-wavelength one: if the Bohr radius $a_B = 1/me^2$ is of the order of the atomic radius, then $ka_B \ll 1$. The matrix elements in Eq. (5.37) can then be expanded in power series of the small parameter ka_B. Retaining the terms of the second order of smallness, we write operator (5.32) as

$$\hat{P}_i\left(\vec{k}\right) = \hat{p}_i + \frac{\imath}{2}\left[\left(\hat{\vec{J}} + \hat{\vec{S}}\right) \times \vec{k}\right]_i + \imath k_j\hat{B}_{ij} - k_jk_k\hat{C}_{i,jk}, \tag{5.42}$$

where

$$\hat{\vec{p}} = \sum_a \hat{\vec{p}}_a, \quad \hat{\vec{S}} = \sum_a \hat{\vec{s}}_a, \quad \hat{\vec{J}} = \hat{\vec{S}} + \hat{\vec{L}}, \quad \hat{\vec{L}} = \sum_a \hat{\vec{l}}_a, \quad \hat{\vec{l}}_a = \vec{r}_a \times \hat{\vec{p}}_a,$$

$$\hat{B}_{ij} = \frac{1}{2}\sum_a \left(\hat{p}_{ai} r_{aj} + r_{ai}\hat{p}_{aj}\right) = \hat{B}_{ji},$$

$$\hat{C}_{i,jk} = \frac{1}{4}\sum_a \left(\hat{p}_{ai} r_{aj} r_{ak} + r_{aj} r_{ak}\hat{p}_{ai} - 4e_{ijl} r_{ak}\hat{s}_{al}\right) = \delta_{ik}\hat{C}_j^{(1)}$$

$$-\delta_{jk}\left[\hat{C}_i^{(1)} - \hat{C}_i^{(2)} + \left(1 - 3\delta_{jz}\delta_{kz}\right)\hat{C}_i^{(3)}\right] + \hat{\hat{C}}_{i,jk},$$

$$\hat{\vec{C}}^{(1)} = \frac{1}{2}\sum_a \vec{r}_a \times \hat{\vec{s}}_a,$$

$$\hat{\vec{C}}^{(2)} = \frac{1}{12}\sum_a \left(\hat{\vec{p}}_a r_a^2 + r_a^2 \hat{\vec{p}}_a\right),$$

$$\hat{\vec{C}}^{(3)} = \frac{1}{20}\sum_a \left[\vec{r}_a\left(\vec{r}_a \cdot \hat{\vec{p}}_a\right) + \left(\hat{\vec{p}}_a \cdot \vec{r}_a\right)\vec{r}_a - \frac{1}{3}\left(r_a^2 \hat{\vec{p}}_a + \hat{\vec{p}}_a r_a^2\right)\right], \tag{5.43}$$

and $\hat{\hat{C}}_{i,jk}$ are transformed via the representation $D^{(3)} + 2D^{(2)} + D^{(0)}$ and give no contribution to $A^{l,tr}_{J'n',Jn}(k)$ with the adopted accuracy (on the order of $(ka_B)^2$). Note that

$$\hat{\vec{p}} = m\frac{d\vec{r}}{dt}, \quad \vec{r} = \sum_a \vec{r}_a = -\frac{\vec{d}}{e},$$

$$\hat{B}_{ij} = \frac{m}{2}\frac{d}{dt}R_{ij}, \tag{5.44}$$

$$R_{ij} = \sum_a r_{ai} r_{aj} = \frac{1}{3}\delta_{ij}R_{kk} - \frac{1}{3e}Q_{ij},$$

where $\vec{d}$ and Q_{ij} are the dipole and quadrupole electric moments of the atom, respectively [21].

It is convenient to calculate the matrix elements of operators (5.42)–(5.44) by introducing the corresponding spherical tensors and using the Wigner–Eckart theorem and the standard properties of the 3*j* symbols [21]. As a result, we obtain

$$
\begin{aligned}
A^{l,\,tr}_{J'n',\,Jn}(k) &= A^{l,\,tr(d)}_{J'n',\,Jn}(k) + A^{l,\,tr(m)}_{J'n',\,Jn}(k) + A^{l,\,tr(q)}_{J'n',\,Jn}(k) + A^{l,\,tr(s)}_{J'n',\,Jn}(k),\\
A^{l,\,tr(d)}_{J'n',\,Jn}(k) &= \frac{m^2\omega^2_{J'n',\,Jn}}{3e^2}\left|\left\langle Jn\left\|\vec{P}\right\|J'n'\right\rangle\right|^2 + \frac{3}{2}k^2\,\mathrm{Re}\left|\left\langle Jn\left\|\vec{P}\right\|J'n'\right\rangle\right|\left|\langle J'n'\|C^{l,\,tr}\|Jn\rangle\right|,\\
\hat{\vec{C}}^{(l)} &= -\left(\hat{\vec{C}}^{(2)} + 2\hat{\vec{C}}^{(3)}\right),\\
\hat{\vec{C}}^{tr} &= \hat{\vec{C}}^{(1)} - \hat{\vec{C}}^{(2)} + \hat{\vec{C}}^{(3)},\\
A^{l(m)}_{J'n',\,Jn}(k) &= 0,\\
A^{tr(m)}_{J'n',\,Jn}(k) &= \frac{1}{12}k^2|\langle Jn\|J+S\|J'n'\rangle|^2,\\
A^{l(q)}_{J'n',\,Jn}(k) &= \frac{4}{3}A^{tr(q)}_{J'n',\,Jn}(k) = \frac{1}{270e^2}m^2\omega^2_{J'n',\,Jn}k^2|\langle Jn\|Q\|J'n'\rangle|^2,\\
A^{l(s)}_{J'n',\,Jn}(k) &= \frac{1}{36}m^2\omega^2_{J'n',\,Jn}k^2|\langle Jn\|R_0\|J'n'\rangle|^2,\\
A^{tr(s)}_{J'n',\,Jn}(k) &= 0,
\end{aligned}
\tag{5.45}
$$

where $\langle Jn\|\ldots\|J'n'\rangle$ are the reduced matrix elements; $R_{00} = R_{ii}$. Since $\hat{\vec{J}}$ commutes with the Hamiltonian of the atom, then

$$
\langle Jn\|J\|J'n'\rangle = \delta_{J'J}\delta_{n'n}\sqrt{J(J+1)(2J+1)}. \tag{5.46}
$$

In the zero approximation in spin–orbit interaction, we can calculate the matrix elements $\langle Jn\|S\|J'n'\rangle$ to the end [32]. Denote the atomic term unsplit by the spin–orbit interaction by LSn, so that $J = L + S, L + S - 1, \ldots, |L - S|$. Then,

$$
\begin{aligned}
\langle J'L'S'n'\|S\|JLSn\rangle &= \frac{1}{2}\delta_{L'L}\,\delta_{S'S}\,\delta_{n'n}\left\{\delta_{J'J}\sqrt{\frac{2J+1}{J(J+1)}}[J(J+1)+S(S+1)-L(L+1)]\right.\\
&-\delta_{J',J+1}\sqrt{\frac{(J+2+L+S)(J+1+S-L)(J+1+L-S)(L+S-J)}{J+1}}\\
&\left.-\delta_{J',J-1}\sqrt{\frac{(J+L+S+1)(J+S-L)(J+L-S)(L+S+1-J)}{J}}\right\},
\end{aligned}
\tag{5.47}
$$

i.e., the matrix elements of the spin $\hat{\vec{S}}$ are non-zero only for transitions inside the fine structure of the term.

Formulas (5.40) and (5.45) give the final expressions for $\varepsilon^{l}(\omega,k)$ and $\varepsilon^{\mathrm{tr}}(\omega,k)$. Below, we restrict our analysis to moderately high temperatures: if ω_0 is the energy interval between the ground and the first excited atomic levels, then $T \ll \omega_0$. Given relation (5.33), formula (5.40) then takes the form

$$\varepsilon^{l,\,\mathrm{tr}}(\omega,k) = 1 - \frac{\omega_p^2}{\omega^2} - \frac{4\pi N e^2}{m^2\omega^2(2J_0+1)} \sum_{Jn} \Phi_{Jn,\,J_0\,n_0}(\omega,k;T)\, A^{l,\,\mathrm{tr}}_{Jn,\,J_0\,n_0}(k). \quad (5.48)$$

We mark the states of the ground atomic level by the subscript $0 : |J_0 M_0 n_0\rangle$ and $E_{J_0\,n_0}$. If $L_0 = 0$ or $S_0 = 0$, then the ground term has no fine structure, with $\omega_0 \sim \omega_R = 1/(m a_B)^2$. If L_0 and $S_0 \neq 0$, then we have [21]

$$J_0 = |L_0 - S_0|, \quad \omega_0 = |A_{L_0 S_0}|(J_0+1),$$

for the normal multiplet, and

$$J_0 = L_0 + S_0, \quad \omega_0 = |A_{L_0 S_0}| J_0,$$

for the inverted multiplet. Here, the constant $|A_{L_0 S_0}| \sim \omega_R \left(Ze^2/c\right)^2$.

Below, we will have to use a quantity proportional to the difference $\varepsilon^{\mathrm{tr}}(\omega,k) - \varepsilon^{l}(\omega,k)$. We specially denote it by

$$1 - \frac{1}{\mu(\omega,k)} = \frac{\omega^2}{c^2k^2}\left[\varepsilon^{\mathrm{tr}}(\omega,k) - \varepsilon^{l}(\omega,k)\right] = -\frac{\pi e^2 N}{3(mc)^2(2J_0+1)} \sum_{Jn} \Phi_{Jn,\,J_0\,n_0}(\omega,k;T)\, M_{Jn,\,J_0\,n_0}, \quad (5.49)$$

where

$$\begin{aligned}
M_{Jn,\,J_0\,n_0} &= M^{(d)}_{Jn,\,J_0\,n_0} + M^{(m)}_{Jn,\,J_0\,n_0} + M^{(q)}_{Jn,\,J_0\,n_0} + M^{(s)}_{Jn,\,J_0\,n_0},\\
M^{(d)}_{Jn,\,J_0\,n_0} &= 8\,\mathrm{Re}\,\langle J_0\,n_0\|p\|Jn\rangle\langle Jn\|C^{\mu}\|J_0\,n_0\rangle,\\
\vec{\hat{C}}^{\mu} &= \vec{\hat{C}}^{(1)} + 3\vec{\hat{C}}^{(3)},\\
M^{(m)}_{Jn,\,J_0\,n_0} &= |\langle J_0\,n_0\|J+S\|Jn\rangle|^2,\\
M^{(q)}_{Jn,\,J_0\,n_0} &= -\frac{m^2\omega^2_{Jn,\,J_0\,n_0}}{90e^2}|\langle J_0\,n_0\|Q\|Jn\rangle|^2,\\
M^{(s)}_{Jn,\,J_0\,n_0} &= -\frac{m^2\omega^2_{Jn,\,J_0\,n_0}}{3}|\langle J_0\,n_0\|R_0\|Jn\rangle|^2.
\end{aligned} \quad (5.50)$$

Now we will derive expressions for $\varepsilon^{\mathrm{tr},\,l}(\omega,k)$ in various frequency ranges from formula (5.48). Let us first consider the case where ω is close to the frequency of a particular atomic transition: the detuning $\delta_{Jn} = \omega - \omega_{Jn,\,J_0\,n_0}$ satisfies the condition

$$\begin{gathered}|\delta_{Jn}| \ll |\omega_{Jn,J'n'}|,\\(Jn) \neq (J_0 n_0), \quad (J'n') \neq (Jn).\end{gathered} \tag{5.51}$$

Since we disregard the (natural and collisional) width ν_{Jn} of the excited level, the detuning in (5.51) cannot be very small: $|\delta_{Jn}| \gg \nu_{Jn}$. When condition (5.51) is satisfied, it will suffice to retain only the second term in function (5.35). Then, as follows from expressions (5.48),

$$\begin{aligned}\varepsilon^{l,\,\mathrm{tr}}(\omega,k) &= \tilde{\varepsilon}(\omega)\\&+\frac{4\pi e^2 N}{m^2\omega^2(2J_0+1)\left(\delta_{Jn}-k^2/2M\right)} I_+\left(\frac{\delta_{Jn}-k^2/2M}{\sqrt{2}k\bar{v}}\right) A^{l,\,tr}_{Jn,\,J_0n_0}(k),\end{aligned} \tag{5.52}$$

where $\tilde{\varepsilon}(\omega)$ is a smooth function of the frequency in the range (5.51). Note that $k\bar{v}$ is identical to the Doppler width of the spectral line which corresponds to the $J_0n_0 \leftrightarrow Jn$ transition [25, 33]. Expression (5.49) takes the form

$$1-\frac{1}{\mu(\omega,k)}=\frac{\pi e^2 N}{3(mc)^2(2J_0+1)\left(\delta_{Jn}-k^2/2M\right)} I_+\left(\frac{\delta_{Jn}-k^2/M}{\sqrt{2}k\bar{v}}\right) M_{Jn,\,J_0n_0}. \tag{5.53}$$

Let us now consider the case where ω is far from the resonance frequencies:

$$|\delta_{Jn}| \gg k\bar{v}, k^2/M, \quad (Jn) \neq (J_0n_0). \tag{5.54}$$

In this case, according to expressions (5.35) and (5.41), we have with the adopted accuracy

$$\begin{aligned}\Phi_{Jn,\,J_0n_0}(\omega,k;T) &= \frac{2\omega_{Jn,\,J_0n_0}}{\omega^2-\omega^2_{Jn,\,J_0n_0}}\\\times&\left[1+\frac{k^2}{M\left(\omega^2-\omega^2_{Jn,\,J_0n_0}\right)}\times\left(\frac{\omega^2+\omega^2_{Jn,\,J_0n_0}}{2\omega_{Jn,\,J_0n_0}}+\frac{3\omega^2+\omega^2_{Jn,\,J_0n_0}}{\omega^2-\omega^2_{Jn,\,J_0n_0}}T\right)\right].\end{aligned} \tag{5.55}$$

With the same accuracy, from Eqs. (5.45), (5.48), (5.49), and (5.50) we obtain

$$\varepsilon^{l,\mathrm{tr}}(\omega,k)=1-\frac{\omega_p^2}{\omega^2}+\frac{8\pi e^2N}{m^2\omega^2(2J_0+1)}\times\sum_{Jn}{}'\frac{\omega_{Jn,J_0n_0}}{\omega^2_{Jn,J_0n_0}-\omega^2}$$

$$\times\left[A^{l,tr}_{Jn,J_0n_0}(k)+\frac{m^2k^2\omega^2_{Jn,J_0n_0}}{3Me^2\left(\omega^2-\omega^2_{Jn,J_0n_0}\right)}\times|\langle J_0n_0\|d\|Jn\rangle|^2\left(\frac{\omega^2+\omega^2_{Jn,J_0n_0}}{2\omega_{Jn,J_0n_0}}+\frac{T\left(3\omega^2+\omega^2_{Jn,J_0n_0}\right)}{\omega^2-\omega^2_{Jn,J_0n_0}}\right)\right.$$

$$-\frac{4\pi e^2N}{m^2\omega^2(2J_0+1)}A^{l,tr}_{J_0n_0,J_0n_0}(k)\Phi_0(\omega,k;T),$$

$$1-\frac{1}{\mu(\omega,k)}=\frac{\pi e^2N}{3(mc)^2}\left[\frac{2}{2J_0+1}\sum_{Jn}\frac{\omega_{Jn,J_0n_0}}{\omega^2_{Jn,J_0n_0}-\omega^2}\times M_{Jn,J_0n_0}-J_0(J_0+1)g_0^2\Phi_0(\omega,k;T)\right],\tag{5.57}$$

where

$$\sum_{Jn}{}'(\ldots)=\sum_{(Jn)\neq(J_0n_0)}{}'(\ldots),$$

$$\Phi_0(\omega,k;T)=\Phi_{J_0n_0,J_0n_0}(\omega,k;T),\qquad g_0=g_{J_0L_0S_0},$$

and

$$g_{JLS}=1+\frac{J(J+1)+S(S+1)-L(L+1)}{2J(J+1)}\tag{5.58}$$

is the Lande factor of the *JLSn* atomic level [21]. In expressions (5.56) and (5.57), we take into account the fact that, as follows from relations (5.45)–(5.47),

$$\begin{aligned}A^{l}_{Jn,Jn}(k)&=0,\\ A^{\mathrm{tr}}_{Jn,Jn}(k)&=\frac{1}{12}k^2J(J+1)(2J+1)g^2_{JLS}.\end{aligned}\tag{5.59}$$

If, in addition to conditions (5.54), conditions (5.51) are also satisfied, then the expressions derived from Eqs. (5.56) and (5.57) are identical to the expressions derived from Eqs. (5.52) and (5.53) when it is considered that in this case, $\omega^2-\omega^2_{Jn,J_0n_0}\approx 2\omega_{Jn,J_0n_0}\delta_{Jn}$ and [see expression (5.41)]

$$\frac{1}{z^{(-)}_{Jn,J_0n_0}}I_+\left(z^{(-)}_{Jn,J_0n_0}\right)\approx-\frac{\sqrt{2}k\bar{v}}{\delta_{Jn}}\left[1+\frac{k^2}{M\delta_{Jn}}\left(\frac{1}{2}+\frac{T}{\delta_{Jn}}\right)\right],\tag{5.60}$$

$$\varepsilon^{l,\,\mathrm{tr}}(\omega,k)=\tilde{\varepsilon}(\omega)-\frac{4\pi e^2N}{m^2\omega^2(2J_0+1)\delta_{Jn}}\times\left[A^{l,\,tr}_{Jn,\,J_0n_0}(k)+\frac{m^2k^2\omega^2_{Jn,\,J_0n_0}}{3e^2M\delta_{Jn}}\times|\langle J_0n_0\|d\|Jn\rangle|^2\left(\frac{1}{2}+\frac{T}{\delta_{Jn}}\right)\right],\tag{5.61}$$

$$1-\frac{1}{\mu(\omega,k)}=-\frac{\pi e^2N}{3(mc)^2(2J_0+1)\delta_{Jn}}M_{Jn,\,J_0n_0}.\tag{5.62}$$

The case of low frequencies, $\omega\ll\omega_0$, is contained in formulas (5.56) and (5.57). The terms proportional to k^2 in the expression for $A^{l,\,tr}_{Jn,\,J_0n_0}(k)$ at $Jn\neq J_0n_0$ are of the order of $m\omega_R(ka_B)^2$, while the remaining terms in square brackets in Eq. (5.56) at $\omega\ll\omega_0$ are of the order of $m\omega_R(ka_B)^2m/M$. We disregard these terms (of the order of m/M) from the outset. Therefore, at $\omega\ll\omega_0$,

$$\varepsilon^{l,\,\mathrm{tr}}(\omega,k)=1-\frac{\omega_p^2}{\omega^2}-\frac{8\pi e^2N}{m^2\omega^2(2J_0+1)}\times\sum_{Jn}{}'\frac{\omega_{Jn,\,J_0n_0}}{\omega^2-\omega^2_{Jn,\,J_0n_0}}A^{l,\,tr}_{Jn,\,J_0n_0}(k)-\frac{4\pi e^2N}{m^2\omega^2(2J_0+1)}A^{l,\,tr}_{J_0n_0,\,J_0n_0}(k)\Phi_0(\omega,k;T),\tag{5.63}$$

We know how the pole at $\omega=0$ in the expression for $\varepsilon_{ij}\left(\omega,\vec{k}\right)$ can be eliminated if the contribution of only dipole transitions is considered [27]. A similar procedure can also be performed in Eq. (5.63). First, note that

$$\frac{1}{\omega^2\left(\omega^2-\omega^2_{Jn,\,J_0n_0}\right)}=\frac{1}{\omega^2_{Jn,\,J_0n_0}}\left(\frac{1}{\omega^2-\omega^2_{Jn,\,J_0n_0}}-\frac{1}{\omega^2}\right).\tag{5.64}$$

The sums (proportional to $1/\omega^2$) that result from the substitution of relation (5.64) into Eq. (5.63) can be calculated by using formulas (5.43)–(5.45), the Wigner–Eckart theorem, and the standard properties of the $3j$ and $6j$ symbols [21]. As a result, we obtain

$$\sum_{Jn}{}'\frac{1}{\omega_{Jn,\,J_0n_0}}A^{l}_{Jn,\,J_0n_0}(k)=\frac{Zm}{2}(2J_0+1),$$

$$\sum_{Jn}{}'\frac{A^{tr}_{Jn,\,J_0n_0}(k)}{\omega_{Jn,\,J_0n_0}}=\frac{Zm}{2}(2J_0+1)+\frac{k^2}{12}\left[\sum_{Jn}{}'\frac{|\langle J_0n_0\|S\|Jn\rangle|^2}{\omega_{Jn,\,J_0n_0}}-\frac{m(2J_0+1)}{\sqrt{2L_0+1}}\langle J_0n_0\|R_0\|J_0n_0\rangle\right].\tag{5.65}$$

The final expressions can be derived from Eqs. (5.59) and (5.63)–(5.65):

$$\varepsilon^l(\omega,k) = 1 - \frac{8\pi e^2 N}{m^2(2J_0+1)} \times \sum_{Jn}{}' \frac{1}{\omega_{Jn,J_0n_0}\left(\omega^2 - \omega^2_{Jn,J_0n_0}\right)} A^l_{Jn,J_0n_0}(k), \quad (5.66)$$

$$\varepsilon^{\text{tr}}(\omega,k) = 1 - \frac{8\pi e^2 N}{m^2(2J_0+1)} \times \sum_{Jn}{}' \frac{1}{\omega_{Jn,J_0n_0}\left(\omega^2 - \omega^2_{Jn,J_0n_0}\right)} A^{tr}_{Jn,J_0n_0}(k)$$
$$+ \frac{2\pi e^2 N k^2}{3m^2\omega^2}\left[-\frac{m\langle L_0n_0\|R_0\|L_0n_0\rangle}{\sqrt{2L_0+1}} + \frac{1}{2J_0+1}\sum_{Jn}{}' \frac{|\langle J_0n_0\|S\|Jn\rangle|^2}{\omega_{Jn,J_0n_0}} - \frac{J_0(J_0+1)}{2} g_0^2 \Phi_0(\omega,k;T)\right], \quad (5.67)$$

$$1 - \frac{1}{\mu(\omega,k)} = -\frac{2\pi e^2 N\omega^2}{3(mc)^2(2J_0+1)} \times \sum_{Jn}{}' \frac{1}{\omega_{Jn,J_0n_0}\left(\omega^2 - \omega^2_{Jn,J_0n_0}\right)} M_{Jn,J_0n_0}$$
$$+ \frac{2\pi e^2 N}{3(mc)^2} \times \left[-\frac{m\langle L_0n_0\|R_0\|L_0n_0\rangle}{\sqrt{2L_0+1}} + \frac{1}{2J_0+1}\sum_{Jn}{}' \frac{|\langle J_0n_0\|S\|Jn\rangle|^2}{\omega_{Jn,J_0n_0}} - \frac{J_0(J_0+1)}{2} g_0^2 \Phi_0(\omega,k;T)\right]. \quad (5.68)$$

Formula (5.68) can also be derived from Eq. (5.57) by using a transformation similar to the transformation used when passing from Eq. (5.63) to Eqs. (5.66) and (5.67).

The longitudinal permittivity $\varepsilon^l(\omega,k)$ is a regular function of ω. Using relation (5.45), from Eq. (5.66), we obtain

$$\varepsilon^l(\omega,0) = \varepsilon^l(\omega,k)\big|_{\frac{ka_B}{\omega/\omega_R}\to 0} = 1 + \frac{8\pi N}{3(2J_0+1)} \sum_{Jn}{}' \frac{\omega_{Jn,J_0n_0}|\langle J_0n_0\|d\|Jn\rangle|^2}{\omega^2_{Jn,J_0n_0} - \omega^2}, \quad (5.69)$$

$$\varepsilon^l(0,k) = \varepsilon^l(\omega,k)\big|_{\frac{\omega/\omega_R}{ka_B}\to 0} = 1 + \frac{8\pi e^2 N}{m^2(2J_0+1)} \sum_{Jn}{}' \frac{A^l_{Jn,J_0n_0}(k)}{\omega^3_{Jn,J_0n_0}}. \quad (5.70)$$

Thus, the limiting value

$$\varepsilon^l_0 = \varepsilon^l(\omega,0)\big|_{\omega/\omega_R\to 0} = \varepsilon^l(0,k)\big|_{ka_B\to 0}$$
$$= 1 + \frac{8\pi N}{3(2J_0+1)} \sum_{Jn} \frac{|\langle J_0n_0\|d\|Jn\rangle|^2}{\omega_{Jn,J_0n_0}} \quad (5.71)$$

does not depend on the order of the passage to the limit: first $ka_B \to 0$ and then $\omega/\omega_R \to 0$, or vice versa, first $\omega/\omega_R \to 0$ and then $ka_B \to 0$.

In contrast to $\varepsilon^l(\omega,k)$, the transverse permittivity $\varepsilon^{\text{tr}}(\omega,k)$ and the permeability $\mu(\omega,k)$ have singularities at $\omega = 0$. The function $\Phi_0(\omega,k;T)$ has the following limiting expressions [see relations (5.35) and (5.41)]:

$$\Phi_0(\omega,k;T)\approx\frac{k^2}{M\omega^2}\quad\text{for}\quad\omega\gg\frac{k^2}{M},k\bar{v},\tag{5.72}$$

$$\Phi_0(\omega,k;T)\approx-\frac{4M}{k^2}\quad\text{for}\quad k\gg\sqrt{M\omega},M\bar{v},\tag{5.73}$$

$$\Phi_0(\omega,k;T)\approx-\frac{1}{T}\quad\text{for}\quad\sqrt{M\omega}\ll k\ll M\bar{v}.\tag{5.74}$$

The conditions in relation (5.72) are satisfied for $ka_B \ll \omega/\omega_R$. Therefore, using relations (5.45) and (5.69), from Eqs. (5.67) and (5.68), we obtain

$$\varepsilon^{\rm tr}(\omega,0)=\varepsilon^{\rm tr}(\omega,k)|_{\frac{ka_B}{\omega/\omega_R}\to0}=\varepsilon^{l}(\omega,0),\quad\varepsilon^{\rm tr}(\omega,0)|_{\frac{\omega}{\omega_R}\to0}=\varepsilon^{l}_0,\tag{5.75}$$

$$1-\frac{1}{\mu(\omega,0)}=1-\frac{1}{\mu(\omega,k)}\bigg|_{\frac{ka_B}{\omega/\omega_R}\to0}=\frac{2\pi N}{3(2J_0+1)}\left(\frac{e\omega}{mc}\right)^2{\sum_{Jn}}'\frac{M_{Jn,J_0n_0}}{\omega_{Jn,J_0n_0}\left(\omega^2_{Jn,J_0n_0}-\omega^2\right)}$$
$$+\frac{2\pi e^2N}{3(mc)^2}\left(-\frac{m}{\sqrt{2L_0+1}}\langle L_0n_0\|R_0\|L_0n_0\rangle+\frac{1}{2J_0+1}{\sum_{Jn}}'\frac{|\langle J_0n_0\|S\|Jn\rangle|^2}{\omega_{Jn,J_0n_0}}\right),\tag{5.76}$$

$$1-\frac{1}{\mu(\omega,0)}\bigg|_{\frac{\omega}{\omega_R}\to0}=\frac{2\pi e^2N}{3(mc)^2}\times\left(-\frac{m}{\sqrt{2L_0+1}}\langle L_0n_0\|R_0\|L_0n_0\rangle+\frac{1}{2J_0+1}{\sum_{Jn}}'\frac{|\langle J_0n_0\|S\|Jn\rangle|^2}{\omega_{Jn,J_0n_0}}\right).\tag{5.77}$$

If the condition $k\gg\sqrt{M\omega}$ in expressions (5.73) and (5.74) is satisfied, then the condition $ka_B \gg \omega/\omega_R$ is also satisfied. In this case, from Eqs. (5.67) and (5.68), we obtain

$$\varepsilon^{\rm tr}(\omega,k)|_{\frac{\omega}{k^2/M}\to0}=1+\frac{2\pi e^2Nk^2}{3m^2\omega^2}$$
$$\times\left[-\frac{m}{\sqrt{2L_0+1}}\langle L_0n_0\|R_0\|L_0n_0\rangle+\frac{1}{2}J_0(J_0+1)\,g_0^2\Phi_0+\frac{1}{2J_0+1}{\sum_{Jn}}'\frac{|\langle J_0n_0\|S\|Jn\rangle|^2}{\omega_{Jn,J_0n_0}}\right],\tag{5.78}$$

$$1-\frac{1}{\mu(0,k)}=1-\frac{1}{\mu(\omega,k)}\bigg|_{\frac{\omega}{k^2/M}\to0}=\frac{2\pi e^2N}{3(mc)^2}$$
$$\times\left[-\frac{m}{\sqrt{2L_0+1}}\langle L_0n_0\|R_0\|L_0n_0\rangle+\frac{1}{2}J_0(J_0+1)g_0^2\Phi_0+\frac{1}{2J_0+1}{\sum_{Jn}}'\frac{|\langle J_0n_0\|S\|Jn\rangle|^2}{\omega_{Jn,J_0n_0}}\right].\tag{5.79}$$

Function $\Phi_0=\Phi_0(\omega,k;T)=-4M/k^2$ if $k\gg M\bar{v}$, and $\Phi_0=\Phi_0(\omega,k;T)=-1/T$ if $k\ll M\bar{v}$[see expressions (5.73) and (5.74)].

In the approach to the electrodynamics of material media in which, apart from the vectors $\vec{E}$ and $\vec{B}$, the vectors $\vec{D}$ and $\vec{H}$ are also introduced, the properties of an isotropic medium are characterized by the permittivity $\varepsilon(\omega)$ and the permeability $\mu(\omega)$ (Sect. 1.2):

$$\vec{D}(\omega) = \varepsilon(\omega)\vec{E}(\omega), \quad \vec{B}(\omega) = \mu(\omega)\vec{H}(\omega). \tag{5.80}$$

They are related to the longitudinal, $\varepsilon^l(\omega, k)$, and transverse, $\varepsilon^{\mathrm{tr}}(\omega, k)$, permittivities considered here by (Sect. 1.2)

$$\varepsilon(\omega) = \varepsilon^l(\omega, k), \quad \mu(\omega) = \mu(\omega, k), \tag{5.81}$$

where $\mu(\omega, k)$ is the function introduced in Eq. (5.49). It is clear from these relations that $\varepsilon(\omega)$ and $\mu(\omega)$ have physical meaning only when the k dependences of $\varepsilon^l(\omega, k)$ and $\mu(\omega, k)$ may be ignored. Thus, as follows from the above results, $\varepsilon(\omega)$ loses its physical meaning (as the coefficient that relates $\vec{D}(\omega)$ and $\vec{E}(\omega)$) not only at ω close to the frequency of any permitted transition [25], but also in all of the ω ranges in which electric quadrupole and completely symmetric transitions contribute appreciably to $\varepsilon^l(\omega, k)$.

According to (5.81), the magnetic permeability $\mu(\omega)$ at low ($\omega \ll \omega_R$) frequencies is defined by formulas (5.76) and (5.77). If we introduce the magnetic susceptibility (1.38), $\chi(\omega) = [\mu(\omega) - 1]/4\pi$, then, according to (5.77), its low-frequency limit will be defined by

$$\chi(0) = \frac{Ne^2}{6(mc)^2} \times \left(-\frac{m}{\sqrt{2L_0+1}} \langle L_0 n_0 \| R_0 \| L_0 n_0 \rangle + \frac{1}{2J_0+1} \sum_{Jn}{}' \frac{|\langle J_0 n_0 \| S \| Jn \rangle|^2}{\omega_{Jn,\, J_0 n_0}} \right). \tag{5.82}$$

In order of magnitude, $|\chi(0)| \approx N a_B^3 \omega_R / mc^2 \ll 1$, because $N a_B^3 < 10^{-4}$ and $\omega_R / mc^2 \approx 10^{-4}$. The static susceptibility in a uniform field can be obtained from Eq. (5.79) for $\Phi_0 = -1/T$:

$$\chi_{st}(0) = \chi(0) + \frac{1}{3T} N \mu_B^2 J_0 (J_0 + 1) g_0^2, \tag{5.83}$$

where $\chi(0)$ is given by expression (5.82). If the g factor of the ground atomic level is non-zero, then the last term in expression (5.83) is much larger than $\chi(0)$; it is of the order of $\chi(0)\omega_R/T$; however, $\chi_{st}(0) \ll 1$ in this case as well. As must be the case, expression (5.83) for $\chi_{st}(0)$ matches the Van Vleck standard formula [29]. The T-independent diamagnetic and paramagnetic terms are also retained at high frequencies [see Eqs. (5.76) and (5.82)]. Therefore, there is no reason to believe that, in contrast to $\varepsilon(\omega)$, $\mu(\omega)$ loses its physical meaning with increasing ω relatively early and we should set $\mu(\omega) = 1$. However, at optical frequencies, the contribution of electric quadrupole transitions to $\mu(\omega)$ is comparable to the contribution of magnetic dipole and completely symmetric transitions [see Eqs. (5.50) and (5.76)]. Thus, the vector

$$\vec{M} = \frac{1}{4\pi}\left(\vec{B} - \vec{H}\right) = \frac{1}{4\pi}[\mu(\omega) - 1]\vec{H}$$

at these frequencies loses the meaning of magnetic moment per unit volume of the gas.

In conclusion, let us consider the question of whether a transverse electromagnetic wave with a group velocity antiparallel to its phase velocity can propagate through a monatomic gas [34]. Besides, the propagation of an electromagnetic wave with antiparallel group and phase velocities through a transparent isotropic medium has interesting features [35].[7]

The dispersion relation for a transverse electromagnetic wave in an isotropic non-gyrotropic medium is [see Eq. (1.116)]

$$\frac{c^2k^2}{\omega^2} = \varepsilon^{\mathrm{tr}}(\omega, k). \tag{5.84}$$

When the wave damping is absent (i.e., for the wave vector k to be real), the following conditions should be satisfied:

$$\varepsilon^{tr''}(\omega, k) \approx 0, \quad \varepsilon^{tr'}(\omega, k) > 0. \tag{5.85}$$

For the phase, $\vec{u}_f = \omega\vec{k}/k^2$, and group, $\vec{u}_{gr} = \left(\vec{k}/k\right)d\omega/dk$, velocities of the wave, we find from Eq. (5.84) that

$$\begin{aligned} &\text{if} \quad \frac{\omega^2}{c^2}\frac{\partial\varepsilon^{\mathrm{tr}\prime}}{\partial k^2} < 1, \quad \text{then} \quad \vec{u}_{gr}\|\vec{u}_f; \\ &\text{if} \quad \frac{\omega^2}{c^2}\frac{\partial\varepsilon^{\mathrm{tr}\prime}}{\partial k^2} > 1, \quad \text{then} \quad \vec{u}_{gr} \text{ and } \vec{u}_f \text{ are antiparallel.} \end{aligned} \tag{5.86}$$

Thus, the group and phase velocities can be antiparallel only if spatial dispersion is taken into account. In the frequency range where $\varepsilon(\omega)$ and $\mu(\omega)$ [see relations (5.49) and (5.81)] have physical meaning, Eq. (5.84) is equivalent to the equation

[7]It is clear from symmetry considerations that the group and phase velocities of any wave in an isotropic medium are either parallel or antiparallel.

$$\frac{c^2 k^2}{\omega^2} = \varepsilon(\omega)\mu(\omega), \tag{5.87}$$

while conditions (5.85) and (5.86) are equivalent to the following conditions [36]:

$$\varepsilon'' \approx 0, \quad \mu'' \approx 0, \quad \varepsilon'\mu' > 0, \tag{5.88}$$

$$\text{if} \quad \varepsilon', \mu' > 0, \quad \text{then} \quad \vec{u}_{gr} \| \vec{u}_f, \tag{5.89}$$

$$\text{if} \quad \varepsilon', \mu' < 0, \quad \text{then} \quad \vec{u}_{gr} \ \text{and} \ \vec{u}_f \ \text{are antiparallel.}$$

The first condition in relations (5.85) is satisfied only if ω is not too close to any of the eigen frequencies of the atom: the detuning $|\delta_{Jn}| \gg \tilde{\nu}_{Jn}$, where $\tilde{\nu}_{Jn}$ is the total width of the Jn level, includes the natural and collisional width ν_{Jn} and the Doppler width $k\bar{v}$. In this case [see expressions (5.45), (5.56), and (5.59)],

$$\begin{aligned}\frac{\omega^2}{c^2}\frac{\partial \varepsilon^{\text{tr}}(\omega,k)}{\partial k^2} &= \frac{8\pi e^2 N}{3(mc)^2(2J_0+1)} \times \Bigg\{ -\frac{1}{8} J_0(J_0+1) g_0^2 \Phi_0 \\ &+ \sum_{Jn}{}' \frac{\omega_{Jn,J_0n_0}}{\omega^2_{Jn,J_0n_0} - \omega^2}\Bigg[\frac{1}{4}|\langle J_0 n_0 \| S \| Jn\rangle|^2 + \frac{m^2\omega^2_{Jn,J_0n_0}}{120 e^2}|\langle J_0n_0\|Q\|Jn\rangle|^2 \\ &+ \frac{m^2 \omega^2_{Jn,J_0n_0}}{Me^2\left(\omega^2 - \omega^2_{Jn,J_0n_0}\right)}|\langle J_0n_0\|d\|Jn\rangle|^2 \left(\frac{\omega^2 + \omega^2_{Jn,J_0n_0}}{2\omega_{Jn,J_0n_0}} + T\frac{3\omega^2 + \omega^2_{Jn,J_0n_0}}{\omega^2 - \omega^2_{Jn,J_0n_0}}\right)\Bigg]\Bigg\}.\end{aligned} \tag{5.90}$$

Assuming that ω and $|\delta_{Jn}| \sim \omega_R$ in Eq. (5.90) and using relation (5.72), we find that

$$\frac{\omega^2}{c^2}\left|\frac{\partial \varepsilon^{\text{tr}}}{\partial k^2}\right| \sim Na_B^3 \frac{\omega_R}{mc^2} \ll 1. \tag{5.91}$$

It follows from this relation and from condition (5.86) that $\vec{u}_{gr} \| \vec{u}_f$ for waves with frequencies far from the atomic transition frequencies.

At low frequencies ($\omega \ll T$), equality (5.90) reduces to [see Eq. (5.78)]

$$\begin{aligned}\frac{\omega^2}{c^2}\frac{\partial \varepsilon^{\text{tr}}(\omega,k)}{\partial k^2} &= \frac{2\pi e^2 N}{3(mc)^2}\Bigg[-\frac{m}{\sqrt{2L_0+1}}\langle L_0n_0\|R_0\|L_0n_0\rangle \\ &+ \frac{1}{2T}J_0(J_0+1)g_0^2 + \frac{1}{2J_0+1}\sum_{Jn}{}' \frac{|\langle J_0n_0\|S\|Jn\rangle|^2}{\omega_{Jn,J_0n_0}}\Bigg].\end{aligned} \tag{5.92}$$

If $g_0 = 0$, then relation (5.91) remains valid; if, however, $g_0 \neq 0$, then

$$\frac{\omega^2}{c^2}\frac{\partial \varepsilon^{\text{tr}}(\omega,k)}{\partial k^2}=\frac{4\pi\mu_B^2 N}{3T}g_0^2 J_0(J_0+1)\sim Na_B^3\frac{\omega_R}{mc^2}\frac{\omega_R}{T}\ll 1. \tag{5.93}$$

Consequently, $\vec{u}_{gr}\|\vec{u}_f$ for low-frequency waves as well.

It remains to consider a wave with a frequency close to ω_{Jn,J_0n_0} of a particular atomic transition: condition (5.51) is satisfied, but $|\delta_{Jn}|\gg\tilde{\nu}_{Jn}$ as before. In this case, equality (5.90) reduces to [see Eqs. (5.45) and (5.61)]

$$\frac{\omega^2}{c^2}\frac{\partial \varepsilon^{\text{tr}}(\omega,k)}{\partial k^2}=-\frac{4\pi e^2 N}{3(mc)^2(2J_0+1)\delta_{Jn}}\times\left\{\frac{1}{4}|\langle J_0n_0\|S\|Jn\rangle|^2+2\,\mathrm{Re}\,\langle J_0n_0\|p\|Jn\rangle\right.$$
$$\left.\times\langle Jn\|C^{\text{tr}}\|J_0n_0\rangle+\frac{m^2\omega_{Jn,J_0n_0}^2}{e^2}\times\left[\frac{1}{120}|\langle J_0n_0\|Q\|Jn\rangle|^2+\frac{1}{M\delta_{Jn}}|\langle J_0n_0\|d\|Jn\rangle|^2\left(\frac{1}{2}+\frac{T}{\delta_{Jn}}\right)\right]\right\}. \tag{5.94}$$

For the Jn levels to which the magnetic dipole and electric quadrupole transitions are permitted, we find that

$$\frac{\omega^2}{c^2}\left|\frac{\partial \varepsilon^{\text{tr}}}{\partial k^2}\right|\sim Na_B^3\frac{\omega_R}{mc^2}\frac{\omega_R}{|\delta_{Jn}|}\ll 1, \tag{5.95}$$

because even $\omega_R/\nu_{Jn}<10^7$ and $|\delta_{Jn}|\gg\tilde{\nu}_{Jn}>\nu_{Jn}$.

For the Jn levels to which an electric dipole transition is permitted, from Eq. (5.94), we obtain

$$\frac{\omega^2}{c^2}\frac{\partial \varepsilon^{\text{tr}}(\omega,k)}{\partial k^2}=-\frac{2\pi N}{3Mc^2(2J_0+1)}\frac{\omega_{Jn,J_0n_0}^2}{\delta_{Jn}^2}|\langle J_0n_0\|d\|Jn\rangle|^2\left(1+\frac{2T}{\delta_{Jn}}\right). \tag{5.96}$$

The contribution of the first term on the right-hand side is of the order of $Na_B^3(\omega_R/mc^2)(m/M)(\omega_R/\delta_{Jn})^2$. Since $m/M<10^{-3}$, the first term is of the order of unity only for $|\delta_{Jn}|<10^{-6}\omega_R$. If, however, $|\delta_{Jn}|\gg\tilde{\nu}_{Jn}$, then the first term on the right-hand side of Eq. (5.96) is smaller than unity in order of magnitude. The second term on the right-hand side of Eq. (5.96) differs in order of magnitude from the first term by a factor of $((T/\omega_R)(\omega_R/|\delta_{Jn}|))$. Therefore, it can become approximately equal to unity even for $|\delta_{Jn}|\le 10^{-4}\omega_R$. If, in this case, $\delta_{Jn}<0$, then the second condition (5.85) is also satisfied, and, according to condition (5.86), the group and phase velocities of the wave can be antiparallel.

Thus, the group velocity of the transverse electromagnetic wave in a monatomic gas at all frequencies coincides in direction with its phase velocity, except for frequencies slightly detuned from the frequencies of electric dipole transitions toward longer wavelengths. In this case [see Eqs. (5.61) and (5.62)],

$$\varepsilon^l(\omega,k)=\varepsilon^{\text{tr}}(\omega,k),\quad \mu(\omega,k)=1,$$

and conditions (5.88) and (5.89) are inapplicable.

5.3 Problems

5.3.1 The charged oscillator with the eigen frequency of vibrations ω_0 and the damping constant γ is initially at rest. Then it is affected by an external electric field $\vec{E}(t)$, which depends on time in an arbitrary way. The wavelengths of field variations are larger than the amplitude of the oscillator vibrations. Calculate the dipole moment of the particle $\vec{p}(t)=e\,\vec{r}(t)$ with respect to the center of vibrations as the integral containing the external field, and calculate the response function (which characterizes the oscillator polarizability) [37].

Solution We use the following equation of motion for the oscillator:

$$\ddot{\vec{r}}+\gamma\dot{\vec{r}}+\omega_0^2\vec{r}=\frac{e}{m}\vec{E}(t). \tag{5.97}$$

We ignore the effect of the magnetic field on the oscillator because the factor $v/c\ll 1$ is small. We do not take into account the field inhomogeneity due to the condition $r\ll\lambda$. The solution of the equation can be written in the integral form

$$\vec{r}(t)=\frac{e}{m}\int_{-\infty}^{\infty}G(t-t')\vec{E}(t')dt', \tag{5.98}$$

where $G(t-t')$ is the Green's function satisfying the equation

$$\ddot{G}+\gamma\dot{G}+\omega_0^2G=\delta(t-t'), \tag{5.99}$$

Let us find the particular solution of the latter equation, corresponding to the delta function on the right-hand side, by the method of variation of constants, that is, in the form

$$G(\tau)=A(\tau)e^{s_1\tau}+B(\tau)e^{s_2\tau},\quad \tau=t-t', \tag{5.100}$$

where s_1 and s_2 are two roots of the characteristic equation $s^2+\gamma s+\omega_0^2=0$:

$$s_{1,2}=\pm\imath\sqrt{\omega_0^2-\frac{\gamma^2}{4}}-\frac{\gamma}{2}. \tag{5.101}$$

The functions $A(\tau)$ and $B(\tau)$ are determined from the system of equations

$$\dot{A}e^{s_1\tau} + \dot{B}e^{s_2\tau} = 0, \quad s_1\dot{A}e^{s_1\tau} + s_2\dot{B}e^{s_2\tau} = \delta(\tau), \tag{5.102}$$

Integrating this system under the assumption, that at $\tau \to -\infty$ the oscillator was at rest (adiabatic switch on of the field), we obtain

$$B(\tau) = -A(\tau) = \frac{\imath\Theta(\tau)}{2\sqrt{\omega_0^2 - \gamma^2/4}}, \tag{5.103}$$

where $\Theta(\tau)$ is the step function. As a result, we have

$$G(\tau) = \left(\omega_0^2 - \gamma^2/4\right)^{-1/2}\Theta(\tau)e^{-\gamma\tau/2}\sin\left(\sqrt{\omega_0^2 - \gamma^2/4}\tau\right). \tag{5.104}$$

With the help of Eqs. (5.98) and (5.104), we present the dipole oscillator moment $\vec{p}(t) = e\,\vec{r}(t)$ in the form

$$\vec{p}(t) = \int_{-\infty}^{\infty} f(t-t')\vec{E}(t')dt' = \int_{-\infty}^{t} f(t-t')\vec{E}(t')dt', \tag{5.105}$$

where the response function is

$$f(t-t') = \frac{e^2}{m}G(t-t') \tag{5.106}$$

Due to the presence of the step function $\Theta(\tau)$ in Eq. (5.104), Eq. (5.105) describes the causal connection of the oscillator dipole moment with an external field: $\vec{p}(t)$ is determined by the field values at preceding instants of time $t' \leq t$.

5.3.2 The semi-classical dispersion model is based on treating atomic electrons as classical oscillators, with eigen frequency ω_0 and damping constant γ. In the simplest model, all oscillators may be considered to be equal, and the difference of the local field from the average one is neglected. Calculate the dielectric permeability $\varepsilon(\omega)$ of the medium with an average number density of electrons n. Find the dependence of $\varepsilon'(\omega) =$ Re $\varepsilon(\omega)$ and $\varepsilon''(\omega) =$ Im $\varepsilon(\omega)$ on frequency for the transparent medium ($\gamma \ll \omega$). Use the response function obtained in the problem 5.3.1 [37].

Solution The medium polarization vector is $\vec{P}(t) = n\vec{p}(t)$, and the electric induction vector is $\vec{D}(t) = \vec{E}(t) + 4\pi n\vec{p}(t)$, where $\vec{p}(t)$ is given by Eqs. (5.105) and (5.106) of the previous problem. Calculating $\varepsilon(\omega)$ from

$$\varepsilon(\omega) = 1 + 4\pi\int_0^{\infty} \alpha(\tau)e^{\imath\omega\tau}d\tau,$$

where $\alpha(\tau)$ is the response function, and using the explicit form of the Green's function (5.104) of previous problem, we find

$$\varepsilon(\omega) = 1 + \frac{\omega_{0e}^2}{\omega_0^2 - \omega^2 - \imath\gamma\omega}, \quad \omega_{0e}^2 = \frac{4\pi n e^2}{m}, \tag{5.107}$$

that is,

$$\varepsilon'(\omega) = 1 + \frac{\omega_{0e}^2\left(\omega_0^2 - \omega^2\right)}{\left(\omega_0^2 - \omega^2\right)^2 + \gamma^2\omega^2}, \quad \varepsilon''(\omega) = \frac{\gamma\omega_{0e}^2\omega}{\left(\omega_0^2 - \omega^2\right)^2 + \gamma^2\omega^2}. \tag{5.108}$$

The frequency dependence of the real and imaginary parts of ε is shown in Fig. 5.1. The imaginary part, ε'', determining absorption of electromagnetic energy, noticeably differs from zero only in the vicinity of the eigen frequency ω_0 of oscillations in the medium. It is positive everywhere (at $\omega > 0$). In the frequency range near ω_0, the quantity ε' decreases with frequency increasing (the anomalous dispersion). Otherwise, ε' increases with frequency increasing (the normal dispersion).

5.3.3 Quantum theory of dispersion far from resonances. Let the substance consist of neutral atoms, whose number density is *N*. Using quantum theory, calculate the dipole moment of a separate atom, induced by a weak external long-wave field. Then obtain the dielectric permeability of the medium neglecting the difference of the local field from the average one. Compare the result with that from the previous problem, obtained on the basis of semi-classical model [37].

Solution The wavelength of the electromagnetic field is much larger than the atomic size. Therefore, the operator of the atom-field interaction can be written, assuming the field to be a classical object:

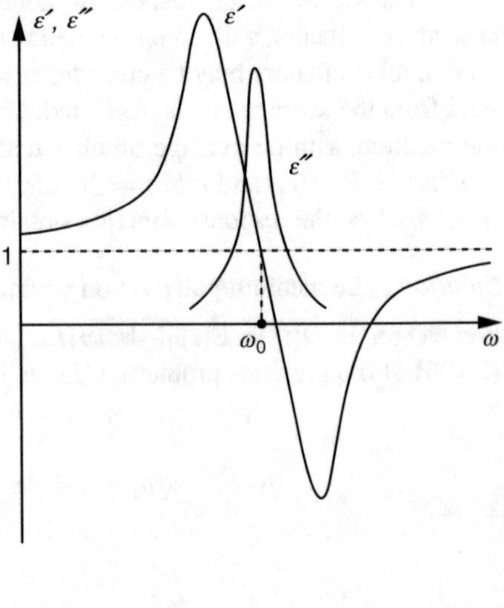

Fig. 5.1 Schematic plot of the real and imaginary parts of dielectric permittivity versus frequency in the model of oscillator with one eigen frequency ω_0

$$\hat{V}(t) = -\hat{\vec{d}} \cdot \vec{E}(t), \quad \hat{\vec{d}} = e\sum_{a=1}^{Z} \hat{\vec{r}}_a, \tag{5.109}$$

Here, $\vec{E}(t)$ is the real vector of the electric field strength, $\hat{\vec{d}}$ is the operator of the atomic dipole moment, and the summation is over all electrons. Using the perturbation theory, we find the first-order correction $\Phi^{(1)}\left(\vec{r}_1, \ldots \vec{r}_Z, t\right)$ to the wave function $\Phi^{(0)}\left(\vec{r}_1, \ldots \vec{r}_Z, t\right)$ of the non-disturbed atom, satisfying the Schrödinger equation

$$\imath\hbar \frac{\partial \Phi^{(0)}}{\partial t} = \hat{\mathcal{H}}_0 \Phi^{(0)}, \tag{5.110}$$

In our problem, the electrons can be considered particles without spin, since the effect of an external field on the spin is weak and spin states of the atom do not change. Substituting

$$\Phi\left(\vec{r}_1, \ldots \vec{r}_Z, t\right) = \Phi^{(0)}\left(\vec{r}_1, \ldots \vec{r}_Z, t\right) + \Phi^{(1)}\left(\vec{r}_1, \ldots \vec{r}_Z, t\right)$$

in the Schrödinger equation with the Hamiltonian $\hat{\mathcal{H}} + \hat{\mathcal{H}}_0 + \hat{V}(t)$, we obtain the approximate equation for $\Phi^{(1)}$:

$$\imath\hbar \frac{\partial \Phi^{(1)}}{\partial t} - \hat{\mathcal{H}}_0 \Phi^{(1)} = \hat{V}(t)\Phi^{(0)}, \tag{5.111}$$

where the term, bilinear in perturbation, is omitted. Let us switch on the field at $t \to -\infty$ slowly (adiabatically):

$$\vec{E}(t) = \begin{cases} \vec{E}_0 e^{\alpha t} \cos \omega t, & t \le 0, \quad \alpha > 0, \\ \vec{E}_0 \cos \omega t, & t \ge 0, \end{cases} \tag{5.112}$$

where we should set $\alpha \to 0$ after all integrations over time. This is done to eliminate transient process of setting-up the stationary state. According to this choice, we write the initial conditions for Eq. (5.111) as:

$$\Phi|_{t\to-\infty} \to \varphi_0\left(\vec{r}_1, \ldots \vec{r}_Z\right) e^{-\imath E_0 t/\hbar}, \quad \Phi^{(1)}\Big|_{t\to-\infty} \to 0. \tag{5.113}$$

Here, and further on, by φ_s we denote the wave functions of stationary states of the non-disturbed atom; $s = 0$ corresponds to the ground state. Let us find the solution of Eq. (5.111) by expansion over eigen functions of the non-disturbed atom:

$$\Phi^{(1)} = \sum_s c_s(t)\left(\vec{r}_1, \ldots \vec{r}_Z\right) e^{-\imath E_0 t/\hbar}. \tag{5.114}$$

Using orthogonality of the functions φ_s, we obtain from Eq. (5.111):

$$\frac{dc_s}{dt} = -\frac{1}{\imath\hbar}\left(\vec{d}_{s0} \cdot \vec{E}(t)\right) e^{\imath\omega_{s0} t}, \tag{5.115}$$

where $\vec{d}_{s0}$ is the matrix element of the atomic dipole moment. Notice that $\vec{d}_{00} = 0$, because the wave functions of the stationary states possess definite parity.

Integrating over time in Eq. (5.115) and using Eqs. (5.112) and (5.113), we obtain at finite t:

$$c_s(t) = \frac{\vec{d}_{s0} \cdot \vec{E}}{2\hbar}\left(\frac{e^{\imath(\omega_{s0}+\omega)\,t}}{\omega_{s0}+\omega} + \frac{e^{\imath(\omega_{s0}-\omega)\,t}}{\omega_{s0}-\omega}\right). \tag{5.116}$$

This result can be used only for non-resonance frequencies, when the denominators in the latter expression are not small. The average quantum mechanical value of the induced dipole moment of the atom can be calculated from

$$\vec{d}(t) = \int \Phi^* \hat{\vec{d}}\, \Phi dq = \int \left(\Phi^{(0)*} \hat{\vec{d}}\, \Phi^{(1)} + \Phi^{(1)*} \hat{\vec{d}}\, \Phi^{(0)}\right) dq, \tag{5.117}$$

where $d_q = d^3 r_1 \ldots d^3 r_Z$, the quadratic correction is neglected, and the relation $\vec{d}_{00} = 0$ is used. With the help of Eqs. (5.114), (5.116), and (5.117) we find

$$\vec{d}(t) = \frac{1}{2\hbar}\sum_s \left\{ \left(\frac{e^{\imath\omega t}}{\omega_{s0}+\omega} + \frac{e^{-\imath\omega t}}{\omega_{s0}-\omega}\right)\left(\vec{d}_{s0} \cdot \vec{E}_0\right)\vec{d}_{0s} \right.$$
$$\left. \times + \left(\frac{e^{-\imath\omega t}}{\omega_{s0}+\omega} + \frac{e^{\imath\omega t}}{\omega_{s0}-\omega}\right)\left(\vec{d}_{s0} \cdot \vec{E}_0\right)^* \vec{d}_{s0} \right\}. \tag{5.118}$$

As clear from symmetry arguments, the macroscopic polarization vector is directed along the field vector $\vec{E}_0$, which, in our case, is real. Therefore, it is sufficient to calculate the projection of the vector $\vec{d}$ in the field direction. Aligning the field along the ox-axis and using the Hermitian character of the matrix $(d_x)_{0s} = (d_x)^*_{s0}$, we find from the general formula (5.118) the atomic polarization,

$$d_x = \beta(\omega) E_0, \tag{5.119}$$

where

$$\beta(\omega) = \sum_s \frac{2\omega_{s0}\left|(d_x)_{s0}\right|^2}{\hbar\left(\omega_{s0}^2 - \omega^2\right)}.$$

From here, we obtain the dielectric permittivity:

$$\varepsilon(\omega) = 1 + 4\pi N\beta(\omega) = 1 + \frac{4\pi N}{\hbar}\sum_s \frac{2\omega_{s0}\left|(d_x)_{s0}\right|^2}{\hbar\left(\omega_{s0}^2 - \omega^2\right)}. \tag{5.120}$$

Let us compare Eq. (5.120) with the results of semi-classical theory [Eq. (5.107) of problem 5.3.2]. The oscillator model requires the following modifications: (1) the introduction of many oscillators with different eigen frequencies, $\omega_0 \to \omega_{s0}$; (2) summation over all such oscillators; and (3) as far as the damping is concerned, we can set $\gamma = 0$, if $\gamma \ll \omega_0$, far from resonances.

Introducing the ratio $f_s = n_s/s$, which is the fraction of electrons, belonging to an oscillator s, we present the semi-classical formula (5.107) from the problem 5.3.2 in the form of the quantum formula (5.120),

$$\varepsilon(\omega) = 1 + \frac{4\pi Ne^2}{m}\sum_s \frac{f_s}{\omega_{s0}^2 - \omega^2}, \tag{5.121}$$

if the quantities f_s are determined in terms of matrix elements of the dipole moment:

$$f_s = \frac{2m\omega_{s0}}{e^2\hbar}\left|(d_x)_{s0}\right|^2, \quad (d_x)_{s0} = e\sum_{a=1}^{Z} x_{s0}^a. \tag{5.122}$$

The quantities f_s are referred to as the *oscillator strengths*; the oscillator strengths satisfy the sum rule $\sum_s f_s = Z$. Therefore, at $\omega \gg \omega_{s0}$ the quantum formulas (5.120) and (5.121) take the universal form; the average number of electrons per unit volume is $n = NZ$. The dielectric permittivity, derived in this problem, is not applicable near the frequencies of spectral lines.

5.3.4 Quantum theory of dispersion near the frequencies of atomic spectral lines: Under the conditions of problem 5.3.3 take into account the radiation broadening of spectral lines and calculate the dielectric permeability at the frequencies close or coinciding with the frequencies of spectral lines [37].

Solution Near the resonance, $\omega \approx \omega_{s0}$, the population of state s by an external field increases infinitely [see Eq. (5.116) of Problem 5.3.3]. This is because an atom absorbs the quanta with energy $\hbar\omega$ from the external field. However, in addition to transitions of atoms in excited states, there will also exist inverse processes of spontaneous quantum emissions by excited atoms. Absorptions and emissions will have tendency to be balanced; the population of state s will take then some stationary value. Along with the interaction with the external field $\vec{E}(t)$, we will include the interaction of the atom with the quantized electromagnetic field of photons.

The operator of this interaction will be denoted here as ^$\hat{V}'$:

$$\hat{V}' = -\hat{\vec{d}} \cdot \hat{\vec{E}} . \tag{5.123}$$

In this case, we will take into account only vacuum and single-photon states of the quantized field. The state vector, describing the system of the atom in the external field + quantized field, can be written in the form

$$|\Phi(t)\rangle = c_0(t)\varphi_0|0\rangle\, e^{-\imath E_0 t/\hbar} + \sum_k c_k(t)\varphi_0|1_k\rangle\, e^{-\imath(\omega_k+E_0/\hbar)t} + c_s(t)\varphi_s|0\rangle\, e^{-\imath E_s t/\hbar}, \tag{5.124}$$

where $\hbar\omega_k$ is the energy of the photon emitted by the atom, $k = \left(\vec{k}, \sigma\right)$ are photon's quantum numbers, and $c_s(t)$ are the amplitudes of the single-photon states. The sum over k actually reduces to the integration over frequencies and directions of the photon propagation, along with the summation over polarizations. The sum over s in Eq. (5.124) is absent, since we consider the two-level system, and the index s relates to the upper level. Substituting the expansion (5.124) in the Schrödinger equation, we obtain the system of equations for the coefficients:

$$\frac{dc_k}{dt} = -\frac{\imath}{\hbar}\left\langle 01_k|\hat{V}'|s0\right\rangle e^{\imath(\omega_k-\omega_{s0})\,t} c_s, \quad \omega_{s0} = \frac{1}{\hbar}(E_s - E_0), \tag{5.125}$$

$$\frac{dc_0}{dt} = -\frac{\imath}{\hbar}\langle 00|\hat{V}|s0\rangle e^{-\imath\omega_{s0}\,t} c_s, \tag{5.125a}$$

$$\frac{dc_s}{dt} = -\frac{\imath}{\hbar}\langle s0|\hat{V}|00\rangle e^{\imath\omega_{s0}\,t} c_0 - \frac{\imath}{\hbar}\sum_k\left\langle s0|\hat{V}'|01_k\right\rangle e^{-\imath(\omega_k-\omega_{s0})\,t} c_k. \tag{5.126}$$

Here, we have used the relations $\vec{d}_{ss} = \vec{d}_{00} = 0$ and $\langle 0|1_k\rangle = 0$. Since we take into account only the spontaneous transitions from the excited level, the transient process will be damped out. Accordingly, in this case there is no need for the adiabatic switching-on of the external field. We assume that it is switched on at $t = 0$, and then the initial conditions are $c_s(0) = c_k(0) = 0$ and $c_0(0) = 1$. Let us present Eqs. (5.125) and (5.125a) in the integral form:

$$c_0(t) = 1 - \frac{\imath}{\hbar}\int_0^t \langle 00|\hat{V}|s0\rangle e^{-\imath\omega_{s0}\,t'} c_s(t')dt',$$

$$c_k(t) = -\frac{\imath}{\hbar}\left\langle 01_k|\hat{V}'|s0\right\rangle \int_0^t e^{\imath(\omega_k-\omega_{s0})\,t'} c_s(t')dt'.$$

The first equation implies that the difference of $c_0(t)$ from unity is of the second order of smallness in perturbation, which we neglect, setting $c_0(t) = 1$. The second equation is rearranged by integrating by parts,

$$c_k(t) = -\frac{1}{\hbar}\left\langle 01_k|\hat{V}'|s0\right\rangle e^{\imath(\omega_k-\omega_{s0})\,t}\int\limits_0^t \frac{1-e^{\imath(\omega_k-\omega_{s0})(t'-t)}}{\omega_k-\omega_{s0}}\frac{dc_s}{dt'}dt', \qquad (5.127)$$

Substitution of the obtained result in Eq. (5.126) enables us to derive the inhomogeneous integro-differential equation for $c_s(t)$:

$$\frac{dc_s}{dt'} = -\frac{\imath}{\hbar}\langle s0|\hat{V}|00\rangle e^{\imath\omega_{s0}\,t} + \frac{\imath}{\hbar^2}\sum_k\left|\langle s0|\hat{V}'|01_k\rangle\right|^2\int\limits_0^t\frac{1-e^{\imath(\omega_{s0}-\omega_k)(t-t')}}{\omega_k-\omega_{s0}} \times\frac{dc_s}{dt'}dt'. \qquad (5.128)$$

We consider a rather large interval of time, during which the excitation of atoms by an external field and their spontaneous radiation evolve to a stationary state. At large t, the interval $t - t'$ in the exponent in the integrand will be also large. Let present the integrand in the form

$$\left.\frac{1-e^{\imath(\omega_{s0}-\omega_k)(t-t')}}{\omega_k-\omega_{s0}}\right|_{t-t'\to\infty} \to \frac{\mathcal{P}}{\omega_k-\omega_{s0}} + \imath\pi\delta(\omega_k-\omega_{s0}). \qquad (5.129)$$

After substituting the right-hand side of this equality in Eq. (5.128), the first term (principal value) will describe a small radiation frequency shift of the transition (Lamb shift); we will ignore it. The term with the delta function is the one-half constant of damping, induced be spontaneous radiation:

$$\frac{\gamma_s}{2} = \frac{\pi}{\hbar^2}\sum_\sigma\int\left|\langle s0|\hat{V}'|01_k\rangle\right|^2\delta(\omega_k-\omega_{s0})\frac{\mathcal{V}\omega_k^2 d\omega_k d\Omega}{(2\pi c)^3}. \qquad (5.130)$$

Here, the summation over the discrete modes is replaced by the integration over photon frequencies and over directions of photon propagation, only the summation over polarizations remains. We will also simplify the first term on the right-hand side of Eq. (5.128), retaining the principal (resonance) term alone. Then Eq. (5.128) takes the form

$$\frac{dc_s}{dt} + \frac{\gamma_s}{2}c_s = -\frac{\imath}{2\hbar}\vec{d}_{s0}\cdot\vec{E}_0 e^{\imath(\omega_{s0}-\omega_k)\,t}. \qquad (5.131)$$

The solution of the derived equation with $t \gg \gamma_s^{-1}$ s is

$$c_s(t) = \vec{d}_{s0} \cdot \vec{E}_0 \frac{e^{\imath(\omega_{s0}-\omega_k)\,t}}{2\hbar(\omega_{s0} - \omega_k - \imath\gamma_s/2)}. \tag{5.132}$$

In comparison with Eq. (5.116) of Problem 5.3.3, here the resonance value of the atom excitation amplitude is finite. Using Eq. (5.132), we find the dielectric permittivity at frequencies close to one of the resonances:

$$\varepsilon(\omega) = 1 + \frac{4\pi N}{\hbar} \frac{\left|(d_x)_{s0}\right|^2}{\omega_{s0} - \omega - \imath\gamma_s/2}, \tag{5.133}$$

where the index k at the frequency is dropped now. The effect of radiation damping is responsible for the complexity of the dielectric permeability. At arbitrary frequencies (and at $\gamma_s \ll \omega_{s0}$), it is possible to use the approximate interpolation formula of the form

$$\varepsilon(\omega) = 1 + \frac{4\pi Ne^2}{m} \sum_s \frac{f_s}{\omega_{s0}^2 - \omega^2 - \imath\gamma_s\omega}. \tag{5.134}$$

Far from the resonance it reproduces the quantum Eq. (5.121) of problem 5.3.3, while near one of the resonances it gives the result close to Eq. (5.133).

5.3.5 Calculate the dielectric permittivity of a conducting sphere, assuming that the ions are at rest and the dielectric susceptibility $\alpha_i(\omega)$ of ion medium is given. The energy dissipation should be included by introducing the friction force $-\eta\,\dot{\vec{r}}$, acting on the conduction electrons of number density N. Relate the coefficient η to the statistic electric conductivity σ [37].

Solution The equation of motion for a conduction electron will be written in the form

$$m\ddot{\vec{r}} + \eta\dot{\vec{r}} = e\vec{E}_0 e^{-\imath\omega t}. \tag{5.135}$$

Its particular solution, corresponding to eigen modes, is

$$\vec{r} = -\frac{e\vec{E}_0 e^{-\imath\omega t}}{m(\omega^2 + \imath\gamma\omega)}, \tag{5.136}$$

where $\gamma = \eta/m$. We obtain the dipole moment per unit volume multiplying $\vec{r}$ by an electron charge e and by a number of particles per unit volume, N. Then we determine the polarizability of the medium $\alpha(\omega)$ and the dielectric permittivity $\varepsilon(\omega)$, induced by conduction electrons:

$$\varepsilon(\omega) = 1 + 4\pi\alpha(\omega) = 1 - \frac{\omega_{0e}^2}{\omega^2 + \imath\gamma\omega}, \quad \omega_{0e}^2 = \frac{4\pi e^2 N}{m}. \tag{5.137}$$

Using Eq. (5.135) and Ohm's law for a constant current, we will relate the specific resistance ρ to a constant current and the coefficient η:

$$\rho \equiv \frac{1}{\sigma} = \frac{\eta}{Ne^2}. \tag{5.138}$$

Rewriting Eq. (5.137) in the form

$$\varepsilon(\omega) = 1 + \imath \frac{4\pi}{\omega} \frac{Ne^2}{m(\gamma - \imath\omega)}, \tag{5.139}$$

we find the dependence of the electric conductivity on frequency:

$$\sigma(\omega) = \frac{Ne^2}{m(\gamma - \imath\omega)}. \tag{5.140}$$

To include the effect of bound electrons, one should add to the right-hand side of Eq. (5.137) the term $4\pi\,\alpha_i(\omega)$ containing the ion polarizability:

$$\varepsilon(\omega) = 1 + 4\pi\alpha_i(\omega) + \imath \frac{4\pi\sigma(\omega)}{\omega}. \tag{5.141}$$

At sufficiently low frequencies, we can neglect dispersion of the ion permeability and rewrite Eq. (5.141) as

$$\varepsilon(\omega) = \varepsilon_0 + \imath \frac{4\pi\sigma(\omega)}{\omega}, \qquad \varepsilon_0 = 1 + 4\pi\alpha_i(0). \tag{5.142}$$

Let us estimate $\gamma = \eta/m$ for copper (the static conductivity $\sigma = 5 \cdot 10^{17}\,\mathrm{s}^{-1}$). From Eq. (5.138), we obtain:

$$\gamma = \frac{Ne^2}{\sigma m} = \frac{N_0 e^2 d}{\sigma m A},$$

where $N_0 \approx 6 \cdot 10^{23}\,\mathrm{mol}^{-1}$ is the Avogadro number, $A \approx 63.5\,\mathrm{g/mol}$ is the atomic weight and $d \approx 8.9\,\mathrm{g/cm}^3$ is the copper mass density. The estimate gives $\gamma \approx 10^{+14}\,\mathrm{s}^{-1}$; let us note for comparison, that optical spectrum refers to frequencies $\approx 10^{15}\,\mathrm{s}^{-1}$. Therefore, in this case, it is likely that the conductivity retains its static value, up to frequencies of infrared range. However, one should bear in mind, that at high frequencies, when the mean free path of electrons becomes comparable with the penetration depth of the field into the metal, the effects of the spatial inhomogeneity begin to be important (anomalous skin-effect) and the static conductivity σ becomes inappropriate.

For a semiconductor (germanium), we have $\sigma \approx 2 \times 10^{10}\,\mathrm{s}^{-1}, \gamma \approx 10^{12} - 10^{13}\,\mathrm{s}^{-1}$, and $\varepsilon_0 = 16$. The results of this problem are applicable, in the restricted frequency region, to metals, semiconductors, and ionized gases, if the motion of positive ions can be ignored.

5.3.6 A gaseous dielectric in statistic equilibrium state at a temperature T consists of molecules with number density N, the principal values of the polarization tensor $\beta^{(1)} = \beta$ and $\beta^{(2)} = \beta^{(3)} = \beta'$ (β and β' depend on frequency ω). The dielectric is placed in a constant and uniform electric field $\vec{E}_0$. Find the tensor of the dielectric permittivity for the electric field, harmonically dependent on time, $\vec{E}(t) = \boldsymbol{\mathcal{E}} e^{-\iota\omega t}$ at $\mathcal{E} \ll E_0$ [37].

Solution Molecules in a dielectric are not spherically symmetric. Therefore, the external field $\vec{E}_0$ partially orients them, and the dielectric becomes anisotropic. However, the orienting effect of alternating field can be ignored due to the condition $\mathcal{E} \ll E_0$. Because the anisotropy is caused by the external electric field $\vec{E}_0$, one of the principal axes of the dielectric permittivity tensor will coincide with its direction, and the other two principal axes will be perpendicular to $\vec{E}_0$.

Let us denote the components of the molecule polarization in these axes in terms of β'_{ik} (the values $i, k = 1$ correspond to the axes parallel to $\vec{E}_0$). The components β'_{ik} are expressed through the principal values $\beta^{(i)}$ by the conventional formula:

$$\beta'_{ik} = \alpha_{il}\alpha_{km}\beta_{lm} = (\beta - \beta')\alpha_{i1}\alpha_{k1} + \beta'\delta_{ik},$$

where α_{il} are cosines of angles between the molecule symmetry axes and the principal axes of the dielectric permittivity tensor (we use here the relation $\alpha_{il}\alpha_{kl} = \delta_{ik}$, which follows from orthogonality of the matrix α_{ik}). To calculate the dielectric permittivity tensor per unit volume of the dielectric, it is necessary to use the Boltzmann distribution and find the statistic average values of β'_{ik}, that is, to average the product $\alpha_{i1}\alpha_{k1}$.

If we denote in the primed reference frame the polar angles of the molecule symmetry axis as φ and ϑ, then the quantities α_{1i} could be written as:

$$\alpha_{11} = \cos\vartheta, \quad \alpha_{12} = \sin\vartheta\cos\varphi, \quad \alpha_{13} = \sin\vartheta\sin\varphi.$$

Performing the averaging with the Boltzmann distribution and retaining the terms linear in $a = (\beta_0 - \beta'_0)E_0^2/2T$ we obtain:

$$\overline{\alpha_{11}^2} = \frac{1}{3}\left(1 + \frac{4}{15}a\right),$$
$$\overline{\alpha_{12}^2} = \overline{\alpha_{13}^2} = \frac{1}{3}\left(1 - \frac{2}{15}a\right),$$
$$\overline{\alpha_{i1}\alpha_{k1}} = 0 \quad \text{for} \quad \iota \neq k$$

where β_0 and β'_0 are the static values of the molecule polarization tensor. Then

$$\overline{\beta'_{11}} = \frac{1}{3}(\beta - \beta')\left(1 + \frac{4}{15}a\right) + \beta',$$
$$\overline{\beta'_{22}} = \overline{\beta'_{33}} = \frac{1}{3}(\beta - \beta')\left(1 - \frac{2}{15}a\right) + \beta'.$$

Neglecting the difference between the actual and average fields, we obtain the principal values of the dielectric permittivity tensor:

$$\varepsilon^{(1)} = 1 + 4\pi N\overline{\beta'_{11}}, \quad \varepsilon^{(2)} = \varepsilon^{(3)} = 1 + 4\pi N\overline{\beta'_{22}}.$$

This result shows that in a strong constant electric field, the dielectric becomes anisotropic with respect to high-frequency oscillations (for example, oscillations in optical range). The appearance of the anisotropy under the action of a constant electric field is known as the *Kerr effect*. The inertia of this effect is very low: a time for onset and disappearance of anisotropy is of the order of 10^{-10} s, being determined by a time for onset of statistic equilibrium in dielectrics. The Kerr effect is widely used in the industry for the fast modulation of light intensity.

5.3.7 A gaseous dielectric consists of polar molecules, whose electric dipole moment, in the absence of external fields, is p_0. The principal values of the molecule polarization in an alternating field are equal to $\beta^{(1)} = \beta$ and $\beta^{(2)} = \beta^{(3)} = \beta'$, and the axis x_1 is parallel to $\vec{p}_0$. The dielectric is immersed in the constant electric field $\vec{E}_0$ and alternating field $\vec{E}(t) = \mathcal{E}e^{-\imath\omega t}$. Neglecting the orientation effect of the alternating field and the orientation effect related to the anisotropic polarizability of molecules in a constant field, find the tensor of the dielectric permittivity of the dielectric in an alternating field at a given temperature T and number density of particles N [37].

Solution Considering the parameter $pE_0/T = a$ to be small and keeping the terms of the order a^2, we obtain:

$$\overline{\beta'_{11}} = \frac{1}{3}(\beta - \beta')\left(1 + \frac{2}{15}a^2\right) + \beta',$$
$$\overline{\beta'_{22}} = \overline{\beta'_{33}} = \frac{1}{3}(\beta - \beta')\left(1 - \frac{1}{15}a^2\right) + \beta',$$
$$\varepsilon^{(1)} = 1 + 4\pi N\overline{\beta'_{11}}, \quad \varepsilon^{(2)} = \varepsilon^{(3)} = 1 + 4\pi N\overline{\beta'_{22}}.$$

The notations are the same as in problem 5.3.6.

5.3.8 Find the polarizability of an atom, β_{ik}, in a field of the plane monochromatic wave in the presence of the weak external constant magnetic field $\vec{B}_0$. Use the model of the elastically bound electron (see problem 5.3.2). Apply the method of iterations. Neglect the action of the magnetic field of the plane wave and on losses of electromagnetic energy. Determine also the gyration vector $\vec{g}$ [37].

Solution The equation of motion for an atomic electron, bound to a nucleus by an elastic force, can be written as

$$\ddot{\vec{r}} + \omega_0^2 \vec{r} = \frac{e}{m}\left[\vec{E}_0 e^{-\imath\omega t} + \left(\frac{\mathbf{v}}{c} \times \vec{B}_0\right)\right], \tag{5.143}$$

where ω_0 is the eigen frequency. Solving it by the method of iterations, in the linear over B_0 approximation we obtain:

$$\vec{r} = \frac{e\vec{E}}{m\left(\omega_0^2 - \omega^2\right)} - \imath\frac{e^2\omega}{m^2c\left(\omega_0^2 - \omega^2\right)^2}\left(\vec{E} \times \vec{B}_0\right). \tag{5.144}$$

To obtain the tensor of atomic polarization, we use the form of the vector product containing the anti-symmetric tensor $e_{\imath kl}$. This gives

$$\beta_{ik} = \frac{e^2}{m\left(\omega_0^2 - \omega^2\right)}\delta_{\imath k} - \imath\frac{e^3\omega B_{0l}}{m^2c\left(\omega_0^2 - \omega^2\right)^2}e_{\imath kl}. \tag{5.145}$$

According to the general statement, this tensor is Hermitian. It is clear that Hermitian tensor β_{ik} can be written as

$$\beta_{ik} = \beta^{(i)}\delta_{\imath k} + \imath g_l e_{\imath kl},$$

where $\vec{g}$ is the gyration vector. The gyration vector in this case has the form

$$\vec{g} = -\frac{e^3\omega}{m^2c\left(\omega_0^2 - \omega^2\right)^2}\vec{B}_0 = -\frac{2e^2\omega}{m\left(\omega_0^2 - \omega^2\right)^2}\vec{\omega}_L, \tag{5.146}$$

where $\vec{\omega}_L = e\vec{B}_0/2mc$ is the Larmor frequency.

5.3.9 The crystals with two ions in the elementary cell (*NaCl*, *LiF*, *KBr*, and others) have the cubic symmetry, and their long-wave oscillations ($ka \ll 1$, a being the lattice constant) are isotropic. In such crystals, acoustic and optical oscillations can be excited. In the acoustic oscillations, macroscopic elements of the lattice participate as a whole, that is, positive and negative ions oscillate jointly, in phase. The optical oscillations are those of the sub-lattice of the negative ions relative to the sub-lattice of the positive ions. Such oscillations are similar to the plasma oscillations of the electron gas relative to the ion background. However, in the ion crystals the elastic forces, holding the ions near the lattice sites, are of great importance. Along with the elastic forces, the macroscopic electric field $\vec{E}$ and macroscopic polarization $\vec{P}$, affecting the optical oscillations, can occur.

Find the frequencies of the longitudinal and transverse optical oscillations by taking into account the elastic and electric forces. Consider the oscillations as quasi-static, that is, neglect the delay of electromagnetic disturbances, and describe them by the equations of electrostatics. Express the required frequencies in terms of the given frequency ω_0 of the purely elastic oscillations (neglecting the electric forces), and also in terms of two dielectric permittivities: the static one, ε_0, and that, which includes the polarizability of the electron shells of the ions, but not the mutual displacement of positive and negative ions. The latter permittivity is commonly denoted as ε_∞ and measured at frequencies much greater than the ion oscillation frequency, but much smaller than transition frequencies in the electron shells of the ions [37].

Solution The relative oscillations of ions occur with the reduced mass $m = m_+m_-/(m_+ + m_-)$. We introduce the parameter $\vec{w} = \sqrt{m/V}\left(\vec{r}_+ - \vec{r}_-\right)$, characterizing the relative displacement of ions of both signs from the equilibrium position. Here, V is the volume per one elementary crystal cell.

The oscillations produced only due to elastic forces would be described by the equation

$$\ddot{\vec{w}} = -\omega_0^2\vec{w}, \tag{5.147}$$

The Lagrangian describing these oscillations is

$$L\left(\dot{\vec{w}}, \vec{w}\right) = \dot{\vec{w}}^2/2 - \omega_0^2\vec{w}^2/2. \tag{5.148}$$

On the other hand, if we include the electric forces, we can write down the linear equation, determining the polarization vector of the crystal:

$$\vec{P} = \gamma\vec{w} + \alpha\vec{E}. \tag{5.149}$$

The first term is due to a displacement of ions from equilibrium positions. The nature of the second term and of constant α is simple. In the absence of the relative displacement of ions, the polarization is induced only by deformation of their electron shells. Therefore, α is the corresponding electric susceptibility, related to ε_∞. The latter is introduced in the statement of the problem by equation $\alpha = \alpha_\infty = (\varepsilon_\infty - 1)/4\pi$. The existence of the polarization leads to the additional potential energy

$$U = -\int_0^E \vec{P}\left(\vec{E}\right) \cdot d\vec{E} = -\gamma\vec{r} \cdot \vec{E} - \frac{1}{2}E^2. \tag{5.150}$$

Adding it to the Lagrangian (5.148), we obtain

$$L\left(\vec{w}, \vec{w}\right) = \vec{\dot{w}}^2/2 - \omega_0^2 \vec{w}^2/2 + \gamma \vec{w} \cdot \vec{E} + \frac{1}{2} E^2. \tag{5.151}$$

From the Lagrangian (5.151), we obtain the equation of motion with account for the electric forces:

$$\ddot{\vec{w}} = -\omega_0^2 \vec{w} + \gamma \vec{E}. \tag{5.152}$$

Let us express the constant γ in Eq. (5.152) in terms of the static dielectric permittivity ε_0. For this purpose, we write Eqs. (5.149) and (5.152) in the static limit:

$$\vec{P}_0 = \gamma \vec{w}_0 + \alpha_\infty \vec{E}, \quad \vec{w}_0 = \frac{\gamma}{\omega_0^2} \vec{E}. \tag{5.153}$$

On the other hand, we evidently have $\vec{P}_0 = \alpha_0 \vec{E}$, where α_0 is the static susceptibility.

With the help of Eq. (5.153), we find

$$\gamma = \omega_0 \sqrt{\frac{\varepsilon_0 - \varepsilon_\infty}{4\pi}}. \tag{5.154}$$

After all constants are determined, let us focus on oscillation frequencies. Up to now, we have considered the system homogeneous. Now we take into account that the oscillations in crystals propagate as weakly inhomogeneous waves (small $\vec{k}$). The vectors $\vec{w}$, $\vec{E}$, and $\vec{P}$ may be aligned perpendicular or parallel to $\vec{k}$. Let us represent every vector as a sum of the transverse and longitudinal components: $\vec{w} = \vec{w}_{tr} + \vec{w}_l, \ldots$. Consequently, Eqs. (5.149) and (5.152) will be related either to transverse or longitudinal vectors. For vector $\vec{E}_{tr}$, the Maxwell's equations in the quasi-statistical approximation have the form of $\nabla \times \vec{E}_{tr} = 0$ and $\nabla \cdot \vec{E}_{tr} = 0$. Hence, $\vec{E}_{tr} = 0$, that is, the electric field does not affect transverse oscillations. Equation (5.149) takes the form of $\ddot{\vec{w}}_{tr} = -\omega_0^2 \vec{w}_{tr}$, that is, the frequency of the transverse oscillations is $\omega_{tr} = \omega_0$.

The longitudinal field satisfies the equations of $\nabla \times \vec{E}_l = 0$ and $\nabla \cdot \vec{D} = div\left(\vec{E}_l + 4\pi \vec{P}_l\right) = 0$. From these equations we have

$$\vec{E}_l = -4\pi \vec{P}_l, \tag{5.155}$$

since at $\vec{P} = 0$ in the quasi-classical case there should be $\vec{E} = 0$. Now we exclude from Eqs. (5.149), (5.152), and (5.155) the vectors of $\vec{P}_l$ and $\vec{E}_l$ to find the equation of motion $\ddot{\vec{w}}_l = -\omega_l^2 \vec{w}_l$, from which we obtain

$$\omega_l^2 = \omega_0^2 + \frac{4\pi\gamma^2}{1+4\pi\gamma^2} = \frac{\varepsilon_0}{\varepsilon_\infty}\omega_{tr}^2, \quad \omega_l = \sqrt{\frac{\varepsilon_0}{\varepsilon_\infty}}\omega_{tr}.$$

The table contains experimental values of parameters for some ion crystals.

ε_∞	ε_0	ω_{tr} 10^3 rad/s	Crystal
2.25	5.62	3.09	NaCl
2.13	4.68	2.67	KCl
5.10	31.9	1.61	TiCl
5.07	8.30	5.71	ZnS

5.3.10 Find the dispersion relation for transverse electromagnetic waves near one of the resonance frequencies of a dielectric, whose molecules do not have constant dipole moments (non-polar dielectric). The permittivity of the non-polar dielectric can be obtained in the Lorentz–Lorenz model. Find the region of non-transparency of such a dielectric [37].

Solution In the Lorentz–Lorenz model, the permittivity near the resonance s is (see problem 5.3.4)

$$\varepsilon(\omega) \approx 1 + \frac{\omega_p^2 f_s}{\omega_s^2 - \omega^2 - \imath\gamma_s\omega},$$

where $\omega_p^2 = 4\pi\ Ne^2/m$ and N is the number of atoms or molecules. The electromagnetic field of frequency ω can propagate in the form of transverse waves with the wave number

$$k = \frac{\omega}{c}\sqrt{\frac{\omega_l^2 - \omega^2}{\omega_s^2 - \omega^2}} \quad (\gamma_s \ll \omega_s, \omega_l),$$

where $\omega_l^2 = \omega_s^2 + \omega_p^2 f_s$ is a squared frequency of longitudinal. The wave number (constant of propagation) becomes purely imaginary in the frequency range $\omega_s < \omega < \omega_l$, in which the dielectric is opaque. The damping results not from the dissipation of electromagnetic energy, but from destructive interference produced by vibrations of oscillators. Outside this interval, the dielectric is transparent; at $0 \leq \omega \leq \omega_l$ the dispersion relation is $2\omega_{tr1}(k) = \omega_l^2 + c^2k^2 - \sqrt{(\omega_l^2 + c^2k^2)^2 - \omega_s^2c^2k^2}$, $0 \leq k < \infty$. At frequencies $\omega > \omega_l$, the dispersion relation reads $2\omega_{tr2}(k) = \omega_l^2 + c^2k^2 + \sqrt{(\omega_l^2 + c^2k^2)^2 - \omega_s^2c^2k^2}$,$0 \leq k < \infty$. In the special case in which $ck \ll \omega_l$, we have $\omega_{tr2} \approx \omega_l + (1 - \omega_s^2/\omega_l^2)c^2k^2/2$. At $ck \gg \omega_l$, the dispersion relation $\omega_{tr2} = ck$ is the same as in vacuum. At frequencies of the order of ω_s, ω_l, there appears a strong coupling between the oscillations of the electromagnetic field and atomic oscillators (polariton waves and their quantum excitations–polaritons).

References

1. A.S. Davydov, Theory of light absorption in molecular crystals, Transactions of the Institute of Physics, Academy of Sciences of the Ukrainian SSR [in Russian], Izd. AN UkrSSR, Kiev (1951)
2. A.S. Davydov, J. Exp. Theor. Phys. **19**, 930 (1949)
3. U. Fano, Phys. Rev. **103**, 1202 (1956)
4. S. Pekar, J. Exp. Theor. Phys. **33**, 1022 (1957).; J. Exp. Theor. Phys. **34**, 1176 (1958); J. Exp. Theor. Phys. **35**, 522 (1958); J. Exp. Theor. Phys. **36**, 451 (1959)
5. J.J. Hopfield, Phys. Rev. **112**, 1555 (1958)
6. V.M. Agranovich, J. Exp. Theor. Phys. **35**, 430 (1959)
7. A.S. Davydov, A.F. Lubchenko, J. Exp. Theor. Phys. **35**, 1499 (1958)
8. V.S. Mashkvich, J. Exp. Theor. Phys. **38**, 906 (1960)
9. V.L. Strizhevskii, Russ. Solid State Phys. **2**, 1806 (1960)
10. S.I. Pekar, B.E. Tsekvava, Russ. Solid State Phys. **2**, 211 (1960)
11. B.E. Tsekvava, Russ. Solid State Phys. **2**, 482 (1960)
12. M.S. Brodin, S.I. Pekar, J. Exp. Theor. Phys. **38**, 74 (1960).; J. Exp. Theor. Phys. **38**, 1910 (1960)
13. M. Born, K. Huang, *Dynamical Theory of Crystal Lattices* (Oxford University Press, Oxford, 1954)
14. A.S. Davydov, J. Exp. Theor. Phys. **17**, 1106 (1947)
15. M.V. Volkenshtein, *Molecular Optics* (Gostekhteorizdat, Moscow, 1951)
16. M.V. Volkenshtein, *Structural and Physical Properties of Molecules* (Izd-vo Academy Science of USSR, Moscow, 1955)
17. V.M. Agranovich, Opt. Spectrosc. **1**, 338 (1956)
18. M.I. Kaganov, M.Y. Azbel, Dokl. Acad. Sci. USSR **102**, 49 (1955)
19. O.V. Konstantinov, V.I. Perel, J. Exp. Theor. Phys. **37**, 786 (1959)
20. V.L. Ginzburg, A.A. Rukhadze, V.P. Silin, Sov. Solid State Phys. **3**(5), 1337 (1961)
21. L.D. Landau, E.M. Lifshitz, *Course of Theoretical Physics, Vol. 3: Quantum Mechanics: Non-Relativistic Theory*, 3rd edn. (Nauka, Moscow, 1974; Pergamon, New York, 1977)
22. A.F. Prikhotko, J. Exp. Theor. Phys. **19**, 383 (1949)
23. E.F. Gross, A.A. Kaplianskii, Dokl. Acad. Sci. USSR **132**, 98 (1960)
24. V.M. Agranovich, A.A. Rukhadze, J. Exp. Theor. Phys. **35**, 982 (1958)
25. L.D. Landau, E.M. Lifshitz, *Electrodynamics of Continuous Media*, 2nd edn. (Pergamon, New York, 1984)
26. V.P. Makarov, A.A. Rukhadze, J. Exp. Theor. Phys. **98**(2), 305–315 (2004)
27. V.M. Agranovich, V.L. Ginzburg, *Crystal Optics with Spatial Dispersion, and Excitons*, 2nd edn. (Nauka, Moscow, 1979; Springer, New York, 1984)
28. E.M. Lifshitz, L.P. Pitaevskii, *Physical Kinetics* (Nauka, Moscow, 1979; Pergamon Press, Oxford, 1981)
29. L.D. Landau, E.M. Lifshitz, *Course of Theoretical Physics, Vol. 5: Statistical Physics*, 3rd edn. (Nauka, Moscow, 1976).; Pergamon Press, Oxford, 1980), Part 1
30. E.M. Lifshitz, L.P. Pitaevskii, *Course of Theoretical Physics, Vol. 5: Statistical Physics* (Nauka, Moscow, 1978; Pergamon Press, Oxford, 1980), Part 2
31. L.D. Landau, E.M. Lifshitz, *Course of Theoretical Physics, Vol. 1: Mechanics*, 3rd edn. (Nauka, Moscow, 1973; Pergamon Press, Oxford, 1976)
32. V.B. Berestetskii, E.M. Lifshitz, L.P. Pitaevskii, *Quantum Electrodynamics*, 2nd edn. (Nauka, Moscow, 1980; Pergamon Press, Oxford, 1982)
33. L.A. Vainshteœn, I.I. Sobel'man, E.A. Yukov, *Excitation of Atoms and Broadening of Spectral Lines* (Nauka, Moscow, 1979; Springer, Berlin, 1981)
34. A.A. Houck, J.B. Brock, I.L. Chung, Phys. Rev. Lett. **90**, 137401 (2003)

35. L.I. Mandel'shtam, Zh. Éksp, Teor. Fiz. **15**, 475 (1945); in Complete Works, Akad. Nauk USSR, Moscow (1947), Vol. 2, p. 334; in Complete Works, Akad. Nauk USSR, Moscow, Vol. 5, p. 461 (1950)
36. V.G. Veselago, Usp. Fiz. Nauk **92**, 517 (1967) [Sov. Phys. Usp. **10**, 509 (1968)]
37. I.N. Toptygin, *Electromagnetic Phenomena in Matter: Statistical and Quantum Approaches*, 1st edn. (Wiley-VCH, Weinheim, 2015)

Appendix A: The Main Operators of Field Theory in Orthogonal Curvilinear Coordinate Systems

The main operators of field theory are the gradient, the divergence, the rotation and the Laplacian. In Cartesian coordinates where

$$\vec{r} = \vec{i}x + \vec{j}y + \vec{k}z,$$

and $\vec{i}, \vec{j}, \vec{k}$ being the unit vectors along the coordinate axes, we have

$$\begin{aligned}
&grad\psi(x,y,z) = \nabla\psi = \vec{i}\frac{\partial\psi}{\partial x} + \vec{j}\frac{\partial\psi}{\partial y} + \vec{k}\frac{\partial\psi}{\partial z},\\
&div\vec{A}(x,y,z) = \nabla\cdot\vec{A} = \frac{\partial A_x}{\partial x} + \frac{\partial A_y}{\partial y} + \frac{\partial A_z}{\partial z},\\
&curl\,\vec{A}(x,y,z) = \nabla\times\vec{A} = \begin{vmatrix} \vec{i} & \vec{j} & \vec{k} \\ \frac{\partial}{\partial x} & \frac{\partial}{\partial y} & \frac{\partial}{\partial z} \\ A_x & A_y & A_z \end{vmatrix} = \vec{i}\left(\frac{\partial A_Z}{\partial y} - \frac{\partial A_y}{\partial z}\right) + \vec{j}\left(\frac{\partial A_x}{\partial z} - \frac{\partial A_z}{\partial x}\right) + \vec{k}\left(\frac{\partial A_y}{\partial x} - \frac{\partial A_x}{\partial y}\right),\\
&\nabla^2\psi(x,y,z) = \frac{\partial^2\psi}{\partial x^2} + \frac{\partial^2\psi}{\partial y^2} + \frac{\partial^2\psi}{\partial z^2}.
\end{aligned} \tag{A.1}$$

Thus

$$\nabla = \vec{i}\frac{\partial}{\partial x} + \vec{j}\frac{\partial}{\partial y} + \vec{k}\frac{\partial}{\partial z}, \quad \Delta = \nabla^2 = \frac{\partial^2}{\partial x^2} + \frac{\partial^2}{\partial y^2} + \frac{\partial^2}{\partial z^2}. \tag{A.2}$$

It is not difficult to write these operators in another curvilinear coordinate system (q_1, q_2, q_3), related to the Cartesian system by the transformation formulas

B. Shokri, A. A. Rukhadze, *Electrodynamics of Conducting Dispersive Media*, Springer Series on Atomic, Optical, and Plasma Physics 111,
https://doi.org/10.1007/978-3-030-28968-3

$$x = x(q_1, q_2, q_3), \quad y = y(q_1, q_2, q_3), \quad z = z(q_1, q_2, q_3). \tag{A.3}$$

If the transformation determinant is nonzero:

$$D = \left|\frac{\partial x_i}{\partial q_j}\right| = \begin{vmatrix} \frac{\partial x}{\partial q_1} & \frac{\partial x}{\partial q_2} & \frac{\partial x}{\partial q_3} \\ \frac{\partial y}{\partial q_1} & \frac{\partial y}{\partial q_2} & \frac{\partial y}{\partial q_3} \\ \frac{\partial z}{\partial q_1} & \frac{\partial z}{\partial q_2} & \frac{\partial z}{\partial q_3} \end{vmatrix} \neq 0, \tag{A.4}$$

the following relations are valid:

$$\begin{aligned}
\nabla\psi(q_1, q_2, q_3) &= \frac{e_1}{H_1}\frac{\partial\psi}{\partial q_1} + \frac{e_2}{H_2}\frac{\partial\psi}{\partial q_2} + \frac{e_3}{H_3}\frac{\partial\psi}{\partial q_3}, \\
\nabla\cdot\vec{A}(q_1, q_2, q_3) &= \frac{1}{H_1H_2H_3}\left[\frac{\partial}{\partial q_1}(A_1H_2H_3) + \frac{\partial}{\partial q_2}(A_2H_1H_3) + \frac{\partial}{\partial q_3}(A_3H_1H_2)\right], \\
\nabla\times\vec{A}(q_1, q_2, q_3) &= \frac{\vec{e}_1}{H_2H_3}\left[\frac{\partial}{\partial q_2}(A_3H_3) - \frac{\partial}{\partial q_3}(A_2H_2)\right] + \frac{\vec{e}_2}{H_1H_3}\times \\
&\times\left[\frac{\partial}{\partial q_3}(A_1H_1) - \frac{\partial}{\partial q_1}(A_3H_3)\right] + \frac{\vec{e}_3}{H_1H_2}\left[\frac{\partial}{\partial q_1}(A_2H_2) - \frac{\partial}{\partial q_2}(A_1H_1)\right], \\
\nabla^2\Psi(q_1, q_2, q_3) &= \frac{1}{H_1H_2H_3}\left[H_2H_3\frac{\partial}{\partial q_1}\left(\frac{1}{H_1}\frac{\partial\Psi}{\partial q_1}\right) + H_1H_3\frac{\partial}{\partial q_2}\left(\frac{1}{H_2}\frac{\partial\Psi}{\partial q_2}\right) + H_1H_2\frac{\partial}{\partial q_3}\left(\frac{1}{H_3}\frac{\partial\Psi}{\partial q_3}\right)\right].
\end{aligned} \tag{A.5}$$

Here, $\vec{e}_1$, $\vec{e}_2$, $\vec{e}_3$ are the unit vectors of the curvilinear coordinate system and H_1, H_2, H_3 the so-called Lame coefficients determined by

$$H_i = \sqrt{\left(\frac{\partial x}{\partial q_i}\right)^2 + \left(\frac{\partial y}{\partial q_i}\right)^2 + \left(\frac{\partial z}{\partial q_i}\right)^2}, \tag{A.6}$$

for $i = 1, 2, 3$.

Throughout the book we have frequently used the cylindrical coordinate system where

$$x = r\cos\phi, \quad y = r\sin\phi, \quad z = z, \tag{A.7}$$

and $\vec{e}_1 = \vec{e}_r$, $\vec{e}_2 = \vec{e}_\phi$, $\vec{e}_3 = \vec{e}_z$ are the unit vectors and $q_1 = r$, $q_2 = \phi$, $q_3 = z$. Then we have

$$H_1 = 1, \quad H_2 = r, \quad H_3 = 1. \tag{A.8}$$

Thus, in the cylindrical coordinate system we obtain

$$\nabla\Psi(r,\phi,z) = \vec{e}_r\frac{\partial\Psi}{\partial r} + \frac{\vec{e}_\phi}{r}\frac{\partial\Psi}{\partial\phi} + \vec{e}_z\frac{\partial\Psi}{\partial z};$$
$$\nabla\cdot\vec{A}(r,\phi,z) = \frac{1}{r}\left[\frac{\partial}{\partial r}(rA_r) + \frac{\partial A_\phi}{\partial\phi} + r\frac{\partial A_z}{\partial z}\right] = \frac{1}{r}\frac{\partial}{\partial r}(rA_r) + \frac{1}{r}\frac{\partial A_\phi}{\partial\phi} + r\frac{\partial A_z}{\partial z};$$
$$\nabla\times\vec{A}(r,\phi,z) = \vec{e}_r\left[\frac{1}{r}\frac{\partial A_z}{\partial\phi} - \frac{\partial A_\phi}{\partial z}\right] + \vec{e}_\phi\left[\frac{\partial A_r}{\partial z} - \frac{\partial A_z}{\partial r}\right] + \vec{e}_z\left[\frac{1}{r}\frac{\partial}{\partial r}(rA_\phi) - \frac{1}{r}\frac{\partial A_r}{\partial\phi}\right];$$
$$\nabla^2\Psi(r,\phi,z) = \frac{\partial^2\Psi}{\partial r^2} + \frac{1}{r}\frac{\partial^2\Psi}{\partial\phi^2} + \frac{1}{r}\frac{\partial\Psi}{\partial r} + \frac{\partial^2\Psi}{\partial z^2}. \tag{A.9}$$

A.1 Exercise

A.1.1 Calculate the Lame coefficients in the spherical coordinate system, where

$$x = r\sin\theta\cos\phi,\quad y = r\sin\theta\sin\phi,\quad z = r\cos\theta.$$

Solution We obtain

$$q_1 = r,\quad q_2 = \phi,\quad q_3 = \theta,$$
$$H_1 = 1,\quad H_2 = r\sin\theta,\quad H_3 = r.$$

A.1.2 Assuming the oz axis to be oriented along the vector $\vec{B}$, calculate the operation $\left[\vec{v}\times\vec{B}\right]\cdot\partial/\partial\vec{v}$ in the cylindrical coordinates of the velocity space.

Solution Taking into account that

$$\frac{\partial}{\partial v_x} = \frac{\partial v_\perp}{\partial v_x}\frac{\partial}{\partial v_\perp} + \frac{\partial\phi}{\partial v_x}\frac{\partial}{\partial\phi} = \cos\phi\frac{\partial}{\partial v_\perp} - \frac{\sin\phi}{v_\perp}\frac{\partial}{\partial\phi},$$
$$\frac{\partial}{\partial v_y} = \frac{\partial v_\perp}{\partial v_y}\frac{\partial}{\partial v_\perp} + \frac{\partial\phi}{\partial v_y}\frac{\partial}{\partial\phi} = \sin\phi\frac{\partial}{\partial v_\perp} + \frac{\cos\phi}{v_\perp}\frac{\partial}{\partial\phi}.$$

We easily obtain

$$\left[\vec{v}\times\vec{B}\right]\cdot\frac{\partial}{\partial\vec{v}} = B\left(v_y\frac{\partial}{\partial v_x} - v_x\frac{\partial}{\partial v_y}\right) = -B\frac{\partial}{\partial\phi}.$$

A.1.3 Write the vector $\left(\vec{v}\cdot\nabla\right)\vec{v}$ in cylindrical coordinates.

Solution Using the identity

$$\left(\vec{v}\cdot\nabla\right)\vec{v}=\frac{1}{2}\nabla v^2-\left(\vec{v}\times\nabla\times\vec{v}\right)$$

and the expressions for grad v^2 and rot $\vec{v}$ in the cylindrical coordinates (A.9), we easily obtain

$$\left\{\left(\vec{v}\cdot\nabla\right)\vec{v}\right\}_r=v_r\frac{\partial v_r}{\partial r}+\frac{v_\phi}{r}\frac{\partial v_r}{\partial\phi}+v_z\frac{\partial v_r}{\partial z}-\frac{v_\phi^2}{r},$$
$$\left\{\left(\vec{v}\cdot\nabla\right)\vec{v}\right\}_\phi=v_r\frac{\partial v_\phi}{\partial r}+\frac{v_\phi}{r}\frac{\partial v_\phi}{\partial\phi}+v_z\frac{\partial v_\phi}{\partial z}+\frac{v_r v_\phi}{r},$$
$$\left\{\left(\vec{v}\cdot\nabla\right)\vec{v}\right\}_z=v_r\frac{\partial v_z}{\partial r}+\frac{v_\phi}{r}\frac{\partial v_z}{\partial\phi}+v_z\frac{\partial v_z}{\partial z}.$$

Appendix B: Elements of Tensor Calculus

The concept of a tensor is closely related to the transformations of coordinate systems. Above we have primarily used the three-dimensional orthogonal Cartesian coordinate system *oxyz*, which will be written in the following in the symmetrical form $ox_1x_2x_3$. We consider two Cartesian systems $ox_1x_2x_3$ and $ox'_1x'_2x'_3$ with the common origin *o*. Then the coordinates of a point *M* in the primed and the primeless system are related by

$$x_k = e_{kj}x'_j, \quad x'_k = e^{-1}_{kj}x_j = e_{jk}x_j, \tag{B.1}$$

where e_{ij} are the cosines of the angles between the axes of the primed and primeless system:

	x'_1	x'_2	x'_3
x_1	e_{11}	e_{12}	e_{13}
x_2	e_{21}	e_{22}	e_{23}
x_3	e_{31}	e_{32}	e_{33}

(B.2)

To simplify the notation, the summation from 1 to 3 is assumed to be carried out over repeated (dummy) indices. It is easy to show the relation

B. Shokri, A. A. Rukhadze, *Electrodynamics of Conducting Dispersive Media*, Springer Series on Atomic, Optical, and Plasma Physics 111,
https://doi.org/10.1007/978-3-030-28968-3

$$e_{ik}e_{jk} = e_{ki}e_{kj} = \delta_{ij} = \begin{cases} 1, & \text{for} \quad i = j, \\ 0, & \text{for} \quad i \neq j. \end{cases} \tag{B.3}$$

The transformation (B.1) is called an orthogonal affine transformation (a rotation of the coordinate system) and the matrix

$$e_{ij} = \begin{pmatrix} e_{11} & e_{12} & e_{13} \\ e_{21} & e_{22} & e_{23} \\ e_{31} & e_{32} & e_{33} \end{pmatrix} \tag{B.4}$$

is known as the transformation matrix.

Since the vector $\vec{x}$ with the components x_1, x_2, x_3 or $\vec{x'}$ with x'_1, x'_2, x'_3 in the primeless and primed coordinates, respectively, represent the point M, relations (B.1) constitute the transformation law of vectors when the coordinate system is rotated. Moreover, a set of three quantities transformed by expression (B.1) is a vector.

If two vectors $\vec{a} = (a_1, a_2, a_3)$ and $\vec{b} = (b_1, b_2, b_3)$ are given, then their scalar product defined by

$$\vec{a}.\vec{b} = a_i b_i = a_1 b_1 + a_2 b_2 + a_3 b_3 \tag{B.5}$$

is invariant under the transformations of the coordinate system (B.1), i.e.,

$$a_i b_i = a'_i b'_i. \tag{B.6}$$

Indeed, due to Eq. (B.3) we have $a'_i b'_i = e_{ki} a_k e_{si} b_s = \delta_{ks} a_k b_s = a_s b_s$, which was to be proven.

The invariance of the scalar product Eq. (B.6) is often used as the definition of a vector. Thus, if the vector $\vec{x} = (x_1, x_2, x_3)$ and a set of three quantities $a_i = (a_1, a_2,$ $a_3)$ are given and the linear from

$$F_1 = a_i x_i \tag{B.7}$$

is invariant with respect to transformations of the coordinate system (B.1), then the set forms a vector $(a_i) = \vec{a} = (a_1, a_2, a_3)$.

A second-rank tensor is defined analogously: if two vectors $\vec{x} = (x_1, x_2, x_3)$ and $\vec{y} = (y_1, y_2, y_3)$ are given and the quadratic from

$$F_2 = d_{ij} x_i y_j \tag{B.8}$$

is invariant with respect to transformations of the coordinate system (B.1), then the set of the nine quantities d_{ij} is called a second-rank tensor.

The set δ_{ij} defined by Eq. (B.3) is also a tensor, Actually,

$$\delta_{ij} x_i y_j = x_j y_j = \vec{x}.\vec{y} = \text{const.} \tag{B.9}$$

From Eq. (B.8) there follows the transformation law of second-rank tensors

$$d_{ij} x_i x_j = d'_{ij} x'_i y'_i = d'_{ij} e_{mi} x_m e_{nj} y_n = d'_{sk} e_{is} e_{jk} x_i y_j, \tag{B.10}$$

or

$$d_{ks} = e_{ki} e_{sj} d'_{ij}. \tag{B.11}$$

Analogously we obtain

$$d'_{ks} = e_{ik} e_{js} d_{ij}. \tag{B.12}$$

Thus, a second-rank tensor is transformed like the outer product of two vectors $a_i b_j$. Therefore, one can define such a tensor as a set of nine quantities being transformed like the outer product of two vectors.

Tensors of higher rank are defined analogously. Thus, a third-rank tensor β_{ijk} is a set of 27 quantities, leaving the cubic form

$$F_3 = \beta_{ijk} x_i y_j z_k \tag{B.13}$$

invariant with respect to the transformation (B.1), when $\vec{x}$, $\vec{y}$, $\vec{z}$ are vectors. Equivalently, the set of the quantities β_{ijk} defines a tensor when it is transformed like the outer product of three vectors $a_i b_j c_k$. A scalar quantity can be regarded as a zero-rank tensor and a vector as a first-rank tensor.

The components of a tensor can be both real and complex. Therefore, in general, we have to deal with complex tensors. Then, the concept of the Hermiticity of a tensor is important. A second-rank tensor is called Hermitian if ("$*$" means complex conjugation)

$$\alpha_{ij}^{*H} = \alpha_{ji}^{H}, \tag{B.14}$$

but if

$$\alpha_{ij}^{*a} = -\alpha_{ji}^{a} \tag{B.15}$$

The tensor is called anti-Hermitian. Any tensor can be decomposed into a Hermitian and an anti-Hermitian part.

In the above, tensors have been referred to as a set of complex quantities. The tensor components, however, can be functions of both a scalar [e.g., of the time,

$\alpha_{ij}(t)$] and a vector [e.g., of the coordinate. $\alpha_{ij}(\vec{r})$]. Therefore, in general, we must write:

$\phi\left(t, \vec{r}\right)$ for a scalar (zero-rank tensor)

$a_i\left(t, \vec{r}\right)$ for a vector (first-rank tensor)

$\alpha_{ij}\left(t, \vec{r}\right)$ for a second-rank tensor

$\beta_{ijk}\left(t, \vec{r}\right)$ for a third-rank tensor, etc.

Tensors as functions of more than one scalar or vector or even tensor variable are defined analogously.

Differentiating a tensor with respect to a scalar its rank is not altered; however, the differentiation with respect to a vector enlarges the tensor rank. Hence,

$\frac{\partial\phi\left(t, \vec{r}\right)}{\partial r_i}$ is a vector (a first-rank tensor),

$\frac{\partial a_i\left(t, \vec{r}\right)}{\partial r_j}$ is a second-rank tensor,

$\frac{\partial \alpha_{ij}\left(t, \vec{r}\right)}{\partial r_k}$ is a third-rank tensor, etc.

We must take account of this fact when expanding a tensor in powers of a vector quantity:

$$\alpha_{ij}\left(t, \vec{r} + \Delta\vec{r}\right) = \alpha_{ij}\left(t, \vec{r}\right) + \frac{\partial \alpha_{ij}\left(t, \vec{r}\right)}{\partial r_k}\Delta r_k + \frac{\partial^2 \alpha_{ij}\left(t, \vec{r}\right)}{\partial r_k \partial r_s}\Delta r_k \Delta r_s. \tag{B.16}$$

The expansion of a scalar quantity in a power series is performed in the usual way.

Up to now we have dealt with a rotation of the coordinate system only, and the tensor quantities have been defined by a certain symmetry property with respect to the rotation transformation (B.1). We now consider the mirror reflection of the symmetry axes of the coordinates. With respect to these transformations the tensors can be subdivided into real and pseudo-tensors. A scalar quantity, which is invariant not only with respect to the rotation of the coordinate system but also with respect to the mirror reflection, is called a real scalar. If it is not modified by a rotation but reverses its sign under a mirror reflection it is called a pseudo-scalar. Real and pseudo-tensors of any rank are defined analogously. A real tensor of an even rank does not reverse its sign under the transformation of the mirror reflection. A pseudo-tensor, however, does reverse it. A real tensor of an odd rank reverses its sign under the transformation of the mirror reflection and a pseudo-tensor of this rank conserves it.

Real vectors in the three-dimension space are, e.g., the radius vector $\vec{r}$, the velocity $\vec{v}$ and the momentum $\vec{p}$ vectors, the wave vector $\vec{k}$, the vectors of the electric field strength $\vec{E}$ and the electric induction $\vec{D}$, the vector of the current density

$\vec{j}$, etc. Real scalars are the time t, the charge density ρ, the particle energy $\epsilon\left(\vec{p}\right)$ and the frequency $\omega\left(\vec{k}\right)$. We have

$$\vec{v} = \frac{d\vec{r}}{dt}, \quad \vec{v} = \frac{\partial \epsilon\left(\vec{p}\right)}{d\vec{p}}, \quad \vec{v}_{\text{gr}} = \frac{\partial \omega\left(\vec{k}\right)}{\partial \vec{k}}. \tag{B.17}$$

A pseudo-vector can be obtained as the vector product of two real vectors $\vec{a}$ and $\vec{b}$, i.e., $\left(\vec{a} \times \vec{b}\right)$. On the other hand, the vector product of a real vector $\vec{a}$ and a pseudo-vector $\vec{d}$ is a real vector $\left(\vec{a} \times \vec{d}\right)$. Thus, the magnetic field $\vec{B}$ is a pseudo-vector, since its vector product with the velocity vector $\vec{v}$ constitutes the real vector of the force $\vec{F} \sim \left(\vec{v} \times \vec{B}\right)$.

In electrodynamics of material media, the completely anti-symmetric third-rank unit tensor e_{ijk}, determined by

$$e_{ijk} = \begin{cases} 0 & \text{if two of the indices } i, j, k \text{ coincide,} \\ 1 & \text{if the indices } i, j, k \text{ form a regula succesionof the numbers } 1, 2, 3, \\ -1 & \text{if the indices } i, j, k \text{ form an irregular succession of the numbers } 1, 2, 3, \end{cases} \tag{B.18}$$

Is of special importance. A cyclic succession of the numbers 1, 2, 3 is called regular and a noncyclic one irregular.

Using the unit tensor e_{ijk}, the product of two vectors $\vec{a}$ and $\vec{b}$ can be written as

$$\left(\vec{a} \times \vec{b}\right)_i = e_{ijk} a_j b_k. \tag{B.19}$$

The scalar product of a real vector $\vec{a}$ and a pseudo-vector $\vec{d}$, in contrast to the scalar product of two real vectors, is a pseudo-scalar:

$$\vec{a} \cdot \vec{d} = a_i d_i = a_i e_{ijk} b_j c_k. \tag{B.20}$$

Here, $d_i = e_{ijk} b_j c_k$ is a pseudo-vector, $\vec{b}$ and $\vec{c}$ are real vectors.

The aforementioned properties are given not only for vector fields $a_i\left(\vec{r}\right)$ and tensor fields $a_{ij}\left(\vec{r}\right)$, $\beta_{ijk}\left(\vec{r}\right)$ but also for vector and tensor operators. As stated above, the differentiation with respect to a vector argument increases the rank of the matrix. We can introduce the differentiation operator as the vector $\partial/\partial r_i = \partial/\partial \vec{r} = \nabla_r$ and define the differentiation process by a vector or scalar product:

$$\frac{\partial}{\partial r_i}\phi\left(\vec{r}\right) = \nabla_r\phi\left(\vec{r}\right) = \nabla\phi\left(\vec{r}\right),$$
$$\frac{\partial}{\partial r_i}a_i\left(\vec{r}\right) = \nabla_r\vec{a}\left(\vec{r}\right) = \nabla.\vec{a}\left(\vec{r}\right), \tag{B.21}$$
$$e_{ijk}\frac{\partial}{\partial r_j}a_k\left(\vec{r}\right) = \left[\nabla_r \times \vec{a}\left(\vec{r}\right)\right]_i = \nabla \times \vec{a}\left(\vec{r}\right)\ldots.$$

If $\phi\left(\vec{r}\right)$ is a real scalar and $\vec{r}$ is a real vector, then the first quantity in (B.1) is a real vector. If $\phi\left(\vec{r}\right)$is a pseudo-scalar and $\vec{r}$ a real vector, then it is a pseudo-vector. It will also be a pseudo-vector if $\phi\left(\vec{r}\right)$ is a real scalar and $\vec{r}$ a pseudo-vector. The other quantities appearing in (B.1) and also the variables and operators of a higher rank can be interpreted analogously. For instance,

$$\nabla\cdot\nabla\phi\left(\vec{r}\right) = \frac{\partial}{\partial r_i}\frac{\partial}{\partial r_i}\varphi\left(\vec{r}\right) = \nabla^2\phi\left(\vec{r}\right) = \left(\frac{\partial^2}{\partial x^2}+\frac{\partial^2}{\partial y^2}+\frac{\partial^2}{\partial z^2}\right)\phi\left(\vec{r}\right),$$
$$\nabla\times\nabla\times\vec{a}\left(\vec{r}\right) = \left[\nabla_r\times\left[\nabla_r\times\vec{a}\left(\vec{r}\right)\right]\right]_i = e_{imn}e_{nkj}\frac{\partial}{\partial r_j}\frac{\partial}{\partial r_m}a_k$$
$$= \frac{\partial}{\partial r_i}\frac{\partial}{\partial r_j}a_j\left(\vec{r}\right) - \frac{\partial^2}{\partial r_j \partial r_j}a_i\left(\vec{r}\right) = \nabla\nabla\cdot\vec{a}\left(\vec{r}\right) - \nabla^2\vec{a}\left(\vec{r}\right)\ldots. \tag{B.22}$$

The described theory of three-dimensional tensors can be easily generalized to the four-dimensional case. In the four-dimensional space of the time and space coordinates $\left(t, \vec{r}\right)$ the Lorentz transformations are rotation transformations and from the basis for the definition of four-dimensional vectors and tensors. Besides, $\left(t, \vec{r}\right)$, the current and charge densities $\left(\rho, \vec{j}\right)$, the wave vector and frequency $\left(\omega, \vec{k}\right)$, etc., are four-dimensional vectors. However, we do not expand on the theory of four-dimensional tensors, since, in fact, they not been used in the book.

B.1 Exercises

B.1.1 Using the identity

$$e_{ikl}e_{mnl} = \delta_{im}\delta_{kn} - \delta_{in}\delta_{km},$$

Verify the equality

$$\left(\vec{A}\times\left(\vec{B}\times\vec{C}\right)\right)=\vec{B}\left(\vec{A}.\vec{C}\right)-\vec{C}\left(\vec{A}.\vec{B}\right).$$

Solution We apply the tensor notation

$$\left[\vec{A}\times\left(\vec{B}\times\vec{C}\right)\right]=e_{ikl}A_kB_mC_ne_{lmn}=A_kB_mC_n(e_{ikl}e_{lmn})=$$
$$=A_kB_mC_n(\delta_{im}\delta_{kn}-\delta_{in}\delta_{km})=B_i(A_nC_n)-C_i(A_nB_n),$$

which was to be proven.

B.1.2 Compose the general second-rank tensor ε_{ij} with $\varepsilon_{ij}\left(\vec{k}\right)=\varepsilon_{ij}\left(-\vec{k}\right)$ for real $\vec{k}$. For $\vec{k}=0$ reduce ε_{ij} over the indices, i.e. take the sum ε_{ii}.

Solution

$$\varepsilon_{ij}\left(\vec{k}\right)=\alpha_1\delta_{ij}+\alpha_2k_ik_j=\left(\delta_{ij}-\frac{k_ik_j}{k^2}\right)\varepsilon^{\mathrm{tr}}+\frac{k_ik_j}{k^2}\varepsilon^l,$$

i.e.,

$$\alpha_1=\varepsilon^{\mathrm{tr}},\quad \alpha_2=\frac{\varepsilon^l-\varepsilon^{\mathrm{tr}}}{k^2}.$$

In the limit $\vec{k}\to 0$ we have $\varepsilon_{ij}(0)=\alpha_1\delta_{ij}$ and $\varepsilon^l=\varepsilon^{\mathrm{tr}}=\varepsilon$. For vanishing $\vec{k}$, $\varepsilon_{ij}(0)=\varepsilon\delta_{ij}$ is the general second-rank tensor. The reduction over the indices reads

$$\varepsilon_{ij}\left(\vec{k}\right)=(3-1)\varepsilon^{\mathrm{tr}}+\varepsilon^l=2\varepsilon^{\mathrm{tr}}+\varepsilon^l,\quad \varepsilon^l_{ii}(0)=3\varepsilon.$$

B.1.3 Compose the second-rank tensor $\varepsilon_{ij}\left(\vec{B}\right)=\varepsilon_{ij}\left(-\vec{B}\right)$ from a pseudo-vector $\vec{B}$ and reduce it over the indices.

Solution

$$\varepsilon_{ij}\left(\vec{B}\right)=\alpha_1\delta_{ij}+\alpha_2b_ib_j+\alpha_3e_{ijk}b_k=\varepsilon_\perp\delta_{ij}+\left(\varepsilon_\parallel-\varepsilon_\perp\right)b_ib_j+\imath g e_{ijk}b_k,$$

where $\vec{b}=\vec{B}/B$. This tensor has the following matrix from

$$\varepsilon_{ij}\left(\vec{B}\right)=\begin{pmatrix}\varepsilon_\perp & \imath g & 0\\ -\imath g & \varepsilon_\perp & 0\\ 0 & 0 & \varepsilon_\parallel\end{pmatrix}.$$

Here, the ox_3-axis is oriented along the vector $\vec{B}$. For vanishing B we have

$$\varepsilon_{ij}(0) = \varepsilon\delta_{ij}, \quad \varepsilon_{\perp} = \varepsilon_{\|} = \varepsilon, \quad g = 0.$$

Finally, the reduction over the indices is

$$\varepsilon_{ii}\left(\vec{B}\right) = 2\varepsilon_{\perp} + \varepsilon_{\|}, \quad \varepsilon_{ii}(0) = 3\varepsilon.$$

B.1.4 Compose the second-rank tensor $\varepsilon_{ij}\left(\vec{k}, \vec{B}\right) = \varepsilon_{ji}\left(-\vec{k}, -\vec{B}\right)$ from a real vector $\vec{k}$ and a pseudo-vector $\vec{B}$ and reduce it over the indices.

Solution

$$\varepsilon_{ij}\left(\vec{k}, \vec{B}\right) = \alpha_1\delta_{ij} + \alpha_2 k_i k_j + \alpha_3 b_i b_j + \alpha_4 e_{ijm} b_m +$$
$$+\alpha_5 e_{imn} e_{jrs} k_m b_n k_r b_s + \alpha_6\left(e_{imn} k_m b_n k_j - e_{jmn} k_m b_n k_i\right),$$

where $\vec{b} = \vec{B}/B$. Orienting the $ox_3 = oz$-axis along the vector $\vec{b}$ and the $ox_1 = ox$-axis so that the vector $\vec{k}$ takes the form $\vec{k} = \left(k_{\perp}, 0, k_{\|}\right)$, we obtain

$$\varepsilon_{ij}\left(\vec{k}, \vec{B}\right) = \begin{pmatrix} \varepsilon_{11} & \varepsilon_{12} & \varepsilon_{13} \\ -\varepsilon_{12} & \varepsilon_{22} & \varepsilon_{23} \\ \varepsilon_{31} & -\varepsilon_{23} & \varepsilon_{33} \end{pmatrix},$$

with

$$\varepsilon_{\|} = \alpha_1 + \alpha_2 k_{\perp}^2, \quad \varepsilon_{22} = \alpha_1 + \alpha_5 k_{\perp}^5,$$
$$\varepsilon_{12} = \alpha_4 + \alpha_6 k_{\perp}^5, \quad \varepsilon_{23} = -\alpha_6 k_{\perp} k_{\|},$$
$$\varepsilon_{13} = \alpha_2 k_{\perp} k_{\|}, \quad \varepsilon_{33} = \alpha_1 + \alpha_2 k_{\|}^2 + \alpha_3.$$

Reducing the derived tensor over the indices yields

$$\varepsilon_{ii}(k, B) = \varepsilon_{11} + \varepsilon_{12} + \varepsilon_{33} = 3\alpha_1 + \alpha_2 k^2 + \alpha_3 + \alpha_5 k_{\perp}^2.$$

B.1.5 Write down the general relation $B_i = \mu_{ij}\left(\vec{H}\right) H_j$ between the pseudo-vectors $\vec{B}$ and $\vec{H}$.

Solution Evidently, $\mu_{ij}\left(\vec{H}\right)$ must be a real tensor. Therefore,

$$\mu_{ij}\left(\vec{H}\right) = \alpha_1\delta_{ij} + \alpha_2 h_i h_j + \alpha_3 e_{ijk} h_k,$$

where $\vec{h} = \vec{H}/H$. The general relation under discussion thus obtains the form

$$\vec{B} = (\alpha_1 + \alpha_2)\vec{H} = \mu\left(\vec{H}\right)\vec{H}.$$

Appendix C: Generalization of Kramers-Kronig Relations with Account of Finiteness of Speed of Light

Taking into account the finiteness of light speed, we find the generalized Kramer-Kronig relations. For a homogenous medium at rest which is in the stationary state, the relation between the electric induction $\vec{D}\left(t, \vec{r}\right)$ and electric field strength $\vec{E}\left(t, \vec{r}\right)$ (material equation) is written in the form

$$D_i\left(t, \vec{r}\right) = E_i\left(t, \vec{r}\right) + \int dV' \int_{-\infty}^{t} dt' \tilde{F}_{ij}\left(t - t', \vec{r} - \vec{r}'\right) E_j\left(t', \vec{r}'\right). \quad \text{(C.1)}$$

According to the causality principle (the field $\vec{E}$ is the cause that precedes the effect which is the induction $\vec{D}$), integration over dt' in (C.1) is limited from above by time moment t. Representing the field and induction in (C.1) in the form of Fourier integral, we find the relation between their Fourier components

$$D_{i\omega\vec{k}} = \varepsilon_{ij}\left(\omega, \vec{k}\right) E_{j\omega\overleftarrow{k}}, \quad \text{(C.2)}$$

where dielectric permittivity tensor is

$$\varepsilon_{ij}\left(\omega, \vec{k}\right) = \delta_{ij} + F_{ij}\left(\omega, \vec{k}\right), \quad F_{ij}\left(\omega, \vec{k}\right) = \int_{0}^{\infty} dt \int dV \tilde{F}_{ij}\left(t, \vec{r}\right) e^{\imath\left(\omega t - \vec{k}\cdot\vec{r}\right)}. \quad \text{(C.3)}$$

Strictly speaking, equality (C.1) does not take into account the finiteness of light speed and the causality principle yet: cause and effect are connected by a time-like interval; hence, integration over dt' in (C.1) is limited by the condition $t - t' \geq \frac{\left|\vec{r} - \vec{r}'\right|}{c}$:

B. Shokri, A. A. Rukhadze, *Electrodynamics of Conducting Dispersive Media*, Springer Series on Atomic, Optical, and Plasma Physics 111,
https://doi.org/10.1007/978-3-030-28968-3

$$D_i\left(t,\vec{r}\right)=E_i\left(t,\vec{r}\right)+\int dV'\int\limits_{-\infty}^{t-\frac{\left|\vec{r}-\vec{r}'\right|}{c}}dt'\tilde{F}_{ij}\left(t-t',\vec{r}-\vec{r}'\right)E_j\left(t',\vec{r}'\right). \tag{C.4}$$

Consequently, in formula (C.3) for dielectric permittivity tensor, we have

$$F_{ij}\left(\omega,\vec{k}\right)=\int dV\int\limits_{r/c}^{\infty}dt\tilde{F}_{ij}\left(t,\vec{r}\right)e^{\imath\left(\omega t-\vec{k}\cdot\vec{r}\right)}. \tag{C.5}$$

This circumstance was first pointed out by M. A. Leontovich in [1]. Taking into account the finiteness of light speed, he also obtained relations similar to the Kramers-Kronig relations. For isotropic non-gyrotropic media, they are of the form

$$\varepsilon^{l,\mathrm{tr}}(\omega,k)-1=\frac{1}{\pi\imath}\int\limits_{-\infty}^{\infty}d\xi\mathcal{P}\frac{\varepsilon^{l,\mathrm{tr}}(\xi,k+\beta(\xi-\omega)/c)-1}{\xi-\omega}, \tag{C.6}$$

where β is an arbitrary parameter which is limited only by condition $|\beta|\leq 1$; but the only correct value is $\beta=1$. To prove this, we note that the dispersion relation is usually derived from the material equation. However, the converse assertion is also true: the dispersion relation corresponds to a single correct material equation and the latter can be obtained from the dispersion relation. Below it is shown that the correct dispersion relation leads to the material equation (C.4) only when $\beta=1$. Differentiating relation (C.6) by β and substituting $\beta=0$ into it, we find

$$\frac{1}{\pi\imath c}\frac{\partial}{\partial k}\int\limits_{-\infty}^{\infty}\varepsilon^{l,\mathrm{tr}}(\omega,k)d\omega=0. \tag{C.7}$$

This indicates that Leontovich relations for arbitrary $|\beta|\leq 1$ are incorrect because for $\omega\to 0$, $\varepsilon^{l,\ \mathrm{tr}}(\omega,k)\to 0$, then the integral in (C.7) diverges and comes to an incorrect result.

Now we study the correspondence of relations (C.6) to the material equation (C.4). For isotropic non-gyrotropic medium, tensor $\tilde{F}_{ij}\left(t,\vec{r}\right)$ in (C.5) can be written as

$$\tilde{F}_{ij}\left(t,\vec{r}\right)=\left(\delta_{ij}-\frac{r_ir_j}{r^2}\right)\tilde{F}^{\mathrm{tr}}(t,r)+\frac{r_ir_j}{r^2}\tilde{F}^{l}(t,r), \tag{C.8}$$

where $\tilde{F}^{\mathrm{tr}}$ and $\tilde{F}^{l}$ are scalar functions of time t and absolute value of radius vector $\vec{r}$. In this case, for $\varepsilon_{ij}\left(\omega,\vec{k}\right)$, we find

$$\varepsilon_{ij}\left(\omega,\vec{k}\right)=\delta_{ij}+\int_0^\infty r^2dr\int_{r/c}^\infty e^{\imath\omega t}dt\left\{\delta_{ij}\tilde{F}^{\rm tr}(t,r)\int d\Omega e^{-\imath\vec{k}\cdot\vec{r}}+\frac{1}{r^2}\left[\tilde{F}^l(t,r)-\tilde{F}^{\rm tr}(tr)\right]\int d\Omega r_i r_j e^{-\imath\vec{k}\cdot\vec{r}}\right\}, \tag{C.9}$$

where $d\Omega$ is the solid angle element in the direction of radius vector $\vec{r}$.

Next, to simplify the calculations, we write the dispersion relation for effective dielectric permittivity

$\varepsilon(\omega,k)=\varepsilon_{ii}\left(\omega,\vec{k}\right)=2\varepsilon^{\rm tr}(\omega,k)+\varepsilon^l(\omega,k)$. For these quantities, we have

$$\varepsilon\left(\omega,\vec{k}\right)=1+F\left(\omega,\vec{k}\right),\quad F\left(\omega,\vec{k}\right)=\int_0^\infty dt\int dV\tilde{F}\left(t,\vec{r}\right)e^{\imath\left(\omega t-\vec{k}\cdot\vec{r}\right)}, \tag{C.10}$$

where $\tilde{F}(\omega,k)=2\tilde{F}^{\rm tr}(\omega,k)+\tilde{F}^l(\omega,k)$.

Introducing function

$$S\left(t,\vec{r}\right)=\tilde{F}\left(t,\vec{r}\right)\theta(t),\quad \theta(t)=\begin{cases}1, & t>0,\\ 0, & t<0.\end{cases} \tag{C.11}$$

$$S\left(t,\vec{r}\right)=\frac{1}{(2\pi)^4}\int_{-\infty}^\infty d\omega\int d^3ke^{\imath\left(\vec{k}\cdot\vec{r}-\omega t\right)}F(\omega,k), \tag{C.12}$$

after integrating over angular variables, and taking into account that $F(\omega,k)$ is an even function of k, we reduce equality (C.12) to the following form

$$S(t,r)=-\frac{\imath}{(2\pi)^3r}\int_{-\infty}^\infty d\omega e^{-\imath\omega t}\int_{-\infty}^\infty k\,dk\,e^{\imath kr}F(\omega,k). \tag{C.13}$$

Substituting $F(\omega,k)$ from relation (C.3) into the right hand side of Eq. (C.13), after trivial change of variables, and taking into account the integral representation of function sgn$t$, we reduce equality (C.12) to the following form:

$$\tilde{F}(t,r)=\tilde{F}(t,r)\,\mathrm{sgn}\,(t-\beta r/c)+I_\beta(r)\delta(t-\beta r/c), \tag{C.14}$$

where

$$I_\beta(r)=\frac{2\beta/c}{(2\pi)^3}\int_{-\infty}^\infty d\omega\int_{-\infty}^\infty dkF^l(\omega,k)e^{\imath r(k-\omega\beta/c)}. \tag{C.15}$$

For all values of t and r, not connected by condition $ct = \beta r$, solution of Eq. (C.14) is of the form $\tilde{F}^{l}(t,r)\theta(t-\beta r/c)$, not excluding negative values of time (the effect is ahead of the cause!). The correct solution should have the form of $\tilde{F}^{l}(t,r)\theta(t-r/c)$, which corresponds to the material equation (C.4) for the longitudinal field which was to be proved. Therefore, $\beta = 1$ is the only correct choice of the value of the parameter β in (C.6).

Finally, we show that factor $I_1(r)$ for $\delta(t - r/c)$ in (C.14) is equal to zero. For this aim, substituting $\beta = 1$ into (C.14), we find:

$$F^{l}(\omega,k) = -\frac{2\pi\imath}{k}\int_{0}^{\infty} dt \int_{0}^{ct} rdr\tilde{F}^{l}(t,r)\left[e^{\imath(\omega t+kr)} - e^{\imath(\omega t-kr)}\right]. \tag{C.16}$$

After simple calculations, we find

$$I_1(r) = \frac{1}{c}\int_{0}^{r} r'dr'\tilde{F}(r/c,r')\frac{1}{\pi\imath}\int_{-\infty}^{\infty}\frac{dk}{k}\left[e^{\imath k(r+r')} - e^{\imath k(r-r')}\right]. \tag{C.17}$$

Since integrand and integration over dk do not have any singularity at $k = 0$, using

$$\frac{1}{\pi\imath}\int_{-\infty}^{\infty}\mathcal{P}\frac{d\xi}{\xi}e^{\imath t\xi} = \operatorname{sgn} t = \begin{cases} +1, & t > 0, \\ -1, & t < 0, \end{cases}$$

we can simply calculate the integral as

$$\frac{1}{\pi\imath}\int_{-\infty}^{\infty}\frac{dk}{k}\left[e^{\imath k(r+r')} - e^{\imath k(r-r')}\right] = \frac{1}{\pi\imath}\int_{-\infty}^{\infty}\mathcal{P}\frac{dk}{k}\left[e^{\imath k(r+r')} - e^{\imath k(r-r')}\right]$$
$$= \operatorname{sgn}(r+r') - \operatorname{sgn}(r-r'). \tag{C.18}$$

Integration over dr' in (C.17) is taken at $r \geq r' \geq 0$. Therefore, the right-hand side of (C.18) is strictly equal to zero and, consequently, $I_1(r) = 0$.

Thus we have proved that dispersion relations for longitudinal and transverse dielectric permittivities when considering the finiteness of light speed are written in the form of

$$\varepsilon^{l,\mathrm{tr}}(\omega,k) - 1 = \frac{1}{\pi\imath}\int_{-\infty}^{\infty} d\xi\mathcal{P}\frac{\varepsilon^{l,\mathrm{tr}}(\omega+\xi, k+\xi/c) - 1}{\xi}, \tag{C.19}$$

only for $\beta = 1$.

Besides,

$$\int_{-\infty}^{\infty} \left[\varepsilon^{l,\mathrm{tr}}(\omega, k + \omega/c) - 1\right] d\omega = 0. \tag{C.20}$$

As it should be, in the limit $c \to \infty$, relations (C.19) pass to the well-known Kramers-Kronig relations.

In conclusion, we note that above, as well as in [1], we considered the field as the cause, and the induction as the effect, and wrote the Kramers-Kronig relations by taking into account the finite speed of light for dielectric permittivities (C.19). Quite analogously, the Kramers-Kronig relations are written by taking into account the finite speed of light in the case when the current is considered as the cause, and the induction as the effect [2]. In this case, they have the same form of Eq. (C.19) but only for quantities

$$\frac{1}{\varepsilon^{l}(\omega, k)} \text{ and } \frac{1}{k^2c^2 - \omega^2\varepsilon^{\mathrm{tr}}(\omega, k)} \tag{C.21}$$

From the first expression for $\beta = 1$, it follows that $1/\,\mathrm{Re}\,\varepsilon(0, k) < 1$. This means that the real part of longitudinal dielectric permittivity of an equilibrium medium can be larger than unity or less than zero.

References

1. M.A. Leontovich, J. Exp. Theor. Phys. **40**, 907 (1961)
2. D.A. Kirzhnits, Physics-Uspekhi **19**, 530 (1976).; **30**, 575 (1987)

Index

B. Shokri, A. A. Rukhadze, *Electrodynamics of Conducting Dispersive Media*,
Springer Series on Atomic, Optical, and Plasma Physics 111,
https://doi.org/10.1007/978-3-030-28968-3

M

Q

Printed in the United States
By Bookmasters